Algebra

Ernest Shult · David Surowski

Algebra

A Teaching and Source Book

Springer

Ernest Shult
Department of Mathematics
Kansas State University
Manhattan, KS
USA

David Surowski
Manhattan, KS
USA

David Surowski is deceased.

ISBN 978-3-319-19733-3   ISBN 978-3-319-19734-0   (eBook)
DOI 10.1007/978-3-319-19734-0

Library of Congress Control Number: 2015941161

Springer Cham Heidelberg New York Dordrecht London

Printed on acid-free paper

Springer International Publishing AG Switzerland is part of Springer Science+Business Media (www.springer.com)

Preface

This book is based on the notes of both authors for a course called "Higher Algebra," a graduate level course. Its purpose was to offer the basic abstract algebra that any student of mathematics seeking an advanced degree might require. Students may have been previously exposed to some of the basic algebraic objects (groups, rings, vector spaces, etc.) in an introductory abstract algebra course such as that offered in the classic book of Herstein. But that exposure should not be a hard requirement as this book proceeds from first principles. Aside from the far greater theoretical depth, perhaps the main difference between an introductory algebra course, and a course in "higher algebra" (as exemplified by classics such as Jacobson's *Basic algebra* [1, 2] and Van der Waerden's *Modern Algebra* [3]) is an emphasis on the student understanding how to construct a mathematical proof, and that is where the exercises come in.

The authors rotated teaching this one-year course called "Higher Algebra" at Kansas State University for 15 years—each of us generating his own set of notes for the course. This book is a blend of these notes.

Listed below are some special features of these notes.

1. (Combinatorial Background) Often the underlying combinatorial contexts—partially ordered sets etc.—seem almost invisible in a course on modern algebra. In fact they are often developed far from home in the middle of some specific algebraic context. Partially ordered sets are the natural context in which to discuss the following:

    (a) Zorn's Lemma and the ascending and descending chain conditions,
    (b) Galois connections,
    (c) The modular law,
    (d) The Jordan Hölder Theorem,
    (e) Dependence Theories (needed for defining various notions of "dimension").

    The Jordan Hölder Theorem asserts that in a lower semimodular semilattice, any semimodular function from the set of covers (unrefinable chains of length one) to a commutative monoid extends to an interval measure on all algebraic intervals (those intervals containing a finite unrefinable chain from the bottom to the top). The extension exists because the multiset of values of the function on the covers in any two unrefinable chains connecting a and b must be the same. The proof is quite easy, and the applications are everywhere. For example, when G is a finite group and P is the poset of subnormal subgroups, one notes that P is a semimodular lower semilattice and the reading of the simple group A/B of a cover $A < B$, is a *semimodular function* on covers by a fundamental theorem of homomorphisms of groups. By the theorem being described, this function extends to an interval measure with values in the additive monoid of multisets on the isomorphism classes of simple groups. The conclusion of the combinatorial Jordan-Hölder version in this context becomes the classical Jordan-Hölder Theorem for finite groups. One needs no "Butterfly Lemma" or anything else.

2. (Free Groups) Often a free group on generators X is presented in an awkward way—by defining a "multiplication" on 'reduced words' $r(w)$, where w is a word in the free monoid $M(X \cup X^{-1})$. 'Reduced' means all factors of the form xx^{-1} have been removed. Here are the complications: First the reductions, which can often be performed in many ways, must lead to a common reduced word. Then one must show $r(w_1 \circ w_2) = r(r(w_1) \circ r(w_2))$ to get "multiplication" defined on reduced words. Then one needs to verify the associative law and the other group axioms. In this book the free group is *defined* to be the automorphism group of a certain labelled graph, and the universal mapping properties of the free group are easily derived from the graph. Since full sets of automorphisms of an object always form a group, one will not be wasting time showing that an akwardly-defined multiplication obeys the axioms of a group.

3. (Universal Mapping Properties) These are always instances of the existence of an initial or terminal object in an appropriate category.

4. (Avoiding Determinants of Matrices) Of course one needs matrices to describe linear transformations of vector spaces, or to record data about bilinear forms (the Grammian). It is important to know when the rows or columns of a matrix are linearly dependant. One can calculate what is normally called the determinant by finding the invariant factors. For an $n \times n$ matrix, that process involves roughly n^3 steps, while the usual procedure for evaluating the determinant using Lagrange's rule, involves exponentially many steps.

One of the standard proofs that the trace mapping $\text{tr} : K \to F$ of a finite separable field extension $F \subseteq K$ is nonzero proceeds as follows: First, one forms the normal closure L of the field K. One then invokes the theorem that $L = F(\theta)$, a simple extension, with the algebraic conjugates of θ as an F-basis of L. And then one reaches the conclusion by observing that a van der Monde determinant is non-zero. Perhaps it is an aesthetic quibble, but one does not like to see a nice "soft" algebraic proof about "soft" algebraic objects reduced to a matrix calculation. In Sect. 11.7 the proof that the trace is non-trivial is accomplished using only the Dedekind Independence Lemma and an elementary fact about bilinear forms.

In general, in this book, the determinant of a transformation T acting on an n-dimension vector space V is defined to be the scalar multiplication it induces on the n-th exterior product $\wedge^n(V)$. Of course there are historical reasons for making a few exceptions to any decree to ban the usual formulaic definition of determinants altogether. Our historical discussion of the *discriminant* on page 395 is such an exception.

In addition, we have shaped the text with several pedagogical objectives in mind.

1. (Catch-up opportunities) Not infrequently, the teacher of a graduate course is expected to accommodate incoming transfer students whose mathematical preparation is not quite the same as that of current students of the program, or is even unknown. At the same time, this accommodation should not sacrifice course content for the other students. For this this reason we have written each chapter at a gradient—with simplest explanations and examples first, before continuing at the level the curriculum requires. This way, a student may "catch up" by studying the introductory material more intensely, while a more brief review of it is presented in class. Students already familiar with the introductory material have merely to turn the page.

2. (Curiosity-driven Appendices) The view of both authors has always been that a course in Algebra is not an exercise in cramming information, but is instead a way of inspiring mathematical curiosity. Real learning is basically curiosity-driven self-learning. Discussing what is is already known is simply there to guide the student to the real questions. For that reason we have inserted a number of appendices which are largely centered around incites connected with proofs in the text. Similarly, in the exercises, we have occasionally wandered into open problems or offered avenues for exploration. Mathematics education is not a catechism.

3. (Planned Redundancy) Beside its role as a course guide, a textbook often lives another life as a source book. There is always the need of a student or colleague in a nearby mathematical field to check on some algebraic fact—say, to make sure of the hypotheses that accompany that fact. He or she does not need to read the whole book. But occasionally one wanders into the following scenario: one looks up topic A in the index, and finds, at the indicated page, that A is defined by further words B and C whose definition can be deciphered by a further visit to the index, which obligingly invites one to further pages at which the frustration may be enjoyed once again. It becomes a tree search. In order to intercept this process, we have tried to do the following: when an earlier-defined key concept re-inserts itself in a later discussion, we simply recall the definition for the reader at that point, while offering a page number where the concept was originally defined.[1] Nevertheless we are introducing a redundancy. But in the

---

[1] If we carried out this process for the most common concepts, pages would be filled with re-definitions of rings, natural numbers, and what the containment relation is. Of course one has to limit these reminders of definitions to new key terms.

view of the authors' experience in Kansas, redundancy is a valuable tool in teaching. Inspiration is useless if the student cannot first understand the words—and no teacher should apologize for redundancy. Judiciously applied, it does not waste class time; it actually saves it.

Of course there are many topics—direct offshoots of the material of this course—that cannot be included here. One cannot do justice to such topics in a brief survey like this. Thus one will not find in this book material about (i) Representation Theory and Character Theory of Groups, (ii) Commutative Rings and the world of Ext and Tor, (iii) Group Cohomology or other Homological Algebra, (iv) Algebraic Geometry, (v) Really Deep Algebraic Number Theory and (vi) many other topics. The student is better off receiving a full exposition of these courses elsewhere rather than being deceived by the belief that the chapters of this book provide such an expertise. Of course, we try to indicate some of these points of departure as we meet them in the text, at times suggesting exterior references.

A few words are inserted here about how the book can be used.

As mentioned above, the book is a blend of the notes of both authors who alternately taught the course for many years. Of course there is much more in this book than can reasonably be covered in a two-semester course. In practice a course includes enough material from each chapter to reach the principle theorems. That is, portions of chapters can be left out. Of course the authors did not always present the course in exactly the same way, but the differences were mainly in the way focus and depth were distributed over the various topics. We did not "teach" the appendices to the chapters. They were there for the students to explore on their own.

The syllabus presented here would be fairly typical. The numbers in parenthesis represent the number of class-hours the lectures usually consume. A two-semester course entails 72 class-hours. Beyond the lectures we normally allowed ourselves 10–12 h for examinations and review of exercises.

1. Chapter 1: (1 or 2) [This goes quickly since it involves only two easy proofs.]
2. Chapter 2: (6, at most) [This also goes quickly since, except for three easy proofs, it is descriptive. The breakdown would be: (a) 2.2.1–2.2.9 (skip 2.2.10), 2.2.10–2.2.15 (3 h), (b) 2.3 and 2.5 (2 h) and (c) 2.6 (1 h).]
3. Chapter 3: (3)
4. Chapter 4: (3) [Sometimes omitting 4.2.3.]
5. Chapter 5: (3 or 4) [Sometimes omitting 5.5.]
6. Chapter 6: (3) [Omitting the Brauer-Ree Theorem [6.4] but reserving 15 minutes for Sect. 6.6.]
7. Chapter 7: (3) Mostly examples and few proofs. Section 7.3.6 is often omitted.]
8. Chapter 8: (7 or 8) [Usually (a) 8.1 (2 or 3 h), and (b) 8.2–8.4 (4 h). We sometimes omitted Sect. 8.3 if behind schedule.]
9. Chapter 9: (4) [One of us taught only 9.1–9.8 (sometimes omitting the local characterization of UFDs in 9.6.3) while the other would teach all of 9.9–9.12 (Dedekind's Theorem and the ideal class group.]

10. Chapter 10: (6) [It takes 3 days for 10.1–10.5. The student is asked to read 10.6, and 3 days remain for Sect. 10.7 and autopsies on some of the exercises.]
11. Chapter 11: (11) [The content is rationed as follows: (a) 11.1–11.4 (2 h), sometimes omitting 11.4.2 if one is short a day, (b) 11.5–11.6 (3 h) (c) 11.7 [One need only mention this.] (d) 11.8–11.9 (3 h) (e) [One of us would often omit 11.10 (algebraic field extensions are often simple extensions). Many insist this be part of the Algebra Catechism. Although the result is vaguely interesting, it is not needed for a single proof in this book.] (f) 11.11 (1 h) (g) [Then a day or two would be spent going through sampled exercises.]]
12. Chapter 12: (5 or 6) [Content divided as (a) 12.1–12.3 (2 h) and (b) 12.4–12.5 (2 h) with an extra hour wherever needed.]
13. Chapter 13: (9 or 10) [Approximate time allotment: (a) 13.1–13.2 (1 h) (only elementary proofs here), (b) 13.3.1–13.3.2 (1 h), (c) 13.3.3–13.3.4 (adjunct functors) (1 h), (d) 13.4–13.5 (1 h), (e) 13.6–13.8 (1 or 2 h), (f) 13.9 (1 h), 13.10 (2 h) and 13.8 (3 h).]

The list above is only offered as an example. The book provides ample "wiggle room" for composing alternative paths through this course, perhaps even re-arranging the order of topics. The one invariant is that Chap. 2 feeds all subsequent chapters.

Beyond this, certain groups of chapters may serve as one semester courses on their own. Here are some suggestions:

GROUP THEORY: Chaps. 3–6 (invoking only the Jordan Holder Theorem from Chap. 2).

THEORY OF FIELDS: After an elementary preparation about UFD's (their maximal ideals, and homomorphisms of polynomial rings in Chap. 6), and Groups (their actions, homomorphisms and facts about subgroup indices from Sects. 3.2, 3.3 and 4.2) one could easily compose a semester course on Fields from Chap. 11.

ARITHMETIC: UFD's, including PID's with applications to Linear Algebra using Chaps. 7–10.

BASIC RING THEORY: leading to Wedderburn's Theorem. Chapters 7, 8 and 12.

RINGS AND MODULES, TENSOR PRODUCTS AND MULTILINEAR ALGEBRA: Chaps. 7, 8 and 13.

References

1. Jacobson N (1985) Basic algebra I, 2nd edn. W. H. Freeingsman and Co., New York
2. Jacobson N (1989) Basic algebra II, 2nd edn. W. H. Freeman and Co., New York
3. van der Waerden BL (1991) Algebra, vols I–II. Springer, New York Inc., New York
   September 2014

In Memory

My colleague and co-author David D. Surowski expired in a Shanghai hospital on March 2, 2011. He had recently recovered from surgery for pancreatic cancer, although this was not the direct cause of his death.

David was a great teacher of incite and curiosity about the mathematics that he loved. He was loved by his graduate students and his colleagues. But most of all he loved and was deeply loved by his family.

He was my best friend in life.

Eight years ago (2004), David and I agreed that this book should be lovingly dedicated to our wives:

Jiang Tan Shult and Susan (Yuehua) Zhang.

Ernest Shult

Contents

Chapter 1
Basics

Abstract The basic notational conventions used in this book are described. Composition of mappings is defined the standard left-handed way: $f\ g$ means mapping g was applied first. But things are a little more complicated than that since we must also deal with both left and right operators, binary operations and monoids. For example, right operators are sometimes indicated exponentially—that is by right superscripts (as in group conjugation)—or by right multiplication (as in right R-modules). Despite this, the "∘"-notation for composition will always have its left-handed interpretation. Of course a basic discussion of sets, maps, and equivalence relations should be expected in a beginning chapter. Finally the basic arithmetic of the natural and cardinal numbers is set forth so that it can be used throughout the book without further development. (Proofs of the Schröder-Bernstein Theorem and the fact that $\aleph_0 \cdot \aleph_0 = \aleph_0$ appear in this discussion.) Clearly this chapter is only about everyone being on the same page at the start.

1.1 Presumed Results and Conventions

1.1.1 Presumed Jargon

Most abstract algebraic structures in these notes are treated from first principles. Even so, the reader is assumed to have already acquired some familiarity with groups, cosets, group homomorphisms, ring homomorphisms and vector spaces from an undergraduate abstract algebra course or linear algebra course. We rely on these topics mostly as a source of familiar examples which can aid the intuition as well as points of reference that will indicate the direction various generalizations are taking.

The Abstraction of Isomorphism Classes

What do we mean by saying that object A is isomorphic to object B? In general, in algebra, we want objects A and B to be isomorphic if and only if one can obtain a complete description of object B simply by changing the names of the operating

E. Shult and D. Surowski, *Algebra*, DOI 10.1007/978-3-319-19734-0_1

parts of object A and the names of the relations among them that must hold—and vice versa. From this point of view we are merely dealing with the same situation, but under a new management which has renamed everything. This is the "alias" point of view; the two structures are really the same thing with some name changes imposed.

The other way—the "alibi" point of view—is to form a one-to-one correspondence (a bijection) of the relevant parts of object A with object B such that a relation holds among parts in the domain (A) if and only if the corresponding relation holds among their images (parts of set B).[1]

There is no logical distinction between the two approaches, only a psychological one.

Unfortunately "renaming" is a subjective human conceptualization that is awkward to define precisely. That is why, at the beginning, there is a preference for describing an isomorphism in terms of bijections rather than "re-namings", even though many of us secretly think of it as little more than a re-baptism.

It is a standing habit in abstract mathematics for one to assert that mathematical objects are "the same" or even "equal" when one only means that the two objects are isomorphic. It is an abuse of language when we say that "two manifolds are the same", "two groups are the same", or that "A and B are really the same ring". We shall meet this over and over again; for this is at the heart of the "abstractness" of Abstract Algebra.[2]

1.1.2 Basic Arithmetic

The integers are normally employed in analyzing any finite structure. Thus for reference purposes, it will be useful to establish a few basic arithmetic properties of the integers. The integers enjoy the two associative and commutative operations of addition and multiplication, connected by the distributive law, that every student is familiar with.

There is a natural (transitive) order relation among the integers: thus

$$\cdots -4 < -3 < -2 < -1 < 0 < 1 < 2 < 3 < \cdots.$$

If $a < b$, in this ordering, we say "integer a is less than integer b". (This can also be rendered by saying "b is greater than a".) In the set $\mathbb{Z}$ of integers, those integers greater than or equal to zero form a set

---

[1] We are deliberately vague in talking about parts rather than "elements" for the sake of generality.

[2] There is a common misunderstanding of this word "abstract" that mathematicians seem condemned to suffer. To many, "abstract" seems to mean "having no relation to the world—no applications". Unfortunately, this is the overwhelming view of politicians, pundits of Education, and even many University Administrators throughout the United States. One hears words like "Ivory Tower", "Intellectuals on welfare", etc. On the contrary, these people have it just backwards. A concept is "abstract" precisely because it has *more than one* application—not that it hasn't *any* application. It is very important to realize that two things introduced in distant contexts are in fact the same structure and subject to the same abstract theorems.

$$\mathbb{N} = \{0, 1, 2, \ldots\}$$

called the *natural numbers*. Obviously, for every integer a that is not zero, exactly one of the integers a or $-a$ is positive, and this one is denoted $|a|$, and is called the *absolute value* of a. We also define 0 to be the absolute value of itself, and write $0 = |0|$.

Of course this subset $\mathbb{N}$ inherits a total ordering from $\mathbb{Z}$, but it also possesses a very important property not shared by $\mathbb{Z}$:

(The well-ordering property)   *Every non-empty subset of $\mathbb{N}$ possesses a least member.*

This property is used in the Lemma below.

Lemma 1.1.1 (The Division Algorithm) *Let a, b be integers with $a \neq 0$. Then there exist unique integers q (quotient) and r (remainder) such that*

$$b = qa + r, \quad where \ 0 \leq r < |a|.$$

Proof Define the set $R := \{b - qa \mid q \in \mathbb{Z}, b - qa \geq 0\}$; clearly $R \neq \emptyset$. Since the set of non-negative integers $\mathbb{N}$ is well ordered (See p. 34, Example 1), the set R must have a least element, call it r. Therefore, it follows already that $b = qa + r$ for suitable integers q, r and where $r \geq 0$. If it were the case that $r \geq |a|$, then setting $r' := r - |a|$, one has $r' < r$ and $r' \geq 0$, and yet $b = qa + r = qa + (r' + |a|) = (q \pm 1)a + r'$ (depending on whether a is positive or negative). Therefore, $r' = b - (q \pm 1)a \in R$, contrary to the minimality of r. Therefore, we conclude the existence of integers q, r with

$$b = qa + r, \quad where \ 0 \leq r < |a|,$$

as required.

The uniqueness of q, r turns out to be unimportant for our purposes; therefore we shall leave that verification to the reader. $\square$

If n and m are integers, and if $n \neq 0$, we say that n *divides* m, and write $n \mid m$, if $m = qn$ for some integer (possibly 0) q. If a, b are integers, not both 0, we call d a *greatest common divisor* of a and b if

(i) $d > 0$,
(ii) $d \mid a$ and $d \mid b$,
(iii) for any integer c satisfying the properties of d in (i), (ii), above, we must have $c \mid d$.

Lemma 1.1.2 *Let a, b be integers, not both 0. Then a greatest common divisor of a and b exists and is unique. Moreover, if d is the greatest common divisor of a and b, then there exist integers s and t such that*

$$d = sa + tb \quad (\text{The Euclidean Trick}).$$

Proof Here, we form the set $D := \{xa + yb \mid x, y \in \mathbb{Z}, xa + yb > 0\}$. Again, it is routine to verify that $D \neq \emptyset$. We let d be the smallest element of D, and let s and t be integers with $d = sa + tb$. We shall show that $d \mid a$ and that $d \mid b$. Apply the division algorithm to write

$$a = qd + r, \ \ 0 \leq r < d.$$

If $r > 0$, then we have

$$r = a - qd = a - q(sa + tb) = (1 - qs)a - qtb \in D,$$

contrary to the minimality of $d \in D$. Therefore, it must happen that $r = 0$, i.e., that $d \mid a$. In an entirely similar fashion, one proves that $d \mid b$. Finally, if $c \mid a$ and $c \mid b$, then certainly $c \mid (sa + tb)$, which says that $c \mid d$. $\square$

As a result of Lemma 1.1.2, when the integers a and b are not both 0, we may speak unambiguously of their *greatest common divisor* d and write $d = \text{GCD}(a, b)$. When $\text{GCD}(a, b) = 1$, we say that a and b are *relatively prime*.

One final simple, but useful, number-theoretic result:

Corollary 1.1.3 *Let a and b be relatively prime integers with $a \neq 0$. If for some integer c, $a \mid bc$, then $a \mid c$.*

Proof By the Euclidean Trick, there exist integers s and t with $sa + tb = 1$. Multiplying both sides by c, we get $sac + tbc = c$. Since a divides bc, we infer that a divides $sac + tbc$, which is to say that $a \mid c$. $\square$

1.1.3 Sets and Maps

1. **Sets**: Intuitively, a set A is a collection of objects. If x is one of the objects of the collection we write $x \in A$ and say that "x is a member of set A".
   The reader should have a comfortable rapport with the following set-theoretic concepts: the notions of membership, containment and the operations of intersection and union over arbitrary collections of subsets of a set. In order to make our notation clear we define these concepts:

   (a) If A and B are sets, the notation $A \subseteq B$ represents the assertion that every member of set A is necessarily a member of set B. Two sets A and B are considered to be the *same* set if and only if every member of A is a member of B and every member of B is a member of A—that is, $A \subseteq B$ and $B \subseteq A$. In this case we write $A = B$.[3]

---

[3]Of course the sets A and B might have entirely different descriptions, and yet possess the same collection of members.

(b) There is a set called the empty set which has no members. It is denoted by the universal symbol ∅. The empty set is contained in any other set A. To prove this assertion one must show that $x \in \emptyset$ implies $x \in A$. By definition, the hypothesis (the part preceding the word "implies") is false. A false statement implies any statement, in particular our conclusion that $x \in A$. (This recitation reveals the close relation between sets and logic.)

In particular, since any empty set is contained in any other, they are all considered to be "equal" as sets, thus justifying the use of one single symbol "∅".

(c) Similarly, if A and B are sets, the symbol $A - B$ denotes the set $\{x \in A \,|\, x \notin B\}$, that is, the set of elements of A which are not members of B. (The reader is warned that in the literature one often encounters other notation for this set—for example "$A \backslash B$". We will stick with "$A - B$".)

(d) If $\{A_\sigma\}_{\sigma \in I}$ is a collection of sets indexed by the set I, then either of the symbols

$$\cap_{\sigma \in I} A_\sigma \text{ or } \cap \{A_\sigma \,|\, \sigma \in I\}$$

denotes the set of elements which are members of each A_σ and this set is called the *intersection* of the sets $\{A_\sigma \,|\, \sigma \in I\}$.

Similarly, either one of the symbols

$$\cup_{\sigma \in I} A_\sigma \text{ or } \cup \{A_\sigma \,|\, \sigma \in I\}$$

denotes the *union* of the sets $\{A_\sigma \,|\, \sigma \in I\}$—namely the set of elements which are members of at least one of the A_σ.

Beyond this, there is the special case of a union which we call a *partition*. We say that a collection $\pi := \{A_\sigma \,|\, \sigma \in I\}$ of subsets of a set X is a *partition of set X* if and only if

  i. each A_σ is a non-empty subset of X (called a *component of the partition*), and

  ii. Each element of X lies in a unique component A_σ—that is, $X = \cup\{A_\sigma \,|\, \sigma \in I\}$ and distinct components have an empty intersection.

2. **The Cartesian product construction,** $A \times B$: That would be the collection of all ordered pairs (a, b) ("ordered" in that we care which element appears on the left in the notation) such that the element a belongs to set A and the element b is a member of set B. Similarly for positive integer n we understand the *n-fold Cartesian product of the sets* $B_1, \ldots, B_n$ to be the collection of all ordered sequences (sometimes called "n-tuples"), $(b_1, \ldots, b_n)$ where, for $i = 1, 2, \ldots, n$, the element b_i is a member of the set B_i. This collection of n-tuples is denoted

$$B_1 \times \cdots \times B_n.$$

3. **Binary Relations**: The student should be familiar with the device of viewing *relations between objects* as subsets of a Cartesian product of sets of these

objects. Here is how this works: Suppose R is a subset of the Cartesian product $A \times B$. One uses this theoretical device to provide a setting for saying that object "a" in set A is *related* to object "b" in set B: one just says "element a is related to element "b" if and only if the pair (a, b) belongs to the subset R of $A \times B$.[4] (This device seems adequate to handle any relation that is understood in some other sense. For example: the relation of being "first cousins" among members of the set P of living U.S. citizens, can be described as the set C of all pairs (x, y) in the Cartesian product $P \times P$, where x is the first cousin of y.)

The phrase "a relation on a set A" is intended to refer to a subset R of $A \times A$. There are several useful species of such relations, such as equivalence relations, posets, simple graphs etc.

4. **Equivalence relations**: Equivalence relations behave like the equal sign in elementary mathematics. No one should imagine that any assertion that x is equal to y (an assertion denoted by an "equation $x = y$") is saying that x *really is* y. Of course that is impossible since one symbol is one side of the equation and the other is on the other side. One only means that *in some respect* (which may be limited by an observer's ability to make distinctions) the objects x and y do not appear to differ. It may be two students in class with the same amount of money on their person, or it may be two presidential candidates with equally fruitless goals. What we need to know is how this notion that things "are the same" operates. We say that the relation R (remember it is a subset of $A \times A$) is an *equivalence relation* if an only if it obeys these three rules:

   (a) (Reflexive Property) For each $a \in A$, $(a, a) \in R$—that is, every element of A is R-related to itself.
   (b) (Symmetric Property) If $(a, b) \in R$, then $(b, a) \in R$—that is, if element a is related to b, then also element b is related to a.
   (c) (Transitive property) If a is related to b and b is related to c then one must have a related to c by the specified relation R.

Suppose R is an equivalence relation on the set A. Then, for any element $a \in A$, the set $[a]$ of all elements related to a by the equivalence relation R, is called the equivalence class containing a, and such classes possess the following properties:

   (a) For each $a \in A$, one has $a \in [a]$.
   (b) For each $b \in [a]$, one has $[a] = [b]$.
   (c) No element of $A - [a]$ is R-related to an element of $[a]$.

---

[4]This is not just a matter of silly grammatical style. How many American Calculus books must students endure which assert that a "function" (for example from the set of real numbers to itself) is a "rule that assigns to each element of the domain set, a unique element of the "codomain" set? The "rules" referred to in that definition are presumably instructions in some language (for example in American English) and so these instructions are strings of symbols in some finite alphabet, syllabary, ideogramic system or secret code. The point is that such a set is at best only countably infinite whereas the collection of subsets R of $A \times B$ may well be uncountably infinite. So there is a very good logical reason for viewing relations as subsets of a Cartesian product.

It follows that for an equivalence relation R on the non-empty set A, the equivalence classes form the components of a partition of the set A in the sense defined above. Conversely, if $\pi := \{A_\sigma | \sigma \in I\}$ is a partition of a set A, then we obtain a corresponding equivalence relation R_π defined as follows: the pair (x, y) belongs to the subset $R_\pi \subseteq A \times A$—that is, x and y are R_π-related—if and only if they belong to the same component of the partition π. Thus there is a one-to-one correspondence between equivalence classes on a set A and the partitions of the set A.

5. **Partially ordered sets**: Suppose a relation R on a set A, satisfies the following three properties:

   (a) (Reflexive Property) For each element a of A, $(a, a) \in R$.
   (b) (Transitive property) If a is related to b and b is related to c then one must have a related to c by the specified relation R.
   (c) (Antisymmetric property) If (a, b) and (b, a) are both members of R, then $a = b$.

A set A together with such a relation R is called a *partially-ordered set* or *poset*, for short. Partially ordered sets are endemic throughout mathematics, and are the natural home for many basic concepts of abstract algebra, such as chain conditions, dependence relations or the statement of "Zorn's Lemma". Even the famous Jordan-Hölder Theorem is simply a theorem on the existence of interval measures in meet-closed semi-modular posets.

One often denotes the poset relation by writing $a \leq b$, instead of $(a, b) \in R$. Then the three axioms of a partially ordered set $(A, \leq)$ read as follows:

   (a) $x \leq x$ for all $x \in A$.
   (b) If $x \leq y$ and $y \leq z$, then $x \leq z$.
   (c) If $x \leq y$ and $y \leq x$, then $x = y$.

Note that the third axiom shows that the relations $x_1 \leq x_2 \leq \cdots \leq x_n \leq x_1$ imply that all the x_i are equal.

A simple example is the relation of "being contained in" among a collection of sets. Note that our definition of equality of sets, realizes the anti-symmetric property. Thus, if set A is contained in set B, and set B is contained in set A then the two sets are the same collection of objects—that is, they are equal as sets.

6. **Power sets**: Given a set X, there is a set 2^X of all subsets of X, called the *power set of X*. In many books, for example, Keith Devlin's *The Joy of Sets* [16], the notation $\mathcal{P}(X)$ is used in place of 2^X. In Example 2 on p. 35, we introduce this notation when we regard 2^X as a partially ordered set with respect to the containment relation between subsets—at which point it is called the "power poset". But in fact, virtually every time one considers the set 2^X, one is aware of the pervasive presence of the containment relation, and so might as well regard it as a poset. Thus in practice, the two notations 2^X and $\mathcal{P}(X)$ are virtually interchangeable. Most of the time we will use $\mathcal{P}(X)$, unless there is some reason not to be distracted by the containment relation or for the reason of a previous commitment of the symbol "$\mathcal{P}$".

7. **Mappings**: The word "mapping" is intended to be indistinguishable from the word "function" as it is used in most literature. We may define a *mapping* R : $X \to Y$ as a subset $R \subseteq X \times Y$ with the property that for every $x \in X$, there exists exactly one $y \in Y$ such that the pair (x, y) is in R.[5] Then the notation $y = R(x)$ simply means $(x, y) \in R$ and we metaphorically express this fact by saying "the function R *sends* element x to y"—as if the function was actively doing something. The suggestive metaphors continue when we also render this same fact—that $y = R(x)$—by saying that y *is the image of element* x or equivalently that x *is a preimage of* y.

8. **Images and range**: If $f : X \to Y$ is a mapping, the collection of all "images" $f(x)$, as x ranges over X, is clearly a subset of Y which we call the *image* or *range of the function* f and it is denoted $f(X)$.

9. **Equality of mappings**: Two mappings are considered equal if they "do the same things". Thus if f and g are both mappings (or functions) from X to Y we say that *mapping* f *is equal to mapping* g if and only if $f(x) = g(x)$ for all x in X. (Of course this does not mean that f and g are *described* or *defined* in the same way. Asserting that two mappings are equal is often a non-obvious Theorem.)

10. **Identity mappings**: A very special example is the following: The mapping $1_X : X \to X$ which takes each element x of X to itself—i.e. $f(x) = x$—is called the *identity mapping on set* X. This mapping is very special and is uniquely defined just by specifying the set X.

11. **Domains, codomains, restrictions and extensions of mappings**: In defining a mapping $f : X \to Y$, the sets X and Y are a vital part of the definition of a mapping or function. The set X is called the *domain* of the function; the set Y is called the *codomain* of the mapping or function.

A simple manipulation of both sets allows us to define new functions from old ones. For example, if A is a subset of the domain set X, and $f : X \to Y$ is a mapping, then we obtain a mapping

$$f|_A : A \to Y$$

which sends every element $a \in A$ to $f(a)$ (which is defined *a fortiori*). This new function is called *the restriction of the function* f *to the subset* A. If $g = f|_A$, we say that f *extends function* g.

Similarly, if the codomain Y of the function $f : X \to Y$ is a subset of a set B (that is, $Y \subseteq B$), then we automatically inherit a function $f|^B : X \to B$ just from the definition of "function". When $f : X \to Y$ is the identity mapping $1_X : X \to X$, the replacement of the codomain X by a larger set B yields a mapping $1_X|^B : X \to B$ called the *containment mapping*.[6]

A mapping $f : X \to Y$ is said to be *one-to-one* or *injective* if any two distinct elements of the domain are not permitted to yield the same image element.

---

[5]Note that there is no grammatical room here for a "multivalued function".

[6]Unlike the notion of "restriction", this construction does not seem to enjoy a uniform name.

This is another way of saying, that the fibre $f^{-1}(y)$ of each element y of Y is permitted to have *at most one* element.

The reader should be familiar with the fact that the composition of two injective mappings is injective and that the composition of two surjective mappings (performed in that chronological order) is surjective. She or he should be able to prove that the restriction $f|_A : A \to Y$ of any injective mapping $f : X \to Y$ (here A is a subset of X) is an injective mapping.

12. **One-to-one mappings (injections) and onto mappings (surjections):**

A mapping $f : X \to Y$ is called *onto* or *surjective* if and only if $Y = f(X)$ as sets. That means that every element of set Y is the image of some element of set X, or, put another way, the fibre $f^{-1}(y)$ of each element y of Y is nonempty.

A mapping $f : X \to Y$ is said to be *one-to-one* or *injective* if any two distinct elements of the domain are not permitted to yield the same image element. This is another way of saying, that the fibre $f^{-1}(y)$ of each element y of Y is permitted to have *at most one* element.

The reader should be familiar with the fact that the composition of two injective mappings is injective and that the composition of two surjective mappings (performed in that chronological order) is surjective. She or he should be able to prove that the restriction $f|_A : A \to Y$ of any injective mapping $f : X \to Y$ (here A is a subset of X) is an injective mapping.

13. **Bijections:**

A mapping $f : X \to Y$ is a *bijection* if and only if it is both injective and surjective—that is, both one-to-one and onto.

When this occurs, the fibre $f^{-1}(y)$ of every element $y \in Y$ contains a unique element which can unambiguously be denoted $f^{-1}(y)$. This notation allows us to define the unique function $f^{-1} : Y \to X$ which we call the *inverse of the bijection* f. Note that the inverse mapping possesses these properties,

$$f^{-1} \circ f = 1_X, \text{ and } f \circ f^{-1} = 1_Y,$$

where 1_X and 1_Y are the identity mappings on sets X and Y, respectively.

14. **Examples using mappings:**

(a) Indexing families of subsets of a set X with the notation $\{X_\alpha\}_I$ (or $\{X_\alpha\}_{\alpha \in I}$) should be understood in its guise as a mapping $I \longrightarrow 2^X$.[7]

(b) The construction $\hom(A, B)$ as the set of all mappings $A \longrightarrow B$. (This is denoted B^A in some parts of mathematics.) The reader should see that if A is a finite set, say with n elements, then $\hom(A, B)$ is just the n-fold Cartesian product of B with itself

$$B \times B \times \cdots \times B \text{ (with exactly } n \text{ factors).}$$

---

[7]Note that in the notation, the "α" is ranging completely over I and so does not itself affect the collection being described; it is what logicians call a "bound" variable.

Also if $\mathbb{N}$ denotes the collection of all natural numbers (including zero), then $\hom(\mathbb{N}, B)$ is the set of all *sequences* of elements of B.

(c) Recall from an earlier item (p. 5) that a *partition* of a set X is a collection $\{X_\alpha\}_I$ of non-empty subsets X_α of X such that (i) $\cup_{\alpha \in I} X_\alpha = X$, and (ii) the sets X_α are *pairwise disjoint*—that is, $X_\alpha \cap X_\beta = \emptyset$ whenever $\alpha \neq \beta$. The sets X_α of the union are called the *components of the partition*.

A partition may be described in another way: as a surjection $\pi : X \longrightarrow I$. Then the collection of *fibers*—that is, the sets $\pi^{-1}(\alpha) := \{x \in X | \pi(x) = \alpha\}$ as α ranges over I—form the components of a partition. Conversely, if $\{X_\alpha\}_I$ is a partition of X, then there is a well-defined surjection $\pi : X \longrightarrow I$ which takes each element of X to the index of the unique component of the partition which contains it.

15. **A notational convention on partitions**: In these lecture notes if A and B are sets, we shall write $X = A + B$ (rather than $X = A \dot\cup B$ or $X = A \uplus B$) to express the fact that $\{A, B\}$ is a partition of X with just two components A and B. Similarly we write

$$X = X_1 + X_2 + \cdots + X_n$$

when X possesses a partition with n components X_i, $i = 1, 2, \ldots, n$. This notation goes back to Galois' rendering of a partition of a group by cosets of a subgroup. The notation is very convenient since one doesn't have to "doctor up" a "cup" (or "union") symbol. Unfortunately similar notation is also used in more algebraic contexts with a different meaning—for example as a set of sums in some additive group. We resolve the possible ambiguity in this way:

When a partition (rather than, say, a set of sums) is intended, the partition will simply be introduced by the words "partition" or "decomposition".

1.1.4 Notation for Compositions of Mappings

There is a perennial awkwardness cast over common mathematical notation for the composition of two maps. Mappings are sometimes written as left operators and sometimes as right operators, and the awkwardness is not the same for both choices due to the asymmetric fact that English is read from left to right. Because of this, right operators work much better for representing the action of sets with a binary operation as mappings with compositions.

Then why are left operators used at all? There are two answers: Suppose there is a division of the operators on X into two sets—say A and B. Suppose also that if an operator $a \in A$ is applied first in *chronological order* and the operator $b \in B$ is applied afterwards; that the result is always the same had we applied b first and *then* applied a later. Then we say that the two operations "commute" (at least in the time scale of their application, if not the temporal order in which the operators are read from left to right). This property can often be more conveniently rendered by having one set, say A, be the left operators, and the other set B be the "right" operators, and then expressing the "commutativity"as something that looks like an "associative law". So one needs both kinds of operators to do that.

The second reason for using left operators is more anthropological than mathematical. The answer comes from the sociological accident of English usage. In the English phrase "function of x", the word "function" comes first, and *then* its argument. This is reflected in the left-to-right notation "$f(x)$" so familiar from calculus.

Then if the composition $\alpha \circ \beta$ were to mean "α is applied first and then β is applied", one would be obligated to write $(\alpha \circ \beta)(x) = \beta(\alpha(x))$. That is, we must reverse their "reading" order.

On the other hand, if we say $\alpha \circ \beta$ means β is applied first and α is applied second—so $(\alpha \circ \beta)(x) = \alpha(\beta(x))$—then things are nice as far as the treatment of parentheses are concerned, but we still seem to be reading things in the reverse chronological order (unless we compensate by reading from right to left). Either way there is an inconvenience.

In the vast majority of cases, the Mathematical Literature has already chosen the latter as the least of the two evils. Accordingly, we adopt this convention:

Notation for Composition of Mappings: If $\alpha : X \to Y$ and $\beta : Y \to Z$, then $\beta \circ \alpha$ denotes the result of *first* applying mapping α to obtain an element y of Y, and *then* applying mapping β to y. Thus if the mappings α and β are regarded as left operators of X, and Y, respectively, we have, for each $x \in X$,

$$(\beta \circ \alpha)(x) := \beta(\alpha(x)).$$

But if α and β are right operators on X and Y, respectively, we have, for each $x \in X$,

$$x(\beta \circ \alpha) = (x\alpha)\beta.$$

But right operators are also very useful. A common instance is when (i) the set F consists of functions mapping a set X into itself, and (ii) F itself possesses an associative binary operation "$*$" (see Sect. 1.4) and (iii) composition of two such functions is the function representing the binary operation of them—that is $f * g$ acts as $g \circ f$. In this case, it is really handier to think of the functions as right operators, so that we can write

$$(x^f)^g = x^{f*g},$$

for all $x \in X$ and f and g in F. For this reason we tend to view the action of groups or rings as induced mappings which are *right operators*.

Finally, there are times when one needs to discuss "morphisms" which commute with all right operators in F. It is then easier to think of these morphisms as left operators, for if the function α commutes with the right operator g, we can express this by the simple equation

$$\alpha(x^g) = (\alpha(x))^g, \text{ for all } x \in X \text{ and } g \text{ in } F.$$

So there are cases where both types of operators must be used.

The convention of these notes is to indicate right operators by exponential notation or with explicit apologies (as in the case of right R-modules), as right multiplications. Thus in general we adopt these rules:

Rule #1:   The symbol $\alpha \circ \beta$ denotes the composition resulting from first applying β and then α in chronological order. Thus

$$(\alpha \circ \beta)(x) = \alpha(\beta(x)).$$

Rule #2:   Exponential notation indicates right operators. Thus compositions have these images:
$$x^{\alpha \circ \beta} = (x^{\beta})^{\alpha}.$$

Exception to Rule #2:   For right R-modules we indicate the right operators by right "multiplication", that is right juxtaposition. Ring multiplication still gets represented the right way since we have $(mr)s = m(rs)$ for module element m and ring elements r and s (it looks like an associative law). (The reason for eschewing the exponential notation in this case is that the law $m^{r+s} = m^r + m^s$ for right R-modules would then not look like a right distributive law.)

1.2 Binary Operations and Monoids

It is not our intention to venture into various algebraic structures at such an early stage in this book. But we are forced to make an exception for monoids, since they are always lurking around so many of the most basic definitions (for example, the definition of *interval measures* on posets).

Suppose X is a set and let $X^{(n)}$ be the n-fold Cartesian product of X with itself. For $n > 0$, a mapping
$$X^{(n)} \to X$$

is called an *n-ary operation on* X. If $n = 1$ such an operation is just a mapping of X into itself. There are certain concepts that are brought to bear at the level of 2-ary (or binary) operations that fade away for larger n.

We say that set X *admits a binary operation* if there exists a 2-ary function $f : X \times X \to X$. In this case, it is possible to indicate the operation by a constant symbol (say "$*$") inserted between the elements of an ordered pair—thus one might write "$x * y$" to indicate $f((x, y))$ (which we shall write as $f(x, y)$ to rid ourselves of one set of superfluous parentheses).

Indeed one might even use the empty symbol in this role—that is $f(x, y)$ is represented by "juxtaposition" of symbols: that is we write "xy" for $f(x, y)$. The juxtaposition convention seems the simplest way to describe properties of a general but otherwise unspecified binary operation.

The binary operation on X is said to be *commutative* if and only if

$$xy = yx, \text{ for all } x \text{ and } y \text{ in } X.$$

The operation is *associative* if and only if

$$x(yz) = (xy)z \text{ for all } x, y, z \in X.$$

Again let us consider an arbitrary binary operation on X. Do not assume that it is associative or commutative. The operation admits a *left identity element* if there exists an element—say e_L in X such that $e_L x = x$ for all $x \in X$. Similarly, we say the operation admits a *right identity element* if there exists an element e_R such that $x e_R = x$ for all $x \in X$. However, if X admits both a left identity element and a right identity element, say e_L and e_R, respectively, then the two are equal for

$$e_R = e_L e_R = e_L.$$

(The first equality is from e_L being a left identity, and the second is from e_R being a right identity.) We thus have

Proposition 1.2.1 *Suppose X is a set admitting a (not necessarily associative) binary operation indicated by juxtaposition. Suppose this operation on X admits at least one left identity element and at least one right identity element (they need not be distinct elements). Then all right identity elements and all left identity elements are equal to a unique element e for which $ex = xe = x$ for all $x \in X$.* (Such an element e is called an *identity element* or a *two-sided identity* for the given binary operation on X.)

A set admitting an associative binary operation is called a *semigroup*. For example if X contains more than two elements and the binary operation is defined by $xy = y$ for all $x, y \in X$, then with respect to this binary operation, X is a semigroup with many left identity elements and no right identity elements.

A semigroup with a two-sided identity is called a *monoid*. A semigroup (or monoid) with respect to a commutative binary operation is simply called a *commutative semigroup* (or *commutative monoid*).

We list several commonly encountered monoids.

1. The set $\mathbb{N}$ of non-negative integers (natural numbers) under the operation of ordinary addition.
2. Let X be any set. We let $M(X)$ be the set of all finite strings (including the empty string) of the elements of X. A string is simply a sequence of elements of X. It becomes a monoid under the binary operation of concatenation of strings. The

concatenation of strings is the string obtained by extending the first sequence by adjoining the second one as a "suffix". Thus if $s_1 = (x_1, x_2, x_3)$ and $s_2 = (y_1, y_2)$, then the concatenation would be $s_1 * s_2 = (x_1, x_2, x_3, y_1, y_2)$. (Note that concatenation is an associative operation, and that the empty string is the two sided identity of this monoid.) For reasons that will be made clear in Chaps. 6 and 7, it is called the *free monoid on the set X*.

3. A *multiset* of the set X is a function $f : X \to \mathbb{N}$ that assigns to each element of X a non-negative integer. In this sense a multiset represents a sort of inventory of objects of various types drawn from a set of types X. The collection of all multisets on set X, denoted $\mathcal{M}(X)$, admits a commutative binary operation which we call "addition". If $f, g : X \to \mathbb{N}$ are two multisets, their sum $f + g$ is defined to be the function that sends x to $f(x) + g(x)$. In the language of inventories, addition of two multisets is just the merging of inventories. Since this addition is associative and the empty multiset (the function with all values zero) is a two-sided identity, the multisets on X form a monoid $(\mathcal{M}(X), +)$, with respect to addition. This monoid is also called the *free commutative monoid on set X*, an appellation fully justified in Chap. 7.

A multiset f is finite if the set of elements x at which the function f assumes a positive value (called the *support* of the function) is a finite set. By setting $f(i) = a_i$, $i \in \mathbb{N}$, we can write any finite multiset as a countable sequence $(a_0, a_1, \ldots)$ of natural numbers which has only finitely many non-zero entries. Addition of two multisets $(a_0, \ldots)$ and $(b_0, \ldots)$ is then performed coordinate-wise. We denote this submonoid of finite multisets of elements chosen from X by the symbol, $(\mathcal{M}_{<\infty}(X), +)$. If the set X is itself finite, then, of course, all elements of $\mathcal{M}(X)$ are finite multisets.

1.3 Notation for Special Structures

There are certain sets and structures which are encountered over and over, and these will have a special fixed notation throughout this book.

Standard Sets

1. $\mathbb{N}$. The system of *natural numbers*, $\{0, 1, 2, \ldots\}$. It is important for the student to realize that this term is understood here to include the integer zero. It is a well-ordered poset with the descending chain condition (see Chap. 2).
2. $\mathbb{Z}$. This is the collection of all integers, $\{\ldots, -2, -1, 0, 1, 2, \ldots\}$. It forms an integral domain under the operations of addition and multiplication (Chap. 7).
3. $\mathbb{Q}$. The field of rational numbers.
4. $\mathbb{R}$. The field of real numbers.
5. $\mathbb{C}$. The field of complex numbers.

1.4 The Axiom of Choice and Cardinal Numbers

1.4.1 The Axiom of Choice

The student ought to be acquainted with and to be able to use

THE AXIOM OF CHOICE: *Suppose* $\{X_\alpha\}_I$ *is a family of pairwise disjoint nonempty sets. Then there exists a subset* R *of the union of the* X_α, *which meets each component* X_α *in exactly a one-element set.*

This assertion is about the existence of systems of representatives. If only finitely many of the X_α are infinite sets this can be proved from ordinary set theory. But the reader should be aware that in its full generality it is independant of set theory, and yet, is consistent with it. The reader is thus encouraged to think of it as an adjunct axiom to set theory, to make a note of each time it is used, and to quietly produce the appropriate guilt feelings when using it.

The Axiom of Choice has many uses. For example it guarantees the existence of a system of coset representatives for any subgroup of any group. In fact we shall see an application of the axiom of choice in the very next subsection on cardinal numbers.

In the presence of set theory (which is actually ever-present for the purposes of these notes) the Axiom of Choice is equivalent to another assertion called Zorn's Lemma, which should also be familiar to the reader. Since it appears in the setting of partially ordered sets, a full discussion of Zorn's Lemma is deferred to a section of the next chapter.

The reader is not required to know a proof of the equivalence of the Axiom of Choice and Zorn's Lemma, or a proof of their consistency with set theory. At this stage, all that is required to read these notes is the psychological assurance that one cannot "get into trouble" by using these principles. For a good development of the many surprising equivalent versions of the Axiom of Choice the curious student is encouraged to peruse Sect. 2.2.7 of the book by Keith Devlin entitled *The Joy of Sets* [16].

1.4.2 Cardinal Numbers

Not all collections of things are actually amenable to the axioms of set theory, as Russell's paradox illustrates. Nonetheless certain operations and constructions on such collections can still exist. It is still possible that they may possess equivalence relations and that is true of the collection of all sets.

We have mentioned that a mapping $f : A \longrightarrow B$ which is both an injection and a surjection is called a *bijection* or a *one-to-one correspondence* in a slightly older language.[8] In that case the partition of A defined by the surjection f (see above) is the trivial partition of A into its one-element subsets. This means the fibering f^{-1}

---

[8] A "one-to-one correspondence" is not to be confused with the weaker notion of a "one-to-one mapping" introduced on p. 9. The latter is just an injective mapping which may or may not be a bijection.

defines an *inverse mapping* $f^{-1} : B \longrightarrow A$ which is also a bijection satisfying $f \circ f^{-1} = 1_A$ and $f^{-1} \circ f = 1_B$ as described in item 15 on p. 9. It is clear, using obvious facts about compositions of bijections and identity mappings, that the relation of two sets having a bijection connecting them is an equivalence relation. The resulting equivalence classes are called *cardinal numbers* and the equivalence class containing set X is denoted $|X|$ and is called the *cardinality of* X.

Cardinal numbers possess an inherent partial ordering. One writes $|X| \leq |Y|$ if and only there exists an injection $f : X \longrightarrow Y$.[9] We are obligated to show three things:

1. The relation "$\leq$" is well-defined—that is, if, for sets X, Y and Z, we have that $|X| \leq |Y|$ and $|Y| = |Z|$, then $|X| \leq |Y|$.
2. The relation "$\leq$" is transitive.
3. The relation "$\leq$" is anti-symmetric.

First we observe that our definitions force the transitive law (item 2). Suppose, for sets X, Y and Z, one has $|X| \leq |Y|$ and $|Y| \leq |Z|$. Then from our definition of "$\leq$", there exist injective mappings $f : X \to Y$, and $g : Y \to Z$. Then $g \circ f$ is an injective mapping from X to Z, and so by definition, $|X| \leq |Y|$.

Next the student may observe that $|Y| = |Z|$ implies $|Y| \leq |Z|$, for the former statement implies a bijection $g : Y \to Z$, and as any bijection is injective, $|Y| \leq |Z|$ by definition. This observation together with the transitive law implies the statement of item 2.

For the anti-symmetric law we appeal to a famous Theorem:

Theorem 1.4.1 (The Schröder-Bernstein Theorem) *If, for two sets X and Y, one has*
$$|X| \leq |Y| \text{ and } |Y| \leq |X|,$$
then $|X| = |Y|$.

Proof By hypothesis there are two injective (one-to-one) mappings $f : X \to Y$ and $g : Y \to X$. Our task is to use this data to devise a bijective (one-to-one onto) mapping $h : X \to Y$.

We may assume that neither of the injective mappings f or g is surjective (onto), for otherwise either f or g^{-1} will serve as our desired mapping h.

As a result, $f(X)$ is a proper subset of Y and, as g is injective, $g(f(Y)) = (g \circ f)(Y)$ is a proper subset of $g(Y)$. In this way, one obtains a properly descending chain of subsets:

$$X \supset g(Y) \supset (g \circ f)(X) \supset (g \circ f \circ g)(Y) \supset \cdots . \tag{1.1}$$

Transposing the roles of f and g presents a second properly descending chain:

---

[9]It should be clear to the student that this partial ordering is on the collection of cardinal numbers. It is not a relation between the sets themselves.

$$Y \supset f(X) \supset (f \circ g)(Y) \supset (f \circ g \circ f)(X) \supset \cdots . \tag{1.2}$$

An element $a \in X \cup Y$ is said to be an *ancestor* of element $b \in Y \cup Y$ if and only if there is a finite string of alternate applications of the mappings f and g which, when applied to a, yields b. For example, if $a, b \in X$, and $(g \circ \cdots \circ f)(a) = b$, for some finite string $g \circ \cdots \circ f$, then a is an ancestor of b. Thus, the non-empty set $X_0 := X - g(Y)$ are the members of X which have no ancestors; the set $X_1 := g(Y) - (g \circ f)(X)$ comprise the members of X with exactly one ancestor. We let X_k be denote the set of elements of X with *exactly* k ancestors—namely, the non-empty set

$$X_k = g \circ (f \circ g)^{(k-1)/2}(Y) - (g \circ f)^{(k+1)/2}(X), \text{ if } k \text{ is odd, or}$$
$$X_k = (g \circ f)^{k/2}(X) - g \circ (f \circ g)^{k/2}(Y), \text{ if } k \text{ is even.}$$

The symbol X_∞ will denote the set of elements of X which possess infinitely many ancestors. If the intersection of the sets in the tower of Eq. (1.1) is empty, then there are no elements with infinitely many ancestors. Thus, unlike the X_k, the set X_∞ could be empty.

Next we similarly define the non-empty sets Y_k, as members of Y with exactly k ancestors, and let Y_∞ denote the set of those members of Y with infinitely many ancestors. Now if an element z of X (or Y) possesses infinitely many ancestors, then so does $g^{-1}z$ and $f(z)$ (or $f^{-1}(z)$ and $g(z)$ when $z \in Y$). Thus f restricted to X_∞ induces a bijection $h_\infty : X_\infty \to Y_\infty$ whether the sets are empty or not. It remains only to devise a bijection $X - X_\infty \to Y - Y_\infty$.

We now have two partitions (into non-empty sets):

$$X - X_\infty = X_0 + X_1 + X_2 + \cdots$$
$$Y - Y_\infty = Y_0 + Y_1 + Y_2 + \cdots$$

If k is even, define $h_k : X_k \to Y_{k+1}$ as the *restriction* of the mapping f to the subset X_k. Note that h_k is surjective, and so is a bijection. If k is odd, then X_k lies in $g(Y)$ and so the inverse mapping g^{-1} may be applied to it, to produce a mapping $h_k : X_k \to Y_{k-1}$, that is surjective and injective since g was a mapping. Thus h_k is a bijection in this case as well.

Now $h : X \to Y$ is defined by $h : x \mapsto h_k(x)$ if $x \in X_k$, and $x \mapsto h_\infty(x)$ if $x \in X_\infty$. Since the h_k are all bijections and the codomains of the h_k reproduce the partition of $Y - Y_\infty$ given above, h is our desired bijection. $\square$

Remark This elementary proof due to J. König may be found in the book of P.M. Cohn entitled *Algebra, vol. 2* [10, p. 11] and in the book of Birkhoff and McLane [8].

These lecture notes presume and use two further results concerning cardinalities of sets:

Theorem 1.4.2 *If there exists a surjection $f : X \longrightarrow Y$ then $|Y| \le |X|$.*

Proof With the axiom of choice one gets a subset X' containing exactly one element x_y from each fiber $f^{-1}(y)$ as y ranges over Y. One then has an injection $Y \to X'$ taking y to x_y. We can then compose this with the inclusion mapping $i : X' \to X$ to obtain the desired injection from Y into X. $\square$

As usual, let $\mathbb{N}$ denote the set of *natural numbers*, that is, the set $\{0, 1, 2, \ldots\}$ of all integers which are positive or zero (non-negative).

A *cardinal number* is defined to be the name of a cardinality equivalence class. For a finite set F, the cardinality $|F|$ is simply a natural number. Thus, the cardinality of the empty set $\emptyset$ is 0, and for non-empty finite sets, this association with natural numbers seems perfectly natural since any finite set is bijective with some finite initial segment—say $(1, 2, \ldots n)$—of the positive integers listed in their natural ordering. Indeed, producing that bijection is what we usually refer to as "counting".

One can define a product of two cardinal numbers, in the following way: If $a = |A|$ and $b = |B|$ are two cardinal numbers, (A and B chosen representatives from the equivalence classes of sets denoted by a and b, respectively), then one writes $ab = |A \times B|$, the cardinality of the Cartesian product of A and B. This product is well-defined, for if one selected other representatives A' and B' of these classes, there are then bijections $\alpha : A \to A'$ and $\beta : B \to B'$ which can be used to define a bijection $A \times B \to A' \times B'$ defined by

$$(a, b) \mapsto (\alpha(a), \beta(b)), \text{ for all } (a, b) \in A \times B.$$

Similarly, the mapping

$$((a, b), c) \mapsto (a, (b, c)), \text{ for all } (a, b, c) \in A \times B \times C$$

defines a bijection $(A \times B) \times C \to A \times (B \times C)$. Thus we see that taking the product among cardinal numbers is an associative operation.

The reader will appreciate that this definition of product is completely compatible with the definition of multiplication of positive integers familiar to most children. If A contains three elements, and B contains seven, then the Cartesian product $A \times B$ contains twenty-one distinct pairs. However now, our definition can be applied to cardinalities of infinite sets, as well.

The simplest infinite set familiar to the young student is the set of natural numbers itself. Custom has assigned the rather unique symbol $\aleph_0$ for the cardinality of $\mathbb{N}$. Any set is said to be *countably infinite*, if its cardinality is $\aleph_0$—or equivalently, there is a bijection taking such a set to $\mathbb{N}$.

Now a property which distinguishes infinite sets from finite sets is that an infinite set can be bijective with a proper subset of itself. For example, $\mathbb{N}$ is bijective with the non-negative even integers. It is also bijective with all of the natural numbers that are perfect squares. Similarly, the integers $\mathbb{Z}$ are bijective with $\mathbb{N}$ by the mapping defined by

$$n \mapsto 2n \text{ for } n \in \mathbb{N}$$
$$-n \mapsto 2n - 1 \text{ for } n \in \mathbb{N}, n > 0.$$

It is also easy to see this:

Theorem 1.4.3 *If k is a positive integer, then the cardinality of the union of k disjoint copies of $\mathbb{N}$ is $|\mathbb{N}| = \aleph_0$.*

Proof This is left as an exercise.

Now something peculiar happens:

Theorem 1.4.4 $\aleph_0 \cdot \aleph_0 = \aleph_0$

Proof We must produce a bijection $f : \mathbb{N} \to \mathbb{N} \times \mathbb{N}$. First we assign the pair $(a, b) \in \mathbb{N} \times \mathbb{N}$ to the point in the real (Cartesian) plane with those coordinates. All the integral coordinates in and on the boundary of the first quadrant now represent points of $\mathbb{N} \times \mathbb{N}$. Now partition the points into non-empty finite sets according to the sum of their coordinates. First comes $(0, 0)$, then $\{(0, 1), (1, 0)\}$, then $\{(0, 2), (1, 1), (2, 0)\}$, and so on. Having ordered the points within each component of the partition by the natural ordering of its first coordinate, we obtain in this way, a sequence S indexed by $\mathbb{N}$. Mapping the n-th member of this sequence to $n - 1$ produces a bijection $g : \mathbb{N} \times \mathbb{N} \to \mathbb{N}$. $\square$

Corollary 1.4.5 $|\mathbb{N}| = |\mathbb{Z} \times \mathbb{Z}|$.

It is left as an exercise, to prove that $|\mathbb{N}| = |\mathbb{Q}|$, where, as usual, $\mathbb{Q}$ is the set of rational numbers—the fractions formed from the integers.

In the next chapter, we shall generalize Theorem 1.4.4 by showing that for any infinite cardinal number a, one has $\aleph_0 \cdot a = a$.

References

1. Cohn PM (1977) Algebra, vol 2. Wiley, London
2. Devlin K (1997) The joy of slets. Springer, New York

Chapter 2
Basic Combinatorial Principles of Algebra

Abstract Many basic concepts used throughout Algebra have a natural home in Partially Ordered Sets (hereafter called "posets"). Aside from obvious poset residents such as Zorn's Lemma and the well-ordered sets, some concepts are more wider roaming. Among these are the ascending and descending chain conditions, the general Jordan-Hölder Theorem (seen here as a theorem on interval measures of certain lower semilattices), Galois connections, the modular laws in lattices, and general independence notions that lead to the concepts of dimension and transcendence degree.

2.1 Introduction

The reader is certainly familiar with examples of sets X possessing a transitive relation "$\leq$" which is antisymmetric. Such a pair $(X, \leq)$ is called a *partially ordered set*—often abbreviated as *poset*. For any subset Y of the poset X there is an induced partial ordering $(Y, \leq)$ imposed on Y by the partially ordered set $(X, \leq)$ which surrounds it: one merely restricts the relation "$\leq$" to the pairs of $Y \times Y$. We then call $(Y, \leq)$ an *induced poset* of $(X, \leq)$.

Certainly one example of a partially ordered set familiar to most readers is the poset $\mathcal{P}(X) := (2^X, \subseteq)$ called the *power poset* of *all* subsets of X under the containment relation.

Throughout this algebra course one will encounter sets X which are closed under various n-ary operations subject to certain axioms—that is, some species of "algebraic object". Each such object X naturally produces a partially ordered set whose members are the subsets of X closed under these operations—that is, the poset of algebraic subobjects of the same species. For example, If X is a group, then the poset of algebraic subobjects is the poset of all subgroups of X. If R is a ring and M is a right R-module, then we obtain the poset of submodules of M. Special cases are the posets of vector subspaces of a right vector space V and the poset of right ideals of a ring R.

In turn these posets have special induced posets: Thus the poset $\mathbf{L}_{<\infty}(V)$ of all finite-dimensional vector subspaces of a (possibly infinite-dimensional) vector space

© Springer International Publishing Switzerland 2015

E. Shult and D. Surowski, *Algebra*, DOI 10.1007/978-3-319-19734-0_2

V is transparently a subposet of the poset of all vector subspaces of V. For example from a group G one obtains the poset of normal subgroups of G, or more generally, the poset of subgroups invariant under any fixed subgroup A of the automorphism group of G. Similarly there are posets of invariant subrings of polynomial rings, and invariant submodules for an R-module admitting operators. Finally, there are induced posets of algebraic objects which are closed with respect to a closure operator on a poset (perhaps defined by a Galois connection).

All of these examples will be made precise later. The important thing to note at this stage is that

1. *Partially ordered sets underly all of the algebraic structures discussed in this book.*
2. *Many of the crucial conditions which make arguments work are basically properties of the underlying posets alone and do not depend on the particular algebraic species within which one is working: Here are the main examples:*

    (a) *Zorn's Lemma,*
    (b) *the ascending and descending chain conditions,*
    (c) *Galois connections and closure operators,*
    (d) *interval measures on semimodular semilattices (The General Jordan-Hölder Theorem), and*
    (e) *dependence theory (providing the notion of "dimension").*

The purpose of this chapter is to introduce those basic arguments that arise strictly from the framework of partially ordered sets, ready to be used for the rest of this book.

2.2 Basic Definitions

2.2.1 Definition of a Partially Ordered Set

A *partially ordered set*, $(P, \leq)$, hereafter called a *poset*, is a set P with a transitive antireflexive binary relation $\leq$. This means that for all elements x, y and z of P,

1. (transitivity) $x \leq y$ and $y \leq z$ together imply $x \leq z$, and
2. (antireflexivity or antisymmetry)[1] the assertions $x \leq y$ and $y \leq x$ together imply $x = y$.

It is often useful to view elements of a poset pictorially, as if they were vertices placed in vertical plane. Thus we say "*x is below y*" or "*y is above x*" if $x \leq y$ in some poset.[2]

---

[1] In the literature on binary relations, the term "antisymmetric" often replaces its equivalent "antireflexive".

[2] This is just metaphorical language, nothing more.

We write $y \geq x$ if $x \leq y$ and write $x < y$ if $x \leq y$ and $x \neq y$. In the latter case we say that x is *properly below* y, or, equivalently, that y *is properly above* x.

2.2.2 Subposets and Induced Subposets

Now let $(P, \leq)$ be a poset, and suppose X is a subset of P with a relation $\leq_X$ for which $(X, \leq_X)$ is a partially ordered set. In general, there may be no relation between $(X, \leq_X)$ and the order relation $\leq$ that the elements of X inherit from $(P, \leq)$. But if it is true that for any x and y in X,

$$x \leq_X y \text{ implies } x \leq y, \tag{2.1}$$

then we say that $(X, \leq_X)$ is a *subposet of* $(P, \leq)$. Thus in a general subposet it might happen that two elements x_1 and x_2 of $(X, \leq_X)$ are incomparable with respect to the ordering $\leq_X$ even though one is bounded by the other (say, $x_1 \leq x_2$) in the ambient poset $(P, \leq)$.

However, if the converse implication in Eq. (2.1) holds for a subposet, we say that poset is an *induced subposet*. Formally, $(X, \leq_X)$ is defined to be an *induced subposet of* $(P, \leq)$, if and only if $X \subseteq P$ and for any x and y in X,

$$x \leq_X y \text{ if and only if } x \leq y. \tag{2.2}$$

Suppose $(P, \leq)$ is a poset, and X is a subset of P. Then we can agree to induce the relation $\leq$ on the subset X—that is, for any two elements x and y of the subset X, we agree to say that $x \leq_X y$ if and only if $x \leq y$ in the poset $(P, \leq)$. Thus an induced subposet of $(P, \leq)$ is entirely determined once its set of elements is specified.

The empty set is considered to be an induced subposet of any poset.

Let X and Y be two subsets of P where $(P, \leq)$ is a poset. We make these observations:

1. If $(X, \leq_X)$ is a partial ordering on X and if $(Y, \leq_Y)$ is a partial ordering on Y such that $(X, \leq_X)$ is an induced subposet of $(Y, \leq_Y)$ and $(Y, \leq_Y)$ is an induced subposet of $(P, \leq)$, then $(X, \leq_X)$ is also an induced subposet of $(P, \leq)$. This fact means that if X is any subset of P, and $(P, \leq)$ is a poset, we don't need those little subscripts attached to the relation "$\leq$" any more: we can unambiguously write $(X, \leq)$ to indicate the subposet induced by $(P, \leq)$ on X. In fact, when it is clear that we are speaking of *induced subposets*, we may write X for $(X, \leq)$ and speak of "the induced poset X".

2. If $(X, \leq)$ and $(Y, \leq)$ are induced subposets of $(P, \leq)$ then we can form the *intersection of induced posets* $(X \cap Y, \leq)$. In fact, since this notion depends only on the underlying *sets*, one may form the intersection of induced posets

$$\left(\bigcap_{\sigma \in I} X_\sigma, \leq \right),$$

from any family $\{(X_\sigma, \leq) | \sigma \in I\}$ of induced posets of a poset $(P, \leq)$.

Perhaps the most important example of an induced poset is the *interval*. Suppose $(P, \leq)$ is a poset and that $x \leq y$ for elements $x, y \in P$. Consider the induced poset

$$[x, y]_P := \{z \in P | x \leq z \leq y\}.$$

This is called the *interval in* $(P, \leq)$ *from x to y*. If $(P, \leq)$ is understood, we would write $[x, y]$ in place of $[x, y]_P$. But we must be very careful: there are occasions in which one wishes to discuss *intervals within an induced poset* $(X, \leq)$, in which case one would write $[x, y]_X$ for the elements z of X between x and y. Clearly one could mix the notations and write $[x, y]_X := [x, y] \cap X$, the intersection of two induced posets.

2.2.3 Dual Posets and Dual Concepts

Suppose now that $(P, \leq)$ is given. One may obtain a new poset, $(P^*, \leq^*)$ whose elements are exactly those of P, but in which the partial ordering has been reversed! Thus $a \leq b$ in $(P, \leq)$ if and only if $b \leq^* a$ in $(P^*, \leq^*)$. In this case, $(P^*, \leq^*)$ is called the *dual poset of* $(P, \leq)$.

We might as well get used to the idea that for every definition regarding $(P, \leq)$, there is a "dual notion"—that is, the generically-defined property or set in $(P, \leq)$ resulting from defining the same property or set in $(P^*, \leq^*)$. Examples will follow.

2.2.4 Maximal and Minimal Elements of Induced Posets

An element x is said to be *maximal* in $(X, \leq)$ if and only if there is no element in X, which is strictly larger than x—that is $x \leq y$ for $y \in X$ implies $x = y$.

Note that this is quite different than the more specialized notion of a *global maximum* (over X) which would be an element x in X for which $x' \leq x$ for *all* elements x' of X. Of course defining something does not posit its existence; $(X, \leq)$ may not even possess maximal elements, or if it does, it may not contain a global maximum.

We let max X denote the full (but possibly empty) collection of all maximal elements of $(X, \leq)$.

By replacing the symbol "$\leq$" by "$\geq$" in the preceding definitions, we obtain the dual notions of a *minimal element* and a *global minimum* of an induced poset $(X, \leq)$.

Of course some posets may contain no minimal elements whatsoever.

2.2.5 Global Maxima and Minima

If a poset $(P, \leq)$ possesses a global minimum, then, by the antisymmetric property, that element is the unique global minimum, and so deserves a special name. We call it *the zero-element of the poset*.

Dually, there may exist a global maximum (a *"one-element"*, denoted 1_P, or something similar) which is an element in $(P, \leq)$ with the property that all other elements are less than or equal to it. Obviously this 1_P is the zero-element of the dual poset, $(P^*, \leq^*)$.

Some posets have a "zero", some have a "one", some have both, and some have neither.

But whether or not a "zero" is present in P, one can always adjoin a new element declared to be properly below all elements of P to obtain a new poset $0(P)$. For example, if P contains just one element p, then $0(P)$ consists of just two elements $\{0_1, p\}$ with $0_1 < p$—called a *chain of length one*. Iterating this construction with the same meaning of P, we see that $0^2(P)$ introduces a "new zero element", 0_2, to produce a 3-element poset with $0_2 < 0_1 < p$—that is, a *chain of length two*. Clearly $0^k(P)$ would be a poset with elements (up to a renaming of the elements) arranged as

$$0_k < 0_{k-1} < \cdots < 0_1 < p,$$

which we call a *chain of length k*. Of course this construction can also be performed on any poset P so that $0^k(P)$ in general becomes the poset P with a tail of length $k - 1$ adjoined to it from below—a sort of attached "kite-tail".

Also, by dually defining $1^k(P)$ one can attach a "stalk" of length $k - 1$ above an arbitrary poset P.

2.2.6 Total Orderings and Chains

A poset $(P, \leq)$ is said to be *totally ordered* if $\{c, d\} \subseteq C$ implies $c \leq d$ or $d \leq c$. Sometimes this notion is referred to as *"simply ordered"*. Obviously any induced poset of a totally ordered set is also totally ordered. Any maximal (or minimal element) of a totally ordered set is in fact a global maximum (or global minimum). Familiar examples of totally ordered sets are obtained as induced posets of the real numbers under its usual ordering, for example (1) the rational numbers, (2) the integers, (3) the positive integers (4) the open, half-closed or closed intervals (a, b), $[a, b)$, $(a, b]$, or $[a, b]$ of the real number system, or (5) the intersections of sets in (4) with those in (1)–(3).

A *chain* of poset $(P, \leq)$ is an induced subposet $(C, \leq)$ which happens to be totally ordered.

2.2.7 Zornification

Although an induced poset $(X, \leq)$ of a poset $(P, \leq)$ may not have a maximum, it *might* have an *upper bound*—that is, an element m in P which is larger than anything in the subset X (precisely rendered by "$m \geq x$ for all elements x of X"). (Of course if it happened that such an element m were already in X then it would be a global maximum of X).

The dual notion of "lower bound" should be easy to formulate.

The existence upper bounds on a class of induced posets of $(P, \leq)$ is connected with a criterion for asserting that maximal elements exist in P.

Zorn's Lemma: Suppose $(P, \leq)$ *is a poset for which any chain has an upper bound. Then any element of* $(P, \leq)$ *lies below a maximal element.*

However, Zorn's Lemma is not a Lemma or even a Theorem. It does not follow from the axioms of set theory (Zermelo-Fraenkel), nor does it contradict them. That is why we called it "a criterion for an assertion". Using it can never produce a contradiction with formal set theory. But since its denial also cannot produce such a contradiction, one can apparently have it either way. The experienced mathematician, though not always eschewing its use, at least prudently reports each appeal to "Zorn's Lemma".[3]

Zorn's Lemma is used in the next subsection. After that, it is used only very sparingly in this book.

2.2.8 Well-Ordered Sets

A *well-ordered set* is a special kind of totally ordered set. A poset $(X, \leq)$ is said to possess the *well-ordered property* (or is said to be *well-ordered*) if and only if

(a) $(X \leq)$ is totally ordered.
(b) every non-empty subset of X possesses a (necessarily unique) minimal member.

---

[3] Even an appropriate feeling of guilt is not discouraged. Who knows? Each indulgence in Zornification might revisit some of you in another life.

It should be clear from this definition that every induced subposet of a well-ordered set is also well-ordered.[4]

Any subset X of a well-ordered set A is called an *initial segment* if it possesses the property that if $x \in X$ and $y \leq x$, then $y \in X$. (In the next subsection, we shall meet sets with this property in the context of general posets. There they are called *order ideals*.) An example of an initial segment of a well-ordered set would be the set $L(a) := \{x \in A | x < a\}$. (Note that $x < a$ means $x \leq a$ while $x \neq a$.)

Lemma 2.2.1 *Suppose X is an initial segment of the well-ordered set $(A, \leq)$. Suppose $X \neq A$. Then X has the form $L(a)$, for some element $a \in A$.*

Proof By hypothesis, the set $A - X$, being non-empty, possesses a minimal element a. All elements of $L(a)$ belong to X by the minimality of a. Conversely, all elements of X are properly less than a by the definition of a. Thus $X = L(a)$. □

Theorem 2.2.2 *Any set A possesses a total ordering $\leq$ with respect to which it is well-ordered.*

Proof This proof is a classic application of Zorn's lemma. Let $\mathcal{W}$ denote the full collection of possible well-ordered posets $(W, \leq_W)$, where W is a subset of A. (Note that the same subset W may possess many possible well-orderings $(W, \leq)$, each representing a distinct element of $\mathcal{W}$.) If $(W_1, \leq_1)$ and $(W_2, \leq_2)$ are two elements of $\mathcal{W}$, we write

$$(W_1, \leq_1) \preceq (W_2, \leq_2)$$

if and only if $(W_1, \leq_1)$ is an *initial segment* of the well-ordered poset $(W_2, \leq_2)$. (Specifically, this means that there is an element $x \in W_2$ such that $W_1 = \{z \in W_2 | z \leq_2 x\}$ and the relation $\leq_1$ is just $\leq_2$ restricted to $W_1 \times W_1$.) Since an initial segment of an initial segment is an initial segment, the relation $\preceq$ is transitive and reflexive. It is clearly antisymmetric. In this way the collection of well-ordered sets $\mathcal{W}$ itself becomes a partially-ordered set with respect to the relation $\preceq$.

Now consider a chain $C = \{w_\lambda = (W_\lambda, \leq_\lambda) | \lambda \in I\}$, in the poset $(\mathcal{W}, \preceq)$. Form the set-theoretic union $W_C := \cup_{\lambda \in I} W_\lambda$. W_C inherits a natural total ordering $\leq$ derived from the $\leq_\lambda$. If x and y are elements of W_C, then $x \in W_\lambda$ and $y \in W_\mu$ for some indices $\lambda, \mu \in I$. Since C is a chain, one of these W's is contained in the other, so we may assume $(W_\lambda, \leq_\lambda) \preceq (W_\mu, \leq_\mu)$. Since both x and y lie in the totally ordered set W_μ, we write $x \leq y$ or $y \leq x$ according as $x \leq_\mu y$ or $y \leq_\mu x$. In other words, in comparing two elements of W_C, we utilize the comparison that works in any of the posets $(W_\lambda, \leq_\lambda)$ or $(W_\mu, \leq_\mu)$ that may contain both of them. The comparisons will always be consistent since each poset is an initial segment of any poset above it in the chain.

Next, we must show that the poset $(W_C, \leq)$ is well-ordered. For that purpose, consider a non-empty subset S of W_C. Choose any $x \in S$. Then $x \in W_\lambda$, for some

---

[4]We shall see very soon that a well-ordered poset is simply a chain with the descending chain condition (see p. 44 and Corollary 2.3.6).

$\lambda \in I$. Now, $(W_\lambda, \leq_\lambda)$ is a well ordered set, and $\leq_\lambda$ is the global relation $\leq$ of W_C restricted to W_λ. Since $S \cap W_\lambda$ is non-empty, it contains a minimal element m. We claim that m is a minimal element of the induced poset $(S, \leq)$. If this were false, one could find a second element $m_0 \in S$, with $m_0 \leq m$ but $m_0 \neq m$. But as $m_0 \in S \subset W_C$, m_0 lies in some W_μ. Now if $W_\mu \subseteq W_\lambda$, then $m_0 \in S \cap W_\lambda$, against the minimality of m. Thus, as C is a chain, we must have that $(W_\lambda, \leq_\lambda) \preceq (W_\mu, \leq_\mu)$. But in that case, W_λ is an initial segment of W_μ so that $m_0 < m$ implies that m_0 is in W_λ. We have just seen that this is impossible as that contradicts the minimality of m in $S \cap W_\lambda$. Thus no such m_0 exists, and m is the unique minimal element of S.

At this point, $(W_C, \leq)$ is a member of $(\mathcal{W}, \preceq)$ that is an upper bound of all members of the chain C. Since the chain C is arbitrary, we have achieved the conditions necessary for applying Zorn's Lemma. Thus we may assume there exists a maximal element $(W_m, \leq_m)$ in the poset $(\mathcal{W}, \preceq)$.

If x were a point of $A - W_m$, one could extend the relation $\leq_m$ to $(\{x\} \cup W_m) \times (\{x\} \cup W_m)$ by declaring $w \leq_m x$ for all $w \in \{x\} \cup W_m$. Then W_m would become an initial segment of $(\{x\} \cup W_m, \leq_m)$, and the latter is again a well-ordered set. Thus one obtains $(W_m, \leq_m) \prec (\{x\} \cup W_m, \leq_m)$, against the maximality of $(W_m, \leq_m)$ in $(\mathcal{W}, \preceq)$. So no such x exists, $W_m = A$, and we have obtained a well-ordering, $(A, \leq_m)$. $\square$

It is time to examine the actual structure of a well-ordered set. First, any well-ordered set $(A, \leq)$ inherits a simple partition into equivalence classes. Let us say that two elements of x and y of A are *near*, if and only if there are only a finite number of elements *between* x and y—that is, if $x \leq y$, the set $\{z \in A | x \leq z \leq y\}$ is finite, and if $y \leq x$ then $\{z \in A | y \leq z \leq x\}$ is finite.

Now, using only the fact that A is a *totally-ordered set*, we can conclude that the nearness relation between points, is transitive. For, given any three points, $\{a, b, c\}$, they possess some order—say $a \leq b \leq c$. Now if two of the intervals $[a, b]$, $[b, c]$, $[a, c]$ are finite, then so is the third interval. It follows that the relation of nearness is transitive. It is obviously symmetric and reflexive, and so the "nearness" relation is an equivalence relation on the elements of A. We let $\{A_\lambda\}$ denote the collection of nearness-equivalence classes of the well-ordered set A.

A well-ordered set A may or may not contain a maximal element. For example, any well-ordered finite set contains a maximal element, while the infinite set of natural numbers $\mathbb{N}$, under its natural ordering, is a well-ordered set with no maximal member. If A contains a maximal element m_A, let A_{max} denote the near-ness equivalence class containing m_A. In that case, there are only finitely many elements between the minimal element of the set A_{max} and the maximal element m_A, forcing A_{max} to be a finite set in this case. Otherwise, if there is no maximal element, let A_{max} be the empty set. Whether or not a maximal element exists in A, let A^* denote the set of non-maximal members of A.

The well-ordered property produces an injective mapping

$$\sigma : A^* \to A$$

which takes each non-maximal element a to the least member $\sigma(a)$ of the set $\{x \in A | x > a\}$. We call a the *predecessor* of $\sigma(a)$. Note that $a < \sigma(a)$ and that no elements properly lie between them. Similarly, for any $n \in \mathbb{N}$, if $\sigma^n(a)$ is not maximal, there are exactly $n - 1$ elements lying properly between a and $\sigma^n(a)$. Thus the elements of $\{\sigma^n(a) | \sigma^{n-1}(a) \in A^*\}$ all belong to the same nearness class.

Lemma 2.2.3 *Let A_λ be any nearness equivalence class of the well-ordered set A.*

1. *There is a least member a_λ, of the set A_λ. It has no predecessors.*
2. *Conversely, is x is an element of A that has no predecessor, then x is the least member of the near-ness equivalence class that contains it.*
3. *If $A_\lambda = A_{max}$—that is, it contains an element that is maximal in A—then it is a finite set.*
4. *If $A_\lambda \neq A_{max}$, then $A_\lambda = \{\sigma^n(a_\lambda) | n \in \mathbb{N}\}$, where a_λ is the least member of the set A_λ. In this case A_λ is an infinite countable set.*

Proof Part 1. If $a_\lambda = \sigma(x)$, then by the definition of σ, x is near a_λ, while being properly less than it. In that case, a_λ could not be the least member of its nearness class.

Part 2. If x has no predecessor and lies in A_λ, then x is near a_λ, forcing $x = \sigma^n(a_\lambda)$ for some natural number n. But since x has no predecessor, $n = 0$, and so $x = a_\lambda$.

Part 3. If m_A were a maximal element of $(A, \leq)$, then m_A would be near the least member a_{max} of its nearness equivalence class A_{max}, forcing $m_A = \sigma^n(a_{max})$, for a natural number n. Since m_A is maximal in A_{max}, $|A_{max}| = n$.

Part 4. Suppose $A_\lambda \neq A_{max}$. Then each element x of A_λ is non-maximal, and so has a successor $\sigma(x)$ that is distinct from it. Thus if a_λ is the least member of A_λ, $\{\sigma^n(a_\lambda) | n \in \mathbb{N}\}$ is an infinite set. Clearly, $A \subseteq \{\sigma^n(a_\lambda) | n \in \mathbb{N}\} \subseteq A$. $\square$

This analysis of well-ordered sets has implications for cardinal numbers in general.

Corollary 2.2.4 *Let A be any infinite set. Then A is bijective with a set of the form $\mathbb{N} \times B$. In other words, any infinite cardinal number a has the form $a = \aleph_0 b$, for some cardinal number b.*[5]

Proof By Theorem 2.2.2 one may impose a total ordering on the set A to produce a well-ordered set $(A, \leq)$. By Lemma 2.2.3 the set A_{max} of elements near a maximal element, is either finite or empty. Since A is infinite, the set $A - A_{max}$ is non-empty and has a partition

$$A - A_{max} = \bigcup \{A_\lambda | \lambda \in I\}.$$

where I indexes the nearness classes distinct from A_{max}. Each class A_λ contains a unique minimal element a_λ which has no predecessor, and each element of A_λ can

---

[5]The definition of cardinal number appears on p. 18, and $\aleph_0$ is defined to be the cardinality of the natural numbers in the paragraphs that follow.

be written as $\sigma^n(a_\lambda)$ for a unique natural number $\mathbb{N}$. Thus we have a bijection

$$A - A_{max} \to \mathbb{N} \times P^-$$

where $P^- = \{a_\lambda | \lambda \in I\}$ is the set of all elements of A which have no predecessor and are not near a maximal element.. The mapping takes an arbitrary element x of $A - A_{max}$ to the pair $(n, a_\lambda) \in \mathbb{N} \times P^-$ where a_λ is the unique minimal element of the nearness-class containing x and $x = \sigma^n(a_\lambda)$.

Now one can adjoin the finite set A_{max} to just one of the classes A_λ without changing the cardinality of that class. This produces an adjusted bijection $A \to \mathbb{N} \times P^-$, as desired. $\square$

Corollary 2.2.5 *Suppose A is an infinite set. Then A is bijective with $\mathbb{N} \times A$. For cardinal numbers, if a is any infinite cardinal number, then $a = \aleph_0 \cdot a$.*

Proof It suffices to prove the statement in the language of cardinal numbers. By Corollary 2.2.4, we may write $a = \aleph_0 \cdot b$, for come cardinal number b. Now

$$\aleph_0 \cdot a = \aleph_0 \cdot (\aleph_0 \cdot b) = (\aleph_0 \cdot \aleph_0) \cdot b = \aleph_0 \cdot b = a,$$

by Theorem 1.4.4 and the associative law for multiplying cardinal numbers.$\square$

The above Corollary is necessary for showing that any two bases of an independence theory (or matroid) have the same cardinality when they are infinite (see Sect. 2.6). That result in turn is ultimately utilized for further dimensional concepts, such as dimensions of vector spaces and transcendence degrees of field extensions.

2.2.9 Order Ideals and Filters

For this subsection fix a poset $(P, \leq)$. An *order ideal of P* is an induced subposet $(J, \leq)$, with this property:

If $y \in J$ and x is an element of P with $x \leq y$, then $x \in J$.

In other words, an order ideal is a subset J of P with the property that once some element belongs to J, then all elements of P below that element are also in J. Note that the empty subposet is an order ideal.

The reader may check that the intersection of any family of order ideals is an order ideal. (Since order ideals are a species of induced posets, we are using "intersection" here in the sense of the previous Sect. 2.1.2 on induced subposets.) Similarly, any set-theoretic union of order ideals is an order ideal.

Then there is the dual notion. Suppose $(F, \leq^*)$ were an order ideal of the dual poset $(P, \leq^*)$. Then what sort of induced poset of $(P, \leq)$ is F? It is characterized by being a subset of $(P, \leq)$ with this property:

If $x \in F$ and y is any element of P with $x \leq y$, then $y \in F$.

Any subset of P with this property is called a *filter*.

There is an easy way to construct order ideals. Take any subset X of P. Then define

$$P_X := \{y \in P | y \leq x \text{ for some element } x \in X\}.$$

Note that always $X \subseteq P_X$. In fact the order ideal P_X is "generated" by X in the sense that it is the intersection of all order ideals that contain X. Also, we understand that $P_\emptyset = \emptyset$.

In the particular case that $X = \{x\}$ we write P_x for $P_{\{x\}}$, and call P_x the *principal order ideal generated by x* (or just *a principal order ideal* if x is left unspecified).

Of course any order ideal has the form P_X (all we have to do is set $X = P_X$) but in general, we do not need all of the elements of X. For example if $x_1 < x_2$ for $x_i \in X$, and if we set $X' := X - \{x_1\}$, then $P_X = P_{X'}$. Thus if we throw out elements of X each of which is below an element left behind, the new set defines the same order ideal that X did. At first the student might get the idea that we could keep throwing out elements until we are left with an antichain. That is indeed true when X is a finite set, or more generally if every element of X is below some member of $\max(X, \leq)$. But otherwise it is generally false.

There is another kind of order ideal defined by a subset X of poset P. We set

$$\bigwedge P_X := \bigcap_{x \in X} P_x = \{y \in P | y \leq x \text{ for every } x \in X\}.$$

Of course this order ideal may be empty. If it is non-empty we say that the set X *has a lower bound in P*—that is, there exists an element $y \in P$ which is below every element of the set X.

The dual notion of the *filter generated by X* should be transparent. It is the set P^X of elements of P which bound from above at least one element of X. It could be described as

$$P^X := \{y \in P | P_y \cap X \neq \emptyset\}.$$

If $X = \{x\}$, then P^X is called a *principal filter*. By duality, the intersection and union of any collection of filters is a filter.

Then there is also the filter

$$\bigwedge P^X := \{y \in P | x \leq y \text{ for all } x \in X\},$$

which may be thought of as the set of all "upper bounds" of the set X. Of course it may very well be the empty set.

An induced subposet $(X, \leq)$ of $(P, \leq)$ is said to be *convex* if, whenever x_1 and x_2 are elements of X with $x_1 \leq x_2$ in $(P, \leq)$, then in fact the entire interval $[x, y]_P$ is contained in X. Any intersection of convex induced subposets is convex. All order ideals, all filters, and all intersections and unions thereof are convex.

2.2.10 Antichains

The reader is certainly familiar with many posets which are not totally ordered, such as the set $\mathcal{P}(X)$ of all subsets of a set X of cardinality at least two. Here the relation "$\leq$" is containment of sets. Again there are many structures that can be viewed as a collection of sets, and thus become a poset under this same containment relation: for example, subspaces of a vector space, subspaces of a point-line geometry, subgroups of a group, ideals in a ring and R-submodules of a given R-module and in general nearly any admissible subobject of an object admitting some specified set of algebraic properties.

Two elements x and y are said to be *incomparable* if both of the statements $x \leq y$ and $y \leq x$ are false. A set of pairwise incomparable elements in a poset is called an *antichain*.[6]

The set $\max(X, \leq)$ where $(X, \leq)$ is an induced subposet of $(P, \leq)$, is always an antichain.

2.2.11 Products of Posets

Suppose $(P_1, \leq)$ and $(P_2, \leq)$ are two posets.[7] The *product poset* $(P_1 \times P_2, \leq)$ is the poset whose elements are the elements of the Cartesian product $P_1 \times P_2$, where element (a_1, a_2) is declared to be less-than-or-equal to (b_1, b_2) if and only if

$$a_1 \leq b_1 \text{ and also } a_2 \leq b_2.$$

It should be clear that this notion can be extended to any collection of posets $\{(P_\sigma, \leq)|\sigma \in I\}$ to form a *direct product of posets*.. Its elements are the elements of the Cartesian product $\Pi_{\sigma \in I} P_\sigma$—that is, the functions $f : I \to U$, where U is the disjoint union of the sets P_σ with the property that at any σ in I, f always assumes a

---

[6]In a great deal of the literature, sets of pairwise incomparable elements are called *independent*. Despite this convention, the term "independent" has such a wide usage in mathematics that little is served by employing it to indicate the property of belonging to what we have called an antichain. However, some coherent sense of the term "independence" is exposed in Sect. 2.6 later in this chapter.

[7]Usually authors feel that the two poset relations should *always* have distinguished notation—that is, one should write $(P_1, \leq_1)$ and $(P_2, \leq_2)$ instead of what we wrote. At times this can produce intimidating notation that would certainly finish off any sleepy students. Of course that precaution certainly seems to be necessary if the two underlying sets P_1 and P_2 are identical. But sometimes this is a little over-done. Since we already have posets denoted by pairs consisting of the set P_i and a symbol "$\leq$", the relation "$\leq$" is assumed to be the one operating on set P_i and we have no ambiguity except possibly when the ground sets P_i are equal. Of course in the case the two "ground-sets" are equal we do not hesitate for a moment to adorn the symbol "$\leq$" with further distinguishing emblems. This is exactly what we did in defining the *dual poset*. But even in the case that $P_1 = P_2$ one could say that in the notation, the relation "$\leq$" is determined by the *name* P_i of the set, rather then the *actual set*, so even then the "ordered pair" notation makes everything clear.

value in P_σ. Then for functions f and g we have $f \leq g$ if and only if $f(\sigma) \leq g(\sigma)$ for each $\sigma \in I$.

Suppose now, each poset P_σ contains its own "zero element", 0_σ less than or equal to all other elements of P_σ. We can define a *direct sum of posets* $\{(P_\sigma, \leq)\}$ as the induced subposet of the direct product consisting only of the functions $f \in \prod_{\sigma \in I} P_\sigma$

for which $f(\sigma)$ differs from 0_σ for only finitely many σ. This poset is denoted $\coprod_{\sigma \in I} P_\sigma$.

2.2.12 Morphisms of Posets

Let $(P, \leq_P)$ and $(Q, \leq_Q)$ be two posets. A mapping $f : P \to Q$ is said to be *order preserving* (or is said to be a *poset morphism*) if and only if

$$x \leq_P y \text{ implies } f(x) \leq_Q f(y).$$

Evidently, the composition $g \circ f$ of two poset morphisms $f : (P, \leq_P) \to (Q, \leq_Q)$ and $g : (Q, \leq_Q) \to (R, \leq_R)$ is also a poset morphism $(P, \leq_P) \to (R, \leq_R)$. Clearly the identity mapping $1_P : P \to P$ is a morphism. If $f : P \to Q$ is a poset morphism as above, then $f \circ 1_P = 1_Q \circ f = f$. The student should be aware that if x and y are incomparable elements of P, it is still quite possible that $f(x) \leq_Q f(y)$ or $f(y) \leq_Q f(x)$ in the poset $(Q, \leq_Q)$.

To clarify this point a bit further, using the morphism $f : (P, \leq_P) \to (Q, \leq_Q)$, let us form the *image poset* $f(P) := (f(P), \leq_f)$ whose elements are the images $f(p), p \in P$, and we write $f(x) \leq f(y)$ if and only if there is a pair of elements $(x', y') \in f^{-1}(f(x)) \times f^{-1}(f(y))$, the Cartesian product of the fibers above $f(x)$ and $f(y)$, such that $x' \leq_P y'$. Then the image poset $(f(P), \leq_f)$ is a subposet of $(Q, \leq_Q)$.

We say that the morphism f is *full* if and only if the image poset $(f(P), \leq)$ is an *induced poset* of $(Q, \leq_Q)$. Thus for a full morphism, we have $a \leq_Q b$ in the image, if and only if there exist elements x and y in P such that $x \leq_P y$ and $f(x) = a$ and $f(y) = b$.

A bijection $f : P \to Q$ is an *isomorphism of posets* if and only if it is also a full poset morphism. Thus if f is an isomorphism, we have $f(x) \leq_Q f(y)$ if and only if $x \leq_P y$. In this case the inverse mapping $f^{-1} : Q \to P$ is also an isomorphism. Thus an isomorphism really amounts to changing the names of the elements and the name of the relation but otherwise does not change anything. Isomorphism of posets is clearly an equivalence relation and we call the corresponding equivalence classes *isomorphism classes of posets*.

If the order preserving mapping $f : P \to Q$ is *injective* then the posets $(P, \leq)$ and $(f(P), \leq_f)$ are isomorphic, and we say that f *is an embedding of poset* $(P, \leq)$ *into* $(Q, \leq)$. The following result, although not used anywhere in this book, at least displays the fact that one is interested in cases in which the image poset of an embedding is *not* an induced subposet of the co-domain.

Any poset $(P, \leq)$ can be embedded in a totally ordered set.[8]

2.2.13 Examples

Example 1 (*Examples of totally ordered sets*) Isomorphism classes of these sets are called *ordinal numbers*. The most familiar examples are these:

1. *The natural numbers* This is the set of non-negative integers,

$$\mathbb{N} := \{0, 1, 2, 3, 4, 5, 6, \ldots\},$$

with the usual ordering

$$0 < 1 < 2 < 3 < 4 < 5 \cdots .$$

Rather obviously, as an ordered set, it is isomorphic to the chain

$$1 < 2 < 3 < \cdots$$

or even

$$k < k + 1 < k + 2 < k + 3 < \cdots ,$$

k any integer, under the shift mapping $n \to n + k - 1$.[9]

Recall that a poset $(X, \leq)$ is said to possess the *well-ordered property* (or is said to be *well-ordered*) if and only if (i) $(X, \leq)$ is totally ordered, and (ii) every non-empty subset of X possesses a (necessarily unique) minimal member. It should be clear from this definition that every induced subposet of a well-ordered poset is also well-ordered. The point here is that the natural numbers is a well-ordered poset under the usual ordering. This fundamental principle is responsible for some of the basic properties concerning greatest common divisors (see Chap. 3, p. 2).

2. The system of *integers*

$$\mathbb{Z} := \{\ldots < -2 < -1 < 0 < 1 < 2 < \ldots\}.$$

---

[8] Many books present an equivalent assertion "any poset has a linear extension". The proof is an elementary induction for finite posets. For infinite posets it requires some grappling with Zorn's Lemma and ordinal numbers.

[9] This isomorphism explains why it is commonplace to do an induction proof with respect to the second of these examples beginning with 1 rather than the first, which begins with 0.

  In enumerative combinatorics, for example, the "natural numbers" $\mathbb{N}$ are defined to be all non-negative integers, not just the positive integers (see *Enumerative Combinatorics, vol 1*, p. 1. by R. Stanley) [1].

One notes that any subset of $\mathbb{Z}$ which possesses a lower bound, forms a well-ordered induced subposet of $\mathbb{Z}$.

3. The system $\mathbb{Q}$ of *the rational numbers* with respect to the usual ordering—that is $a/b > c/d$ if and only if $ad > bd$, an inequality of integers.
4. The real number system $\mathbb{R}$.
5. Any induced subposet of a totally ordered set. We have already mentioned intervals of the real line. (Remark: the word "interval" here is used for the moment as it is used in Freshman College Algebra, open, closed, and half-open intervals such as $(a, b]$ or $[a, \infty)$. In this context, the intervals of posets that we have defined earlier, become the *closed* intervals, $[a, b]$, of the real line, with a consistency of notation.

Here is an example: Consider the induced poset of the rational numbers $(\mathbb{Q}, \leq)$ consisting of those positive fractions less than or equal to $1/2$ which (in lowest terms) have a denominator not exceeding the positive integer d in absolute value. For $d = 7$ this is the chain

$$\frac{1}{7} < \frac{1}{6} < \frac{1}{5} < \frac{1}{4} < \frac{2}{7} < \frac{1}{3} < \frac{2}{5} < \frac{3}{7} < \frac{1}{2}.$$

This is called a *Farey series*. A curiosity is that if $\frac{a}{b}$ and $\frac{c}{d}$ are adjacent members from left to right in such a series, then $bc - ad = 1$!

Example 2 (Examples of the classical locally finite (or finite) posets which are not chains) A poset $(P, \leq)$ is said to be a *finite poset* if and only if it contains only finitely many elements—that is, $|P| < \infty$. It is said to be *locally finite* if and only if every one of its intervals $[x, y]$ is a finite poset.

1. The *Boolean poset* $B(X)$ of all finite subsets of a set X, with the containment relation $(\subseteq)$ between subsets as the partial-ordering. (There is, of course, the *power poset* $\mathcal{P}(X)$, the collection of *all* subsets of X, as well as the *cofinite poset* which is the collection $B^*(X)$ of all subsets of X whose complement in X is finite—both collections being partially ordered by the inclusion relation. Of course, these two posets $\mathcal{P}(X)$, and $B^*(X)$ are not locally finite unless X is finite.)
2. The *divisor poset* $\mathbb{D}$ of all positive integers $\mathbb{N}^+$ under the divisor relation:—that is, we say $a|b$ if and only if integer a *divides* integer b evenly—i.e. b/a is an integer.[10]

---

[10]There are variations on this theme: In an integral domain a non-unit a is said to be *irreducible* if and only if $a = bc$ implies one of b or c is a unit. Let D be an integral domain in which each non-unit is a product of finitely many irreducible elements, and let U be its group of units. Let D^*/U be the collection of all non-zero multiplicative cosets Ux. Then for any two such cosets, Ux and Uy, either every element of Ux divides every element of Uy or else no element of Ux divides any element of Uy. In the former case write $Ux \leq Uy$. Then $(D^*/U, \leq)$ is a poset. If D is a unique factorization domain, then, as above, $(D^*/U, \leq)$ is locally finite for it is again a product of chains (one factor in the product for each association class Up of irreducible elements).

One might ask what this poset looks like when D is *not* a unique factorization domain. Must it be locally finite? It's something to think about.

3. *Posets of vector subspaces.* The partially ordered set $\mathbf{L}_{<\infty}(V; q)$ of all finite-dimensional vector subspaces of a vector space V over a finite field of q elements is a locally finite poset. (There is a generalization of this: the poset $\mathbf{L}_{<\infty}(M)$ of all finitely generated submodules of a right R-module M and in particular the poset $\mathbf{L}_{<\infty}(V)$ of all finite-dimensional subspaces of a right-vector space V over some division ring. But of course these are not locally finite in general.)[11]

4. *The partition set:* Π_n. Suppose X is a set of just n elements. Recall that a partition of X is a collection $\pi := \{Y_1, \ldots, Y_k\}$ of non-empty subsets Y_j whose join is X but which pairwise intersect at the empty set. The subsets Y_j are called the *components of the partition* π.

   Suppose $\pi_1 := \{Y_i | i \in I\}$ is a partition of X and $\pi' = \{Z_k | k \in K\}$ is a second partition. We say that *partition* π' *refines partition* π if and only if there exists a partition $I = J_1 + \cdots J_k$ of the index set, such that

$$Y_i := \bigcup_{\ell \in J_i} Z_\ell.$$

   [We can state this another way: A partition π can be associated with a surjective function $f_\pi : X \to I$ where the preimages of the points are the fibers partitioning X: the same being true of π' and an surjective function $f_{\pi'} : X \to K$. We say that *partition* π' *refines a partition* π if and only if there exists a surjective mapping $\phi : K \to I$, such that $f_\pi = \phi \circ f_{\pi'}$—that is, for each $x \in X$, $f_\pi(x) = \phi(f_{\pi'}(x))$.]
   For example $\{6\}\{4, 9\}\{2, 3, 8\}\{1, 5, 10\}\{9, 10\}$, with five components refines $\{1, 5, 6, 9, 10\}$, $\{2, 3, 4, 8, 9\}$ with just two components.
   Then Π_n is the partially ordered set of all partitions of the n-set X under the refinement relation.[12]

5. *The Poset of Finite Multisets:* Suppose X is any non-empty set. A *multiset* is essentially a sort of inventory whose elements are drawn from X. For example: if $X = \{$oranges, apples, and bananas$\}$ then $m = \{$three oranges, two apples$\}$ is an inventory whose elements are multiple instances of elements from X. Letting $O =$ oranges, $A =$ apples, and $B =$ bananas, one may represent the multiset m by the symbol $3 \cdot O + 2 \cdot A + 0 \cdot B$ or even the sequence $(3, 2, 0)$ (where the order of the coordinates corresponds to a total ordering of X). But both of these notations can become inadequate when X is an infinite set. The best way is to think of a multiset as a mapping. Precisely, a *multiset* is a mapping

$$f : X \to \mathbb{N},$$

   from X into the set $\mathbb{N}$ of *non-negative* integers.

---

[11] In Aigner's book (see references), $\mathbf{L}_{<\infty}(V, q)$ is denoted $\mathcal{L}(\infty, q)$ in the case that V has countable dimension over the finite field of q elements. This makes sense when one's plan is to relate the structure to certain types of generating functions (the q-series). But of course, it is a well-defined locally finite poset whatever the dimension of V.

[12] Its cardinality $|\Pi_n|$ is called the *nth Bell number* and will reappear in Chap. 4 in the context of permutation characters.

A multiset f is *dominated* by a multiset g (written $f \leq g$) if and only if

$$f(x) \leq g(x) \text{ for all } x \in X$$

(where the "$\leq$" in the presented assertion reflects the standard total ordering of the integers).

The collection $\hom(X, \mathbb{N})$ of all multisets of a set X forms a partially ordered set under the dominance relation. Since the multisets are actually mappings from X to $\mathbb{N}$ the dominance relation is exactly that used in comparing mappings in products. We are saying that the definition of the poset of multisets shows us that

$$(\hom(X, \mathbb{N}), \leq) = \prod_{x \in X} \mathbb{N}, \tag{2.3}$$

—that is a product in which each "factor" P_σ in the definition of product of posets is the constant poset $(\mathbb{N}, \leq)$ of non-negative integers.

The multiset f is said to be a *finite multiset of magnitude* $|f|$ if and only if

$$f(x) > 0 \text{ for only finitely many values of } x, \text{ and} \tag{2.4}$$

$$|f| = \sum_{x \in X, f(x) > 0} f(x), \tag{2.5}$$

where the sum in the second equation is understood to be the integer 0 when the range of summation is empty (i.e. $f(x) = 0$ for all $x \in X$).

Thus in the example concerning apples, bananas, and oranges above, the multiset m is finite of magnitude $3 + 2 = 5$.

In this way the collection $\mathcal{M}_{<\infty}(X)$ of all finite multisets forms an induced poset of $(\hom(X, \mathbb{N}), \leq)$. Next one observes that a mapping $f : X \to \mathbb{N}$ is a finite multiset if and only if $f(x) = 0$ for all but a finite number of instances of $x \in X$. This means Eq. (2.3) has a companion with the product replaced by a sum:

$$\mathbf{M}_{<\infty}(X) = \coprod_{x \in X} \mathbb{N}. \tag{2.6}$$

The above list of examples shall continue in the subsequent sections.

2.2.14 Closure Operators

We need a few other definitions related to poset mappings.

An order preserving mapping $f : P \to P$ is said to be *monotone non-decreasing* if $p \leq f(p)$ for all elements p of P.

A *closure operator* is a monotone non-decreasing poset homomorphism $\tau : P \to P$ which is *idempotent*. In other words, τ possesses the following three properties:

(i) (Monotonicity) If $x \in P$, then $x \le \tau(x)$
(ii) (Homomorphism) If $x \le y$, then $\tau(x) \le \tau(y)$.
(ii) (Idempotence) $\tau(x) = \tau(\tau(x))$ for all $x \in P$.

We call the images of τ the *closed elements* of P.

There are many contexts in which closure operators arise, and we list a few.

1. The ordinary topological closure in the poset of subsets of a topological space.
2. The mapping which takes a subset of a group (ring or R-module) to the subgroup (subring or submodule, resp.) generated by that set in the poset of all subsets of a group (ring or R-module).
3. The mapping which takes a set of points to the subspace which they generate in a point-line geometry $(\mathcal{P}, \mathcal{L})$.[13]

2.2.15 Closure Operators and Galois Connections

One interesting context in which closure operators arise are Galois connections. Let $(P, \le)$ and $(Q, \le)$ be posets. A mapping $f : P \to Q$ is said to be *order reversing* if and only $x \le y$ implies $f(x) \ge f(y)$.

Example 3 This example displays a common context that produces order reversing mappings between posets. Suppose $X = \cup_{\sigma \in I} A_\sigma$, a union of non-empty sets indexed by I. Now there is a natural mapping among power posets:

$$\alpha : \mathcal{P}(X) \to \mathcal{P}(I),$$

which takes each subset Y of X, to

$$\alpha(Y) := \{\sigma \in I | Y \subseteq A_\sigma\}.$$

For example, if Y is contained in no A_σ, then $\alpha(Y) = \emptyset$. Now if $Y_1 \subseteq Y_2 \subseteq X$ we see that $\alpha(Y_2) \subseteq \alpha(Y_1)$—that is, as the Y_i get larger, there are generally fewer A_σ that contain them. Thus the mapping α is order-reversing.

Let $(P, \le)$ and $(Q, \le)$ be posets. A *Galois connection* (P, Q, α, β) is a pair of order-reversing mappings $\alpha : P \to Q$ and $\beta : Q \to P$, such that the two compositions $\beta \circ \alpha : P \to P$ and $\alpha \circ \beta : Q \to Q$ are both monotone non-decreasing.

---

[13] Here, the set of lines, $\mathcal{L}$, is simply a family of subsets of the set of points, $\mathcal{P}$. A *subspace* is a set S of points, with the property that if a line $L \in \mathcal{L}$ contains at least two points of S, then $L \subseteq S$. Thus the empty set and the set $\mathcal{P}$ are subspaces. From the definition, the intersection over any family of subspaces, is a subspace. The subspace *generated* by a set of points X is defined to be the intersection of all subspaces which contain X.

Galois connections arise in several contexts, especially when some algebraic object acts on something.

Example 4 (Groups acting on sets) Suppose that G is a group of bijections of a set X into itself. The group operation is composition of bijections and the identity element is the identity mapping $1_X : X \to X$ defined by $x \mapsto x$ for all $x \in X$. For $g \in G$, the corresponding bijection or permutation is described as an exponential operator by $x \mapsto x^g$ for all $x \in X$.

Consider the two posets $\mathcal{P}(X)$ and $\mathcal{P}(G)$, the power posets of X and G. Define

$$C_G : \mathcal{P}(X) \to \mathcal{P}(G)$$

by setting $C_G(U) := \{g \in G \mid g(u) = u^g = u \text{ for all } u \in U\}$, for each subset U of X. This mapping is order reversing: if $U \subseteq V$, then $C_G(U) \supseteq C_G(V)$. Conversely, if H is a subset of G, set $\text{Fix}(H) := \{x \in X \mid x = x^h \text{ for all } h \in H\}$.

Then Fix is an order-reversing mapping $\mathcal{P}(G) \to \mathcal{P}(X)$. Therefore, $(\mathcal{P}(X), \mathcal{P}(G), C_G, \text{Fix})$ becomes a Galois connection upon verifying that the compositions $C_G \circ \text{Fix} : \mathcal{P}(G) \to \mathcal{P}(G)$ and $\text{Fix} \circ C_G : \mathcal{P}(X) \to \mathcal{P}(X)$ are monotone non-decreasing.

Example 5 In the above example, the set X might have extra structure that is preserved by G. For example X might itself be a group, ring, field, or a vector space, and G is a group of automorphisms of X. This situation arises in the classical Galois theory studied in Chap. 11, where X is a field, and where G is a group of automorphisms fixing a subfield Y of X.

Example 6 A ring R may act as a ring of endomorphisms of an abelian group A with the (multiplicative) identity element inducing the identity mapping on A. One can then form a Galois connection $(\mathcal{P}(A), \mathcal{P}(R), C_R, \text{Fix})$ with

$$C_R := \{r \in R \mid u^r = u \text{ for all } u \in U\},$$

$$\text{Fix}(S) := \{a \in A \mid a^s = a \text{ for all } s \in S\}$$

for all $U \in \mathcal{P}(A)$ and for all $S \in \mathcal{P}(R)$.

Example 7 Another example arises in algebraic geometry. We say that a polynomial $p(x_1, \ldots, x_n)$ *vanishes* at a vector $v = (a_1, \ldots, a_n)$ in the vector space $F^{(n)}$ of n-tuples over the field F if and only if $p(v) := p(a_1, \ldots, a_n) = 0 \in F$. Let $(P, \leq)$ be the poset of ideals in the polynomial ring $F[x_1, \ldots, x_n]$ with respect to the containment relation and for each ideal I let $\alpha(I)$ be the set of vectors in $F^{(n)}$ at which each polynomial of I vanishes. Let $(Q, \leq)$ be the poset of all subsets of $F^{(n)}$ with respect to containment. For any subset X of $F^{(n)}$, let $\beta(X)$ be the set of all polynomials which vanish simultaneously at every vector in X. Then (P, Q, α, β) is a Galois connection.

The Corollary following the Lemma below shows how closure operators can arise from Galois connections.

Lemma 2.2.6 *Suppose (P, Q, α, β) is a Galois connection. Then for any elements p and q of P and Q respectively*

$$\alpha(\beta(\alpha(p))) = \alpha(p) \text{ and } \beta(\alpha(\beta(q))) = \beta(q)$$

Proof For $p \in P$, $p \leq \beta(\alpha(p))$ since $\beta \circ \alpha$ is monotone non-decreasing. Thus $\alpha(\beta(\alpha(p))) \leq \alpha(p)$ since α is order reversing.

On the other hand

$$\alpha(p) \leq \alpha(\beta(\alpha(p))) = (\alpha \circ \beta)(\alpha(p))$$

since $(\alpha \circ \beta)$ is monotone non-decreasing. By antisymmetry, we have the first equation of the statement of the Lemma.

The second statement then follows from the symmetry of the definition of Galois connection—that is (P, Q, α, β) is a Galois connection if and only if (Q, P, β, α) is. $\square$

Corollary 2.2.7 *If (P, Q, α, β) is a Galois connection, then $\tau := \beta \circ \alpha$ is a closure operator on $(P, \leq)$.*

Proof Immediate upon taking the β-image of both sides of the first equation of the preceding Lemma. $\square$

Example 8 Once again consider the order reversing mapping $\alpha : \mathcal{P}(X) \to \mathcal{P}(I)$ of Example 3, where X was a union of non-empty sets A_σ indexed by a parameter σ ranging over a set I. For every subset Y of X, $\alpha(Y)$ was the set of those σ for which $Y \subseteq A_\sigma$.

There is another mapping $\beta : \mathcal{P}(I) \to \mathcal{P}(X)$ defined as follows: If $J \subseteq I$ set $\beta(J) := \cap_{\sigma \in J} A_\sigma$, with the understanding that if $J = \emptyset$, then $\beta(J) = X$. Then β is easily seen to be an order-reversing mapping between the indicated power posets.

Now the mapping $\tau = \beta \circ \alpha : \mathcal{P}(X) \to \mathcal{P}(X)$ takes each subset Y of X to the intersection of all of the A_σ which contain it (with the convention that an intersections over an empty family of sets denotes the set X itself). Thus τ is a nice closure operator.

Similarly, $\rho = \alpha \circ \beta : \mathcal{P}(I) \to \mathcal{P}(I)$ takes each subset J to the set

$$\rho(J) = \{\sigma \in I | A_\sigma \supseteq \cap_{j \in J} A_j\}.$$

2.3 Chain Conditions

2.3.1 Saturated Chains

Recall that a *chain C of poset* $(P, \leq)$ is simply a (necessarily induced) subposet $(C, \leq)$ which is totally ordered. In case C is a finite set, $|C| - 1$ is called the *length* of the chain.

We say that a chain C_2 *refines* a chain C_1 if and only if $C_1 \subseteq C_2$. The chains of a poset $(P, \leq)$ themselves form a partially ordered set under the inclusion relation (the dual of the refinement relation) which we denote $(ch(P, \leq), \subseteq)$.

Suppose now that

$$C_0 \subseteq C_1 \subseteq \cdots$$

is an ascending chain in the poset $(ch(P, \leq), \subseteq)$. Then the set-theoretic union $\bigcup C_i$ is easily seen to be totally ordered and so is an upper bound in $(ch(P, \leq), \subseteq)$ of this chain. An easy application of Zorn's Lemma then shows that every element of the poset $(ch(P, \leq), \subseteq)$ lies below a maximal member. These maximal chains are called *unrefinable* chains. Thus

Theorem 2.3.1 *If C is a chain in any poset* $(P, \leq)$, *then C is contained in an unrefinable chain of* $(P, \leq)$.

Of course we can restrict this in special ways to an interval. In a poset $(P, \leq)$, a *chain from x to y* is a chain $(C, \leq)$ with x as its unique minimal element and y as it unique maximal element. Thus $x \leq y$ and $\{x, y\} \subseteq C \subseteq [x, y]$.

The collection of all chains of $(P, \leq)$ from x to y itself becomes a poset $(ch[x, y], \subseteq)$ under the containment (or "corefinement") relation. A chain from x to y is said to be *saturated* if it is a maximal element of $(ch[x, y], \subseteq)$.[14]

We can apply Theorem 2.3.1 for $P = [x, y]$ and the chains C in it that *do* contain $\{x, y\}$ to obtain:

Corollary 2.3.2 *If C is a chain from x to y in a poset* $(P, \leq)$, *then there exists a saturated chain C' from x to y which refines C. In particular, given an interval* $[x, y]$ *of a poset* $(P, \leq)$, *there exists an unrefinable chain in P from x to y.*

2.3.2 Algebraic Intervals and the Height Function

Let $(P, \leq)$ be any poset. An interval $[x, y]$ of $(P, \leq)$ is said to be *algebraic* (or *of finite height*) if and only if *there exists* a saturated chain from x to y of finite

---

[14] Note that $(ch[x, y], \subseteq)$ is not quite the same as $(ch([x, y], \leq)$ since the latter may contain chains which, although lying in the interval $[x, y]$, do not contain x or y.

length.[15] Note that to assert that $[a, b]$ is an algebraic interval, does not preclude the simultaneous existence of infinite chains from a to b. The *height* of an algebraic interval $[x, y]$ is then defined to be the *minimal length* of a saturated chain from x to y.[16]

If $[a, b]$ is an algebraic interval, its *height* is denoted $h(a, b)$, and is always a non-negative integer. We denote the collection of all algebraic intervals of $(P, \leq)$ by the symbol $\mathcal{A}_P$.

Proposition 2.3.3 *The following hold:*

(i) *If $[a, b]$ and $[b, c]$ are algebraic intervals of poset $(P, \leq)$, then $[a, c]$ is also an algebraic interval.*

(ii) *The height function $h : \mathcal{A}_P \to \mathbb{N}$ from the non-empty algebraic intervals of $(P, \leq)$ to the non-negative integers, satisfies this property: If $[a, b]$ and $[b, c]$ are algebraic intervals, then*

$$h(a, c) \leq h(a, b) + h(b.c)$$

2.3.3 The Ascending and Descending Chain Conditions in Arbitrary Posets

Let $(P, \leq)$ be any partially ordered set. We say that P satisfies the *ascending chain condition* or *ACC* if and only if every ascending chain of elements of P stabilizes after a finite number of steps—that is, for any chain

$$p_1 \leq p_2 \leq p_3 \leq \cdots,$$

there exists a positive integer N, such that

$$p_N = p_{N+1} = \cdots = p_k, \text{ for all integers } k \text{ greater than } N.$$

Put another way, $(P, \leq)$ has the ACC if and only if every properly ascending chain

$$p_1 < p_2 < \cdots$$

---

[15]This adjective "algebraic" does not enjoy uniform usage. In Universal Algebras, elements which are the join of finitely many atoms are called *algebraic elements* (perhaps by analogy with the theory of field extensions). Here we are applying the adjective to an *interval*, rather than an element of a poset.

[16]Since the adjective "algebraic" entails the existence of a finite unrefinable chain, the height of a algebraic interval is always a natural number. The term "height" is used here instead of "length" which is appropriate when *all* unrefineable chains have the same length, as in the semimodular lower semilattices that appear in the Jordan-Hölder Theorem.

terminates after some finite number of steps. Finally, it could even be put a third way: $(P, \leq)$ satisfies the ascending chain condition if and only if there is *no* countable sequence $\{p_n\}_{n=1}^{\infty}$ with $p_n < p_{n+1}$ for all natural numbers n.

Lemma 2.3.4 *For any poset $P = (P, \leq)$ the following assertions are equivalent:*

(i) *$(P, \leq)$ satisfies the ascending chain condition (ACC).*

(ii) *(The Maximum Condition) Every non-empty subset X of P the induced subposet $(X, \leq)$ contains a maximal member.*

(iii) *(The Second form of the Maximum Condition) In particular, for every induced poset X of $(P, \leq)$, and every element $x \in X$, there exists an element $y \in X$ such that x is bounded by y and y is maximal in $(X, \leq)$. (To just make sure that we understand this: for every $x \in X$ there exists an element $y \in X$ such that*

(a) *$x \leq y$.*

(b) *If $u \in X$ and $y \leq u$ then $u = y$.)*

Remark The student is reminded: to say that "*x* is a *maximal member* of a subset X of a poset P" simply means x is an element of X which is not properly less than any other member of X.

Proof of Lemma 2.3.4:

1. (The ACC implies the Maximum Condition). Let X be any nonempty subset of P. By way of contradiction, assume that X contains no maximal member. Choose $x_1 \in X$. Since x_1 is not maximal in X, there exists an element $x_2 \in X$ with $x_1 < x_2$. Suppose now, that we have been able to extend the chain $x_1 < x_2$ to $x_1 < \cdots < x_n$. Since x_n is not maximal in X, there exists an element $x_{n+1} \in X$, such that $x_n < x_{n+1}$. Thus, by mathematical induction, for every positive integer n, the chain $x_1 < \cdots < x_n$, can be extended to $x_1 < \cdots < x_n < x_{n+1}$. Taking the union of these extensions one obtains an infinite properly ascending chain

$$x_1 < x_1 < x_3 < \cdots,$$

contrary to the assumption of ACC.[17]

2. (The first version of the Maximum Principle implies the second.) Now assume only the first version of the maximum condition. Take a subset X and an element $x \in X$. Then set $X' = X \cap P^x$ where $P^x := \{z \in P | x \leq z\}$ is the principal filter

---

[17] The graduate student has probably encountered arguments like this many times, where a sequence with certain properties is said to exist because after the first n members of the sequence are constructed, it is always possible to choose a suitable $n + 1$-st member. This has an uncomfortable feel to it, for the sequence alleged to exist must exemplify infinitely many of these choices—at least invoking the Axiom of Choice in choosing the x_i. But in a sense it appears worse. The sets are not just sitting there as if we had prescribed non-empty sets of socks in closets lined up in an infinite hallway (the traditional folk-way model for the Axiom of Choice). Here, it as if each new closet was being defined by our choice of sock in a previous closet, so that it is really a statement about the existence of infinite paths in trees having no vertex of degree one. All we can feebly tell you is that it is basically equivalent to the Axiom of Choice.

generated by x in the induced poset $(X, \leq)$. Then X' is non-empty (it contains x) and so we obtain an element y maximal in X' from the first version of the maximum condition. Now if y were not a maximal member of X there would exist an element $y' \in X$ with $y < y'$. But in that case $x < y'$ so $y' \in X \cap P^x = X'$. But that would contradict y being maximal in X'. Thus y is in in fact maximal in X and dominates x as desired.

3. (The second version of the Maximum Principle implies the ACC.) Assume the second version of the Maximum Principle. Suppose the ACC failed. Then there must exist an infinite properly ascending chain $x_0 < x_1 < \cdots$. Setting $X = \{x_i | i \in \mathbb{N}\}$, and $x = x_0$, we see there is no maximal member of X dominating x, contrary to the statement of the second version of the Maximum Principle. $\square$

Of course, by replacing the poset $(P, \leq)$ by its dual $P^* := (P, \geq)$ and applying the above, we have the dual development:

We say a poset $(P, \leq)$ possesses the *descending chain condition* or *DCC* if and only if every descending chain

$$p_1 \geq p_2 \geq \cdots \geq p_j \in P,$$

stabilizes after a finite number of steps. That is, there exists a positive integer N such that $p_N = p_{N+1} = \cdots p_{N+k}$ for all positive integers k.

We say that x is a *minimal member* of the subset X of P if and only if $y \geq x$ for $y \in X$ implies $x = y$. A poset $(P, \leq)$ is said to possess the *minimum condition* if and only if

(**Minimum Condition**)   *Every nonempty subset of elements of the poset $P = (P, \leq)$ contains a minimal member.*

Then we have:

Lemma 2.3.5 *A poset $P = (P, \leq)$ has the descending condition (DCC)*

(i) *if and only if it satisfies the minimum condition or*
(ii) *if and only if it satisfies this version of the minimum condition: for any induced poset X any $x \in X$, there exists an element y minimal in X which is bounded above by x.*

Proof: Just the dual statement of Lemma 2.3.4. $\square$

Corollary 2.3.6 *Every non-empty totally ordered poset with the descending chain condition (DCC), is a well-ordered set.*

Proof Let $(P, \leq)$ be a non-empty totally-ordered poset with the DCC. By Lemma 2.3.5, the minimum condition holds. The latter implies that any non-empty subset X contains a minimal member, say m. Since $(X, \leq)$ is totally ordered, m is a global minimum of $(X, \leq)$. Thus $(P, \leq)$ is well-ordered.[18] $\square$

---

[18]This conclusion reveals the incipient presence of the Axiom of Choice/Zorn's Lemma in the argument of the first paragraph of the proof of Lemma 2.3.4.

Corollary 2.3.7 *(The chain conditions are hereditary) If X is any induced poset of* $(P, \leq)$, *and P satisfies the ACC (or DCC), then X also satisfies the ACC (or DCC, respectively).*

Proof This is not really a Corollary at all. It follows from the definition of "induced poset" and the chain-definitions directly. Any chain in the induced poset is *a fortiori* a chain of its ambient parent. We mention it only to have a signpost for future reference.

In the next section, Theorem 2.4.2 will have a surprising consequence for lower semilattices with the DCC.

Any poset satisfying both the ACC and the DCC also satisfies

(FC) *Any unrefinable chain of P has finite length.*

This is because patently one of the two chain conditions is violated by an infinite unrefinable chain. Conversely, if a poset P satisfies condition (FC) then there can be no properly ascending or descending chain of infinite length since by Theorem 2.3.1 it would then lie in a saturated chain which was also infinite against (FC). Thus

Lemma 2.3.8 *The condition (FC) is equivalent to the assumption of both DCC and ACC.*

2.4 Posets with Meets and/or Joins

2.4.1 Meets and Joins

Let W be any subset of a poset $(P, \leq)$. The *join of the elements of* W is an element v in $(P, \leq)$ with these properties:

1. $w \leq v$ for all $w \in W$.
2. If v' is an element of $(P, \leq)$ such that $w \leq v'$ for all $w \in W$, then $v \leq v'$.

Similarly there is the dual notion: the *meet of the elements of* W in P would be an element m in P such that

1. $m \leq w$ for all $w \in W$.
2. If m' is an element of $(P, \leq)$ such that $m' \leq w$ for all $w \in W$, then $m' \leq m$.

Of course, P may or may not possess a meet or a join of the elements of W. But one thing is certain: if the meet exists it is unique; if the join exists, it is unique. Because of this uniqueness we can give these elements names. We write $\bigwedge_P(W)$ (or just $\bigwedge(W)$ if the ambient poset P is understood) for the meet of the elements of W in P (if it exists). Similarly, we write $\bigvee_P(W)$ (or $\bigvee(W)$) for the join in P of all of the elements of W (if that exists).

In the case that W is the set $\{a, b\}$, we render the meet and join of a and b by $a \wedge b$ and $a \vee b$, respectively. The reader may verify

$$a \wedge b = b \wedge a \tag{2.7}$$

$$a \wedge (b \wedge c) = (a \wedge b) \wedge c \tag{2.8}$$

$$(a \wedge b) \vee (a \wedge c) \le a \wedge (b \vee c) \tag{2.9}$$

and the three dual statements when the indicated meets and joins exist.

We say that $(P, \le)$ is a *lower semilattice* if it is "meet-closed"—that is, the meet of any two of its elements exists. In this case it follows that the meet of any finite subset $\{a_1, a_2, \ldots, a_n\}$ exists (see Exercise (9) in Sect. 2.7.2). We denote this meet by $a_1 \wedge a_2 \wedge \cdots \wedge a_n$. We then have

Lemma 2.4.1 *Suppose P is a lower semilattice, containing elements $x, a_1, a_2, \ldots,$ $a_n \in P$ such that $x \le a_i$ for $i = 1, 2, \ldots, n$. Then*

$$x \le a_1 \wedge a_2 \wedge \cdots \wedge a_n.$$

Dually we can define an *upper semilattice* (it is "join closed").

A *lattice* is a poset $(P, \le)$ that is both a lower semilattice and an upper semilattice—that is, the meet and join of any two of its elements both exist in P. Thus, a lattice is a self-dual concept: If $P = (P, \le)$ is a lattice, then so is its dual poset $P^* = (P, \le^*)$.

If every non-empty subset U of P has a meet (join) we say that *arbitrary meets (joins) exists*. If both arbitrary meets and joins exist we say that P is a *complete lattice*.

Example 9 Here are some familiar lattices:

1. Any totally ordered poset is a lattice. The meet of a finite set is its minimal member; its join is its maximal member. Considering the open real interval $(0, 1)$ with its induced total ordering from the real number system, it is easy to see

   (a) that there are lattices with no "zero" or "one",
   (b) that there can be (infinite) subsets of a lattice with no lower bound or no upper bound.

2. The *power set* $\mathcal{P}(X)$ is the poset of all subsets of a set X under the containment relation. It is a lattice with the intersection of two sets being their meet, and the union of two sets being their join. This lattice is self dual and is a *complete lattice*, meaning that *any* subset of elements of $\mathcal{P}(X)$, whether infinite or not, has a least upper bound and greatest lower bound—i.e. a meet and a join. Of course the lattice has X as its "one" and the empty set as its "zero".

2.4.2 Lower Semilattices with the Descending Chain Condition

The proof of the following theorem is really due to Richard Stanley, who presented it for the case that P is finite [36, Proposition 3.3.1, p. 103].

Theorem 2.4.2 *Suppose P is a meet-closed poset (that is, a lower semilattice) with the DCC condition.*
   Then the following assertions hold:

   (i) *If X is a meet-closed induced poset of P, then X contains a unique minimal member.*
  (ii) *Suppose, for some subset X of P, the filter of upper bounds*

$$\bigwedge P^X := \{y \in P | x \le y \text{ for all } x \in X\}$$

   is non-empty. Then the join $\bigvee (X)$ exists in P.
   In particular, if $(P, \le)$ possesses a one-element $\hat{1}$, then for any arbitrary subset X of P, there exists a join $\bigvee (X)$. Put more succinctly, if $\hat{1}$ exists, unrestricted joins exist.
 (iii) *For any non-empty subset X of P, the universal meet $\bigwedge (X)$ exists and is expressible as a meet of a finite number of elements of X.*
  (iv) *The poset P contains a $\hat{0}$.*
   (v) *If P contains a $\hat{1}$, then P is a complete lattice.*

Proof (i) The set of minimal elements of X is non-empty (Lemma 2.3.5 of the section on chain conditions.) Suppose there were at least two distinct minimal members of X, say x_1 and x_2. Then $x_1 \wedge x_2$ is also a member of X by the meet-closed hypothesis. But by minimality of each x_i, one has

$$x_1 = x_1 \wedge x_2 = x_2.$$

Since any two minimal members of the set X are now equal (and the set of them is non-empty) there exists a unique minimal member. The proof of Part (i) is complete.
   (ii) Let X be any subset of P for which the filter $\bigwedge P^X$ is non-empty. One observes that $\bigwedge P^X$ is a meet-closed induced poset and so by part 1, contains a unique minimal member $j(X)$. Then, by definition, $j(X)$ is the join $\bigvee (X)$. If $\hat{1} \in P$ then of couse $\bigwedge P^X$ is non-empty for all subsets X and the result follows.
   (iii) Let $W(X)$ be the collection of all elements of P which can be expressed as a meet of finitely many elements of X, viewed as an induced poset. Then $W(X)$ is meet-closed, and so has a unique minimal member x_0 by (i). In particular $x_0 \le x$ for all $x \in X$. Now suppose $z \le x$ for all $x \in X$. Then by Lemma 2.4.1, z is less than or equal to any finite meet of elements of X, and so is less than or equal to x_0. Thus $x_0 = \bigwedge (X)$, by definition of a global meet.
   (iv) By (i), P contains minimal non-zero elements (often called "atoms"). Suppose m is such an atom. If $m \le x$ for all $x \in P$ then x would be the element $\hat{0}$, against our

definition of "minimal element". Thus there exists an element $y \in P$ such that y is not greater than or equal to m. Then $y \wedge m$ (which exists by meet-closure) is strictly less than m. Since m is an atom, we have $y \wedge m = \hat{0} \in P$.

(v) If $\hat{1}$ exists, P enjoys both unrestricted joins by (ii). But by (iii), unrestricted meets exist, and so now P is a complete lattice.

The proof is complete. $\square$

Example 10 The following posets are all lower semilattices with the DCC:

1. $\mathbb{N}^+ = \{1 < 2 < \cdots\}$ of all positive integers with the usual ordering. This is a chain.
2. $\mathbb{D}$, the positive integers under the partial ordering of "dividing"—that is, $a \le b$ if and only if integer a divides the integer b.
3. $B(X)$, the poset of all finite subsets of a set X.
4. $L_{<\infty}(V)$, the poset of finite-dimensional subspaces of a vector space V.
5. $\mathcal{M}_{<\infty}(I)$, the finite multisets over a set I.

It follows that arbitrary meets exist and a $\hat{0}$ exists.

Remark Of course, we could also adjoin a global maximum $\hat{1}$ to each of these examples and obtain complete lattices in each case.

2.4.3 Lower Semilattices with both Chain Conditions

Recall from Sect. 2.4.3, Lemma 2.3.8, that a poset has both the ACC and the DCC if and only if it possesses condition

(FC) *Every proper chain in P has finite length.*

An observation is that if P satisfies (FC), then so does its dual P^*. Similarly, if P has finite height, then so does its dual. This is trivial. It depends only on the fact that rewriting a finite saturated chain in descending order produces a saturated chain of P^*.

We obtain at once

Lemma 2.4.3 *Suppose P is a lower semilattice satisfying* (FC).

  (i) *P contains a $\hat{0}$. If P contains a $\hat{1}$ then P is a complete lattice.*
 (ii) *Every element of $P - \{\hat{1}\}$ is bounded by a maximal member of $P - \{\hat{1}\}$ (that is, an element of $\max(P)$).*
(iii) *Every element of $P - \{\hat{0}\}$ is above a minimal element of $P - \hat{0}$ (that is, an element of $\min(P)$).*
(iv) *The meet of all maximal elements of P—that is $\bigwedge(\max(P))$ (called the Frattini element or radical of $(P, \le)$) exists and is the meet of just finitely many elements of $\max(P)$.*

(v) *If* $\hat{1} \in P$, *the join of all atoms,* $\bigvee(\min(P))$ *(called the* socle of P *and denoted* soc(P)) *exists and is the join of just finitely many atoms.*

Proof Parts (i) and (iv) follow from Theorem 2.4.2. Parts (ii) and (iii) are immediate consequences of the hypothesis (FC), and Part (v) follows from Parts (i) and (iii). $\square$

Remark Hopefully the reader will notice that the hypothesis that P contained the element $\hat{1}$ played a role—even took a bow—in Parts (i) and (v) of the above Lemma. Is this necessary? After all, we have both the DCC *and* the ACC. Well, there is an asymmetry in the hypotheses. P is a lower semilattice, but not an upper semilattice (though this symmetry is completely restored once $\hat{1}$ exists in P, because then we have arbitrary joins).

Example 11 Consider any one of the posets **D**, $B(X)$, $L_{<\infty}(V)$. These are ranked posets, with the rank of an element being (1) the number of prime factors, (2) the cardinality of a set, or (3) the vector-space dimension of a subspace, respectively. Select a positive integer r and consider the induced poset $\mathrm{tr}_r(P)$ of all elements of rank at most r in P where $P = \mathbf{D}, B(X), L_{<\infty}(V)$. Then $\mathrm{tr}_r(P)$ is still a lower semilattice with both the DCC and the ACC. But it has no $\hat{1}$.

2.5 The Jordan-Hölder Theory

In the previous two sections we defined the notions of "algebraic interval", "height of an algebraic interval" and the meet-closed posets which we called "lower semilattices". They shall be used with their previous meanings without further comment.

This section concerns a basic theorem that emerges when a certain property, that of *semimodularity*, is imposed on lower semilattices. Many important algebraic objects give rise to lower semilattices which are semimodular (for example the posets of subnormal subgroups of a finite group, or the submodule poset of an R-module) and each enjoys its own "Jordan-Holder Theorem"—it always being understood there is a general form of this theorem. It is in fact a very simple theorem about extending "semimodular functions"[19] on the set of covers of a semimodular lower semilattice to an interval measure on that semilattice. It sounds like a mouthful, but it is really quite simple.

Fix a poset $(P, \leq)$. An unrefinable chain of length one is called a *cover* and is denoted by (a, b) (which is almost the name of its interval $[a, b]$—altered to indicate that we are talking about a cover). Thus (a, b) is a cover if and only if $a < b$ and there is no element c in P with $a < c < b$—i.e. a is a maximal element in the induced

---

[19]The prefix "semi-" is justified for several reasons. The term "modular function" has quite another meaning as a certain type of meromorphic function of a complex variable. Secondly the function in question is defined in the context of a semimodular lower semilattice. So why not put in the "semi"? We do not guarantee that every term coined in this book has been used before.

poset of all elements properly below b (the principal order ideal P_b minus $\{b\}$). The collection of all covers of $(P, \leq)$ is denoted $\mathcal{C}ov_P$. Note that $\mathcal{C}ov_P$ is a subset of $\mathcal{A}_P$, the collection of all algebraic intervals of $(P, \leq)$.

2.5.1 Lower Semi-lattices and Semimodularity

A lower semilattice $(P, \leq)$ is said to be *semimodular* if, whenever (x_1, b) and (x_2, b) are both covers with $x_2 \neq x_2$, then both $(x_1 \wedge x_2, x_1)$ and $(x_1 \wedge x_2, x_2)$ are also covers.

Lemma 2.5.1 *Suppose $(P, \leq)$ is a semimodular lower semilattice. Suppose $[x, a]$ is algebraic and (b, a) is a cover, where $x \leq b$. Then $[x, b]$ is algebraic.*

Proof Since $[x, a]$ is algebraic, there exists a finite unrefinable chain from x to a, say $A = (x = a_0, a_1, \ldots, a_n = a)$. Clearly each interval $[x, a_j]$ is algebraic. See Fig. 2.1.

By hypothesis, $x \leq b$ and so there is a largest subscript i such that $a_i \leq b$. Clearly $i < n$. If $i = n - 1$ then $a_{n-1} = b$ so $[x, b]$ is algebraic by the previous paragraph. Thus we may assume that $b \neq a_{n-1}$. (See Fig. 2.1.) Since both (a_{n-1}, a) and (b, a) are covers, then by semimodularity, both $(a_{n-1} \wedge b, a_{n-1})$ and $(a_{n-1} \wedge b, b)$ are covers. Continuing in this way we obtain that $(a_k \wedge b, a_k)$ and $(a_k \wedge b, a_{k-1} \wedge b)$ are covers for all k larger than i. Finally, as $a_i \leq b$, this previous statement yields $a_{i+1} \wedge b = a_i$ and (by semimodularity) $(a_i, a_{i+2} \wedge b)$ must be a cover). Note that

$$(x = a_0, \ldots, a_i, a_{i+2} \wedge b, a_{i+3} \wedge b, \ldots, a_{n-1} \wedge b, b)$$

is an finite unrefinable chain since its successive pairs are covers. This makes $[x, b]$ algebraic, completing the proof. $\square$

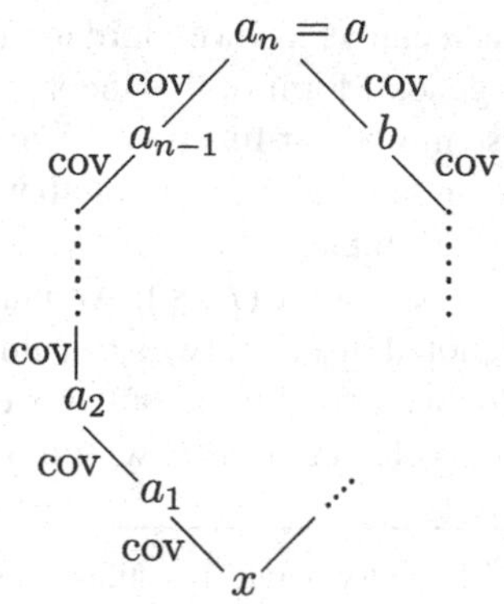

Fig. 2.1 The poset showing $[x, b]$ is algebraic. The symbol "cov" on a depicted interval indicates that it is a cover

2.5.2 Interval Measures on Posets

Let M be a *commutative monoid*. This means M possesses a binary operation, say "$*$," which is commutative and associative, and that M contains an identity element, say "e", such that $m = e * m = m * e$ for all $m \in M$.

There is one commutative monoid that plays an important role in the applications of our main theorem and that is the monoid $\mathcal{M}(X)$ of all finite multisets over a set X. We have met this object before in the guise of a locally finite poset $\mathcal{M}_{<\infty}(X)$ (see the last item under Example 2 of this chapter). We have seen that any finite multiset over X can be represented as a function $f : X \to \mathbb{N}$ from X to the non-negative integers $\mathbb{N}$ whose "support" is finite—i.e. the function achieves a non-zero value in only finitely many instances as x wanders over X. Now, if f and g are two such functions, we may let "$f+g$" denote the function that takes $x \in X$ to the non-negative integer $f(x) + g(x)$, the sum of two integers. Clearly $f + g$ has finite support, and so $(\mathcal{M}_{<\infty}(X), +)$ (under this definition of "plus") becomes a commutative semigroup. But as the constant function $0_X : X \to \{0\}$ is an identity element with respect to this operation, $\mathcal{M}(X) := (\mathcal{M}_{<\infty}(X), +)$ is actually a commutative monoid. We call this the *commutative monoid of finite multisets over* X.

An *interval measure* μ of a poset $(P, \leq)$ is a mapping $\mu : \mathcal{A}_P \to M$ from the set of algebraic intervals of P into a commutative monoid $(M, *)$ with identity element e such that

$$\mu(a, a) = e \text{ for all } a \in P. \tag{2.10}$$

$$\mu(a, b) * \mu(b, c) = \mu(a, c) \text{ whenever } [a, b] \text{ and } [b, c] \text{ are in } \mathcal{A}_P \tag{2.11}$$

[Notice that we have found it convenient to write $\mu(a, b)$ for $\mu([a, b])$.]

Here are some examples of interval measures on posets:

Example 12 Let M be the multiplicative monoid of positive integers. Let $(P, \leq)$ be the set of positive integers and write $x \leq y$ if x divides y evenly. Then every interval of $(P, \leq)$ is algebraic. Define μ by setting $\mu(a, b) := b/a$ for every interval (a, b).

Example 13 Let $(P, \leq)$ be as in Example 14, but now let M be the *additive* monoid of all non-negative integers. Now if we set $\mu(a, b); =$ the total number of prime divisors of b/a, then μ is a measure.

Example 14 Let $(P, \leq)$ be the poset of all finite-dimensional subspaces of some (possibly infinite dimensional) vector space V, where "$\leq$" is the relation of "contained in".

(i) Let M be the additive monoid of all non-negative integers. If we define $\mu(A, B) := \dim(B/A)$, then μ is a measure on $(P, \leq)$.

(ii) If M is the multiplicative group $\{1, -1\}$, then setting $\mu(A, B) := (-1)^{\dim(A/B)}$ for every algebraic interval (A, B) also defines a measure.

2.5.3 The Jordan-Hölder Theorem

Now we can prove

Theorem 2.5.2 (The Jordan-Hölder Theorem) *Let* $(P, \leq)$ *be a semimodular lower semilattice. Suppose* $\mu : \mathcal{C}ov_P \longrightarrow (M, +)$ *is a mapping from the set of covers of* P *to a commutative monoid* $(M, +)$, *and suppose this mapping is "semimodular" in the sense that*

$$\mu(b, y) = \mu(a \wedge b, a) \text{ whenever } (a, y) \text{ and } (b, y) \text{ are distinct covers.} \quad (2.12)$$

Then for any two finite unrefinable chains $U = (u = u_0, \ldots, u_n = v)$ *and* $V = (u = v_0, \ldots, v_m = v)$ *from* u *to* v, *we have*

$$\sum_{i=0}^{n-1} \mu(u_i, u_{i+1}) = \sum_{i=0}^{m-1} \mu(v_i, v_{i+1}) \quad (2.13)$$

and $n = m$ *(the finite summation is taking place in the additive monoid* $(M.+)$*). In particular,* μ *extends to a well-defined interval measure* $\hat{\mu} : \mathcal{A}_P \longrightarrow M$.

Proof If $n = 0$ or 1, then $U = V$, and the conclusion holds. We therefore proceed by induction on the minimal length $h(u, v)$ of an unrefinable chain from u to v for any algebraic interval $[u, v]$—that is, the height of $[u, v]$. It suffices to assume $U = (u = u_0, u_1, \ldots, u_n = v)$ is such a minimal chain (so that $n = h(u, v)$) and prove Eq. (2.13), and $n = m$ for any other unrefinable chain $V = (u = v_0, v_1, \ldots, v_m = v)$.

If $u_{n-1} = v_{m-1}$, $h(u, v_{n-1}) = n - 1$, so by induction, $\sum_{i=0}^{n-2} \mu(u_i, u_{i+1}) = \sum_{i=0}^{m-2} \mu(v_i, v_{i+1})$ and $m - 1 = n - 1$ and the conclusion follows.

So assume $u_{n-1} \neq v_{m-1}$. Set $z = u_{n-1} \wedge v_{m-1}$. Since (u_{n-1}, v) and (v_{m-1}, v) are both covers, by semi-modularity, so also are (z, u_{n-1}) and (z, v_{m-1}) (see Fig. 2.2).

Fig. 2.2 The main figure for the Jordan Hölder Theorem

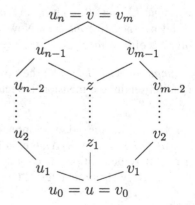

Since $u \leq z$, (z, v_{n-1}) is a cover, and $[u, v_{n-1}]$ is algebraic, we see that Lemma 2.5.1 implies that $[u, z]$ is also algebraic. So a finite unrefinable chain

$$Z = (u = z_0, z_1, \ldots, z_r = z)$$

from u to z exists. Since $h(u, u_{n-1}) = n - 1 < n$, by induction

$$\left(\sum_{i=0}^{r-1} \mu(z_i, z_{i+1}) \right) + \mu(z, u_{n-1}) = \sum_{i=0}^{n-1} \mu(u_i, u_{i+1}) \qquad (2.14)$$

and $r + 1 = n - 1$. But $h(u, v_{m-1}) \leq r + 1 = n - 1$ so induction can also be applied to the algebraic interval $[u, v_{m-1}]$ to yield

$$\left(\sum_{i=0}^{r-1} \mu(z_i, z_{i+1}) \right) + \mu(z, v_{m-1}) = \sum_{i=0}^{m-1} \mu(v_i, v_{i+1}). \qquad (2.15)$$

and the fact that $r + 1 = m - 1$. Thus $n = m$.

But by (2.12), $\mu(z, u_{n-1}) = \mu(v_{m-1}, v)$ and $\mu(z, v_{m-1}) = \mu(u_{n-1}, v)$. The result (2.13) now follows from (2.14) and the commutativity of $(M, +)$. The proof is complete. $\square$

To see how this theorem works in a semimodular lattice in which not all intervals are algebraic, the reader is referred to Example 15 on p. 53 and the remark following.

There are many applications of this theorem. In the case of finite groups (R-modules) we let $(M, +)$ be the commutative monoid of multisets of all iso-morphism classes of finite simple groups (or all isomorphism classes of irreducible R-modules, resp.). These are the classical citations of the Jordan-Hölder Theorem for Groups and R-modules.[20]

2.5.4 Modular Lattices

Consider for the moment the following example:

Example 15 The poset $(P, \leq)$ contains as elements $P = \{a, b, c\} \cup \mathbb{Z}$, where $\mathbb{Z}$ is the set of integers, with ordering defined by these rules:

---

[20]We beg the reader to notice that in the case of groups there is no need for Zassenhaus' famous "butterfly lemma", nor the need to prove that subnormal subgroups of a finite group form a lattice. A lower semilattice will do. One of the classic homomorphism theorems, provides the semimodular function from Cov_P to finite simple groups. The result is then immediate from Theorem 2.4.2, Eq. (2.13), where the interval measure displays the multiset of "chief factors" common to all satu-rated chains in P.

1. $a \leq p$ for all $p \in P$—i.e. a is the poset "zero",
2. $c \geq p$ for all $p \in P$, so c is the poset "one",
3. b is not comparable to any integer in $\mathbb{Z}$, and
4. the integers $\mathbb{Z}$ are totally ordered in the natural way.

Notice that the meet or join of b and any integer are a and c respectively. Thus $(P, \leq)$ is a lattice with its "zero" and "one" connected by an unrefinable proper chain of length 2 and at the same time by an infinite unrefinable chain. The lattice is even lower semimodular since the only covers are (a, b), (b, c) and $(n, n+1)$, for all integers n.

Remark For the purposes of the Jordan-Hölder theory, such examples did not bother us, for the J-H theory was phrased as an assertion about measures which take values on algebraic intervals—that is to say, non-algebraic intervals could be ignored— and the calculation of the measure used only finite unrefinable chains. In Example 15, the only intervals of the semimodular lattice $(P, \leq)$ which are not algebraic are those of the form $[n, c]$, $n \in \mathbb{Z}$. The Jordan-Hölder Theorem is valid here: In fact if f is any function from the set of covers, into a commutative monoid M, then by Theorem 2.5.2, f extends to an interval measure $\mu : \mathcal{A}_P \rightarrow M$. Note that $\mu(a, c) = f(a, b) + f(b, c)$.

However, unlike Example 15, many of the most important posets in Algebra are actually lattices with a property called "the modular law", which *prevents* elements x and y from being connected by both a finite unrefinable chain and an infinite chain. This modular law always occurs in posets of congruence subalgebras which are subject to one of the so-called "Fundamental Theorems of Homomorphisms".

Recall the definition of lattice (p. 46). A lattice L is called *a modular lattice* or L is said to be *modular* if and only if

(M) for all elements a, b, c of L with $a \geq b$,

$$a \wedge (b \vee c) = b \vee (a \wedge c).$$

The dual condition is

(M*) for all elements a, b, c of L with $a \leq b$,

$$a \vee (b \wedge c) = b \wedge (a \vee c).$$

But by transposing the roles of a and b it is easy to see that the two conditions are equivalent. Thus

Lemma 2.5.3 *A lattice satisfies (M) if and only if it satisfies (M*). Put another way: a lattice L is modular if and only if its dual lattice L^* is.*

An immediate consequence of the modular law is the following

Lemma 2.5.4 (Identification Principle) *Let x, y and n be elements of the modular lattice L. Suppose*

$$x \vee n = y \vee n, \quad and \tag{2.16}$$

$$x \wedge n = y \wedge n. \tag{2.17}$$

Then either $x = y$ or else x and y are incomparable. Equivalently, either $x \leq y$ or $y \leq x$ implies $x = y$.

Proof Assume the hypotheses and assume x and y are comparable. Because x and y play symmetric roles in the hypothesis, we may assume that $x \leq y$. Now by the modular law (M):

$$y \wedge (x \vee n) = x \vee (y \wedge n). \tag{2.18}$$

But by Eq. (2.16), the left side of (2.18) is y. On the other hand, by Eq. (2.17), the right side of (2.18) is x. Thus $x = y$. $\square$

The most important property of a modular lattice is given in the following:

Theorem 2.5.5 (The Correspondence Theorem for Modular Lattices) *Suppose L is a modular lattice. Then for any two elements a and b of L, there is a poset isomorphism*

$$\mu : [a, a \vee b] \to [a \wedge b, b],$$

taking each element x of the domain to $x \wedge b$.

Proof As μ is defined, $\mu(a) = a \wedge b$, $\mu(a \vee b) = (a \vee b) \wedge b = b$, and if $a \leq x \leq y \leq b$, then $\mu(x) = x \wedge b \leq y \wedge b = \mu(y)$. Thus μ is poset homomorphism and takes values in $[a \wedge b, b]$.

Suppose now that $x, y \in [a, a \vee b]$ and $\mu(x) = \mu(y)$. Then $x \wedge b = y \wedge b$ and so $a \vee (x \wedge b) = a \vee (y \wedge b)$. Since $a \leq x$ and $a \leq y$ we may apply the law (M) to each side of the last equation. This yields $x = x \wedge (a \vee b) = y \wedge (a \vee b) = y$, since $a \vee b$ dominates both x and y. Thus μ is injective.

Now suppose x is an arbitrary element of the interval $[a \wedge b, b]$, that is, $a \wedge b \leq x \leq b$. Then by the first inequality, $x = x \vee (b \wedge a)$. Since $x \leq b$, applying (M*) (with x and a in the role of a and c in (M*)), $x = b \wedge (x \vee a) = \mu(x \vee a)$. Thus μ is onto.

Finally, we must show that for any $x, y \in [a, a \vee b]$, one has $\mu(x) \leq \mu(y)$ if and only if $x \leq y$.

Obviously if $x \leq y$, then clearly $x \wedge b \leq y \wedge b$, so $\mu(x) \leq \mu(y)$.

Conversely, suppose $\mu(x) \leq \mu(y)$. Then $x \wedge b \leq y \wedge b$, giving us

$$a \vee (x \wedge b) \leq a \vee (y \wedge b). \tag{2.19}$$

Since $a \leq x$ and $a \leq y$, applying (M*) to each side of (2.19) yields

$$x \wedge (a \vee b) \leq y \wedge (a \vee b)$$

which gives us $x \leq y$, since x and y are in $[a, a \vee b]$.

Thus μ is a poset isomorphism. $\square$

Recognizing the Chain Conditions in Modular Lattices

Lemma 2.5.6 *Assume $a \leq b \leq c$ is a chain in a modular lattice L. Then the interval $[a, c]$ has the ACC (DCC) if and only if both intervals $[a, b]$ and $[b, c]$ have the ACC (DCC).*

Proof Recall from Corollary 2.3.7 that if $[a, c]$ has either of the two chain conditions, then its subintervals $[a, b]$ and $[b, c]$ also possess the same condition.

So we assume that the two intervals $[a, b]$ and $[b, c]$ possess the ascending chain condition (ACC). We must show that $[a, c]$ has the ascending chain condition. By way of contradiction, suppose

$$a = c_0 < c_1 < c_2 < \cdots$$

is an infinite properly ascending chain in the poset $[a, c]$. Then we have ascending chains

$$c_0 \vee b \leq c_1 \vee b \leq \cdots \text{ and } c_0 \wedge b \leq c_1 \wedge b \leq \cdots$$

in posets $[b, c]$ and $[a, b]$, respectively. Since these two posets are assumed to possess the ACC, there exists a natural number k such that for every integer m exceeding k we have

$$c_k \vee b = c_m \vee b, \tag{2.20}$$
$$c_k \wedge b = c_m \wedge b. \tag{2.21}$$

Since $c_k \leq c_m$, the Identification Principle, Lemma 2.5.4, forces $c_k = c_m$. But that is impossible since these are distinct entries in a properly ascending chain.

The argument that the presence of the descending chain condition for both $[a, b]$ and $[b, c]$ implies the same condition for $[a, c]$ now follows by duality from our result for the ACC. (It can also be proved by considering an infinite properly descending chain $c_0 > c_1 > \cdots$, and again obtaining a natural number k such that Eqs. (2.20) and (2.21) hold and again invoking the Identification Principle.) $\square$

Lemma 2.5.7 *Assume L is a modular lattice and $\{a_1, \ldots, a_n\}$ is a finite subset of L.*

(i) *Suppose $c \leq a_i$ for all $i = 1, 2, \ldots, n$. Then $[c, a_1 \vee \cdots \vee a_n]$ has the ACC (DCC) if and only if each interval $[c, a_i]$ possesses the ACC (DCC).*

(ii) *Suppose* $a_i \leq c$ *for all* $i = 1, 2, \ldots, n$. *Then* $[a_1 \wedge \cdots \wedge a_n, c]$ *has the ACC (DCC) if and only if each interval* $[a_i, c]$ *has the ACC (DCC).*

Proof Part (i). We assume $c \leq a_i$ for all i. Of course, by Corollary 2.3.7, if $[c, a_1 \vee a_2 \vee \cdots \vee a_n]$ has the ACC (DCC) then so does any of its intervals $[c, a_i]$. So we need only prove the reverse implication.

Assume each interval $[c, a_i], i = 1, 2, \ldots, n$, possesses the ACC (DCC). A simple induction on n reduces us to the case that $n = 2$. By Theorem 2.5.5, $[a_2, a_1 \vee a_2] \simeq [a_1 \wedge a_2, a_1]$ which has the ACC since it is an interval of $[c, a_1]$ which is hypothesized to have this chain condition. But the interval $[a_1 \wedge a_2, a_2]$ is a subinterval of $[c, a_2]$ which has the ACC (DCC). So we see that both intervals $[a_1 \wedge a_2, a_2]$ and $[a_2, a_1 \vee a_2]$ have the ACC (DCC) and so by Lemma 2.5.6, the interval $[a_1 \wedge a_2, a_1 \vee a_2]$ also possesses the ACC (DCC). Finally, noting that $[c, a_1 \wedge a_2]$ has the ACC (DCC) because it is an interval of $[c, a_1]$ hypothesized to have this chain condition, one more application of Lemma 2.5.6 now yields the fact that $[c, a_1 \vee a_2]$ also enjoys this condition.

Part (ii) follows from Part (i) by duality. $\square$

Corollary 2.5.8 *Any modular lattice is a semi-modular lower semilattice and so is subject to the Jordan-Hölder theory (See Theorem 2.5.2.). In particular, any two finite unrefinable chains that may happen to connect two elements x and y of the lattice, must have the same length, a length which depends only on the pair (x, y).*

Proof Apply the previous Lemma for the case $[a, a \vee b]$ and $[b, a \vee b]$ are both covers. $\square$

The preceeding Corollary only compared two *finite unrefinable chains* from x to y. Could one still have both a finite unrefinable chain from x to y as well as an infinite one as in Example 15 at the beginning of this subsection? The next result shows that such examples are banned from the realm of modular lattices.

Theorem 2.5.9 *Suppose L is a modular lattice and suppose $a = a_0 < a_1 < \cdots < a_n = b$ is an unrefinable proper chain of length n preceding from a to b. Then every other proper chain proceeding from a to b has length at most n.*

Proof Of course, by Zorn's lemma, any proper chain is refined by a proper unrefinable chain. So, if we can prove that all proper chains connecting a and b are finite, such a chain possesses a well-defined measure by the Jordan-Hölder Theory. In particular all such chains possess the same fixed length n. So it suffices to show that any proper chain connecting a and b must be finite.

We propose to accomplish this by induction on the parameter n given in the statement of the theorem. But this means that we shall have to keep track of the bound on length asserted by the induction hypothesis, in order to obtain new intervals $[a', b']$ to which induction may be applied.

So we begin by considering a (possibly infinite) totally ordered subposet X of $(P, \leq)$ having a as its minimal member and b as its maximal member. We must show that X is a finite proper chain. If $n = 1$, then $\{a, b\} = X$ is a cover, and we are done.

If each $x \in X$ satisfies $x \le a_{n-1}$, then $X - \{b\}$ is finite by induction on n, and so X is finite. So we may suppose that for some $x_\beta \in X$, we have $x_\beta \vee a_{n-1} = b$. By Theorem 2.5.5 $(a_{n-1} \wedge x_\beta, x_\beta)$ is a cover.

Now the distinct members of the set

$$Y = \{z \wedge a_{n-1} | z \in X\}$$

form a totally ordered set whose maximal member is $a_{n-1} = b \wedge a_{n-1}$, and whose minimal member is $a = a \wedge a_{n-1}$. By induction on n, Y is a finite set of at most $n - 1$ elements. On the other hand, it is the concatenation of two proper chains

$$Y^+ := \{y \in Y | y \ge a_{n-1} \wedge x_\beta\}$$
$$Y^- := \{y \in Y | y \le a_{n-1} \wedge x_\beta\},$$

whose lengths are non-zero and sum to at most $n - 1$.

Now the poset isomorphism $\mu : [x_\beta, b] \to [a_{n-1} \wedge x_\beta, a_{n-1}]$ takes the members of X dominating x_β to Y^+. Thus

$$X \cap [x_\beta, b]$$

is a chain of the same length as Y^+. It remains only to show that the remaining part of X, the chain $X \cap [a, x_\beta]$ is finite.

Let index i be maximal with respect to the condition $a_i \le x_\beta$. Then for each index j exceeding i but bounded by $n - 1$,

$$a_j = a_{j-1} \vee (a_j \wedge x_\beta) \text{ and } (a_{j-1} \wedge x_\beta, x_j \wedge x_\beta)$$

is a cover or is length zero. It follows that a is connected to x_β by an unrefinable chain of length at most $n - 1$. By induction $X \cap [a, x_\beta]$ is finite. The proof is complete. $\square$

Corollary 2.5.10 *Let L be a modular lattice with minimum element 0 and maximum element 1. The following conditions are equivalent:*

1. *There exists a finite unrefinable chain proceeding from 0 to 1.*
2. *Every proper chain has length below a certain finite universal bound.*
3. *L has both the ascending and descending chain conditions (see p. 48).*

Proof The Jordan-Hölder theory shows all chains connecting 0 to 1 are bounded by the length of any unrefinable one. How this affects all proper chains is left as an exercise. $\square$

Any modular lattice which satisfies the conditions of Corollary 2.5.10 is said to *possess a composition series*.

Example 16 Let L be the lattice $\mathbb{D}$ of all positive integers where $a \leq b$ if and only if a divides b. Then $\mathbb{D}$ is a modular lattice which possesses the DCC but not the ACC. There is a "zero" (the integer 1), but no lattice "one". However, every interval $[a, b]$ possesses a composition series.

2.6 Dependence Theories

2.6.1 Introduction

If we say A "depends" on B, what does this mean? In ordinary language one might intend several things: "A depends *entirely* on B" or that "A depends just a little on B"—a statement so mild that it might suggest only that B has a "slight influence" on A. But *some* syntax seems to be applied all the same: thus if we say that A depends on B and that B in turn depends on C, then A (we suppose to some small degree) depends on C.

In mathematics we also usually use the word "depends" in both senses. When one asserts that $f(x, y)$ depends (*to some degree*) upon the variable x, one means only that f *might* be influenced by x. After all, it could be that f is a "constant" function as far as x is concerned. But on the other hand the mathematician also intends that f is *entirely* determined by the pair (x, y). Thus we may consider the phrase "$f(x, y)$ depends on x" as one borrowed from ordinary everyday speech. In its various guises, the stronger idea of total and entire dependence appears throughout mathematics with such a common strain of syntactical features as to deserve codification.

But as you will see, the theory here is highly special.

2.6.2 Dependence

Fix a set S, and let $\mathcal{F}(S)$ be the set of all finite subsets of the set S. A *dependence relation* on S is a relation $\mathcal{D}$ from S to $\mathcal{F}(S)$—that is, a subset of $S \times \mathcal{F}(S)$—subject to the axioms (D1)–(D3) listed below. We shall say that the element s in S *depends* on the subset $A \in \mathcal{F}(S)$ if and only if $(s, A) \in \mathcal{D}$. The dependence relation must satisfy these conditions:

(D1) (The Reflexive Condition) If $s \in S$ and if $s \in A \in \mathcal{F}(S)$, then s depends on A—that is, s depends on any finite subset of S that contains it.

(D2) (The Transitivity Condition) If the element s depends on the finite set A, and if every member of A depends on the finite set B, then s depends on B.

(D3) (The Exchange Condition) If $s, a_1, \ldots, a_n$ are elements of S such that s depends on the set $\{a_1, \ldots, a_n\}$ but does not depend on the set $\{a_1, \ldots .a_{n-1}\}$, then a_n depends on the set $\{a_1, \ldots, a_{n-1}, s\}$.

Before concluding anything from this, let us consider some examples.

Example 17 (*Linear dependence*) Suppose V is a left vector space over some field or division ring D. Let us say that a vector v depends on a finite set of vectors $\{v_1, \ldots, v_n\}$ if and only if v can be expressed as a linear combination of those vectors,—that is, $\sum_{i=1}^{n} \delta_i v_i = v$ for some choice of δ_i in D. One checks that this defines a dependence relation on V.

Example 18 (*Algebraic dependence*) Let K be a field containing k as a subfield. (For example, one might let K and k respectively be the complex and rational number fields.) We say that an element b of K depends on a finite subset $X = \{a_1, \ldots .a_{n-1}\}$ of K if and only if b is the root of a polymonial equation whose coefficients are expressible as a k-polynomial expressions in X. This means that there exists a polynomial $p(x, x_1, \ldots, x_n)$ in the polynomial ring $k[x, x_1, \ldots, x_n]$ when evaluated at $x = b, x_i = a_i, 1 \le i \le n$ yields 0. We shall see in Chap. 11 that all three axioms of a dependence theory hold for this definition of dependence among elements of the field K.

2.6.3 *Extending the Definition to $S \times \mathcal{P}(S)$*

Our first task will be to extend the definition of a dependence relation $\mathcal{D} \subseteq S \times \mathcal{F}(S)$ to a relation $\overline{\mathcal{D}} \subseteq S \times \mathcal{P}(S)$, where $\mathcal{P}(S)$ is the collection of *all* subsets of S (the power set). Thus, it will be possible to speak of element x depending on a (possibly infinite) subset A of S. Namely, for any subset A of S, we say that the element x *depends* on A if and only there exists a finite subset A_1 of A such that x depends on A_1 (in the original sense). We leave it as an exercise to the reader to prove the following:

Lemma 2.6.1 *In the following, the sets A and B may be infinite sets:*

 (i) *Element x of S depends on any set A which contains it.*
 (ii) *If A and B are subsets of S, if x depends on A and every element of A depends on set B, then x depends on B.*
 (iii) *If $a \in A$, a subset of S, and x is an element of S which depends on A but does not depend on $A - \{a\}$, then a depends on $(A - \{a\}) \cup \{x\}$*

(See Exercise (1) in Sect. 2.7.4.)

2.6.4 Independence and Spanning

Let S be a set with dependence relation $\mathcal{D} \subseteq S \times \mathcal{F}(S)$, and let $\overline{\mathcal{D}} \subseteq S \times \mathcal{P}(S)$ be the extension of this relation to infinite subsets as in Lemma 2.6.1 above. Now let A be any subset of S. The *flat generated by* A is the set $\langle A \rangle := \{y \mid y$ depends on $A\}$. A subset X of S is called a *spanning set* if and only if $\langle X \rangle = S$.

Next, we say that a subset Y of S is *independent* if and only if for each element x of Y, x does not depend on $Y - \{x\}$.

Now the collections of all spanning sets form a partially ordered set under the containment relation. A similar statement holds for the collection of all independent sets. Thus it makes perfect sense to speak of a minimal spanning set and a maximal independent set (should such sets exist). We wish to show that these two concepts coincide.

Theorem 2.6.2 *Let S be a set with dependence relation $\overline{\mathcal{D}} \subseteq S \times \mathcal{P}(S)$ as above.*

(i) Every minimal spanning set is a maximal independent set.

(ii) Every maximal independent set is a minimal spanning set.

Proof We prove (i). Suppose U is a minimal spanning set. If U is not independent then there exists an element u in U such that u depends on a finite subset U_1 of $U - \{u\}$. Thus every element of S depends on U and by Lemma 2.6.1, part 1, and what we know of u, every element of U depends on $U - \{u\}$. Thus by Lemma 2.6.1, part (ii), every element of S depends on $U - \{u\}$. Thus $U - \{u\}$ is a spanning set, against the minimality of U. Thus U must be independent.

If U were not a maximal independent set there would exist an element z in $S - U$ which did not depend on U. But that is impossible as U spans S.

Next we show (ii). If X is a maximal independent set, then by maximality, X spans S. If a proper subset X_1 of X also spanned S, then any element x of $X - X_1$ would depend on X_1 and so would depend on $X - \{x\}$ by the definition of dependence. But this contradicts the fact that X is independent. Thus X is actually minimal as a spanning set. The proof is complete. $\square$

Next, we have the following important result.

Theorem 2.6.3 *Maximal independent subsets of S exist.*

Proof This is a straightforward application of Zorn's Lemma. As remarked above, the collection $\mathcal{J}$ of all independent subsets of S form a partially ordered set $(\mathcal{J}, \leq)$ under the inclusion relation. Now consider any chain $C = \{J_\alpha\}$ of independent sets and form their union $\bar{C} := \bigcup_\alpha J_\alpha$. Then if $\bar{C}$ were not an independent set, there would exist an element x in $\bar{C}$ and a finite subset F of $\bar{C} - \{x\}$ such that x depends on F. But since the chain C is totally ordered and $F \cup \{x\}$ is finite, there exists an index σ such that $F \cup x \subseteq J_\sigma$. But this contradicts J_σ independent. Thus we see that $\bar{C}$ *must* be an independent set.

Thus every chain C in $\mathcal{J}$ possesses an upper bound $\bar{C}$ in $(\mathcal{J}, \leq)$. By the Zorn Principle, $\mathcal{J}$ contains maximal elements. The proof is complete. $\square$

2.6.5 Dimension

The purpose of this section is to establish

Theorem 2.6.4 *If X and Y are two maximal independent sets, then $|X| = |Y|$.*

Proof By the Schröder-Bernstein Theorem (Theorem 1.4.1, p. 16), it suffices to show only that $|X| \geq |Y|$ for any two maximal independent sets X and Y.

The proof has a natural division into two cases: the case X is a finite set and the case that X is infinite.

First assume X is finite. We proceed by induction on the parameter $k := |X| - |X \cap Y|$. If $k = 0$, then $X = X \cap Y \subseteq Y$. But as Y is independent and each of its elements depends on its subset X, we must have $X = Y$, whence $|X| = |Y|$ and we are done. Thus we may assume $k > 0$.

Suppose $X = \{x_1, \ldots, x_n\}$ where the x_i are pairwise distinct elements of S and the indexing is chosen so $\{x_1, \ldots, x_r\} = X \cap Y$, thereby making $k = n - r$. From maximality of the independent set X, every element of $Y - X$ depends on X. Similarly, every element of X depends on Y. If every element of Y depended on $X_0 := \{x_1, \ldots, x_{n-1}\}$, then x_n, which depends on Y, would depend on X_0 by Lemma 2.6.1 (i) above, against the independence of X. Thus there is some element $y \in Y - X$ which depends on X but does not depend on X_0. By the exchange condition, x_n depends on $X_1 := \{x_1, \ldots, x_{n-1}, y\}$. Moreover, if X_1 were not independent, then either y depends on $X_1 - \{y\} = \{x_1, \ldots, x_{n-1}\}$ or some x_i depends on $(X_0 - \{x_i\}) \cup \{y\}$. The former alternative is ruled out by our choice of y. If the latter dependence held, then as the independence of X prevents x_i from depending on $X_0 - \{x_i\}$, the exchange condition would force y to depend on X_0, again contrary to the choice of y. Thus X_1 is an independent set of n distinct elements. But as x_n depends on X_1, each element of X depends on X_1. Thus as all elements of S depend on X, they all depend on X_1 as well. Thus X_1 is a maximal independent set. But since $|X_1 \cap Y| = 1 + |X \cap Y|$, induction on k yields $|X_1| \geq |Y|$. The result now follows from $|X| = |X_1|$.

Now assume X is an infinite set. Since Y is a maximal independent set, it is a spanning set. Thus for each element x in X there is at least one finite non-empty subset Y_x of Y on which it depends. Choose, if possible $y \in (Y - \bigcup_X Y_x)$. Since y depends on X and by construction every element of X depends on $\bigcup_X Y_x$, we see that y depends on $\bigcup_X Y_x \subseteq Y$, contradicting the independence of Y. Thus we see that no such y can exist and so $Y = \bigcup_X Y_x$.

Our goal is to produce an injective mapping

$$\phi : Y \to X \times \mathbb{N},$$

where, as usual, $\mathbb{N}$ denotes the natural numbers. Since, for each $x \in X$, the set Y_x is finite, there exists an injective mapping $\phi_x : Y_x \to \mathbb{N}$. The problem is that the sets Y_x may intersect non-trivially, so that merely combining the ϕ_x does not yield a well-defined ϕ.

We get around this problem as follows: First, by Theorem 2.2.2, the set X can be well-ordered. Next given $y \in Y$, let $S(y) := \{x \in X | y \in Y_x\}$, and let $\ell(x)$ be the least element of the set $S(y)$. Now define $\phi : Y \to X \times \mathbb{N}$, by setting

$$\phi(y) := (\ell(y), \phi_{\ell(y)}(y)) \in X \times \mathbb{N} \tag{2.22}$$

for each $y \in Y$. If $\phi(y_1) = \phi(y_2)$, then $\ell(y_1) = \ell(y_2)$ and then $y_1 = y_2$ follows from the injectivity of $\phi_{\ell(y)}$. Thus the mapping ϕ defined in (2.22) is injective.

It follows that
$$|Y| \leq |X \times \mathbb{N}|.$$

Now by Corollary 2.2.5 near the beginning of this chapter, $|X| = |X \times \mathbb{N}|$, since X is infinite.

Thus $|Y| \leq |X|$ as required. The proof is complete. $\square$

This common cardinality of maximal independent sets is called the *dimension* of the dependence system $(S, \mathcal{D})$.

2.6.6 Other Formulations of Dependence Theory

For the sake of completeness, this subsection surveys a number other views of dependence theory. Since this subsection is not essential for anything further in the book, many results are not proved. Most of the missing proofs can be found in the book by Oxley [1].

Fix a set X and let $\mathcal{I}$ be a family of subsets of X. The pair $(X, \mathcal{I})$ is called a *matroid* if and only if the following axioms hold:

(M1) The family $\mathcal{I}$ is closed under taking subsets.
(M2) (Exchange axiom) If $A, B \in \mathcal{I}$ with $|A| > |B|$, then there exists an element
       $a \in A - B$, such that $\{a\} \cup B \in \mathcal{I}$.
(M3) If every finite subset of a set A belongs to $\mathcal{I}$, then A belongs to $\mathcal{I}$.

Note that if A and B are maximal elements of $(\mathcal{I}, \subseteq)$, then $|A| = |B|$ is an immediate consequence of axiom (M2). However, from axioms (M1) and (M2) alone, it does not follow that maximal elements even exist. One needs (M3) for that.

Lemma 2.6.5 *If $(X, \mathcal{I})$ is a matroid, then every element of $\mathcal{I}$ lies in a maximal element of $\mathcal{I}$.*

Proof This proof utilizes Zorn's lemma. Let A_0 be an arbitrary member of $\mathcal{I}$. We examine the poset of all subsets in $\mathcal{I}$ which contain A_0. Let $A_1 \subseteq A_2 \subseteq \cdots$ be a chain in this poset and let B be the union of all the sets A_i. Consider any finite subset F of B. Then each element $f \in F$ lies in some member $A_{i(f)}$ of the chain. Setting m to be the maximal index in the finite set $\{i(f) | f \in F\}$, we see that $F \subseteq A_m$. Since $A_m \in \mathcal{I}$, (M1) implies $F \in \mathcal{I}$. But since F was an arbitrary finite subset of B, (M3) shows that $B \in \mathcal{I}$. Thus every finite chain in the poset of elements of $\mathcal{I}$ which contain A_0 has an upper bound in that poset, so, by Zorn's lemma, that poset contains a maximal element M. Clearly M is a maximal member of $\mathcal{I}$, since any member of $\mathcal{I}$ which contains M also contains A_0, and so lies in the poset of which M is a maximal member. Thus A_0 lies in a maximal member of $\mathcal{I}$, as required. $\square$

Example 19 Let $\mathcal{D} \subseteq X \times \mathcal{F}(\mathcal{S})$ be a dependence theory. Let $\mathcal{I}$ be the set of independent sets of the dependence theory $\mathcal{D}$. Then $(X, \mathcal{I})$ is a matroid.

Example 20 Suppose (V, E) is a simple graph with vertex set V and edge set E. (The adjective "simple" just means that edges are just certain unordered pairs of distinct vertices.) A *cycle* is a sequence of edges $(e_0, e_1, \ldots, e_n)$ such that e_i shares *just one* vertex with e_{i+1}, and *another* vertex with e_{i-1}, indices i taken modulo n. Thus e_{n-1} shares a vertex with $e_n = e_0$, and the "length" of the cycle, n, cannot be one or two. A graph with no cycles is called a *forest*—its connected components are called *trees*. Now let $\mathcal{I}$ be the collection of all subsets A of E such that the graph (V, A) is a forest. Then $(E, \mathcal{I})$ is a matroid.

Example 21 Let $\mathcal{F}$ be a fixed family of subsets of a set X. A finite subset $\{x_1, x_2, \ldots, x_n\}$ of X is said to be a *partial transversal of* $\mathcal{F}$ if and only if there are *pairwise distinct* subsets $A_1, \ldots A_n$ of $\mathcal{F}$ such that $x_i \in A_i$, $i = 1, \ldots, n$— that is, the set $\{x_1, \ldots, x_n\}$ is a "system of distinct representatives" of the sets $\{A_i\}$. Now let $\mathcal{I}$ be the collection of all subsets of X all of whose finite subsets are partial transversals. Then $(X, \mathcal{I})$ is a matroid.

Example 22 Suppose $(X, \mathcal{I})$ is any given matroid, and Y is any subset of X. Let $\mathcal{I}(Y)$ be all members of $\mathcal{I}$ which happen to be contained in the subset Y. Then it is straightforward that the axioms (M1), M(2) and (M3) all hold for $(Y, \mathcal{I}(Y))$. The matroid $(Y, \mathcal{I}(Y))$ is called the *induced matroid on* Y.

Here is another approach to matroids, using a so-called *rank function*.

Theorem 2.6.6 *Suppose r is a map from the set of all subsets of a set X, into the cardinal numbers, satisfying these three rules:*

(R1) *For each subset Y of X, $0 \le \rho(Y) \le |Y|$.*
(R2) (Monotonicity) *If $Y_1 \subseteq Y_2 \subseteq X$, then $\rho(Y_1) \le \rho(Y_2)$.*
(R3) (The submodular inequality) *If A and B are subsets of X, then*

$$\rho(A \cup B) + \rho(A \cap B) \le \rho(A) + \rho(B).$$

Let $\mathcal{I}$ be the collection of a subsets Y of X with the following property:

(*) *For every finite subset F_Y of Y we have $|F_Y| = r(F_Y)$.*

Then $(X, \mathcal{I})$ is a matroid.

Now suppose we are given a matroid $(X, \mathcal{I})$. Can we recover a dependence theory from this matroid? Clearly we need to have a definition of "dependence" constructed exclusively from matroid notions. Consider this definition:

Definition: Let x be an element of X and let A be a finite subset of X. We say that x *depends on* A if and only if there exists a subset A_0 of A which lies in $\mathcal{I}$, such that either (i) $x \in A_0$, or (ii) $\{x\} \cup A_0$ is not in $\mathcal{I}$.

Theorem 2.6.7 *Given a matroid* $(X, \mathcal{I})$, *let the relation of "dependence" between elements of X and finite subsets of X be given as in the preceding definition. Then this notion satisfies the axioms of a dependence theory.*

It now follows from Example 19, and the assertion of Theorem 2.6.7, that matroids and dependence theories are basically the same thing.

There are purely combinatorial ways to express the gist of a dependence theory, and these are the various formulations of the notion of a matroid in terms of circles, in terms of flats, or in terms of closure operators.

Let us consider flats, for a moment. We have already defined them from the point of view a dependence theory. From a matroid point of view, the flat $\langle A \rangle$ spanned by subset A in matroid $(X, \mathcal{I})$, is the set of all elements x for which $\{x\} \cup A_x \notin \mathcal{I}$ for some finite subset A_x of A. It is an easy exercise (Exercise (6) in Sect. 2.7.4) to show that $\langle \langle A \rangle \rangle = \langle A \rangle$, for all subsets A of X. In fact the reader should be able to prove the following:

Theorem 2.6.8 *The mapping τ which sends each subset A of X to the flat $\langle A \rangle$ spanned by A, is a closure operator on the lattice of all subsets of X. The image sets—or "closed" sets—form a lattice:*

1. *The intersection of two closed sets is closed, and is the meet in this lattice.*
2. *The closure of the set-theoretic union of two sets is a join, that is, a global minimum in the poset of all flats above the two sets.*

The characterization of matroids by closure operators is the following:

Proposition 2.6.9 *Suppose $P = P(X)$ is the poset of all subsets of a set X and $\tau : P \to P$ satisfies*

(Increasing) *For each subset A, $A \subseteq \tau(A)$.*
(Closure) *The mapping τ is a closure operator—that is, it is an idempotent monotone mapping of a poset into itself.*
(The Steinitz-MacLane Exchange Property) *If $X \subseteq P$ and $y, z \in P - \tau(X)$ then $y \in \tau(X \cup \{z\})$ implies $z \in \tau(X \cup \{y\})$.*

Then, setting

$$\mathcal{I} = \{A \subseteq X | x \notin \tau(A - \{x\}) \text{ for all } x \in A\},$$

we have that $M := (X, \mathcal{I})$ is a matroid.

Conversely, if $(X, \mathcal{I})$ is a matroid, then the mapping τ which takes each subset of A to the flat $\langle A \rangle$ which it spans, is a monotone increasing closure operator $P \to P$ possessing the Steinitz-MacLane Exchange Property.

2.7 Exercises

2.7.1 Exercises for Sect. 2.2

1. Suppose $\lesssim$ is a reflexive and transitive relation on a set X. (Such a relation is called a *pseudo-order*.) For any two elements x and y of X, let us write $s \sim y$ if and only if $x \lesssim y$ and also $y \lesssim x$.

   (a) Show that "$\sim$" is an equivalence relation.
   (b) For each element x of X, let $[x]$ be the $\sim$-equivalence class containing element x. Show that if $x \lesssim y$ then $a \lesssim b$ for every element $a \in [x]$ and element $b \in [y]$. (In this case we write "$[x] \leq [y]$".)
   (c) Let $X/\sim$ denote the collection of all $\sim$-classes of X. Show that $(X/\sim, \leq)$ is a poset.

2. Make a full list of the posets defined in Examples 1 and 2 above which

   (a) have a zero element.
   (b) have a one element.
   (c) are locally finite.

3. Let P be a fixed poset. If there is a bijection $f : X \to Y$ prove that there exists an isomorphisms of posets:

$$\prod_{x \in X} P \to \prod_{y \in Y} P,$$

$$\sum_{x \in X} P \to \coprod_{y \in Y} P, \text{ if } P \text{ has a zero element.}$$

   (This means that the index set X has only a weak effect on the definition of a product.)

4. Recall that the elements of the divisor poset $\mathbb{D}$ are the positive integers, with $a \leq b$ if and only if the integer a divides the integer b.

   The poset of all finite multisets on the set of natural numbers consists of all infinite sequences of natural numbers with only finitely many positive entries. This multiset is denoted $\mathbf{M}_{<\infty}(\mathbf{N})$.

   For this exercise, the student is asked to assemble a proof of the following theorem.

Theorem 2.7.1 *The divisor poset* $\mathbb{D}$ *is isomorphic to the poset* $\mathbf{M}_{<\infty}(\mathbf{N})$ *of all finite multisets over the set* $\mathbb{N}^+$ *of positive integers. That is, we have an isomorphism*

$$\epsilon : (\mathbb{D}, \leq) \to \sum_{i \in \mathbb{N}^+} \mathbb{N}_i \tag{2.23}$$

where each $\mathbb{N}_i$ *is a copy of the totally ordered poset of the natural numbers.*

[Hint: We sketch the argument: Notice that the right side of (2.23) consists of all sequences of non-negative integers of which only finitely many are positive (equivalently, all finite sequences of positive integers). The mapping ϵ is easy to define. First place the positive prime numbers in ascending (natural) order $2 = p_1 < p_2 < \cdots$. For example $p_6 = 13$. Now any integer $d \in \mathbb{D}$ greater than one has a unique factorization into positive prime powers, with the primes placed from left to right in ascending order:

$$d = \prod_{\sigma \in J_d} p_\sigma^{a_\sigma},$$

for some (possibly empty) ascending sequence of positive integers J_σ. If one declares $\epsilon(d)$ to be the function $f : \mathbb{N}^+ \to \mathbb{N}_0$ such that

$$f(\sigma) = \begin{cases} a_\sigma & \text{if } \sigma \in J_d, \\ 0 & \text{if } \sigma \notin J_d, \end{cases}$$

and declares $\epsilon(1) := \mathbf{0}$, the constant function with all values 0, then one sees that $f(c) \le f(d)$ in the sum on the right side of Eq. (2.23) if and only if c divides d evenly—i.e. $c \le d$ in $(D, \le)$. Clearly ϵ is a bijection.]

5. Recall that if $(P, \le)$ and $X \subseteq P$, then the symbol P_X denoted the order ideal

$$P_X := \{z \in P | z \le x \text{ for some } x \in X\}.$$

Show that $P_X = \bigcup \{P_x | x \in X\}$.
6. For the filters P^X and P^Y generated by these sets, show that

$$\left(\bigwedge P^X\right) \bigcap \left(\bigwedge P^Y\right) = \bigwedge P^{(X \cup Y)}.$$

7. Let X be a subset of a poset $(P, \le)$, and let P_X be the order ideal of all elements bounded above by at least one element of X (see p. 31 or Exercise (5) in Sect. 2.7.1 in this section). Prove that P_X is the intersection of all order ideals of P which contain X.
8. Give an example of a poset P and subset X for which the order ideal P_X is not generated by an antichain.
9. Give an example of a poset $(P, \le)$ and an order ideal J of P such that J does not have the form P_X for any antichain X.
10. Let X be an infinite set. Recall that a partition of X is a decomposition $\pi = \{X_\sigma\}$ of X into pairwise disjoint non-empty subsets X_σ called the *components* of the partition. (The word "decomposition" is there to indicate that the union of the X_σ is X.) A component X_σ is said to be *trivial* if it contains exactly one element of X. Such a partition π is said to be a *finitary partition* if and only if finitely many of the components are non-trivial and each of these is a finite set.

Let $\mathcal{F}P(X)$ be the full collection of finitary partitions of X. Show that with respect to the refinement relation, $\mathcal{F}P(X)$ is a locally finite poset.

11. Let $\mathcal{F} := \{A_\sigma \mid \sigma \in I\} \subseteq \mathcal{P}(X)$. Let $\alpha : \mathcal{P}(X) \to \mathcal{P}(I)$ be the mapping of posets which sends each subset U of X to the set $\alpha(U) := \{\sigma \in I \mid U \subseteq A_\sigma\}$. (If no such subset A_σ contains U, then $\alpha(U)$ is the empty set.) Similarly, for each subset K of I, set $\beta(K) := \bigcap_{\sigma \in K} A_\sigma$ so that $\beta : \mathcal{P}(I) \to \mathcal{P}(X)$ is a mapping of posets.

   (a) Show that α and β are both order reversing and that $(\mathcal{P}(X), \mathcal{P}(I), \alpha, \beta)$ is a Galois connection.
   (b) Show that the closed elements of X are those subsets expressible as intersections of A_σ's.
   (c) Show that a subset J of I is closed if and only if it has the property: If $A_\tau \supseteq \bigcap_{\sigma \in J} A_\sigma$, then $\tau \in J$.

2.7.2 Exercises for Sects. 2.3 and 2.4

1. We let $[n]$ be the chain $\{1 < 2 < \cdots < n\}$ in the usual total ordering of the positive integers. Show that the infinite union of disjoint chains,

$$[1] \cup [2] \cup [3] \cup \cdots ,$$

   satisfies (FC) but does not possess finite height, and possesses neither a $\hat{1}$ nor a $\hat{0}$.

2. If we adjoin $\hat{0}$ to the poset presented in the previous Exercise, show that the Frattini element exists, and is $\hat{0}$.

3. Show that the product of posets,

$$[1] \times [2] \times [3] \times \cdots ,$$

   does not satisfy (FC).

4. Let P be a lower semilattice with $\hat{0}$ and $\hat{1}$ satisfying (FC). Set $\mathcal{B} := \max(P)$ and let $\bar{P}$ be the induced subposet generated by $\mathcal{B}$—that is, the set of elements of P expressible as a finite meet of elements of $\mathcal{B}$. We understand the empty meet to be the element $\hat{1}$, so the latter is an element of $\bar{P}$.

   (a) Show that $\bar{P}$ has the Frattini element $\phi(P)$ as its zero element, $\hat{0}_{\bar{P}}$.
   (b) For each $x \in P - \{\hat{1}\}$, show that the induced poset $P^x \cap \bar{P}$ has a unique minimal member (which we shall call $\sigma(x)$).
   (c) Defining $\sigma(\hat{1}) = \hat{1}$ show the following:
      i. For all $x \in P$, $x \leq \sigma(x)$. (This is built into the definition of σ.)
      ii. The mapping $\sigma : P \to \bar{P}$ is a surjective morphism of posets.

iii. For each $x \in P$, one has $\sigma(\sigma(x)) = \sigma(x)$.

(Recall that any morphism onto an induced subposet which satisfies these three conditions is called a *closure operator*.)

5. Suppose $P = \mathbb{N} \times \{0, 1\}$. We totally order P (lexicographically) as follows:

   (a) $(a, 0) \le (b, 1)$ for any $a, b \in \mathbb{N}$.
   (b) $(a, 0) \le (b, 0)$ if and only if $a \le b$, $a, b \in \mathbb{N}$.
   (c) $(a, 1) \le (b, 1)$ if and only if $a \le b$, $a, b \in \mathbb{N}$.

   Show that $(P, \le)$ has the Descending Chain Condition (DCC), but that there exist intervals (x, y) with no finite unrefinable chain from x to y.

6. Suppose $(P, \le)$ is a poset with a zero element $\hat{0}$. Recall from Sect. 2.4.2 that in this case an *atom of* $(P, \le)$ is an element a distinct from $\hat{0}$ with the property that there is no element $x \in P$ such that $\hat{0} < x < a$,—that is, $(\hat{0}, a)$ is a cover. Let A be the set of atoms of P. Assume now that $(P, \le)$ has the property that every interval $[\hat{0}, b]$ possesses the DCC. Show that either $P = \{\hat{0}\}$ or that the set A of atoms is non-empty.

7. Give an example of a locally finite poset which does not possess the descending chain condition.

8. Suppose $(P, \le)$ is a locally finite poset which possesses a zero element $\hat{0}$. Show by means of an example, that such a poset need not possess the descending chain condition.

9. Suppose L is a lattice. Give an induction proof showing that for any finite collection $\{a_1, \ldots, a_n\}$ of elements of L, the elements

$$a_1 \vee \cdots \vee a_n \text{ and } a_1 \wedge \cdots \wedge a_n,$$

exist and are respectively the greatest lower bound and greatest upper bound in L of the set $\{a_1, \ldots, a_n\}$.

10. Suppose $(P, \le)$ is a lower semilattice with the descending chain condition (DCC). Show that any principle order ideal is a lattice. Conclude that for any non-empty subset $X \subseteq P$, the order ideal

$$P_x := \{z \in P | z \le x \text{ for all } x \in X\}$$

is always a lattice.

11. Suppose $(P, \le)$ is a locally finite poset.

   (a) Suppose $(P, \le)$ possesses a zero element $\hat{0}$ and at least one other element. Show that the set A of atoms of $(P, \le)$ is not empty. (Note that we are not assuming the Descending Condition, so this is a little different than Problem (6) in Sect. 2.7.2.)
   (b) Now assume that $(P, \le)$ is a lower semilattice. Show that any principle order ideal is a finite lattice.

12. A lattice $L := (L, \leq)$ is said to be *distributive* if and only if, for any elements a and b in L, one always has:

$$a \wedge (b \vee c) = (a \wedge b) \vee (a \wedge c). \tag{2.24}$$

An example of a distributive lattice is the power poset $\mathcal{P}(X)$ of all subsets of a set X.

(a) Prove that if L is a distributive lattice and $(M, \leq)$ is an induced subposet closed under taking pair-wise meets and joins, then M is also a distributive lattice.

(b) Use the result of item (a) to prove the following:
   i. The Boolean poset $B(X)$—of all finite subsets of X is a distributive lattice.
   ii. Let $\mathcal{J}(P)$ be the poset of all order ideals of a poset $(P, \leq)$ under the containment relation. Use item (a) to prove that $\mathcal{J}(P)$ is a distributive lattice. [One must define "meet" and "'join" of order ideals and set $L = \mathcal{P}(P)$ in item (a) of this exercise.]

2.7.3 Exercises for Sect. 2.5

1. (The Jordan-Hölder Theorem implies the Fundamental Theorem of Arithmetic.) Let $(P, \leq)$ be the poset of positive integers where $a \leq b$ if and only if integer a divides integer b evenly (Example 1, of this chapter).

(a) Show that $(P, \leq)$ is a lower semimodular semilattice with all intervals algebraic. Let $\mu : Cov_P \to \mathbb{N}^+$ be the function which records the prime number b/a at every cover (a, b) of $(P, \leq)$. Indicate why μ is a semimodular function in the sense given in Eq. (2.12) of the Jordan-Hölder Theorem.

(b) Suppose integer a properly divides integer b. For every factorization $b/a = p_1 p_2 \cdots p_r$ of b/a into primes show that there exists a finite unrefineable chain $(a = a_0, a_1, a_2, \ldots a_r = b)$ such that $a_i/a_{i-1} = p_i$ for $i = 1, \ldots r$.

(c) Conclude from the Jordan-Hölder Theorem that every positive integer possesses a factorization into prime numbers, and that the multiset of positive prime numbers involved in the factorization is unique.

2. Show that the following lattices are modular:

   (i) The poset of subspaces of a vector space.
   (ii) The poset of subgroups of an abelian group.

   In both cases, the order relationship is that of containment.

3. Show that the product $L_1 \times L_2$ of two modular lattices is modular. If the lattices L_σ, $\sigma \in I$ are modular and possess a "zero", does the modularity extend to the direct sum $\coprod_{\sigma \in I} L_\sigma$?

4. Is the lattice of partitions of a finite set, modular? (Here the order relationship is refinement of one partition by another.)

5. Write out an explicit proof of Corollary 2.5.10.

2.7.4 Exercises for Sect. 2.6

1. Prove the three parts of Lemma 2.6.1 for the extended definition of dependence allowing infinite subsets of S.

2. Verify the axioms of a matroid for Example 22, on the edge-set of a graph.

3. Verify the axioms of a matroid for Example 23, on finite partial transversals.

4. Let $\Gamma = (V, E)$ be a simple graph. Let $M = (E, \mathcal{I})$ be the matroid of Example 20. For any subset F of the edge set E, let

$$(V, F) = \bigcup_{\sigma \in K} (V_\sigma, F_\sigma)$$

be a decomposition of Γ into connected components. For each connected component (V_σ, F_σ), let E_σ be the collection of edges connecting two vertices in V_σ. Thus (V_σ, E_σ) is the subgraph induced on the vertex set V_σ. Prove that in the matroid, the flat spanned by F is the union $\cup_{\sigma \in K} E_\sigma$.

5. Let $M = (X, \mathcal{I})$ be a matroid. A set which is minimal with respect to not lying in $\mathcal{I}$ is called a *circuit*. Show that any circuit is finite. (Remark: There is also a characterization of matroids by circuits. See *Matroid Theory*, by James G. Oxley [1], Proposition 1.3.10.)

6. Let $M = (X, \mathcal{I})$ be a matroid. For each subset A of X, let $\langle A \rangle$ be defined as the set of all elements $x \in X$ for which there exists a finite subset $A_x \subseteq A$ such that $\{x\} \cup A_x \notin \mathcal{I}$ (this is the matroid version of "the flat generated by A" defined on p. 65. Show that $\langle A \rangle = \langle \langle A \rangle \rangle$.

References

1. Oxley J (2011) Matroid theory, Oxford graduate texts. vol 21, Oxford University Press, Oxford
2. Stanley RP (1997) Enumerative Combinatorics, vol 1, Cambridge studies in advanced mathematics, vol 49, Cambridge University Press, Cambridge

Chapter 3
Review of Elementary Group Properties

Abstract Basic properties of groups are collected in this chapter. Here are exposed the concepts of order of a group (any cardinal number) or of a group element (finite or countable order), subgroup, coset, the three fundamental theorems of homomorphisms, semi-direct products and so forth.

3.1 Introduction

Groups are systems of symmetries of objects, in particular mathematical objects. Understanding groups can be useful in classifying objects in a particular class. One uses a group of symmetries to transfer any object of the class to a representative object which is an easily-studied canonical form. For example, the group $GL(n, F) \times GL(n, F)$, generated by elementary row and column operations on the class $M_n(F)$ of $n \times n$ matrices is used to transport an arbitrary matrix to a more easily studied canonical form.

This is just *one* of the reasons that group theory is needed, whatever field of mathematics you might choose to enter. There are of course many other reasons having to do with special uses of groups (such as "Polya counting",[1] quantum physics, invariant theory etc.). But overall, the need to classify things seems the *broadest* reason that group theory is pertinent to all of mathematics.

Groups are defined by a hallowed set of axioms which are useful only for the purpose of banishing all ambiguities from the subject. Somehow, staring at the axioms of a group mesmerizes one away from their basic and natural habitat. *Every group that ever exists in the world is in fact the full group of symmetries of something.* That is what the theory of groups is: the study of symmetries. But it is useful to know this only on a philosophical plane. Knowing that there is a set of objects such that every group—or even every finite group—is the full group of automorphisms of at least one of these objects, is not very helpful for classifying the groups themselves.

---

[1] This has a long pre-Polya history.

© Springer International Publishing Switzerland 2015

E. Shult and D. Surowski, *Algebra*, DOI 10.1007/978-3-319-19734-0_3

3.2 Definition and Examples

This is probably a good place to explain a convention. A *group* is a set G with a binary operation with certain properties—of which there are a good many examples. In some of these examples the binary operation is written as "$+$"; in others, it is written as "$\circ$" or "$\times$". (If one uses "$+$", one might think the operation is commutative. If one uses "$\circ$", the elements seem to be mappings, and if one uses "$\times$", Cartesian products can confuse the issue). How then does one speak generally about all possible groups? The standard solution is this: The group operation of an arbitrary group, will be indicated by *simple juxtaposition*, the act of writing one symbol directly after the other—thus, the group operation, applied to the ordered pair (x, y) will be denoted xy.

A *group* is a set G equipped with a binary operation $G \times G \to G$ (denoted here by juxtaposition) such that the following axioms hold:

1. (The Associative Law) The binary operation is associative—that is $(ab)c = a(bc)$ for all elements a, b, c of G. (The parentheses indicate the temporal order in which operations were performed. This axiom more-or-less says that the *temporal* order of applying the group operations doesn't matter, while, of course, the *left-right* order of the elements being operated on *does* matter.)
2. (Identity Element) There exists an element, say e, such that $eg = g = ge$ for all elements g in G.
3. (The Existence of Inverses) For each element x in G, there exists an element x' such that $x'x = xx' = e$ where e is the element referred to in Axiom 2.[2]

One immediately deduces the following:

Lemma 3.2.1 *For any group G one has:*

1. *The identity element e is the unique element of G possessing the property of Axiom 2.*
2. *Given element g in G, there is at most one element g' such that $gg' = g'g = e$.*
3. *In light of the preceding item 2, we may denote the unique inverse of g by the symbol g^{-1}. Then note that*

   (i) $(ab)^{-1} = (b^{-1})(a^{-1})$ for all elements $a, b \in G$.
   (ii) $(a^{-1})^{-1} = a$, for each element $a \in G$.
   (iii) For any element x and natural number k, the product $xx \ldots x$ (with k factors) is a unique element which we denote as x^k. We have $(x^k)^{-1} = (x^{-1})^k$. (As a notational convention, this element is also written as x^{-k}.).

---

[2]People have figured out how to "improve" these axioms, by hypothesizing only right identities and left inverses and so on. There is no attempt here to make these axioms independent or logically neat. Of course, the axioms indeed over-state their case; but a little redundancy won't hurt at this beginning stage.

Example 23 Familiar Examples of Groups:

(a) The group $(\mathbb{Z}, +)$ of integers under the operation of addition.

(b) The group of non-zero rational numbers under multiplication.

(c) The group of non-zero complex numbers under multiplication.

(d) *The cyclic group* $\mathbb{Z}_n$ *of order* n. This can be thought of as the group of rotations of a regular n-gon. This group consists of the elements $\{1, x, x^2 \ldots, x^{n-1}\}$ where x is a clockwise rotation of $2\pi/n$ radians and "1" is the identity transformation. All multiplications of the powers of x can be deduced solely from the algebraic identity $x^n = 1$.

(e) *The dihedral group* D_{2n}, *of order* $2n$. This is the group of all rigid symmetries of the regular n-gon. In addition to the group of n rotations just considered, it also contains n reflections which are symmetries of the n-gon. If n is an odd number, these reflections are about an axis through a corner or vertex of the polygon and bisecting an opposite side. If n is even, the axes of the reflections are of two sorts: those which go from a vertex through its opposite vertex, and those which bisect a pair of opposite sides. There are then $n/2$ of each type. The elements of the group all have the form $t^i x^j$, where t is any reflection, x is the clockwise rotation by $2\pi/n$ radians and $0 \le i \le 1$ and $0 \le j \le n - 1$. The results of all multiplications of such elements can be deduced entirely from the relations $x^n = 1, t^2 = 1, tx = x^{-1}t$ (and its consequence, $tx^{-1} = xt$).

The isomorphism type of the dihedral group of order $2n$ is denoted D_{2n}. In the special case that $n = 2$, each group in the resulting class D_4 is called a *fours group*. It is distinguished from the cyclic group of order four by the fact that the square of any of its elements is the identity element.

(f) *The symmetric groups* $Sym(X)$. Let X be a set. A *permutation* of the elements of X is a bijective mapping $X \rightarrow X$. This class of mappings is closed under composition of mappings, inverses exists, and it is easy to verify that they form a group with the identity mapping 1_X which takes each element to itself, as the group identity element. This group is called the *symmetric group on* X and is denoted $Sym(X)$. If $|X| = n$ it is well known from elementary counting principles that there are exactly $n!$ permutations of X. In this case one writes $Sym(n)$ for $Sym(X)$, since the names or identities of the elements of X do not really affect the nature of the group of all permutations.

(g) *The group of rigid motions of a (regular) cube.* Just imagine a wooden cube on the desk before you. We consider the ways that cube can be rotated so that it achieves a position congruent to the original one. The result of doing one rotation after another is still some sort of rotation. It should be clear that these rotations can have axes which are situated in three different ways with respect to the cube. The axis of rotation may pass through the centers of opposite faces, it may pass though the midpoints of opposite edges, or it could be passing through opposite vertices.

(h) Let $G = (V, E)$ be a simple graph. This means the edges E are pairwise distinct 2-subsets of the vertex set V. Two (distinct) vertices are said to be *adjacent* if and only if they are the elements of an edge. Now the *group of automorphisms of the graph G* is the set of permutations of the set of vertices V which preserve the adjacency relation. The group operation is composition of the permutations.

(i) Let R be any ring with identity element e.[3] An element u of R is said to be a *unit* if and only there exists a *two-sided multiplicative inverse in R*—that is, an element u' such that $u'u = e = uu'$. (Observe in this case, that u' itself must be a unit.) Then the set $U(R)$ of all units of the ring R forms a group under ring multiplication. Some specific cases:

(i) The group $U(\mathbb{Z})$ of units of the ring of integers $\mathbb{Z}$, is the set of numbers $\{+1, -1\}$ under multiplication. Clearly this group behaves just like the cyclic group of order 2, one of those introduced in part (d) above.

(ii) Let $D = \mathbb{Z} \oplus \mathbb{Z}i$, where $i^2 = -1$, the ring of *Gaussian integers* $\{a + bi\,|\,a, b \in \mathbb{Z}\}$ as a multiplicatively and additively closed subset (that is a *subring*) of the ring of all complex numbers. Then the reader may check that $U(D)$ is the set $\{\pm 1, \pm i\}$ under multiplication. This is virtually identical with the cyclic group of order four as defined in part (d) of this series of examples.

(iii) Let V be a left vector space over a division ring D. Let hom(V, V) be the collection of all linear transformations $f : V \to V$ (viewed as right operators on the set of vectors V). This is an additive group under the operation "+" where, for all $f, g \in \text{hom}(V, V)$, $(f + g)$ is the linear transformation which takes each vector v to $vf + vg$. With composition of such transformations as "multiplication" the set hom(V, V) becomes a ring.[4]

Now the group of units of hom(V, V) would be the set $GL(V)$ of all linear transformations $t : V \to V$ where t is bijective on the set of vectors (i.e. a *permutation* of the set of vectors. (V is not assumed to be finite or even finite-dimensional in this example.) The group $GL(V)$ is called the *general linear group of V*.

(iv) *The group GL(n, F).* Let F be a field, and let $G = GL(n, F)$ be the set of all invertible $n \times n$ matrices with entries in F.[5] This set is closed under multiplication and forms a group. In fact it is the group of units of the ring of all $n \times n$ matrices with respect to ordinary matrix multiplication and matrix addition.

---

[3]It is presumed that the reader has met rings before. Not much beyond the usual definitions are presumed here.

[4]Actually it is more, for hom(V, V) can be made to have the structure of a vector space, and hence an algebra if D possesses an anti-automorphism—e.g. when D is a field. But we can get into this later.

[5]Recall from your favorite undergraduate linear algebra or matrix theory course that if n is a positive integer, then an n-by-n matrix M has a right inverse if and only if it has a left inverse (this is a consequence of the equality of row and column ranks).

3.2.1 Orders of Elements and Groups

Suppose x is an element of a group G. We set $x^0 := e$, $x^1 := x$, $x^2 := xx$, and for each natural number n, inductively define $x^n := xx^{n-1}$. If there exists a natural number n such that $x^n = e$, then x is said to *have finite order*. If n is the *smallest* positive integer such that $x^n = e$ then we write $o(x) = n$, calling n the *order of the element x*. Otherwise, we say that x has *infinite order* and write $o(x) = \infty$. In the latter case, the powers $e, x, x^2, \ldots$ of x are all distinct, for if $x^n = x^m$ for $n < m$, then $e = x^n x^{-n} = x^m x^{-n} = x^{(m-n)}$, and so x would have finite order.

Lemma 3.2.2 *Let G be a group, and let $x \in G$ be an element of finite order n.*

(i) *If $x^m = e$, then $n \mid m$.*
(ii) *If n and m are relatively prime, then $o(x^m) = n$.*

Proof We apply the Division Algorithm (Lemma 1.1.1 of Chap. 1) and write $m = qn + r$ with $0 \le r < n$. From this, one has $r = m - qn$ and so $x^r = x^{m-qn} = x^m x^{-qn} = e$. By definition of $o(x) = n$ together with $0 \le r < n$, we must have $r = 0$, i.e., that $n \mid m$, proving part (i).

Next, let $o(x^m) = k$, and so $x^{km} = e$. By part (i), we infer that $n \mid km$. By Lemma 1.1.3 we have $n \mid k$. Since it is clear that $(x^m)^n = e$, we conclude that, $k \mid n$ as well, forcing $n = k = o(x^m)$. $\square$

Elements of order 2 play a special role in finite group theory, and so are given a special name: any element of order two is called an *involution*.

The *order of a group G* is the number $|G|$ of elements within it.

3.2.2 Subgroups

Let G be a group. For any two subsets X and Y of G, the symbol XY denotes the set of all group products xy where x ranges over X and y ranges independently over Y. (It is not immediately clear just how many group elements are produced in this way, since a single element g might be expressible as such a product in more than one way. At least we have $|XY| \le |X| \cdot |Y|$.)

A second convention is to write X^{-1} for the set of all inverses of elements of the subset X. Thus $X^{-1} := \{x^{-1} \mid x \in X\}$. This time $|X^{-1}| = |X|$, since the correspondence $x \to x^{-1}$ defines a bijection $X \to X^{-1}$ (using Lemma 3.2.1, part 3(ii) here).

A subset H of G is said to be a *subgroup of G* if and only if

1. $HH \subseteq H$, so that by restriction H admits the group operation of G, and
2. with respect to this operation, H is itself a group.

One easily obtains the useful result:

Lemma 3.2.3 (Subgroup Criterion) *For any subset X of G, the following are equivalent:*

1. *X is a subgroup of G.*
2. $X = X^{-1}$ *and* $XX \subset X$.
3. $XX^{-1} \subseteq X$.

A remark on notation Whenever subset X is a *subgroup* of G, we write $X \leq G$ instead of $X \subseteq G$. This is in keeping with the general practice in abstract algebra of writing $A \leq B$ whenever A is a *subobject of the same algebraic species as B*. For example we write this if A is an R-submodule of the R-module B, and its special case: when A is a vector subspace of a vector space B. Usually the context should make it clear what species of algebraic object we are talking about. Here it is groups.

Corollary 3.2.4   1. *The set-intersection over any family of subgroups of G is a subgroup of G.*
2. *If A and B are subgroups of G, then AB is a subgroup of G if and only if $AB = BA$ (an equality of sets).*
3. *For any subset X of G, the set $\langle X \rangle_G$ of all finite products $y_1 y_2 \ldots y_n$, $n \in \mathbb{N}$ where the y_i range independently over $X \cup X^{-1}$, is a subgroup of G, which is contained in any subgroup of G which contains subset X.*
   Thus we can also write

$$\langle X \rangle_G = \cap_{X \subseteq H \leq G} H$$

   where the intersection on the right is taken over all subgroups of G which contain set X.

The proof is left to the reader (see Exercise (2) in Sect. 3.7.1). The subgroup $\langle X \rangle_G$ (which is often written $\langle X \rangle$ when the "parent" group G is understood) is called the *subgroup generated by X*. As is evident from the Corollary, it is the smallest subgroup in the poset of all subgroups of G which contains set X.

Example 24 Examples of Subgroups.

(a) Let $\Gamma(V, E)$ be a graph with vertex set V and edge set E, and set $H := \text{Aut}(\Gamma)$, the group of automorphisms of the graph Γ, as in Example 23, part (h) in the previous subsection. Then H is a subgroup of $\text{Sym}(V)$, the symmetric group on the vertex set.
(b) *Cyclic Subgroups* Let x be an element of the group G. The set $\langle x \rangle := \{x^n | n \in \mathbb{Z}\}$ (where, for negative integers n, we adopt the convention of Lemma 3.2.1 3(iii) that $x^{-n} = (x^{-1})^n$) is clearly a subgroup of G (Corollary 3.2.4). Now one sees that *the order of the element x is the (group) order of the subgroup $\langle x \rangle$.*
(c) Let Y be a subset of X. The set of all permutations of X which map the subset Y onto itself forms a subgroup of $\text{Sym}(X)$ called *the stabilizer of Y*. If G is any subgroup of $\text{Sym}(X)$, then the intersection of the stabilizer of Y with G is called the *stabilizer of Y in G* and is denoted $\text{Stab}_G(Y)$. This is clearly a subgroup of G. Thus it makes sense to speak of the stabilizer of a specified vertex in

the automorphism group Aut(Γ) of a graph Γ. (For sets with a single member, the convention is to write Stab(y) instead of Stab($\{y\}$). Note also that all these definitions apply when $G = \text{Sym}(X)$, as well.

(d) Let us return once more to the example of the group of rotations of a regular cube (Example (g) of the previous subsection). We asserted that each non-identity rigid motion was a rotation in ordinary Euclidean 3-space about an axis symmetrically placed on the cube.

One can see that the group of rotations about an axis through the center of a square face and through the center of its opposite face, is a cyclic group generated by a rotation y of order four. There are 6/2 = 3 such axes: in fact, they can be taken to be the three pair-wise orthogonal coordinate axes of the surrounding Euclidean space. This contributes 6 elements y of order four and 3 elements of order two (such as y^2).

Another type of axis extends from the midpoint of one edge to the midpoint of an opposite edge. Rotations about such an axis form a subgroup of order two. The generating rotation t of order two does not stabilize any face of the cube, and so is not any of the involutions (elements of order two) stabilizing any of the previous "face-to-face" axes. Since there are twelve edges in six opposite pairs, these "edge-to-edge" axes contribute 6 new involutions to the group.

Finally there are 8/2 = 4 "vertex-to-vertex" axes, the rotations about which form cyclic subgroups generated by a rotation of order three. Thus each of these four groups contribute two elements of order three.

Thus the group of rotations of the cube contains 1 identity element, 6 elements of order four, 3 involutions stabilizing a face, 6 involutions not stabilizing a face, and 8 elements of order three—a total of 24 elements.

3.2.3 Cosets, and Lagrange's Theorem in Finite Groups

Suppose H is a subgroup of a group G. If x is any element of G, we write Hx for the product set $H\{x\}$ introduced in the last subsection. Such a set Hx is a *right coset of H in G*. If x and y are elements of G and $H \leq G$, then y is an element of Hx if and only if $Hy = Hx$. This is any easy exercise. It follows that all the elements of G are partitioned into right cosets as

$$G = \cup_{x \in T} Hx \text{ a disjoint union of right cosets, for appropriate } T.$$

Here T is merely a set consisting of one element from each right coset. Such a set is called a *system of right coset representatives of H in G*, or sometimes a *(right) transversal of H in G*.

The *components* of this partition—that is the sets $\{Hx | x \in T\}$ for *any* transversal T—is denoted G/H. It is just the collection of all right cosets themselves. The cardinality of this set is called the *index of H in G* and is denoted $[G : H]$.

One notes that right multiplication by element x induces a bijection $H \rightarrow Hx$. Thus all right cosets have the same cardinality. Since they partition all the elements of G we have the following:

Lemma 3.2.5 (Lagrange's Theorem)

1. *If $H \leq G$, then $|G| = [G : H] \cdot |H|$.*
2. *The order of any subgroup divides the order of the group.*
3. *The order of any element of G divides the order of G.*

We conclude with a useful result.

Lemma 3.2.6 *Suppose A and B are subgroups of the finite group G. Then $|AB| \cdot |A \cap B| = |A| \cdot |B|$.*

Proof Consider the mapping $f : A \times B \rightarrow AB$, which maps every element (a, b) of the Cartesian product to the group product ab. This map is surjective, and the fibre $f^{-1}(ab)$ contains all pairs $\{(ax, x^{-1}b)|x \in A \cap B\}$. (Note that in order for $(ax, x^{-1}b)$ to be in the designated fibre, one must have $ax \in A$, and $x^{-1}b \in B$, forcing $x \in A$ and $x^{-1} \in B$–that is, $x \in A \cap B$.) Thus $|A \times B| \geq |A \cap B| \cdot |AB|$. On the other hand, if $ab = a'b'$ for (a, b) and (a', b') in $A \times B$, then $a^{-1}a' = b(b')^{-1} = x \in A \cap B$. But then $ax = a', x^{-1}b = b'$. So the fibers are no larger than $|A \cap B|$. This gives the inequality in the other direction. $\square$

3.3 Homomorphisms of Groups

3.3.1 Definitions and Basic Properties

Let G and H be groups. A mapping $f : G \rightarrow H$ is called a *homomorphism of groups* if and only if

$$f(xy) = f(x)f(y) \text{ for all elements } x, y \in G. \tag{3.1}$$

Here, as was our convention, we have represented the group operation of both abstract groups G and H by juxtaposition. Of course in actual practice, the operations may already possess some other notation customary for familiar examples.

For any subset X, we set $f(X) := \{f(x)|x \in X\}$. In particular, the set $f(G)$ is called the *homomorphic image* of G.

We have the usual glossary for special properties of homomorphisms. Suppose $f : G \rightarrow H$ is a homomorphism of groups. Then

1. f is an *epimorphism* if f is onto—that is, f is a surjection of the underlying sets of group elements. Equivalently, f is an epimorphism if and only if $f(G) = H$.
2. f is an *embedding of groups* whenever f is an injection of the underlying set of group elements. (It need not be surjective).

3. f is an *isomorphism of groups* if and only if f is a bijection of the underlying set of elements—that is, f is both an embedding and an epimorphism.
4. f is an *endomorphism of groups* if it is a homomorphism of G into itself.
5. f is an *automorphism of a group* G if it is an isomorphism $G \rightarrow G$ of G to itself.

The following is an elementary exercise.

Lemma 3.3.1 *Suppose* $f : G \rightarrow H$ *is a homomorphism of groups. Then*

1. *If* 1_G *and* 1_H *denote the unique identity elements of* G *and* H, *respectively, then* $f(1_G) = 1_H$.
2. *For any element* x *of* G, $f(x^{-1}) = (f(x))^{-1}$.
3. *The* homomorphic image $f(G)$ *is a subgroup of* H.

Lemma 3.3.2 *Suppose* $f : G \rightarrow H$ *and* $g : H \rightarrow K$ *are group homomorphisms. Then the composition* $g \circ f : G \rightarrow K$ *is also a homomorphism of groups. Moreover:*

1. *If* f *and* g *are epimorphisms, then so is* $g \circ f$.
2. *If* f *and* g *are both embeddings of groups, then so is* $g \circ f$,
3. *If* f *and* g *are both isomorphisms, then so is* $g \circ f$, *and the inverse mapping* $f^{-1} : H \rightarrow G$.
4. *If* f *and* g *are both endomorphisms (i.e.* $G = H = K$), *then so is* $g \circ f$.
5. *If* f *and* g *are both automorphisms of* G *then so are* $g \circ f$ *and* f^{-1}.

Thus the set of automorphisms of a group G *form a group under composition of automorphisms.* (This is called the *automorphism group* of G and is denoted $\mathrm{Aut}(G)$).

Finally we introduce an invariant associated with every homomorphism of groups. The *kernel* of the group homomorphism $f : G \rightarrow H$ is the set

$$\ker f := \{x \in G | f(x) = 1_H\}.$$

The beginning reader should use the subgroup criterion to verify that $\ker f$ is a subgroup of G. If $f(x) = f(y)$ for elements x and y in G, then $xy^{-1} \in \ker f$, or equivalently, $(\ker f)x = (\ker f)y$ as cosets. Thus we see

Lemma 3.3.3 *The group homomorphism* $f : G \rightarrow H$ *is an embedding if and only if* $\ker f = 1$, *the identity subgroup of* G.

We shall have more to say about kernels later.

3.3.2 Automorphisms as Right Operators

As noted just above, homomorphisms of groups may be composed when the arrangement of domains and codomains allows this. In that case we wrote $(g \circ f)(x)$ for $g(f(x))$—that is, f is applied first, then g.

As remarked in Chap. 1, that notation is not very convenient if composition of mappings is to reflect a binary operation on the set of mappings itself. We have finally reached such a case. The automorphisms of a group G themselves form a group $\mathrm{Aut}(G)$. To represent how the group operation is realized by composition of the induced mappings, it is notationally convenient to represent then in "*exponential notation*" and view the composition as a composition of right operators.

(*The exponential convention*)   If σ is an automorphism of a group G, and $g \in G$, we rewrite $\sigma(g)$ as g^σ. This way, passing from the group operation (in $\mathrm{Aut}(G)$) to composition of the automorphisms does not entail a reversal in the order of the group arguments.[6]

Thus for automorphisms σ and τ of G and any $x \in G$ we then have,

$$x^{\sigma\tau} = (x^\sigma)^\tau.$$

3.3.3 Examples of Homomorphisms

Symmetries that are induced by group elements on some object X are a great source of examples of group homomorphisms. Where possible in these examples we write these as left operators with ordinary composition "∘"—but we will begin to render these things in exponential notation here and there, to get used to it. In the next chapter on group actions, we will be using the exponential notation uniformly when a group acts on anything.

Example 25 Examples of homomorphisms.

(a) Suppose there is a bijection between sets X and Y. Then there is an isomorphism $\mathrm{Sym}(X) \to \mathrm{Sym}(Y)$. This just amounts to changing the names of the objects being permuted.

(b) Let $\mathbb{R}^*$ be the multiplicative group of all nonzero real numbers, and let $\mathbb{R}^{+*}$ be the multiplicative group of the *positive* real numbers. Then the "squaring" mapping, which sends each element to its square, defines a homomorphism of groups

$$\sigma : \mathbb{R}^* \to \mathbb{R}^{+*},$$

and, by restriction, an embedding

$$\sigma|_{\mathbb{R}^{+*}} : \mathbb{R}^* \to \mathbb{R}^*.$$

The kernel of σ is the multiplicative group consisting of the real numbers ± 1.

---

[6]This is part of a general scheme in which elements of some 'abstract group' G (with its own multiplication) induce a group of symmetries $\mathrm{Aut}(X)$ of some object X so that group multiplication is represented by composition of the automorphisms. These are called "group actions" and are studied carefully in the next chapter.

(c) In any intermediate algebra course (cf. books of Dean or Herstein, for example), one learns that complex conjugation (which sends any complex number $z = a + bi$ to $\bar{z} := a - bi$, a and b real) is an automorphism of the field of complex numbers. The *norm mapping* $N : \mathbb{C}^* \to \mathbb{R}^{+*}$ from the multiplicative group of non-zero complex numbers to the multiplicative group of positive real numbers is defined by setting $N(z) := z \cdot \bar{z}$ for each complex number z. Since complex conjugation is an automorphism of the commutative multiplicative group $\mathbb{C}^*$, it follows that the norm mapping N satisfies $N(z_1 z_2) = N(z_1)N(z_2)$ and hence is a homomorphism of groups. The kernel of the homomorphism is the group $\mathbb{C}_1$ of complex numbers of norm 1—the so-called *circle group*.
(When one considers that $N(a + bi) = a^2 + b^2$, $a, b \in \mathbb{R}$, it is not mysterious that the set of integers which are the sum of two perfect squares is closed under multiplication.)

(d) (Part 1.) Now consider the group of rigid rotations of the (regular) cube. There are four diagonal axes intersecting the cube from a vertex to its opposite vertex. These four axes intersect at the center of the cube, which we take to be the origin of Euclidean 3-space. The angle α formed at the origin by the intersection of any of these two axes, satisfies $\cos(\alpha) = \pm 1/3$. Let us label these four axes 1, 2, 3, 4 in any manner. As we rotate the cube to a new congruent position, the four axes are permuted among themselves. Thus we have a mapping

$$\text{rotations of the cube } \to \text{ permutations of the labeled axes}$$

which defines a group homomorphism

$$\text{rigid rotations of the cube } \to \text{Sym}(4),$$

the symmetric group on the four labels of the axes. The kernel would be a group of rigid motions which stabilizes each of the four axes. Of course it is conceivable that some axes are reversed (sent end-to-end) by such a motion while others are fixed point-wise by the same motion. In fact if we had used the three face-centered axes, it *would* be possible to reverse two of the axes while rigidly fixing the third. But with these four vertex-centered axes, that is not possible. (Can you show why? It has to do with the angle and the rigidity of the motion.) So the kernel here is the identity rotation of the cube. Thus we have an embedding of the rotations of the cube into Sym(4). But we have seen in the previous subsection that both of these groups have order 24. Thus by the "pigeon-hole principle", the homomorphism we have defined is an isomorphism.

(d) (Part 2.) Again G is the group of rigid rotations of the cube. There are exactly three face-centered axes which are at right angles to one another. A 90° rotation about one of these three axes fixes it, but transposes the other two. Thus if we label the three face-centered axes by the letters $\{1, 2, 3\}$, and send each rotation in the group G to the permutation of the labels of the three face-centered axes which it induces, we obtain an epimorphism of groups $G \to \mathrm{Sym}(3)$, or, in view of part 1 of (d), a homomorphism $\mathrm{Sym}(4) \to \mathrm{Sym}(3)$. The kernel is the group K which stabilizes each of the three face-centered axes. This group consists of the identity element, together with three involutions, each being a $180°$ rotation about one of the three face-centered axes. Multiplication in K is commutative.

(e) *Linear groups to matrix groups.* Now let V be a vector space over a field F of finite dimension n. We have seen in the previous subsection that the bijective linear transformations from V into itself form a group which we called $GL(V)$, the *general linear group on* V. Now fix a basis $\mathcal{A} = \{v_1, \ldots, v_n\}$ of V. Any linear transformation $T : V \to V$, viewed as a right operator of V can be represented as a matrix

$$_{\mathcal{A}}T_{\mathcal{A}} := (p_{ij})$$

where

$$(v_i)T = p_{i1}v_1 + p_{i2}v_2 + \cdots + p_{in}v_n$$

(so that the rows of the matrix depict the fate of the vector v_i under T).[7] For composition of the right operators S and T on V let us write

$$v(T * S) = ((v)T)S, v \in V,$$

so that $T * S$ is simply $S \circ T$ in the standard notation for composition. Then we see that

$$_{\mathcal{A}}(T * S)_{\mathcal{A}} = {}_{\mathcal{A}}(T)_{\mathcal{A}} \cdot {}_{\mathcal{A}}(S)_{\mathcal{A}}$$

(where chronologically, the notation intends that T is applied first, then S, being right operators and "$\cdot$" denotes ordinary multiplication of matrices.) This way the symbolism does not transpose the order of the arguments, so in fact the mapping

$$T \to_{\mathcal{A}} T_{\mathcal{A}}$$

defines a group homomorphism

$$f_{\mathcal{A}} : GL(V) \to GL(n, F)$$

---

[7]Thanks to the analysts' notation for functions, combined with our habit of reading from left to right, many linear algebra books make linear transformations left operators of their vector spaces, so that their matrices are then the transpose of those you see here. That is, the *columns* of their matrices record the fates of their basis vectors. However as algebraists are aware, this is actually a very awkward procedure when one wants to regard the composition of these transformations as a binary operations on any set of such transformations.

of the group of linear bijections into the group of $n \times n$ invertible matrices under ordinary matrix multiplication.

What is the kernel? This would be the group of linear transformations which fix each basis element, hence every linear combination of them, and hence every vector of V. Only the identity mapping can do this, so we see that f_A is an isomorphism of groups.

(f) *The determinant homomorphism of matrix groups.* The determinant associates with each $n \times n$ matrix, a scalar which is non-zero if the matrix is invertible. That determinants preserve matrix multiplication is not very obvious from the formulae expressing it as a certain sum over the elements of a symmetric group.[8] Taking it on faith, for the moment, this would mean that the mapping

$$\det : GL(n, F) \to F^*,$$

taking each invertible matrix to its determinant is a group homomorphism into the multiplicative group of non-zero elements of the ground field F. The kernel, then, is the group of all $n \times n$ matrices of determinant 1, which is called the *special linear group* and is denoted $SL(n, F)$.

(g) *Even and odd permutations and the sgn homomorphism.* Now consider the *symmetric group on n letters*. In view of Example (a) above, the symmetric groups $\mathrm{Sym}(X)$ on finite sets X of size n are all isomorphic to one another, and so are given a neutral uniform description: $\mathrm{Sym}(n)$ is the group of all permutations of the set of "letters" $\{1, 2, \ldots, n\}$. Subgroups of $\mathrm{Sym}(n)$ are called *permutation groups on n letters*. Representing an abstract group as such a group of permutations provides an environment for calculating products. Many properties of finite groups are in fact proved by such calculations. In general, the way to transport arguments with symmetric groups to arbitrary groups G is to exploit homomorphisms $G \to \mathrm{Sym}(n)$. These are called "group actions" and are fully described in the next chapter.

Now we can imagine that the neutral set of letters $\Omega_n := \{1, 2, \ldots, n\}$ are formally a basis $\mathcal{A}$ of an n-dimensional vector space over any chosen field F. Then any permutation becomes a permutation of the basis elements of V, which extends to a linear transformation T of V, and can then be rendered as a matrix $_A T_A$ with respect to the basis $\mathcal{A}$ as in Example (f). Thus, by a composition of several isomorphisms that we understand, together with their restrictions to subgroups, we have obtained an embedding of groups

$$\mathrm{Sym}(n) \to GL(n, F)$$

---

[8] The multiplicative properties follow easily from a much nicer definition of determinate which will emerge from the exterior algebras studied in Chap. 13.

which represents each permutation by a matrix possessing exactly one 1 in each row and in each column, all other entries being zero. Such a matrix is called a *a permutation matrix.*

Now even the usual sum formula for the determinant shows that the determinant of a permutation matrix is ± 1. Now if we accept the thesis of Example (f) just above, that the determinant function is in fact multiplicative, we obtain a useful group homomorphism:

$$\text{sgn} : \text{Sym}(n) \rightarrow \{\pm 1\}$$

into the multiplicative group Z_2 of numbers ± 1, which records the determinant of each permutation matrix representing a permutation. The kernel of *sgn* is called the *alternating group*, denoted $\text{Alt}(n)$, and its elements are called *even permutations*. All other permutations are called *odd permutations*. Since *sgn* is a group homomorphism, we see that

An even permutation times an even permutation is even.

An odd permutation times an odd permutation is even.

An even permutation times an odd permutation (in any order) is odd.

Since the argument developed for this example assumed the thesis of part (f) (of this same Example 25)—that the determinant of a product of matrices is the product of their determinants—and since that thesis may not be known from first principles by some students, we shall give an elementary proof of the existence of the *sgn* homomorphism in Sect. 4.2.2 of the next chapter, without any appeal to determinants.

(h) *The automorphism group of a cyclic group of order n.* Finally, perhaps, we should consider an example of an automorphism group of a group. We consider here, the group $\text{Aut}(Z_n)$, the group of automorphisms of the cyclic group of order n, where n is any natural number. Suppose, then, that C is the additive group of integers mod n—that is, the additive group of residue classes modulo n. Thus $\{[j] := j + n\mathbb{Z}\}, \; j = 1, 2, \ldots, n-1, n$. The addition rule becomes

$$[i] + [j] = [i+j] \text{ or } [i+j-n], \text{ whichever does not exceed } n,$$

where $1 \leq i \leq j \leq n$. Then element $[1]$ generates this group. Indeed so does $[m]$ if and only if $\gcd(m, n) = 1$. Thus, if $f : C \rightarrow C$ is an automorphism of C it follows than $f([1]) = [m]$ where $\gcd(m, n) = 1$. Moreover, since f is a homomorphism, $f[k] = [mk]$. Thus the automorphism f is completely determined by the natural number m coprime to and less than n. The number of such numbers is called the *Euler ϕ-function*, and it's value at n is denoted $\phi(n)$. Thus $\phi(n) = |\text{Aut}(Z_n)|$.

It now follows that $\text{Aut}(C)$ is isomorphic to the multiplicative group $\Phi(n)$ of *all residues mod n which consist only of numbers which are relatively prime to n.*[9]

---

[9] We will obtain a more exact structure of this group when we encounter the Sylow theorems.

(i) *The inner automorphism group.* Let G be a general abstract group and let x be a fixed element of G. We define a mapping

$$\psi_x : G \to G,$$

called *conjugation by x*, by the rule $\psi_x(g) := x^{-1}gx$, for all $g \in G$. One can easily verify that $\psi_x(gh) = \psi_x(g)\psi_x(h)$, and as ψ_x is a bijection, it is an automorphism of G. Any automorphism of G which is of the form ψ_x for some x in G, is called an *inner automorphism* of G.

Now if $\{x, y, g\} \subseteq G$, one always has

$$y^{-1}(x^{-1}gx)y = (y^{-1}x^{-1})g(xy) = (xy)^{-1}g(xy) \tag{3.2}$$

which means

$$\psi_y \circ \psi_x = \psi_{xy} \tag{3.3}$$

for all x, y. Thus the set of inner automorphisms is closed under composition of morphisms. Setting $y = x^{-1}$ in Eq. (3.3) we have

$$\psi_{x^{-1}} = \psi_x^{-1}, \tag{3.4}$$

and so the set of inner automorphisms is also closed under taking inverses. It now follows that the set of inner automorphisms of a group G forms a subgroup of the full automorphism group $\text{Aut}(G)$. We call this subgroup the *inner automorphism group of G*, denoted by $\text{Inn}(G)$.

Now Eq. (3.3) would suggest that there is a homomorphism from G to $\text{Aut}(G)$ except for one thing: the arguments of the ψ-morphisms come out *in the wrong order* in the right side of the equation. That is because *the operation "$\circ$" is denoting composition of left operators.*

This reveals the efficacy of using the exponential notation for denoting automorphisms as right operators. We employ the following:

(*Convention of writing conjugates in groups*)   If a and b are elements of a group G, we write

$$a^{-1}ba \text{ in the exponential form } b^a.$$

In this notation Eq. (3.2) reads as follows:

$$(g^x)^y = g^{xy} \tag{3.5}$$

for all $\{g, x, y\} \subseteq G$. What could be simpler?

Then we understand Eq. (3.3) to read

$$\psi_{xy} = \psi_x \psi_y, \tag{3.6}$$

where the juxtaposition on the right hand side of the equation indicates *composition of right operators*—that is the chronological order in which the mappings are performed reads from left to right, ψ_x first and then ψ_y.

Now Eq. (3.6) provides us with a group homomorphism:

$$\psi : G \rightarrow \text{Aut}(G)$$

taking element x to the inner automorphism ψ_x, the automorphisms of $\text{Aut}(G)$ being composed as right operators (as in the exponential convention for isomorphisms on p. 81).

What is the kernel of the homomorphism ψ? This would be the set of all elements $z \in G$ such that $\psi_z = 1_G$, the identity map on G. Thus this is the set $Z(G)$ of elements z of G satisfying any one of the following equivalent conditions:

(a) $\psi_z = 1_G$, the identity map on G,
(b) $z^{-1}gz = g$ for all elements g of G,
(c) $zg = gz$ for all elements g of G.

The subgroup $Z(G)$ is called the *center of G*. The identity element is always in the center. If $Z(G) = G$, then multiplication in G is "commutative"—that is $xy = yx$ for all $(x, y) \in G \times G$. Such a group is said to be *commutative*, and is affixed with the adjective *abelian*. Thus G is abelian if and only if $G = Z(G)$.

A Glossary of Terms Expected to be Understood from the Examples

1. Homomorphism of groups.
2. Epimorphism of groups.
3. Embedding of groups.
4. Isomorphism of groups.
5. Endomorphism of a group.
6. Automorphism of a group.
7. The kernel of a homomorphism, $\ker f$.
8. The automorphism group, $\text{Aut}(G)$, of a group G.
9. The inner automorphism group, $\text{Inn}(G)$, of a group, G.
10. An inner automorphism.
11. The center of a group, $Z(G)$.
12. Abelian groups.

3.4 Factor Groups and the Fundamental Theorems of Homomorphisms

3.4.1 Introduction

A teacher is sometimes obliged to present to a class a Theorem labeled as some sort of "Fundamental Theorem". More often than not, such a theorem is not quite as fundamental as it must have seemed at an earlier time in our history.[10]

At a minimum it would seem that a proposition should be labelled "a fundamental theorem" if it has these properties:

1. It should be used so constantly in the daily life of a scholar of the field, that quoting it becomes repetitive.
2. It's logical distance from the "first principles" of the field should be short enough to bear a short explanation to a puzzled student (that is, the alleged "fundamental theorem" should be "teachable").

We are lucky today! The fundamental theorems of homomorphisms of groups actually meets both of these criteria. They tell us that the homomorphic images of groups, their compositions, and their effects on subgroups, can all be derived from an internal study of the groups themselves.

The custom has been to name these three theorems as the "First-", "Second-", and "Third Fundamental Theorems of Homomorphisms". However a perusal of eight well-known textbooks in algebra shows this nomenclature to be far from uniform.[11] So we have tried to sidestep the ambiguity by naming these three very basic Theorems in a way related to what these theorems are telling us.

3.4.2 Normal Subgroups

The set of all subgroups of a group are permuted among themselves by automorphisms of a group. Explicitly: if $\sigma \in \mathrm{Aut}(G)$, and K is a subgroup of G, then $K^\sigma := \{k^\sigma | k \in K\}$ is again a subgroup of G.[12] Moreover, if H is a subgroup of $\mathrm{Aut}(G)$, then any subgroup of G left invariant by the elements of H is said to be

---

[10] Who on earth decided that the *Fundamental Theorem of Algebra* should be the fact that the Complex Numbers form an algebraically closed field? Who on earth decided that the *Fundamental Theorem of Geometry* should be the fact that an isomorphism between two Desarguesian Projective Spaces of sufficient dimension is always induced by a semilinear transformation of the underlying vector spaces?

[11] One author's "First" theorem is another's "Second", all three ordinal labels are used by one author (Hall) for what another calls the "First" Theorem, and some authors (Michael Artin, for example), seeing the problem, wisely declined to assign ordinal numbers beyond the "First".

[12] Note the convention of regarding automorphisms as right operators whose action is denoted exponentially (see p. 81).

H-invariant. Naturally we have special cases for special choices of *H*. A subgroup *K* which is invariant under the full group Aut(*G*) is said to be a *characteristic subgroup* of *G*. We write this as *K* char *G*. (The *center* which we previously encountered is certainly one of these).

But let us drop down to a larger class of *H*-invariant subgroups by letting *H* descend from the full automorphism group of *G* to the case that *H* is simply the inner automorphism group, Inn(*G*), encountered in the previous Section, Example 25, part (i). A subgroup which is invariant under Inn(*G*) is said to be *normal in G* or to be a *normal subgroup of G*.

Just putting definitions together one has

Lemma 3.4.1 *The following are equivalent conditions for a subgroup K of G:*

1. $\psi_x(K) = K$, *for all* $x \in G$,
2. $x^{-1}Kx = K$ *for every* $x \in G$,
3. $xK = Kx$, *for each* $x \in G$.

(In this Lemma, ψ_x is the inner automorphism induced by the element x: see Example 25, part (i) preceding.)

As a notational convenience, the symbol $K \trianglelefteq G$ will always stand for the assertion: "*K* is a normal subgroup of *G*" or, equivalently, "*K* is normal in *G*". This is always a relation *between* a subgroup *K* and some subgroup *G* which contains it. It is *not* a transitive relation. It is quite possible for a group *G* to possess subgroups *L* and *K*, with $L \trianglelefteq K$ and $K \trianglelefteq G$, for which *L* is not normal in *G*.[13] Two immediate consequences of normality are the following:

Corollary 3.4.2 *Suppose N is a normal subgroup of the group G and suppose H is any subgroup of G. Then the following statements hold:*

1. *N is normal in any subgroup of G which contains it.*
2. $N \cap H$ *is a normal subgroup of H. (As a special case, if H contains N, then N is normal in H).*
3. $NH = HN$ *is a subgroup of G.*
4. *If H is also normal in G, then so is HN.*

The proof is left for the beginning student in Exercise (4) in Sect. 3.7.1 at the end of this chapter.

One should not leave a basic section on normal subgroups without touching on the relationship between the normal subgroups and characteristic subgroups. As noted above, a normal subgroup of *G* is simply a subgroup *N* of *G* which is invariant under all the inner automorphisms of *G*, while a characteristic subgroup is a subgroup

---

[13] In the group *G* of rigid rotations of a cube (Example 23, part (g) of Sect. 3.2), the 180° rotation about one of the three face-centered axes, generates a subgroup *L* which point-wise fixes its axis of rotation, but inverts the other two face-centered axes. This a normal subgroup of the abelian subgroup *K* of all rotations of the cube which leave each of the three face-centered axes invariant. Then *K* is normal in *G*, being the kernel of the homomorphism of Example 25, Part (d) of Sect. 3.3. But clearly *L* is not normal in *G*, since otherwise its unique point-wise fixed face-centered axis, would also be fixed. But it is not fixed as *G/K* induces all permutations of Sym(3) on these axes.

K of G that is invariant under *all* automorphisms of G. Thus every characteristic subgroup of G is already normal, but, in general, being characteristic is a much stronger condition.

Theorem 3.4.3 *Assume N is a normal subgroup of G. If K is a characteristic subgroup of N (i.e. K is invariant under $Aut(N)$), then K is normal in G.*

One more result:

Theorem 3.4.4 *For any group G, $Inn(G) \trianglelefteq Aut(G)$.*

The proofs of these two theorems are left as Exercises 3 and 4 of Sect. 3.7.3 at the end of this chapter.

Almost any subgroup of G that is unique in some respect is a characteristic subgroup of G. For example the identity group 1, the whole group G and the center, $Z(G)$, are all characteristic subgroups of G.

3.4.3 Factor Groups

Let N be a normal subgroup if G. By Part 3. of the above Lemma 3.4.1, one sees that the subset xN is in fact the set Nx, for each $x \in G$. But it also asserts that $Nx \cdot Ny = N(xN)y = NNxy = Nxy$ as subsets of G. Thus there is actually a multiplicative group G/N whose elements are the right cosets of N in G, where multiplication is unambiguously given by the rule

$$(Nx) \cdot (Ny) = Nxy. \tag{3.7}$$

We have a name for this multiplicative group of cosets of a normal subgroup N. It is called the *factor group*, G/N. Its identity element is the subgroup N itself (certainly a right coset $N \cdot n$, for any $n \in N$) since $Nx \cdot N = NNx = Nx$. The inverse in G/N of the element Nx is the element Nx^{-1}. In fact it is now easy to verify that the mapping $G \to G/N$ that sends element x to coset Nx is a group homomorphism (see Theorem 3.4.5, part (i) below).

Now let $f : G \to H$ be a homomorphism of groups. Two special groups associated with a group homomorphism f have been introduced earlier in this chapter: the *range* or *image*, $f(G)$, and the *kernel*, $\ker f$. Now if $y \in \ker f$, and $x \in G$, we see that

$$f(x^{-1}yx) = (f(x))^{-1}f(y)f(x) = (f(x))^{-1} \cdot 1 \cdot f(x) = 1 \in H,$$

since $f(y) = 1$. Thus for all $y \in \ker f$, and $x \in G$, $x^{-1}yx \in \ker f$ so $\ker f$ is always a normal subgroup. What about the corresponding factor group?

Theorem 3.4.5 (The Fundamental Theorem of Homomorphisms)

(i) *If N is a normal subgroup of a group G, the mapping $G \to G/N$ which maps each element x of G to the right coset Nx which contains it, is an epimorphism of groups. Its kernel is N.*

(ii) *If $f : G \to H$ is a homomorphism of groups, there is an isomorphism $\eta :$ $f(G) \to G/(\ker f)$ taking each element $f(x)$, $x \in G$, to $(\ker f)x$. Thus every homomorphic image $f(G)$ is isomorphic to a factor group of G.*

(iii) *In particular, the kernel of the homomorphism is trivial—i.e. $\ker f = 1$—if and only if the homomorphism itself is injective.*

Proof (i) That the mapping $\pi : G \to G/N$ defined by $\pi(x) := Nx$, is a homomorphism follows from $NxNy = Nxy$, for all $(x, y) \in G \times G$. By definition of G/N this map is onto. (The epimorphism $\pi : G \to G/N$ is usually called the *projection homomorphism* or sometimes the *natural homomorphism onto the factor group* G/N.) Since the coset $N = N \cdot 1$ is the identity element of G/N, the kernel of ν is precisely the subset $\{x \in G | Nx = N\} = \{x \in G | x \in N\} = N$. Thus $\ker \pi = N$.

(ii) Now set $\ker f := N$ We propose a mapping $\eta : f(G) \to G/N$ which takes an image element $f(x)$ to the coset Nx. Since this recipe is formulated in terms of a single element x, we must show that the proposed mapping is well-defined. Suppose $f(x) = f(y)$. We wish to show $Nx = Ny$. But the hypothesis shows that $f(x^{-1}y) = f(x^{-1})f(y) = (f(x))^{-1}f(x) = 1 \in G$, so $x^{-1}y \in \ker f = N$, whence $Nx = Ny$, as desired.

Now

$$\eta(f(x)f(y)) = \eta(f(xy)) = Nxy \qquad (3.8)$$

$$= NxNy = \eta(f(x)) \cdot \eta(f(y)). \qquad (3.9)$$

Thus η is a homomorphism. If $f(x)$ were in the kernel of η, then $\eta(f(x)) = Nx = N$, the identity element of G/N. Thus $x \in N$, so $f(x) = f(1)$, the identity element of $f(G)$. Finally, η is onto, since, for any $g \in G$, the coset $Ng = \eta(f(g))$. Thus η is a bijective homomorphism and so is an isomorphism.

(iii) This is obvious from first principles since $xy^{-1} \in \ker f$ if and only if $f(x) = f(y)$. It also follows from (ii). The proof is complete. $\square$

Remark (a) The main idea about (i) is that when you spot a homomorphism shooting off somewhere, you do not have to search all over the Universe to study it. Instead, you can realize the homomorphic image right inside the structure of the group G itself, as one of its factor groups.

(b) We have included statement (iii) in the Fundamental Theorem of Homomorphisms since it is so often implicitly used without any particular reference. Since it is an immediate consequence of part (ii) we have given it a home here.

There are two further consequences of Theorem 3.4.5 which are contained in the following Corollary.

Corollary 3.4.6 (The classical isomorphism theorems)

(i) (The Composition Theorem for Groups) *Suppose K and N are both normal subgroups of G, with K contained in N. Then N/K is a normal subgroup of G/K, and G/N is isomorphic to the factor group $(G/K)/(N/K)$.*

(ii) (The Modularity Theorem for Groups) *If N is a normal subgroup of G, and H is any subgroup of G, HN/N is isomorphic to $H/(H \cap N)$.*

Proof (i) By Corollary 3.4.2, part 1, K is normal in N, so N/K is the multiplicative group of cosets $\{Kn | n \in N\}$. For any coset Kx of G/K, and coset Kn of N/K, $(Kx)^{-1}KnKx = K(x^{-1}nx)$, which is in N/K. Thus $N/K \trianglelefteq G/K$. Thus there is a natural epimorphism $f_2 : G/K \to (G/K)/(N/K)$ onto the factor group as described in the Fundamental Theorem (Theorem 3.4.5, part (i)). By the same token, there is a canonical epimorphism $f_1 : G \to G/K$. By Lemma 3.3.2 of Sect. 3.2. The composition of these epimorphisms is again an epimorphism:

$$f_2 \circ f_1 : G \to (G/K)/(N/K).$$

An element x of G is first mapped to the coset Kx and then to the coset $(Kx)\{Kn | n \in N\}$ (which, viewed as a set of elements of G is just Nx). Thus x maps to the identity element $(N/K)/(N/K)$ of $(G/K)/(N/K)$ if and only if $x \in N$. Thus the kernel of epimorphism $f_2 \circ f_1$ is N. The result now follows from the Fundamental Theorem of Homomorphisms (Theorem 3.4.5 Part (ii)).

(ii) Clearly $N \trianglelefteq HN$ and $N \cap H \trianglelefteq H$ by Corollary 3.4.2, part1. We propose to define a morphism $f : H \to HN/N$ by sending element h of H to coset $Nh \in HN/N$. Clearly hh' is sent to $Nhh' = (Nh)(Nh')$, so f is a group homomorphism. Since every coset of N in $HN = NH$ has the form Nh for some $h \in H$, f is an epimorphism of groups. Now if $h \in H$, then $Nh = N$, the identity of HN/N, if and only if $h \in H \cap N$. Thus ker $f = H \cap N$ and now the conclusion follows from the Fundamental Theorem of Homomorphisms (Theorem 3.4.5 Part (ii)). $\square$

3.4.4 Normalizers and Centralizers

Let X be some non-empty subset of a group G. For each $x \in G$, the set $\psi_x(X) := x^{-1}Xx$ is called a *conjugate (in G) of the set X*. Given X, consider the set

$$N_G(X) := \{x \in G | x^{-1}Xx = X\}.$$

The reader may verify that the set on the right side of the presented equation indeed satisfies any one of the equivalent conditions listed in the Subgroup Criterion (Lemma 3.2.3). The subgroup described by this equation is called the *normalizer (in G) of* the set X and is denoted $N_G(X)$. Also, if H is any subgroup of G, we say that H *normalizes the set X* if and only if H is a subgroup of the normalizer $N_G(X)$.

In practice, X is often itself a subgroup.

Lemma 3.4.7 *The following statements about a group G are true.*

1. *The subgroup H is a normal subgroup of G, i.e. $H \unlhd G$, if and only if $N_G(H) = G$. More generally, H is normal in the subgroup K—i.e. $H \unlhd K$—if and only if $H \leq K \leq N_G(H)$.*
2. *If H and K are subgroups of G, then $N_K(H) = N_G(H) \cap K$.*
3. *If H and K are subgroups of G, and K normalizes H, then $HK = KH$, and KH is a subgroup of G.*
4. *Suppose the subgroup K normalizes the subset X of G. Then K also normalizes the subgroup $\langle X \rangle_G$ generated by X.*

The statements are immediate consequences of the definitions. The beginning student is urged to warm up some nice fall afternoon by devising formal proofs of these statements.

There is another kind of subgroup determined by a subset X of a group G, namely the subgroup

$$C_G(X) := \{g \in G | g^{-1}xg = x, \text{ for all } x \in X\},$$

called the *centralizer (in G) of* X. It consists precisely of those elements in G which commute with every element of X. We say that subgroup H *centralizes the set X* if and only if $H \subseteq C_G(X)$.

At this point it might be useful to compare the centralizer and the normalizer. The normalizer $N_G(X)$ is the set of elements $g \in G$ whose associated inner automorphism ψ_g leaves the subset X invariant as a whole. The centralizer $C_G(X)$ consists of those elements $g \in G$ whose associated inner automorphism ψ_g fixes set X *element-wise*. Now, as with the normalizer, we possess a number of elementary observations.

Lemma 3.4.8 *Suppose G is some fixed group with a designated subset X. The following statements hold.*

1. *We have $C_G(X) \unlhd N_G(X)$.*
2. *If H is a subgroup of G, then $C_H(X) = C_G(X) \cap H$.*
3. *If subgroup H centralizes X, then it also centralizes the subgroup $\langle X \rangle_G$ generated by X.*

Once again the beginning student is invited to spend a few moments some nice fall afternoon assembling formal proofs of the statements in Lemma 3.4.8 This time, in view of part 1, the student is permitted to order a drink.

3.5 Direct Products

A direct product is a formal construction for getting new groups in a rather easy way. First let $\mathcal{G} = \{G_\sigma | \sigma \in I\}$ be a family of groups indexed by the index set I. We can write elements of the Cartesian product of the G_σ as functions

$$f : I \to \bigcup_{\sigma \in I} G_\sigma \text{ such that } f(\sigma) \in G_\sigma.$$

Of course, when I is countable, we can represent elements f in the usual way, as sequences $(f(1), f(2), \ldots)$. Define a binary operation on the Cartesian product by declaring the "product" $f_1 f_2$ of f_1 and f_2 to be the function whose value $(f_1 f_2)(\sigma)$ at $\sigma \in I$ is $f_1(\sigma) f_2(\sigma)$—that is, the values $f_i(\sigma)$ at the "coordinate" σ are multiplied in the group G_σ) to yield the σ-coordinate of the "product". This is termed "coordinate-wise multiplication", since for sequences, the product of $(a_1, a_2, \ldots)$ and $(b_1, b_2, \ldots)$ is $(a_1 b_1, a_2 b_2, \ldots)$, by this definition. The Cartesian product with this coordinate-wise multiplication clearly forms a group, which we call *the direct product over* $\mathcal{G}$ and is denoted by

$$\prod\nolimits_{\sigma \in I}(G_\sigma) \text{ or } \prod\nolimits_{\mathcal{G}} G_\sigma$$

or, when $I = \{1, 2, \ldots, n\}$, simply by

$$G_1 \times G_2 \times \cdots \times G_n.$$

It contains a subgroup consisting of all maps f in the definition of direct product, for which $f(\sigma)$ fails to be the identity element of the group G_σ only finitely many times. This subgroup is called the *weak direct product* or *direct sum* over $\mathcal{G}$ and is denoted

$$\bigoplus\nolimits_{\sigma \in I}(G_\sigma) \text{ or } \bigoplus\nolimits_{\mathcal{G}}(G_\sigma).$$

When I is finite, there is no distinction between the direct product and direct sum.

Consider the group $G = \{\pm 1\}$ under ordinary multiplication of integers. Thus G is a group of order 2, with involution -1. Then $G \times G$ is the group of pairs (u, v), $u, v \in \{\pm 1\}$, with coordinate-wise multiplication. For example, one calculates that $(-1, 1) \cdot (1, -1) = (-1, -1)$. In this case $G \times G$ is a group of order four with an identity element and three involutions. The product of any two distinct involutions is the third involution. Any group in the isomorphism class of this group is called a *fours group*.

For every index τ in I, there is clearly a homomorphism $\pi_\tau : \bigoplus_{\mathcal{G}}(G_\sigma) \to G_\tau$ which takes f to $f(\tau)$, for all f in the direct product. Such an epimorphism is called *a projection onto the coordinate indexed by* τ. This epimorphism retains this name even when it is restricted to the direct sum.

Any permutation of the index set induces an obvious isomorphism of direct products and direct sums. Thus $G_1 \times G_2$ is isomorphic to $G_2 \times G_1$ even though they are not formally the same group.

Similarly, any (legitimate) rearrangement of parentheses involved in constructing direct products yields isomorphisms; specifically

$$G_1 \times G_2 \times G_3 \simeq (G_1 \times G_2) \times G_3 \simeq G_1 \times (G_2 \times G_3).$$

Now suppose A and B are two subgroups of a group G with $A \cap B = 1$, and B normalizing A (that is, $B \leq N_G(A)$). Then together they generate a subgroup AB with order $|A| \cdot |B|$ (Lemma 3.2.6). Now suppose in addition that A normalizes B. Then for any element $a \in A$ and element $b \in B$, the factorizations $(aba^{-1})b^{-1} = a(ba^{-1}b^{-1})$ show that $aba^{-1}b^{-1} \in A \cap B = \{1\}$ and so $ab = ba$. Thus all elements of A commute with all elements of B. In that case the mapping $A \times B \to AB$ which takes $(a, b) \in A \times B$ to ab, is a group homomorphism whose kernel is $\{(x, x^{-1})|x \in A \cap B\}$. Since $A \cap B = \{1\}$, this map is an isomorphism.

These remarks are summarized in the next Lemma.

Lemma 3.5.1

 (i) *For any permutation π in $Sym(n)$, There is an isomorphism*

$$G_1 \times G_2 \times \cdots \times G_n \to G_{\pi(1)} \times G_{\pi(2)} \times \cdots \times G_{\pi(n)}$$

although neither of the groups are necessarily formally the same.

 (ii) *Any two well-formed groupings of the factors of a direct product into parentheses yields groups which are isomorphic to the original direct product and hence are isomorphic to each other.*

(iii) *(Internal Direct Products) Suppose $A_1, A_2, \ldots$ is a countable sequence of subgroups of a group G.*

    (a) *Suppose A_j normalizes A_i whenever $i \neq j$, and*
    (b) *$A_1 A_2 \cdots A_{k-1} \cap A_k = 1$, for all $k \geq 2$.*

Then $A_1 A_2 \cdots A_n$ is isomorphic to the direct product $A_1 \times \cdots \times A_n$ for any finite initial segment $\{A_1, \ldots, A_n\}$ of the sequence of subgroups.

3.6 Other Variations: Semidirect Products and Subdirect Products

3.6.1 Semidirect Products

Suppose we are given a homomorphism

$$f : H \to \text{Aut}(N),$$

for two abstract groups N and H. Then f defines a formal construction of a group $N{:}H$, which is called a *semidirect product of N by H*.[14] The elements of $N{:}H$ are

---

[14]In some older books there is other notation for the semidirect product—some sort of decoration added to the "$\trianglelefteq$" symbol. The notation $N{:}H$ is the notation for a "split extension of groups" used in the *Atlas of Finite Groups* [12].

the elements of the Cartesian product $N \times H$. Multiplication proceeds according to this rule: For any two elements (n_1, h_1) and (n_2, h_2) in $N \times H$:

$$(n_1, h_1)(n_2, h_2) := (n_1 (n_2)^{f(h_1^{-1})}, h_1 h_2).$$

Verification of the associative law for a triple product $(n_1, h_1)(n_2, h_2)(n_3, h_3)$ comes down to the calculation

$$n_3^{f((h_1 h_2)^{-1})} = (n_3^{f(h_2^{-1})})^{f(h_1^{-1})}.$$

Example 26 1. If $f : H \to \mathrm{Aut}(N)$ is the trivial homomorphism—that is, f maps every element of H to the identity automorphism of N—then $N{:}H$ is just the ordinary direct product $N \times H$ of the previous section.
2. If A and B are subgroups of a group G for which A normalizes B and $A \cap B = 1$, then the subgroup AB of G, is isomorphic to the semidirect product $B{:}A$ with respect to the morphism $f : A \to \mathrm{Aut}(B)$ which takes each element a of A to the automorphism of B induced by conjugating the elements of B by a—that is $\psi_a|_B$.
3. Let N be the additive group of integers mod n, and let H be any subgroup of the multiplicative group of residues coprime to n (for example, the quadratic residues coprime to n). Let G be the group of all permutations of N of the form

$$\mu(m, h) : x \to hx + m, \text{ for all } x \in N.$$

Then G is the semidirect product $N{:}H$.
4. Let F be any field, let F^* be the multiplicative group of all nonzero elements of F, and let H be any subgroup of F^*. Then the set of all transformations of F of the form

$$x \to hx + a, a \in F, h \in H,$$

forms a group under the composition of such transformations. This group is isomorphic to the semidirect product $(F, +){:}H$, where $(F, +)$ denotes the group on F whose operation is addition in the field F.
5. (The Frobenius group of order 21.) The group is generated by elements x and y subject only to the relations

$$x^3 = 1, y^7 = 1, x^{-1} y x = y^2.$$

It is a semidirect product $\mathbf{Z}_7 : \mathbf{Z}_3$. (This is an example of a "presented group", which we shall meet in Chap. 6.) Here, the presentation reveals the homomorphism $H = \langle x \rangle \to \mathrm{Aut}(\langle y \rangle)$. This group is isomorphic to a group constructed by the recipe in part 4, where N is the additive group of the field $\mathbb{Z}/(7)$ and H is the multiplicative group of quadratic residues mod 7.

6. Suppose A is an abelian group of order n which is not a direct product of Z_2's. Then the mapping $\sigma : A \to A$ defined by $a^\sigma = a^{-1}$ for all $a \in A$, is an automorphism of order 2. Using the inclusion $\langle \sigma \rangle \subseteq \mathrm{Aut}(A)$ for f, one can form the semidirect product $A\langle\sigma\rangle$ as above. Some authors refer to $A\langle\sigma\rangle$ as a *generalized dihedral group*.

7. A group is said to be a *normal extension of a group N by H* (written $G = H.N$) if and only if

$$N \trianglelefteq G \text{ and } G/N \simeq H.$$

The extension is said to be *split* if and only if there is a subgroup H_1 of G such that

$$G = NH_1, \quad N \cap H_1 = 1.$$

In this case, $H_1 \simeq H$. One can see that

> G is a split extension of G by H if and only if G is a semidirect product of N by H.

First, if the extension is split, every element of G has the form $g = nh$, $(n, h) \in N \times H$. Because $N \cap H_1 = 1$, the expression $g = nh$ is unique for a given element g. We thus have a bijection

$$\sigma : G \to N \times H_1$$

which takes $g = nh$ to $(n, \tau(h))$, where τ is the isomorphism $H_1 \to H = G/N$ (from Corollary 3.4.6, part (ii)). Now if we multiply two elements of G, we typically obtain

$$(n_1 h_1)(n_2 h_2) = n_1 (h_1 n_2 h_1^{-1}) h_1 h_2 = n_1 (n_2)^{(h_1^{-1})} \cdot h_1 h_2,$$

where, by our convention on inner automorphisms, $g^x := x^{-1}gx$. The σ-value of this element is the pair

$$((n_1)(n_2)^{(h_1^{-1})}, \tau(h_1)\tau(h_2)).$$

Thus σ is an isomorphism of G with the semidirect product of N by H, defined by composing τ^{-1} with the homomorphism

$$H_1 \to \mathrm{Aut}(N)$$

induced by conjugation. (The latter is just the restriction to H_1 of the homomorphism ψ of Example 3, part (i).)

Conversely, if G is the semidirect product of N by H defined by some homomorphism $\rho : H \to \mathrm{Aut}(N)$, then, setting

$$N_0 := \{(n, 1)|n \in M\}$$
$$H_0 := \{(1, h)|h \in H\}$$

then

$$N_0 \trianglelefteq G = N_0 H_0 \text{ and } N_0 \cap H_0 = 1,$$

so G is the split extension N_0 by H_0 and conjugation in G by element h_0 of H_0 induces automorphism $\rho(h_0)$ on N_0. (Clearly, $N_0 \simeq N$ and $H_0 \simeq H$.)

3.6.2 Subdirect Products

Something a little less precise is the notion of a subdirect product. We say that a group H is a *subdirect product of the groups* $\{G_\sigma | \sigma \in I\}$ if and only if (i) H is a subgroup of the direct product $\prod_I G_\sigma$ and (ii) for each index τ, the restriction of the projection map π_τ to H is onto—i.e. $\pi_\tau(H) = G_\tau$. There may be many such subgroups, so the isomorphism type of H is not uniquely determined by (i) and (ii) above.

Subdirect products naturally arise in the following situation: Suppose M and N are normal subgroups of a group G. Then $G/(M \cap N)$ is a subdirect product of G/N and G/M. This is because $G/(M \cap N)$ can be embedded as a subgroup of $(G/M) \times (G/N)$ by mapping each coset $(N \cap M)x$ to the ordered pair (Mx, Nx).

In fact something very general happens:

If $\{N_\sigma | \sigma \in I\}$ is a family of normal subgroups of a group G, then

- $G/(\cap_I N_\sigma)$ is the subdirect product of the groups $\{G/N_\sigma | \sigma \in I\}$.
- Suppose $\mathcal{F}$ is a family of groups "closed under taking subdirect products"—that is, any subdirect product of members of $\mathcal{F}$ is in $\mathcal{F}$. (For example the class of abelian groups is closed under subdirect products.) Then for any arbitrary group G, there is a "smallest" normal subgroup $G_\mathcal{F}$ whose associated factor group $G/G_\mathcal{F}$ lies in $\mathcal{F}$. Precisely, $G/G_\mathcal{F} \in \mathcal{F}$ and if N is normal in G with $G/N \in \mathcal{F}$, then $G_\mathcal{F} \leq N$.

A *simple group* is a group G whose only proper normal subgroup is the identity subgroup. (Note that by this definition, the group of order one is not a simple group.)

A *maximal normal subgroup of a group* G is a maximal element in the partially ordered set of *proper normal subgroups* of G (ordered by inclusion, of course).

In Exercise (1) in Sect. 3.7.2 the following is proved:

Lemma 3.6.1 *A factor group G/N of G is a simple group if and only if N is a maximal normal subgroup of G.*

We begin with an elementary Lemma.

Lemma 3.6.2 *Suppose* $\{M_i | i \in I\}$ *is a finite collection of maximal normal subgroups of a group* G. *Then for some subset* J *of* I,

$$G/(\cap_{i \in I} M_i) = G/(\cap_{j \in J} M_j) \simeq \prod_{j \in J} (G/M_j),$$

a direct product of simple groups.

Proof The result is true for $|I| = 1$, as G/M_1 is simple. We use induction on $|I|$. Renumbering the subscripts if necessary, the induction hypothesis allows us to assume that for some subset $J = \{1, \ldots, d\} \subseteq I' = \{1, \ldots, k-1\}$,

$$N := \cap_{i \in I'} M_i = \cap_{j \in J} M_j. \tag{3.10}$$
$$G/N \simeq (G/M_1) \times \cdots \times (G/M_d). \tag{3.11}$$

We set $I := \{1, \ldots, k\} = I' + \{k\}$ and assume, without any loss of generality, that M_k is not contained in N. Then since $N M_k$ is a normal subgroup of G properly containing M_k, we must have $G = N M_k$. Also, since M_k is a simple group, $N \cap M = 1$. Then

$$G/(N \cap M_k) \simeq (G/N) \times (G/M_k),$$

and the result for $|I| = k$ follows upon substitution for G/N in the right-hand side. $\square$

One may conclude.

Corollary 3.6.3

(i) *Suppose* G *is a finite group with no non-trivial proper characteristic subgroups. Then* G *is a direct product of pairwise isomorphic simple groups.*

(ii) *If* N *is a minimal normal subgroup of a finite group* G, *then* N *is a direct product of isomorphic simple groups.*

Proof Part (i) Let M be a maximal normal subgroup of the finite group G. Then G has a finite automorphism group, and so $\{M^\sigma | \sigma \in \text{Aut}(G)\}$ is a finite collection of maximal normal subgroups of G whose intersection N is a characteristic subgroup of G properly contained in G. By hypothesis, $N = 1$. Then by the above Lemma 3.6.2, $G = G/N$ is the direct product of simple groups. If $G = 1$, we have an empty direct product and there is nothing to prove. Otherwise, the product of those direct factors isomorphic to the first direct factor clearly form a non-trivial characteristic subgroup, which, by hypothesis must be the whole group.

Part (ii) Since N is a minimal normal subgroup of the finite group G, N is a non-trivial finite group with only the identity subgroup as a proper characteristic subgroup, so the conclusion of Part (i) holds for N. $\square$

3.7 Exercises

3.7.1 Exercises for Sects. 3.1–3.3

1. Write out formal proofs of Lemmas 3.3.1 and 3.3.2 (all parts).
2. Prove Corollary 3.2.4.
3. Prove that $G/Z(G)$ can never be a cyclic group. [Hint: If false, cosets of the form $Z(G)x^{\pm k}$ partition G, and elements on one of these cosets commute with those in any other such coset. So G can be shown to be abelian.]
4. Prove Corollary 3.4.2. [Hint: Use the subgroup criterion (Lemma 3.2.3) for part 3.]
5. Suppose a group G has *exponent* 2—that is, $g^2 = 1$ for every $g \in G$. Show that G is abelian.
6. Suppose p is a (positive) prime number and k is a positive integer.

   (i) If $p > 2$, show that $\mathrm{Aut}(Z_{p^k}) \simeq Z_{p-1} \times Z_{p^{k-1}}$.
   (ii) If $p = 2$ show that $\mathrm{Aut}(Z_{2^k}) \simeq Z_2 \times Z_{2^{k-2}}$.

3.7.2 Exercises for Sect. 3.4

1. Suppose N is a normal subgroup of the group G.

   (a) Show that there is an isomorphism between the poset of all subgroups H of G which contain N, and the poset of all subgroups of G/N (both posets partially ordered by the containment relation).
   [Hint: The isomorphism takes H in the first poset to H/N in the second.]
   (b) Show that H is normal in G if and only if H/N is normal in G/N. Thus the isomorphism of part (a) and its inverse both preserve the normality relation. (Note that they need not preserve the property of being characteristic in either direction. The group H containing N is called *the inverse image* of H/N.)
   (c) Using the fundamental theorem of homomorphisms conclude that if

   $$f : G \to L$$

   is an epimorphism of groups, then there is a 1-1 correspondence of the subgroups of L with the subgroups of G containing ker f preserving containment and normality. The correspondence takes a subgroup L_1 of L to the subset $\{g \in G | f(g) \in L_1\}$, also called *the inverse image of L_1.*

2. If X is a *subset* of the group G (that is, the subset X is not necessarily a subgroup of G), define the *centralizer in G of X* to be the set of all group elements g of G such that $gx = xg$ for all $x \in X$. This set is denoted $C_G(X)$. Similarly, the

normalizer in G of X is the set of elements $\{g \in G | g^{-1}Xg = X\}$. Such elements may permute the elements of X by conjugation, but they stabilize X as a set. A subgroup H is said to *normalize X* if and only if $H \leq N_G(X)$. If $N_G(X) = G$, then X is called *a normal set in G*.

  (a) Show that $C_G(X)$ and $N_G(X)$ are subgroups of G.
  (b) Show that if X is a normal set in G, then $C_G(X)$ is also normal in G.
  (c) Show that if N is a normal subgroup of G, then the inner automorphism $\psi_g : x \to g^{-1}xg$ induces an automorphism of N.

3. (a) Conclude that a characteristic subgroup of a normal subgroup of G is normal in G, that is, K char $N \trianglelefteq G$ implies $K \trianglelefteq G$.
  (b) Conclude that a characteristic subgroup of a characteristic subgroup of G is characteristic in G, that is, L char K char G implies L char G.
  (c) Suppose N is a normal subgroup of G. Show that the mapping which sends element g to $\psi_g|_N$, the restriction of the inner automorphism conjugation-by-g to N, defines a group homomorphism

$$\psi^N : G \to \operatorname{Aut}(N)$$

  whose kernel is $C_G(N)$. Conclude that the group of automorphisms induced on N by the inner automorphisms of G is isomorphic to $G/C_G(N)$.

4. Show that for any group G, $\operatorname{Inn}(G) \trianglelefteq \operatorname{Aut}(G)$. (The factor group $\operatorname{Out}(G) := \operatorname{Aut}(G)/\operatorname{Inn}(G)$ is called the *outer automorphism group of G*.)
5. Prove the assertions of Corollary 3.4.2.
6. Recall from p. 99 that a group is said to be a *simple group* if and only if its only proper normal subgroup is the identity subgroup. (Note that the definition forbids the identity group to be a simple group.)

  (a) Prove that any group of prime order is a simple group.
  (b) Suppose G is a group. A subgroup M is a *maximal subgroup of G* if and only if it is maximal in the poset of all *proper* subgroups of G. A subgroup M is a *maximal normal subgroup of G* if and only if it is maximal in the poset of all proper normal subgroups of G. (Note that it *does not* mean that it is a maximal subgroup which is normal. A maximal normal subgroup may very well not be a maximal subgroup.) Prove that a factor group G/N of G is a simple group if and only if N is a maximal normal subgroup. [Hint: Use the result of Exercise (1)b in Sect. 3.7.2 just above.]

3.7.3 Exercises for Sects. 3.5–3.7

1. Let V be any vector space, and form the group $GL(V)$ of all invertible linear transformations of V. Let G be any group and let $f : G \to GL(V)$ be a homomorphism of groups. [Such a homomorphism is said to be a *representation of the group G*.] Define multiplication on $V \times G$ by the rule

$$(v_1, g_1) \cdot (v_2, g_2) := (v_1 + v_2^{f(g_1)}, g_1 g_2).$$

   (a) Show that this group is isomorphic to the semidirect product $(V, +){:}G$ upon noting that $GL(V) \leq \mathrm{Aut}(V, +)$.

   (b) Show that the center of $(V, +){:}G$ is the direct product $C_G(V) \times \ker f$.

2. A popular textbook in intermediate algebra (after presenting the Sylow theorems which will appear in the next chapter) offers an exercise requesting that the reader prove that every group of order 75 is abelian. Using the semidirect product construction, show that there exists a non-abelian group of order 75.

   [Hint: Let $V := Z_5 \times Z_5$. One can regard V as a vector space over the field $\mathbf{Z}/(5)$ of integers mod 5. Let $t : V \to V$ be a linear transformation acting with minimal polynomial $x^2 + x + 1$. This means V has a basis $\{v_1, v_2\}$ with $v_1^t = v_2$, and $v_2^t = -v_1 - v_2$. Then t^3 induces the identity transformation 1_V. Thus $\langle t \rangle$, as a subgroup of $GL(V)$, has order 3. One can then form the semidirect product $(V, +)\langle t \rangle$ as described in the previous exercise.]

3. Suppose K is a characteristic subgroup of N and N is a normal subgroup of G. Show that K is a normal subgroup of G.

   [Hint: Conjugation by any element of G induces an automorphism of N and so leaves K invariant, since K is characteristic in N.]

4. Show that the group of inner automorphism of G is a normal subgroup of the group of all automorphisms of G.

   [Hint: Let $\psi(g)$ be conjugation by g. show that

$$\sigma^{-1} \cdot \psi(g) \cdot \sigma$$

is conjugation by $\sigma^{-1}(g)$ and so is an inner automorphism.]

Chapter 4
Permutation Groups and Group Actions

Abstract A useful paradigm for discussing a group is to regard it as acting as a group of permutations of some set. The power of this point of view derives from the flexibility one has in choosing the set being acted on. Odd and even finitary permutations, the cycle notation, orbits, the basic relation between transitive actions and actions on cosets of a subgroup are first reviewed. For finite groups, the paradigm produces Sylow's theorem, the Burnside transfer and fusion theorems, and the calculations of the order of any group of automorphisms of a finite object. Of more special interest are primitive and multiply transitive groups.

4.1 Group Actions and Orbits

As defined earlier, a *permutation* of a set X is a bijective mapping $X \to X$. Because we will be dealing with groups of permutations it will be convenient to use the "exponential" notation which casts permutations in the role of right operators. Thus if π is a permutation of X, we write x^π for the image of element x under the permutation π. Recall that composition of bijections and inverses of bijections are bijections, so that the set of all bijections of set X into itself form a group under composition which we have called the *symmetric group* on X and have denoted $\mathrm{Sym}(X)$. Recall also that any bijection $\nu : X \to Y$ induces a group isomorphism $\bar{\nu} : \mathrm{Sym}(X) \to \mathrm{Sym}(Y)$ by the rule:

$$\nu(x^\pi) = (\nu(x))^{\bar{\nu}(\pi)} .$$

In order to emphasize the distinction between the elements of $\mathrm{Sym}(X)$ and the elements of the set X which are being permuted, we often refer to the elements of X by the neutral name "letters". In view of the isomorphism just recorded, the symmetric group on any finite set of n elements can be thought of as permuting the set of symbols (or 'letters') $\Omega_n := \{1, 2, \ldots, n\}$. This group is denoted $\mathrm{Sym}(n)$ and has order $n!$.

Now suppose H is any subgroup of $\mathrm{Sym}(X)$. Say for the moment that two elements of X are H-related if and only if there exists an element of H which takes one to the other. It is easy to see that "H-relatedness" is an equivalence relation on the elements of X. The equivalence classes with respect to this relation are called the

© Springer International Publishing Switzerland 2015   105
E. Shult and D. Surowski, *Algebra*, DOI 10.1007/978-3-319-19734-0_4

H-orbits of *X*. Within an *H*-orbit, it is possible to move from any one letter to any other, by means of an element of *H*. Thus *X* is partitioned into *H*-orbits. We say *H* is *transitive* on *X* if and only if *X* comprises a single *H*-orbit.

We now extend these notions in a very useful way. We say that a *group G acts on set X* if and only if there is a group homomorphism

$$f : G \to \text{Sym}(X).$$

We refer to both *f* and the image *f(G)* as the *action* of *G*, and we shall borrow almost any adjective of a subgroup *f(G)* of Sym(*X*), and apply it to *G*. Thus we call an *f(G)*-orbit, a *G*-orbit, we say "*G* acts in *k* orbits on *X*" if *X* partitions into exactly *k* such *f(G)*-orbits, and we say "*G* is transitive on *X*", or "acts transitively on *X*" if and only if *f(G)* is transitive on *X*.

If the action *f* is understood, we write x^g instead of $x^{f(g)}$. Also the unique *G*-orbit containing letter *x* is the set $\{x^g \mid g \in G\}$, and is denoted x^G. Its cardinality is called the *length of the orbit*. The group action is said to be *faithful* if and only if the kernel of the homomorphism $f : G \to \text{Sym}(X)$ is the identity subgroup of *G*—that is, *f* is an embedding of *G* as a subgroup of the symmetric group.

The power of group actions derives from the fact that the same group can have several actions, each of which can yield new information about a group. We have already met several examples of group actions, although not in this current language. We list a few.

4.1.1 Examples of Group Actions

1. Recall that each rigid rotation of a regular cube induces a permutation on the following sets:

    (a) the four vertex-centered axes.
    (b) the three face-centered axes.
    (c) the six edge-centered axes.
    (d) the six faces.
    (e) the eight vertices.
    (f) the twelve edges.

    If *Y* is any one of the six sets just listed, and *G* is the full group of rigid rotations of the cube, then we obtain a transitive action $f_Y : G \to \text{Sym}(Y)$. Except for the case *Y* is the three perpendicular face-centered axes (Case (b)), the action is faithful. The action is an epimorphism onto the symmetric group only in Cases (a) and (b).

2. If N is a normal subgroup of G, then G acts on the elements of N by inner automorphisms of G. Thus $n^g := g^{-1}ng$ for each element g of G and each element n of N, to give the group action: $f : G \to \mathrm{Sym}(N)$. Clearly the identity element of G forms one of the G-orbits and has length one.

   If $N = G$, the G-orbits $x^G := \{g^{-1}xg|g \in G\}, x \in G$ are called *conjugacy classes* of G. Now one may consider the transitive action $G \to \mathrm{Sym}(x^G)$ on a single conjugacy class x^G.

3. The image H^σ of any subgroup H of G under an automorphism σ is again a subgroup of G, and this is certainly true of inner automorphisms. Thus $H^g := g^{-1}Hg$ is a subgroup of G called a *G-conjugate of H* or just *conjugate of H*. Thus there is an action $G \to \mathrm{Sym}(\mathcal{S}(G))$ where $\mathcal{S}(G)$ is the collection of all subgroups of G. As in the previous case, this restricts to the transitive action $G \to \mathrm{Sym}(H^G)$ on the set $H^G := \{H^g|g \in G\}$ of all *conjugates of the subgroup H*.

4. Finally we can take any fixed subset X of elements of G and watch the action $G \to \mathrm{Sym}(X^G)$ on the collection $X^G := \{g^{-1}Xg|g \in G\}$ of all G-conjugates of set X.

5. Let $P_k = P_k(V)$ be the collection of all k-dimensional subspaces of the vector space V, where k is a natural number. We assume $\dim V \geq k$ to avoid the possibility that P_k is empty. Since an invertible linear transformation must preserve the dimension of any finite dimensional subspace, we obtain an action

$$f : GL(V) \to \mathrm{Sym}(P_k).$$

   It is necessarily transitive.

6. Similarly, if we have an action $f : G \to \mathrm{Sym}(X)$, where $|X| \geq k$, we also inherit an action $f_Y : G \to \mathrm{Sym}(Y)$ on the following sets Y:

   (a) The set $X^{(k)}$ of ordered k-tuples of elements of X.
   (b) the set $X^{(k)*}$ of ordered k-tuples with pairwise distinct entries.
   (c) the set $X(k)$ of all (unordered) subsets of X of size k.

4.2 Permutations

A permutation $\pi \in \mathrm{Sym}(X)$ is said to *displace* letter $x \in X$ if and only if $x^\pi \neq x$. A permutation is said to be *finitary* if and only if it displaces only a finite number of letters. The products among, and inverses and conjugates of finitary permutations are always finitary, and so the set of all finitary permutations always form a normal subgroup $\mathrm{FinSym}(X)$ of the symmetric group $\mathrm{Sym}(X)$. Any finitary permutation obviously has finite order.

In this subsection we shall study certain arithmetic properties of group actions which are mostly useful for finite groups.

4.2.1 Cycle Notation

Now suppose π is a finitary permutation in $\mathrm{Sym}(X)$. Letting H be the cyclic subgroup of $\mathrm{Sym}(X)$ generated by π, we may partition X into H-orbits. By our self-imposed finitary hypothesis, there are only finitely many H-orbits of length greater than one. Let O be one of these, say of length n. Then for any letter x in O, we see that $O = x^H$ must consist of the set $\{x, x^\pi, x^{\pi^2}, \ldots x^{\pi^{n-1}}\}$. Now x^{π^n} must be a letter in this sequence which, by the injectivity of π, can only be x itself. We represent this permutation of O by the symbol:

$$(x \ x^\pi \ x^{\pi^2} \ldots x^{\pi^{n-1}})$$

called a *cycle*. Stated in this generality, the notation is not very impressive. But in specific instances it is quite useful. For example the notation (1 2 4 7 6 3) denotes a permutation which takes 1 to 2, takes 2 to 4, takes 4 to 7, takes 7 to 6, takes 6 to 3, and takes 3 back to 1. Thus everyone is moved forward one position along a circular trail indicated by the cycle. In particular the cycle notation is not unique since you can begin anywhere in the cycle: thus (7 6 3 1 2 4) represents the same permutation just given. Moreover, writing the cycle with the reverse orientation of the circuit yields the inverse permutation, (4 2 1 3 6 7) in this example.

Now the generator π of the cyclic group $H = \langle \pi \rangle$ acts on each H-orbit as a cycle. This can be restated as the assertion that *any finitary permutation can be represented as a product of disjoint cycles*, that is, cycles which pairwise displace no common letter. Such cycles commute with one another so it doesn't matter in which order the cycles are written. Also, if the set X is known, the permutation is determined only by its list of disjoint cycles of length greater than one—the cycles of length one (indicating letters that are fixed) need not be mentioned. Thus in $\mathrm{Sym}(9)$, the following permutations are the same:

$$(14597)(26) = (62)(14597)(8).$$

Now to multiply two such permutations we simply compose the two permutations—and here the order of the factors *does* matter. (It is absolutely necessary that the student learn to compute these compositions.) Suppose

$$\pi = (7 \ 6 \ 1)(3 \ 4 \ 8)(2 \ 5 \ 9) \tag{4.1}$$

$$\sigma = (7, 10, 11)(9 \ 2 \ 4). \tag{4.2}$$

Then, applying π *first* and σ *second*, the composition $\sigma \circ \pi$ of these *right* operators is

$$\pi\sigma = (1, 10, 11, 7 \ 6)(2 \ 5)(3 \ 9 \ 4 \ 8).$$

We simply compute $\langle \pi \sigma \rangle$-orbits.[1] The computation begins with an involved letter—say "1"—and asks what $\pi \sigma$ did to it? $1^\pi = 7$ and $7^\sigma = 10$, so $1^{\pi \sigma} = 10$. Then it asks what $\pi \sigma$ did to 10 in turn? One gets $10^{\pi \sigma} = 11$, $11^{\pi \sigma} = 7$, and so on, until the $\langle \pi \sigma \rangle$-orbit is completed upon returning to 1. One then looks for a new letter displaced by at least one of the two permutations, but which is not involved in the orbit just calculated—say, "2"—and one then repeats the process.

We make this observation:

Lemma 4.2.1 *The order of a finitary permutation expressed as a product of disjoint cycles is the least common multiple of the lengths of those cycles.*

Proof Suppose x and y are disjoint cycles of lengths a and b respectively, Since x and y commute, we have $(xy)^k = x^k y^k$ for any positive integer k. Thus if m is any common multiple of a and b, then $(xy)^m = 1$. On the other hand, if $(xy)^d = 1$, the fact that the cycles are disjoint yields $x^d = 1 = y^d$. this forces d to be a multiple of both a and b. Thus the order of xy is the least common multiple of a and b.$\square$

4.2.2 Even and Odd Permutations

We begin with a technical result:

Lemma 4.2.2 *If k and l are non-negative integers, then*

(i) $(a\ b)(a\ x_1\ \ldots\ x_k\ b\ y_1\ \ldots\ y_l) = (a\ y_1\ \ldots\ y_l)(b\ x_1\ \ldots\ x_k)$, *and*
(ii) $(a\ b)(a\ y_1\ \ldots\ y_l)(b\ x_1\ \ldots\ x_k) = (a\ x_1\ \ldots\ x_k\ b\ y_1\ \ldots\ y_l)$.

Proof The permutation on the left side of the first equation sends a to y_1, sends y_i to y_{i+1} for $i < l$, and y_l to a. Similarly it sends b to x_1, sends x_j to x_{j+1} for $j < k$, and sends x_k to b. Thus the left side has been expressed as the product of the two disjoint cycles on the right side of the first equation.

The second equation follows from the first my multiplying both sides of the first equation by the 2-cycle $(a\ b)$.$\square$

Let $FinSym(X)$ denote the group of all finitary permutations of the set of 'letters' X. Each element π of FinSym(X) is expressible as a finite product of disjoint cycles of lengths $d_1, \ldots d_k$ greater than one, together with arbitrarily many cycles of length one. These numbers $\{d_j\}$ are uniquely determined by π since they are the orbit-lengths greater than one, of the group $\langle \pi \rangle$. Define the *sign* of π to be the number

$$\prod_{j=1}^{j=k} (-1)^{d_j - 1}.$$

Then the function $sgn : \mathrm{FinSym}(X) \to \{\pm 1\}$ is well-defined.

---

[1] We will follow Burnside in explaining that commas are introduced into the cycle notation only for the specific purpose of distinguishing a 2-or-more-digit entry from other entries.

A permutation of $Sym(X)$ which displaces exactly two letters is called a *transposition*. Thus every transposition can be expressed in cycle notation by $(a\ b)$, for the two displaced letters, a and b. Clearly, the sign of a transposition is $(-1)^{2-1} = -1$.

We now can state

Lemma 4.2.3 *If g is a permutation in $FinSym(X)$, and if t is a transposition, then*

$$sgn(tg) = -sgn(g) = sgn(gt).$$

Proof The permutation g is a product of disjoint cycles c_j of length d_j. Thus $sgn(g) = \prod_j (-1)^{d_j-1}$. Suppose t is the transposition $(a\ b)$. If the letters a and b are not involved in any of the cycles g_i, then the result follows from the formula for the sign of $gt = tg$ since we have simply tacked on the disjoint cycle $(a\ b)$ in forming gt.

Suppose on the other hand, that a and b appear in the same $\langle g \rangle$-orbit, say the one represented by the cycle g_j. Then g_j has the form $(a\ x_1 \ldots x_k\ b\ y_1 \ldots y_l)$, where k and l are non-negative. By Lemma 4.2.2,

$$tg_j = (a\ y_1\ \ldots\ y_l)(b\ x_1 \ldots x_k).$$

Thus $d_j = k + l + 2$ so $sgn(g_j) = (-1)^{k+l+1}$ while $sgn(tg_j) = (-1)^k(-1)^l$, so $sgn(tg_j) = -sgn(g_j)$. Since $tg = g_1 \cdots (tg_j) \cdots g_i \cdots$, the result follows from the formula.

Now suppose a occurs in one cycle, say $g_1 = (a\ x_1, \ldots x_k)$ and b occurs in another cycle, which we can take to be $g_2 = (b\ y_1\ \ldots\ y_l)$. Then by Lemma 4.2.2,

$$tg = (tg_1g_2)g_3 \cdots g_m$$
$$= (a\ y_1\ \ldots\ y_l\ b\ x_1\ \ldots x_k)g_3 \cdots g_m.$$

Thus

$$sgn(tg) = sgn(tg_1g_2)\prod_{i>2}(-1)^{(d_i-1)}$$
$$= (-1)sgn(g_1)sgn(g_2)\prod_{i>2}(-1)^{(d_i-1)}$$
$$= -sgn(g).$$

So the result follows in this case. The proof is complete. □

Now any cycle $(x_1\ \ldots x_m)$ is a product of transpositions

$$(x_1\ x_k)(x_k\ x_{k-1}) \cdots (x_3\ x_2).$$

Since each element of $FinSym(X)$ is the product of disjoint cycles we see that $FinSym(X)$ is generated by its transpositions.

It now follows from Lemma 4.2.3 that no element can be both a product of an even number of transpositions as well as a product of an odd number of transpositions. Thus we have a partition

$$\mathrm{FinSym}(X) = A^+ + A^-$$

of all the elements of $\mathrm{FinSym}(X)$ into the set A^+ of all finitary permutations which are a product of an even number of transpositions, and A^-, all finitary permutations which are a product of an odd number of transpositions. Right multiplication by any finitary permutation at best permutes the two sets A^+ and A^-, and multiplication by a transposition clearly transposes them. Thus we have a transitive permutation action

$$sgn : \mathrm{FinSym}(X) \rightarrow \mathrm{Sym}(\{A^+, A^-\}).$$

The right side is $\mathrm{Sym}(2) \simeq Z_2$, isomorphic to the multiplicative group $\{\pm 1\}$.

The kernel of this action is the normal subgroup $\mathrm{FinAlt}(X) := A^+$, the *finitary alternating group*. Its elements are called the *even permutations*. The elements in the other coset A^-, are called *odd permutations*. These terms are used only for finitary permutations. If X is a finite set, there is no distinction gained by singling out the finitary permutations, and the prefix "Fin" is dropped throughout. So in that case the *alternating group* is denoted $\mathrm{Alt}(X)$ or $\mathrm{Alt}(n)$ when $|X| = n$.

The factorization of a cycle of length n given above shows that any cycle of even length is an odd permutation and any cycle of odd length is an even permutation. It only "sounds" confusing; by the formula, the sign of an n-cycle is $(-1)^{n-1}$.

4.2.3 Transpositions and Cycles: The Theorem of Feit, Lyndon and Scott

Recall that a *transposition* of a symmetric group $\mathrm{Sym}(X)$ is an element which displaces exactly two of the letters of the set X. These two letters must clearly exchange places, otherwise we are dealing with the identity permutation.

Let $\mathcal{T}$ be any collection of transpositions of $\mathrm{Sym}(X)$. We can then construct a simple[2] graph $\mathcal{G}(\mathcal{T}) := (X, \mathcal{T})$ with vertex set X and edge set consisting of the transposed 2-subsets determined by each involution in $\mathcal{T}$.

---

[2]In the context of graphs (as opposed to groups) "simple" actually means something rather simple. A simple graph is one which is undirected, without loops or multiple edges. That is, edges connect only distinct vertices, there is at most one edge connecting two distinct vertices, and no orientation to any edge. In other words, edges are 2-subsets of the vertex set.

Lemma 4.2.4 *Let* T *be a set of transpositions of* $Sym(X)$. *Suppose* $G(T)$, *the subgroup generated by the transpositions of* T, *acts on* X *with orbits* X_σ *of finite length. Then* $G(T)$ *can be expressed as a direct sum:*

$$G(T) \simeq \bigoplus_{\sigma \in I} Sym(X_\sigma)$$

where X_σ *are the* $G(T)$-*orbits on* X.

Proof First of all, since transpositions displace just two letters, $G(T)$ is a finitary permutation group on X. Let X_σ be a connected component of the graph $\mathcal{G}(T)$ and suppose x and y are two of its vertices.. Since there is a sequence of transpositions defining the edges of a path in X_σ, connecting x and y, the product of these transpositions in the order they occur in the path is an element of $G(T)$ taking x to y. On the other hand, there is no such path connecting a vertex in one connected component of $\mathcal{G}(T)$ with a vertex in another connected component, and so no element of the group $G(T)$ can move one of these vertices to the other. Thus the connected components X_σ are actually the $G(T)$-orbits. Let T_σ be the set of transpositions which form an edge of the connected component X_σ of $\mathcal{G}(T)$, and let G_σ be the subgroup they generate. Then G_σ is transitive on X_σ but fixes each other orbit point-wise. It follows that any conjugate $g^{-1} G_\sigma g$ in $G(T)$, displaces only vertices in X_σ, and so is a subgroup G_σ since it is generated by transpositions in T_σ. Thus each G_σ is normal in $G(T)$ and if X_τ and X_σ are distinct orbits, we have $G_\sigma \cap G_\tau = 1$. Since $G(T)$ is finitary, each element of the group is uniquely determined by its action on each orbit X_σ. Conversely, if $G_\sigma \simeq Sym(X_\sigma)$, we see that any product of permutations displacing only finitely many letters, is a product of involutions, and so is an element of $G(T)$. It then follows that $G(T)$ is the direct sum of the groups

$$G_\sigma := \langle T_\sigma \rangle.$$

So all that remains is to show that G_σ acts on X_σ as the full symmetric group on the finite orbit X_σ. Thus, without loss of generality, we assume that $G(T)$ is transitive on the finite set X of cardinality n and that the transpositions T define a connected graph on X. If $|T| = 1$, then $|X| = 2$ and there is nothing to prove. We may now proceed by induction on $|T|$ and conclude that $\mathcal{G}(T)$ is a tree. Now we are free to choose the transposition $t \in T$ so that as an edge of the graph $\mathcal{G}(T)$, t connects an "end point" a of valence one, and a vertex b. Thus $\mathcal{G}(T - \{t\})$ is a tree on the vertices $X - \{a\}$, and so by induction the stabilizer in $\mathcal{G}(T)$ of the vertex a has order $(n - 1)!$, and so as $\mathcal{G}(T)$ is clearly transitive, it has order $n!$. Since it is faithful, (for its elements fix every single letter of every other orbit), we have here the full symmetric group.$\square$

Theorem 4.2.5 (Feit, Lyndon and Scott) *Let* T *be a minimal set of transpositions generating* $Sym(n)$. *Then their product in any order is an* n-*cycle.*

Proof We have seen from above that the minimality of T implies that the graph $\mathcal{G}(T)$ is a tree.

Let $a = t_1t_2 \cdots t_{n-1}$ be the product of these transpositions in any particular order. Consider t_1 and remove the edge t_1 from the graph $\mathcal{G}(T)$ to obtain a new graph $\mathcal{G}'$ which is not connected, but has two connected component trees $\mathcal{G}_1$ and $\mathcal{G}_2$, connected in $\mathcal{G}$ by the "bridging" edge t_1. Now by an obvious induction any order of multiplying the transpositions of

$$T_i := \{s \in T \,|\, s \text{ is an edge in} \mathcal{G}_i\},$$

$i = 1, 2, \ldots$, yields an n-cycle. Thus

$$a = t(x_1, \ldots x_k)(y_1, \ldots y_l),$$

where the second cycle b is the product of the transpositions of T_1 as they are encountered in the factorization $a = t_1 \cdots t_n$, and the third cycle c above is the product of the transpositions of T_2 in the order in which they are encountered in the product given for a. (Recall that every transposition in T_1 commutes with every transposition in T_2.) Now these two cycles involve two separate orbits bridged by the transposition t_1. That the product t_1bc is an n-cycle follows directly from Lemma 4.2.2 Part (ii) above. The proof is complete.$\square$

This strange theorem will bear fruit in the proof of the Brauer-Ree Theorem, which takes a bow in the chapter on generation of groups (Chap. 6).

4.2.4 Action on Right Cosets

Let H be any subgroup of the group G, and let G/H be the collection $\{Hg \,|\, g \in G\}$ of all right cosets of H in G. Note that we have extended our previous notation. G/H used to mean a group of right cosets of a normal subgroup H of G. It is still the same *set* (of right cosets of H in G) even when H is not normal in G. But it no longer has the structure of a group. Nonetheless, as we shall see, it still admits a right G-action.

For any element g of G and coset Hx in G/H, Hxg is also a right coset of H in G. Moreover, $Hxg = Hyg$ implies $Hx = Hy$, while $Hx = (Hxg^{-1})g$. Thus right multiplication of all the right cosets of H in G by the element g induces a bijection $\pi_g : G/H \to G/H$. Moreover for elements $g, h \in G$, the identity $Hx(gh) = ((Hx)g)h$, forced by the associative law, implies $\pi_g\pi_h = \pi_{gh}$ as right operators on G/H. Thus we have a group action

$$\pi_H : G \to \mathrm{Sym}(G/H)$$

which takes g to the permutation π_g, induced by right multiplication of right cosets by g. This action is always transitive, for coset Hx is mapped to coset Hy by $\pi_{x^{-1}y}$.

4.2.5 Equivalent Actions

Suppose a group G acts on sets X and Y—that is, there are group homomorphisms $f_X : G \to \text{Sym}(X)$ and $f_Y : G \to \text{Sym}(Y)$. The two actions are said to be *equivalent actions* if and only if there is a bijection $e : X \to Y$ such that for each letter $x \in X$, and element $g \in G$,

$$e(x^{f_X(g)}) = (e(x))^{f_Y(g)}.$$

That is, the action is the same except that we have used the bijection e to "change the names of the elements of X to those of Y".

4.2.6 The Fundamental Theorem of Transitive Actions

First we make an observation:

Lemma 4.2.6 *Let $f : G \to \text{Sym}(X)$ be a transitive action of the group G on the set X. For each letter $x \in X$, let G_x be the subgroup of all elements of G which leave the letter x fixed—i.e. $G_x := \{g \in G | x^g = x\}$.*

(i) *If $x, y \in X$, then G_x and G_y are conjugate subgroups of G. Precisely, $G_y = g^{-1}G_x g$ whenever $x^g = y$.*

(ii) *The set $\{g \in G | x^g = y\}$ of all elements of G taking x to y, is a right coset of G_x.*

Proof (i) If $x^g = y$, $y^{g^{-1}G_x g} = x^{gg^{-1}G_x g} = x^{G_x g} = x^g = y$, so $g^{-1}G_x g \subseteq G_y$.. But since $x = y^{g^{-1}}$, we have $G_y \subseteq gG_x g^{-1}$, by the same token. Thus the first containment is reversible and $g^{-1}G_x g = G_y$.

(ii) If $x^g = y$ then clearly, $x^h = y$ for all $h \in G_x g$. Conversely, if $x^g = x^h = y$, then $gh^{-1} \in G_x$, so $G_x g = G_x h$.□

Theorem 4.2.7 *(The Fundamental Theorem of Transitive Group Actions) Suppose $f : G \to \text{Sym}(X)$ is a transitive group action. Then for any letter x in X, f is equivalent to the action*

$$\pi_{G_x} : G \to \text{Sym}(G/G_x),$$

of G on the right cosets of G_x by right multiplication.

Proof In order to show the equivalence, we must construct the bijection $e : X \to G/G_x$ compatible with both actions. For each letter $y \in X$ set $e(y) := \{g \in G | x^g = y\}$. Lemma 4.2.6 informs us that the latter is a right coset $G_x h$ of G_x. Now for any element g in G, $e(y^g)$ is the set of all elements of G which take x to y^g. This clearly contains $G_x hg$; yet by Lemma 4.2.6, it must itself be a right coset of G_x. Thus $e(y^g)$ is the right multiple $e(y)g$, establishing the compatibility of the actions.□

Corollary 4.2.8

(i) *If G acts on the set X, then every G-orbit has length dividing the order of G.*
(ii) *If G acts transitively on X, then for any subgroup H of G which also acts transitively on X, one has $G = G_x H$, for any $x \in X$.*

Proof (i) The action of G on any G-orbit O is transitive. Applying the fundamental Theorem 4.2.7, we see that the action on O is equivalent to the action of G on G/G_x, the set of right cosets of G_x, where x is any fixed element of O. Thus $|O| = |G/G_x|$ for any letter x in O.

(ii) To say that subgroup H is transitive on O, says that right multiplication of one such coset by the elements of H yields all such cosets. Since these cosets partition all the elements of G, one obtains $G_x H = G$. $\square$

4.2.7 Normal Subgroups of Transitive Groups

If G acts transitively on a set X, we say that G *acts regularly on X* if and only if for some $x \in X$, $G_x = 1$.

Lemma 4.2.9 *Suppose G acts transitively on X.*

(i) *If N is a normal subgroup of G, then all N-orbits on X have the same length.*
(ii) *In particular, if N is a non-identity normal subgroup of G lying in a subgroup G_x, for some x in X, then N acts trivially on X, and the action is not faithful.*
(iii) *The faithful transitive action of an abelian group is always a regular action.*
(iv) *Suppose N is a normal subgroup of G which acts regularly on X. Then for any $x \in X$, $G = G_x N$, with $G_x \cap N = 1$, and the action of G_x on $X - \{x\}$ (by restriction) is equivalent to that action of G_x on $N - \{1\}$, the set of all non-identity elements of N, induced by conjugation by the elements of G_x.*

Proof (i) For any $g \in G$ and $x \in X$, the mapping

$$x^N \rightarrow (x^N)^g = x^{Ng} = x^{gN} = (x^g)^N$$

is a bijection of N-orbits. Since G is transitive, all N-orbits can be gotten this way.

(ii) This is immediate from (i), since N has an orbit of length 1.

(iii) If A is an abelian subgroup acting transitively on X, for any $x \in X$, A_x acts trivially on X by part (ii) Since A acts faithfully, $A_x = 1$, so A has the regular action.

(iv) That $G = G_x N$ and $G_x \cap N = 1$ follows from Lemma 4.2.8, part (ii), and the fact that N is regular.

Since the normal subgroup N is regular, there is a bijection, $e_x : N \rightarrow X = x^N$ which sends element $n \in N$ to x^n. For each element $h \in G_x$,

$$e(h^{-1}nh) = x^{h^{-1}nh} = x^{nh} = (x^n)^h = e(n)^h.$$

So e defines an equivalence of the action of G_x on N by conjugation and the action of G_x on X obtained by restriction of the action of G. Throwing away the G_x-orbit $\{x\}$ of length 1 yields the last statements of (iv).$\square$

4.2.8 Double Cosets

Let H and K be subgroups of a group G. Any subset of the form HgK is called an (H, K)-*double coset*. Such a set is, on the one hand, a disjoint union of right cosets of H, and on the other, a disjoint union of left cosets of K. So it's cardinality must be a common multiple of $|H|$ and $|K|$. The precise cardinality is given in the following:

Lemma 4.2.10 *Let H and K be subgroups of G.*

(i) $|HgK| = |H| \cdot [K : K \cap g^{-1}Hg] = |H||K|/|gKg^{-1} \cap H| = [H : H \cap gKg^{-1}]|K|$.

(ii) *The (H, K)-double cosets partition the elements of G.*

Proof In the action π_H of G on G/H by right multiplication, the K-orbits are precisely the right cosets of H within an (H, K)-double coset. Since G/K is partitioned by such orbits, and the right cosets of H partition G, part (ii) follows.

The length of a K-orbit on G/H, is the index of its subgroup fixing one of its "letters", say Hg. This subgroup would be $\{k \in K | Hgk = Hg\} = K \cap gHg^{-1}$. Since each, right coset of H has $|H|$ elements of G, the second and third terms of the equations in Part (i) give $|HgK|$. The last term appears from the symmetry of H and K.$\square$

4.3 Applications of Transitive Group Actions

4.3.1 Cardinalities of Conjugacy Classes of Subsets

For any subset X of a group G, the *normalizer of X in G* is the set of all elements $N_G(X) := \{g \in G | g^{-1}Xg = X\}$. Now G acts transitively by conjugation on $X^G = \{g^{-1}Xg | g \in G\}$, the set of distinct conjugates of X, and the normalizer $N_G(X)$ is just the subgroup fixing one of the "letters" of X^G. Thus

Lemma 4.3.1

(i) *Let X be a subset of G. The cardinal number $|X^G|$ of distinct conjugates of X in G is the index $[G : N_G(X)]$ of the normalizer of X in G. This holds when X is a subgroup, as well.*

(ii) *The cardinality of a conjugacy class x^G in G is the index of the centralizer $C_G(x)$ in G.*

4.3.2 Finite p-groups

A *finite p-group* is a group whose order is a power of a prime number p. The trivial group of order 1 is always a p-group.

Lemma 4.3.2

(i) *If a finite p-group acts on a finite set X of cardinality not divisible by the prime p, then it fixes a letter.*

(ii) *A non-trivial finite p-group has a nontrivial center.*

(iii) *If H is a proper subgroup of a finite p-group P, then H is properly contained in its normalizer in P, that is $H < N_P(H)$.*

Proof (i) Let P be a finite p-group, say of order p^n. Then any P-orbit on X has length a power of p. Thus if no orbit had length 1, p would divide $|X|$, since X partitions into such orbits. But as p does not divide $|X|$, some P-orbit must have length 1, implying the conclusion.

(ii) Let P be as in part (i), but of order $p^n > 1$. Now P acts by conjugation on the set $P - \{1\}$ of $p^n - 1$ non-identity elements. Since p does not divide the cardinality of this set, part (i) implies there is a non-identity element z left fixed by this action. That is, $g^{-1}zg = z$, for all $g \in P$. Clearly z is a non-identity element of the center $Z(P)$, our conclusion.

(iii) Suppose H is a proper subgroup of a p-group P. By way of contradiction assume that $H = N_P(H)$. Then certainly, $Z(P) \leq H$. By Part (ii), $Z(P) \neq 1$. If $H = Z(P)$, then $H \trianglelefteq P$, so $P = N_P(H) = H$, a contradiction to H being a proper subgroup. Thus, under the homomorphism

$$f : P \to \bar{P} := P/Z(P),$$

H maps to a non-trivial proper subgroup $f(H) := \bar{H}$ of the p-group $\bar{P}$. By induction on $|\bar{P}|$, $N_{\bar{P}}(\bar{H})$ properly contains $\bar{H}$. Thus the preimage of $N_{\bar{P}}(\bar{H})$, which is $f^{-1}(N_{\bar{P}}(\bar{H}))$, properly contains $H = f^{-1}(\bar{H})$. But this preimage is

$$f^{-1}(N_{\bar{P}}(\bar{H})) = \{x \in P \,|\, x^{-1}(H/Z(P)x = H/Z(P)\} = N_p(H) \leq H,$$

and this contradicts its proper containment of H. The proof is complete.$\square$

4.3.3 The Sylow Theorems

Theorem 4.3.3 (Sylow's Theorem) *Let G be a group of finite order n and let k be the highest power of the prime p such that p^k divides the group order n. Then the following three statements hold:*

(i) *(Existence) G contains a subgroup P of order p^k.*

(ii) (Covering and Conjugacy) *Every p-subgroup R of G is contained in some conjugate of P. In particular, any subgroup of order p^k is conjugate in G to P.*

(iii) (Arithmetic) *The number of subgroups of order p^k is congruent to 1 mod p.*

Proof (i) Let Σ denote the collection of all subsets of G of cardinality p^k. The number of such subsets is $|\Sigma| = \binom{n}{p^k}$. Since, for any natural number $r \leq p^k$, the numbers $n - r$ and $p^k - r$ are both divisible by the same highest power of p, the number $|\Sigma|$ is not divisible by the prime p. Now G acts on Σ by right multiplication— that is, if $X \in \Sigma$ then $Xg \in \Sigma$, for any group element g. Thus Σ partitions into G-orbits under this action, and not all of these orbits can have length divisible by p, since p doesn't divide $|\Sigma|$. Let Σ_1 be such a G-orbit of length not divisible by p. By the Fundamental Theorem of Transitive Group Actions (Theorem 4.2.7), the number of sets in Σ_1 is the index of the subgroup G_X fixing the "letter" X in Σ_1. Since $n = |G| = [G : G_X]|G_X| = |\Sigma_1||G_X|$, we see that p^k divides $|G_X|$. But by definition, $XG_X = X$, so X is a union of left cosets of G_X; so $|G_X|$ divides $|X| = p^k$. It follows that G_X has order p^k exactly. So G_X is the desired subgroup P of statement (i).

(ii) Let R be any p-subgroup of G. Then R also acts on Σ_1 by right multiplication. Since p does not divide $|\Sigma_1|$, Lemma 4.3.2, part (i), shows that R must fix a letter Y in Σ_1. Thus $R \leq G_Y$. But as G is transitive on Σ_1, Lemma 4.2.6 shows that G_Y is conjugate to $P = G_X$. Thus R lies in a conjugate of P as required.

(iii) Now let $\mathcal{S}$ denote the collection of all subgroups of G having order p^k exactly. By parts (i) and (iii), already proved, $\mathcal{S}$ is non-empty, and G acts transitively on $\mathcal{S}$ by conjugation. Thus by Lemma 4.3.1 part (i), $|\mathcal{S}| = [G : N]$, where $N := N_G(P)$, the normalizer in G of a subgroup P in $\mathcal{S}$. Now P itself acts on $\mathcal{S}$ by conjugation, fixing itself, and acting on $\mathcal{S} - \{P\}$ in orbits of p-power length. Suppose $\{R\}$ were such a P-orbit in $\mathcal{S} - \{P\}$ of length one. Then P normalizes R so PR is a subgroup of G. Then $|PR| = |P| \cdot [R : P \cap R]$, (Theorem 3.4.5, part (ii)) a product of p^k and $[R : P \cap R]$, another power of p. Since $|PR|$ divides n and p^k is the largest p-power dividing n, one must conclude that $[R : P \cap R] = 1$, which forces $P = R$, contrary to the choice of R in $\mathcal{S} - \{P\}$. Thus P acts on $\mathcal{S} - \{P\}$ in orbits of lengths divisible by p. This yields $|\mathcal{S}| \equiv 1 \bmod p$. Thus all parts of Sylow's theorem have been proved.$\square$

The subgroups of G of maximal p-power order are called the *p-Sylow subgroups of G*, and the single conjugacy class which they form is denoted $\mathrm{Syl}_p(G)$.

Corollary 4.3.4 (The Frattini Argument) *Suppose N is a normal subgroup of the finite group G, and select $P \in Syl_p(N)$. Then $G = N_G(P)N$.*

Proof We know that conjugation by elements of G induces automorphisms of the normal subgroup N. Since $\mathrm{Syl}_p(N)$ is the full collection of subgroups of N of their order, G acts on $\mathrm{Syl}_p(N)$ by conjugation. But by Sylow's Theorem (Theorem 4.3.3) the subgroup N is already transitive on $\mathrm{Syl}_p(N)$. The result now follows from Lemma 4.2.8, part 2.$\square$

Corollary 4.3.5

(i) *Any normal p-Sylow subgroup of a finite group is in fact characteristic in it.*
(ii) *Any finite abelian group is the direct product of its p-sylow subgroups, that is,*
$A \simeq S_1 \times S_2 \cdots \times S_n$, *where S_i is the p_i-Sylow subgoup of A, and $\{p_1, \ldots, p_n\}$ lists all the distinct prime divisors of $|A|$ exactly once each. Moreover, we have:*

$$Aut(A) \simeq Aut(S_1) \times Aut(S_2) \times \cdots, Aut(S_n).$$

(iii) *The Euler phi-function is multiplicative, that is,*

$$\phi(n) = \phi(a)\phi(b)$$

whenever $n = ab$ and $\gcd(a, b) = 1$.

Proof (i) Any normal p-sylow subgroup is the unique subgroup of its order, and hence is characteristic.

(ii) Since, in a finite abelian group, each p-Sylow subgoup is normal, and has order prime to the direct product of the remaining r-Sylow subgroups ($r \neq p$), A is the internal direct product (see Lemma 3.5.1, part (iii)) Since by part (i) above, each S_i is characteristic in A, each automorphism σ of A induces an automorphism σ_i of S_i. But conversely, if we apply all possible automorphisms σ_i to the direct factor S_i and the identity automorphism to all other p-sylow subgroups $S_j, j \neq i$, we obtain a subgroup B_i of Aut(A). Now the internal direct product characterization of Lemma 3.5.1 applies to the B_i to yield the conclusion.

(iii) Applying part (ii) of this Corollary when A is the cyclic group of order n, the group Z_n, one obtains

$$\phi(n) = \prod_{i=1}^{i=n} \phi(p_i^{a_i})$$

when n has prime factorization $n = p_1^{a_1} p_2^{a_2} \cdots p_n^{a_n}$, upon equating group orders. □

4.3.4 Fusion and Transfer

Lemma 4.3.6 (The Tail-Wags-the-Dog Lemma) *Suppose $\Gamma = (V, E)$ is a bipartite graph with vertices in two parts V_1 and V_2, and each edge of E involving one vertex from V_1 and one vertex from V_2. Suppose G is a group of automorphism of the graph Γ acting on the vertex set V with V_1 and V_2 as its two orbits. Then the following conditions are equivalent.*

(i) *G acts transitively on the edges of Γ.*
(ii) *For each vertex $v_1 \in V_1$, the subgroup G_1 of G fixing v_1 acts transitively on the edges on v_1*

(iii) For each vertex v_2 in V_2, the subgroup G_2 of G fixing v_2 acts transitively on the edges of Γ on vertex v_2.

Proof By the symmetry of the V_i, is suffices to prove the equivalence of (i) and (ii) If G transitively permutes the edges (as in (i)), any edge on v_1 must be moved to any other edge on v_1 by some element g. But since v_1 is the only vertex of the G-orbit V_1 incident with these edges, such an element g fixes v_1. So (ii) holds. Conversely, suppose (ii), so that the subgroup G_1 transitively permutes the edges on v_1. Since G is transitive V_1, any edge meeting V_1 at a single vertex can be taken to any other. But by hypothesis, all edges have this property and (i) follows. The proof is complete. $\square$

Theorem 4.3.7 (The Burnside Fusion Theorem) *Let G be a finite group, and let X_1 and X_2 be two normal subsets of a p-sylow subgroup P of G. Then X_1 and X_2 are conjugate in G if and only if they are conjugate in the normalizer in G of P. In particular, any two elements of the center of P are conjugate if and only if they are conjugate in $N_G(P)$.*

Proof Obviously, if the X_i are conjugate in $N_G(P)$, they are conjugate in G. So assume the X_i belong to $X := X_1^G$. We now form a graph whose vertex set is $X \cup S$ where $S := \mathrm{Syl}_p(G)$. An edge will be any pair $(Y, R) \in X \times S$ for which the subset Y is a normal subset of the p-sylow group R. Then G acts by conjugation on the vertex set V with just two orbits, X and S (the first by construction, the second by Sylow's Theorem). Moreover the conjugation action preserves the normal-subset relationship on $X \times S$, and so G acts as a group of automorphisms of our graph, which is clearly bipartite. Now by Sylow's theorem, the subgroup $G_1 := N_G(X_1)$ acts transitively by conjugation on $\mathrm{Syl}_p(G_1)$. But since X_1 is normal in some p-sylow subgroup (namely P),

$$\mathrm{Syl}_p(G_1) \subseteq \mathrm{Syl}_p(G)$$

Thus $\mathrm{Syl}_p(G_1)$ are the S-vertices of all the edges of our graph which lie on vertex X_1. We now have the hypothesis (ii) of the Tail-Wags-the-Dog Lemma. So the assertion (iii) of that lemma must hold. In this context, it asserts that the stabilizer G_2 of a vertex of S—for example, $N_G(P)$—is transitive on the members of X which are normal in P. But that is exactly the desired conclusion.$\square$

Theorem 4.3.8 (The Thompson Transfer Theorem) *Suppose N is a subgroup of index 2 in a 2-sylow subgroup S of a finite group G. Suppose t is an involution in $S - N$, which is not conjugate to any element of N. Then G contains a normal subgoup M of index 2.*

Proof Let G act on G/N, the set of right cosets of N in G, by right multiplication. In this action, the involution t cannot fix a letter, for if $Nxt = Nx$, then $xtx^{-1} \in N$, contrary to assumption. Thus t acts as $[G : S]$ 2-cycles on G/N. Since this makes t the product of an odd number of transpositions in $\mathrm{Sym}(G/N)$, we see that the image of G in the action $f : G \to \mathrm{Sym}(G/N)$ intersects the alternating group $\mathrm{Alt}(G/N)$ at a subgroup of index 2. Its preimage M is the desired normal subgroup of G.$\square$

4.3.5 Calculations of Group Orders

The Fundamental Theorem of Transitive Group Actions is quite useful in calculating the orders of the automorphism groups of various objects. We peruse a few examples.

(Petersen's Graph) The reader is invited to verify that the two graphs which appear in Fig. 4.1 are indeed isomorphic (vertices with corresponding numerical labels are matched under the isomorphism). Presenting them in these two ways reveals the existence of certain automorphisms.

In the left hand figure, vertex 1 is adjacent to vertices 2, 3, and 4. The remaining vertices form a hexagon, with three antipodal pairs, (5, 8), (6, 9) and (7, 10), which are connected to 2, 4 and 3, respectively. Any symmetry of the hexagon induces a permutation of these antipodal pairs, and hence induces a corresponding permutation of $\{2, 3, 4\}$. Thus, if we rotate the hexagon clockwise 60 degrees, we obtain an automorphism of the graph which permutes the vertices as:

$$y = (1)(2\ 3\ 4)(5\ 10\ 9\ 8\ 7\ 6).$$

But if we reflect the hexagon about its vertical axis we obtain

$$t = (1)(4)(2\ 3)(7\ 5)(8\ 10)(6)(9).$$

Now any automorphism of the graph which fixes vertex 1 must induce *some* symmetry of the hexagon on the six vertices not adjacent to vertex 1. The automorphism inducing the identity, preserves the three antipodal pairs, and so fixes vertices 2,3 and 4, as well as vertex 1—i.e. it is the identity element. Thus if G is the full automorphism group of the Petersen graph, G_1 is the dihedral group $\langle y, t \rangle$ of order 12, acting in orbits, $\{1\}$, $\{2, 3, 4\}$, $\{5, 6, 7, 8, 9, 10\}$.

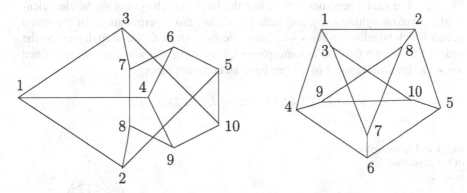

Fig. 4.1 Two views of Petersen's graph. The positions labeled 1–10 are the vertices, and arcs connecting two vertices represent the edges. (Note that since the graph is non-planar, it is drawn with apparent intersections of arcs, which, if unlabeled, do not represent vertices)

On the other hand, the right hand figure in Fig. 4.1 reveals an obvious automorphism of order five, namely

$$u = (1\,2\,5\,6\,4)(3\,8\,10\,7\,9).$$

That G is transitive comes from overlaying the three orbits of G_1, and the two orbits of $\langle u \rangle$ of length five. Together, one can travel from any vertex to any other, by hopping a ride on each orbit, and changing orbits where they overlap. This forces $[G : G_1] = 10$, the number of vertices. Since we have already determined $|G_1| = 12$, we see that the full automorphism group of this graph has order $|G| = [G : G_1]|G_1| = 10 \cdot 12 = 120$.

(The Projective Plane of Order 2.) Now consider a system $\mathcal{P}$ of seven points, which we shall name by the integers $\{0, 1, 2, 3, 4, 5, 6\}$. We introduce a system $\mathcal{L}$ of seven triplets of points, which we call lines:

$$L_1 := [1, 2, 4]$$
$$L_2 := [2, 3, 5]$$
$$L_3 := [3, 4, 6]$$
$$L_4 := [4, 5, 0]$$
$$L_5 := [5, 6, 1]$$
$$L_6 := [6, 0, 2]$$
$$L_0 := [0, 1, 3]$$

Notice that the first line L_1 is the set of quadratic residues mod 7, and that the remaining lines are translates, mod 7, of L_1—that is, each L_i has the form $L_1 + (i - 1)$, with entries in the integers mod 7.

This system is also depicted by the six straight lines and the unique circle in Fig. 4.2. (The reader need only check that the lines in each system are labeled identically.) Automorphisms of the system $(\mathcal{P}, \mathcal{L})$ are those permutations of the seven points which take lines to lines—i.e. preserve the system $\mathcal{L}$. We shall determine the order of the group G of all automorphisms. First, from the way lines were defined above, as translates mod 7 of L_1, we have an automorphism

$$u = (0\,1\,2\,3\,4\,5\,6) : (L_1\,l_2\,L_3\,L_4\,L_5\,L_6\,L_0).$$

Fig. 4.2 The configuration of the projective plane of order 2

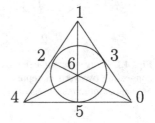

Thus G is certainly transitive on points. Thus the subgroup G_1 fixing point 1 has index 7. There are three lines on point 1, namely, L_0, L_1 and L_5. There is an obvious reflection of the configuration of Fig. 4.2 about the axis formed by line $L_5 = [1, 6, 5]$, giving the automorphism

$$s = (1)(6)(5)(0\ 4)(3\ 2) : (L_0\ L_1)(L_3\ L_6)(L_2)(L_4)(L_5).$$

But not quite so obvious is the fact that $t = (0)(1)(3)(2\ 6)(4\ 5)$ induces the permutation $(L_0)(L_1\ L_5)(L_2\ L_3)(L_4)(L_6)$ and so is an automorphism fixing 1 and L_5 and transposing L_0 and L_1. Thus G_1 acts as the full symmetric group on the set $\{L_0, L_1, L_5\}$ of three lines on point 1. Consequently the kernel of this action is a normal subgroup K of G_1, at index 6. Now K stabilizes all three lines $\{L_0, L_1, L_5\}$, and contains

$$r = (0\ 3)(1)(2\ 4)(5)(6) : (L_0)(L_1)(L_2\ L_4)(L_3\ L_6)(L_5).$$

So K acts as a transposition on the remaining points of L_0 beyond point 1. Thus K in turn has a subgroup K_1 fixing L_0 pointwise, and stabilizing L_1 and L_5. K_1, in turn, contains

$$v = (0)(1)(2\ 4)(3)(5\ 6) : (L_0)(L_1)(L_5)(L_2\ L_3)(L_4\ L_6).$$

It in turn contains a subgroup of index 2 whose elements now fix 0, 1, 3, 5 and 6. One can see that any such element stabilizes every line (for every pair of fixed points determines a fixed line), and hence stabilizes all intersections of such lines, and hence fixes all points. Thus K_1 has order only 2. We now have

$$|G| = [G : G_1][G_1 : K][K : K_1]|K_1| = 7 \cdot 6 \cdot 2 \cdot 2 = 168.$$

4.4 Primitive and Multiply Transitive Actions

4.4.1 Primitivity

Suppose G acts on a set X. A *system of imprimitivity for the action of G* is a nontrivial partition of X which is stabilized by G. Thus X is a disjoint union of two or more proper subsets X_i, not all of cardinality one, such that for each $g \in G$, $X_i^g = X_j$ for some—that is, G permutes the components of the partition among themselves. Of course, if G is transitive in its action on X, the components X_i all have the same cardinality (evidently one dividing the cardinality of X), and are transitively permuted among themselves.

As an example, the symmetric group $\mathrm{Sym}(n)$ acts on all of its elements by right multiplication (the regular representation). But these elements were partitioned as

Sym$(n) = A^+ \cup A^-$, into those elements which were a product of an even number of transpositions (A^+) and those which were a product of an odd number of transpositions (A^-). Right multiplication permutes these sets, and they form a system of imprimitivity when $n > 2$.

We have seen that when N is a non-trivial normal intransitive subgroup of a group G acting faithfully on X, then the N-orbits form a system of imprimitivity. (Lemma 4.2.9 considers the case of N-orbits when G acts transitively on X.)

If a transitive group action preserves no system of imprimitivity, it is said to be a *primitive group action*. (Often one renders this by saying that *G acts primitively*.) We understand this to mean that primitive groups are always transitive. The immediate characterization is this:

Theorem 4.4.1 *Suppose G acts transitively on X, where $|X| > 1$. Then G acts primitively if and only G_x is a maximal subgroup of G for some x in X.*

Proof Since G is transitive, all the subgroups $G_x, x \in X$ are conjugate in G (see Lemma 4.2.6, part 1). By the fundamental theorem of group actions (Theorem 4.2.7), we may regard this action as that of right multiplication of the right cosets of G_x. Now if $G_x < H < G$, then each right coset of H is a union of right cosets of G_x. In this way the rights cosets of H partition G/G_x to form a system of imprimitivity. Thus in the imprimitive case, no such H exists, so G_x is G or is maximal in G. The former case is ruled out by $|X| > 1$.

Conversely, if $X = X_1 + X_2 + \cdots$ is a system of imprimitivity, then all the X_i have the same cardinality and for $x \in X_1, G_x$ stabilizes the set X_1. But by transitivity of G, if $L = \text{Stab}_G(X_1)$ is the stabilizer in G of the component X_1, then L must act transitively on X_1. Since $|X_1| \neq 1$, G_x is a proper subgroup of L. Since $X_1 \neq X$, L is a proper subgroup of G. Thus transitive but imprimitive action, forces G_x not to be maximal. That is, G_x maximal implies primitive action of G on G/G_x and hence X. The proof is complete.$\square$

We immediately have:

Corollary 4.4.2 *If G has primitive action on X, then for any normal subgroup N of G either (i) N acts trivially on X (so either $N = 1$ or the action is not faithful), or (ii) N is transitive on X.*

Proof From Lemma 4.2.9, p. 99, the N-orbits either form a system of imprimitivity or form a trivial partition of X. The former alternative is excluded by our hypothesis. Therefore the partition into N-orbits is trivial—that is, (i) or (ii) holds.$\square$

4.4.2 The Rank of a Group Action

Suppose G acts transitively on a set X. The *rank of the group action* is the number of orbits which the group G induces on the cartesian product $X \times X$. Here an element

g of G takes the ordered pair (x, y) of $X \times X$ to (x^g, y^g). Since G is transitive, one such orbit is always the *diagonal orbital* $D = \{(x, x)|x \in X\}$. The total number of such *"orbitals"* —as the orbits on $X \times X$ are called—is the *rank of the group action.*

Basically, the non-diagonal orbitals mark the possible binary relations on the set X which could be preserved by G. Such a relation is defined by asserting that letter x has relation R_i to letter y (which we write as $x R_i y$) if and only if the ordered pair (x, y) is a member of the non-diagonal orbital O_i (G-orbit on $X \times X$.) We could then represent this relation by a directed graph, $\Gamma_i = (X, E_i)$, where the *ordered* pair (x, y) is a directed edge of E_i if and only if $x R_i y$ holds. In that case G acts as a group of automorphisms of the graph G_i—that is, $G \leq \text{Aut}(\Gamma_i)$.

If O_i is an orbital, there is also an orbital O_i^* consisting of all pairs (y, x) for which (x, y) belongs to O_i. Clearly O_i^* has the same cardinality as O_i.

The orbital O_i is *symmetric* if and only if $(x, y) \in O_i$ implies $(y, x) \in O_i$, that is $O_i = O_i^*$. In this case, the relation R_i is actually a symmetric relation and if $O_i \neq D$, then the graph Γ_i is a simple undirected graph.[3] Thus representing groups as groups of automorphisms of graphs, has a natural setting in permutation groups of rank 3 or more.

Now fix a letter x in X. If (u, v) is in the orbital O_i, there exists, by transitivity, an element g such that $u^g = x$. Thus the orbital O_i possesses a representative (x, v^g), with first coordinate equal to x. Moreover, if (x, v) and (x, w) are two such elements of O_i with first coordinate x, the fact that O_i is a G-orbit, shows that there is an element h in G_x taking (x, u) to (x, w). That is, u and w must belong to the same G_x-orbit on X. Thus there is a one-to-one correspondence between the orbitals of the action of G on X and G_x-orbits on X. The lengths of these orbits are called the *subdegrees* of the permutation action of G on X. By the Fundamental Theorem of Group Actions (Theorem 4.2.7), these G_x-orbits on X (or equivalently, G/G_x) correspond to (G_x, G_x)-double cosets of G, and the length of $G_x g G_x$ is $[G_x : G_x \cap g^{-1} G_x g]$.

We summarize most of this in the following;

Theorem 4.4.3 *Suppose G acts transitively on X.*

(i) *Then the rank of this action is the number of G_x-orbits on X.*

(ii) *The lengths of these G_x-orbits are called* subdegrees *and are the numbers*

$$[G_x : G_x \cap g^{-1} G_x g]$$

as g ranges over a full system of (G_x, G_x)-double coset representatives.

(iii) *The correspondence of an orbital O_i and a (G_x, G_x)-double coset is that $(G_x, G_x h) \in O_i$ if and only coset $G_x h$ is in the corresponding (G_x, G_x)-double coset, namely $G_x h G_x$. Then*

$$|O_i| = |X|[G_x : G_x \cap g^{-1} G_x g].$$

---

[3]For graphs, the word "simple" means is has no multiple edges (of the same orientation) or loops.

Clearly O_i and O_i^* correspond to G_x-orbits of the same size. In particular, if an orbital is not symmetric, then there must be two subdegrees of the same size.

(iv) *If a double coset $G_x g G_x$ contains an involution t, then the corresponding orbital must be symmetric.*

Remark All but item (iv) was developed in the discussion preceding. If t is an involution in $G_x g G_x$, then t inverts the edge $(G_x, G_x t)$ in the graph Γ_i.

Let's look at a few simple examples.

Example 27 (i) (Grassmannian Action) Suppose V is a finite-dimensional vector space of dimension at least 2. As we have done before, let $P_k(V)$ be the collection of all k-dimensional subspaces, where $k \leq (\dim V)/2$. Then the group $GL(V)$ acts transitively on $P_k(V)$, as noted earlier. But because of the inequality on dimensions, a given k-subspace W may intersect other k-subspaces at any dimension from 0 to k. Moreover, the stabilizer H in $GL(V)$ of the subspace W is transitive on all further k-subspaces which meet W at a specific dimension. Thus we see that $GL(V)$ acts on $P_k(V)$ with an action which is rank $k + 1$.

(ii) (Symmetric Groups) If $k \leq n/2$, the group $G = \text{Sym}(n)$, transitively permutes $A(k)$, the k-subsets of $A = \{1, 2, \ldots, n\}$. Again, the stabilizer H of a fixed k-subset U, is transitive on the set of all k-subsets which meet U in a subset of fixed cardinality. Thus $G = \text{Sym}(n)$ has rank $k + 1$ in its action on all k-subsets. For example, $G = \text{Sym}(7)$, acts faithfully on the 35 points of $A(3)$, as a rank four group with subdegrees $1, 12, 18, 4$.

$G = \text{Sym}(6)$, acts as a rank 4 group on the 20 letters of $A(3)$, with subdegrees $1, 9, 9, 1$.

Similarly, $G = \text{Sym}(5)$, acts as a rank 3 group on the 10 letters of $A(2)$, with subdegrees $1, 3, 6$. If O_i is the orbital of 30 ordered pairs corresponding to the subdegree 3, then, as this orbital is symmetric, it contributes 15 undirected edges on the 10 vertices, yielding a graph $\Gamma_i = (A(2), E_i)$ isomorphic to Petersen's graph, whose automorphism group of order 120 we met in the last section.

(iii) We have mentioned that G turns out to be a subgroup of the automorphism group of the graph Γ_i. It need not be the full automorphism group, and whether it is or not depends on which orbital O_i one uses. For example, we have just seen above that the group $G = \text{Sym}(7)$, has a rank 4 action on the 35 3-subsets chosen from the set A of 7 letters, with subdegrees $1, 12, 18, 4$. Now it happens that the graph Γ_3 corresponding to the subdegree 18 has diameter 2, and its full automorphism group is $\text{Sym}(8)$.

(iv) This might be a good time to introduce an example of a rank 3 permutation group with two non-symmetric orbitals. Consider the group $G = AQ$ of all affine transformations,

$$x \to sx + y \text{ where } x \in \mathbf{Z}_p, s \text{ a square,}$$

where $\mathbf{Z}_p$ denotes the additive group of integer residue classes modulo a prime $p \equiv 3 \bmod 4$. Here A denotes the additive group of translations $\{t_y : x \to x + y, x \in \mathbf{Z}_p\}$, and Q denotes the multiplications by quadratic residue classes modulo p. Then G acts faithfully on G/Q, with A as a normal regular abelian subgroup. Now the group G has order $p(p - 1)/2$ which is odd. G is rank three with subdegrees $1, (p - 1)/2, (p - 1)/2$, and the non-diagonal orbitals are not symmetric. By Lemma 4.2.9, the graph Γ can be described as follows: The vertices are the residue classes $[m] \bmod p$. Two of them are adjacent if and only if they differ by a quadratic residue class. Since -1 is not a quadratic residue here, this adjacency relation is not symmetric.

A rank 1 permutation group is so special that it is trivial, for the diagonal orbital D is the only orbital, forcing $|X| = 1$.

A rank 2 group action has just two orbitals, the diagonal and one other. These are the doubly transitive group actions which are considered in the next subsection. In fact

Corollary 4.4.4 *The following are equivalent:*

 (i) *G acts as a rank 2 permutation group.*
 (ii) *G transitively permutes the ordered pairs of distinct letters of X.*
 (iii) *G acts transitively on X and for some letter $x \in X$, G_x acts with two orbits: $\{x\}$ and $X - \{x\}$.*

4.4.3 Multiply Transitive Group Actions

Suppose G acts transitively on a set X. We have remarked in Sect. 3.1 that in that case, for every positive integer $k \le |X|$, that G also acts on any of the following sets:

1. $X^{(k)}$, the set of ordered k-tuples of elements from X,
2. $X^{(k)*}$, the set of ordered k-tuples of elements of X with entries pairwise distinct.
3. $X(k)$, the set of (unordered) subsets of X of size k.

For $k > 1$, the first set is of little interest to us; for the group can never act transitively here, and orbits that exist tend to be equivalent to orbits on the other two sets. A group G is said to *act k-fold transitively on X* if and only if it induces a transitive action on the set $X^{(k)*}$ of k-tuples of distinct letters. It is said to have a *k-homogeneous action* if and only if G induces a transitive action of the set of all k-sets of letters.

Lemma 4.4.5 (i) *If G acts k-transitively on X, then it acts k-homogeneously on X. That is, k-transitivity implies k-homogeneity.*
 (ii) *For $1 < k \le |X|$, a group which acts k-transitively on X acts $(k-1)$-transitively on X.*

(iii) *For* $1 < k < |X| - 1$, *a group* G *which acts transitively on* X, *acts* k-*fold transitively on* X *if and only if the subgroup fixing any* $(k-1)$-*tuple* $(x_1, \ldots, x_{k-1})$ *transitively permutes the remaining letters,* $X - \{x_1, \ldots, x_{k-1}\}$.

Proof The first two parts are immediate from the definitions. The third part follows from Corollary 4.4.4 when $k = 2$. So we assume $2 < k < |X| - 1$, and will apply induction.

Assume $X = \{1, 2, 3, \ldots\}$ (the notation is not intended to suggest that X is countable; only that a countable few elements have been listed first), and suppose G acts transitively on X and that for any finite ordered $(k-1)$-tuple $(1, 2, \ldots, (k-1))$, the subgroup $G_{1,\ldots,(k-1)}$ that fixes the $(k-1)$-tuple point-wise, is transitive on the remaining letters, $X - \{1, 2, \ldots, (k-1)\}$. Clearly, the result will follow if we can first show that G is $(k-1)$-fold transitive on X. Now set $U = X - \{1, \ldots, (k-2)\}$. The subgroup $H := G_{1,\ldots,k-2}$, of all elements of G fixing $\{1, \ldots, k-2\}$ point-wise, properly contains the group $G_{1,\ldots,k-1}$ just discussed, and has this property:

(4.43) For every letter u in U, H_u fixes $\{u\}$ and is transitive on $U - \{u\}$.

Since $k < |X| - 1$ causes $|U| \geq 3$, the presented property implies H acts transitively on U. Thus we have the hypotheses of part (iii) for $k - 1$, and so we can conclude by induction, that G has $(k-1)$-fold action on X. Then the hypothesis on $G_{1,\ldots,k-1}$ implies the k-fold transitive action.

The converse implication in part (iii), that k-fold transitive action causes $G_{1,\ldots,k-1}$ to be transitive on all remaining letters, is straightforward from the definitions. $\square$

Remark The condition that $k < |X| - 1$ is necessary. The group Alt(5) acts transitively on five letters, and any subgroup fixing four of these letters, is certainly transitive on the remaining letters (or more accurately "letter", in this case). Yet the group is not 4-transitive on the five letters: it is only 3-fold transitive.

Lemma 4.4.6 *Suppose* $f : G \to Sym(X)$ *is a 2-transitive group action. Then the following statements hold:*

  (i) G *is primitive in its action on* X.
 (ii) *If* $x \in X$, $G = G_x + G_x t G_x$ *where* t *is congruent to an involution* mod ker f.
(iii) *Suppose* N *is a finite regular normal subgroup of* G. *Then* N *has no proper characteristic subgroups and is a direct product of cyclic groups of prime order* p. *(Such a group is called an* elementary abelian p-group*). It follows that* $|N| = |X| = p^a$ *is a prime power.*

Proof (i) Suppose X_1 is a non-trivial component of a system of imprimitivity, so that $1 \neq |X_1|$ and $X_1 \neq X$. For any $x \in X_1$, the subgroup G_x stabilizes X_1. But G_x acts transitively on $X - \{x\}$, forcing $X_1 = X$, a contradiction. So there is no such system of imprimitivity, as was to be shown.

(ii) Since $n(n - 1) = |X^{(2)*}|$ divides $|G/\ker f|$, the latter number is even. So by Sylow's theorem applied to $G/\ker f$, there must exist an element z of G which acts as an involution on X. Then z must induce at least one 2-cycle, say (a, b) on X.

From transitivity, we can choose $g \in G$ such that $a^g = x$. Then $t := g^{-1}zg$ is the involution displacing x and can be taken as the double coset representative.

(iii) By Lemma 4.2.9 part (iv), G_x transitively permutes the non-identity elements of N by conjugation. Since the conjugation action, induces automorphisms of N, N does not contain any proper characteristic subgroups. Also, all non-identity elements are $G - conjugate$ and so have the same finite order, which must be a prime p. This means $\text{Syl}_r(N) = \{1\}$ if $r \neq p$, so N is a p-group. By Lemma 4.3.2, N has a non-trivial center, and so is abelian. Since all its elements have order p, it is elementary abelian. Since it is regular, the statements about $|X|$ follow.□

Let us discuss a few examples.

Example 28 (i) The groups $\text{FinSym}(X)$ are k-fold transitive for every natural number k not exceeding $|X|$. However, an easy induction proof shows that $\text{Alt}(X)$ is $(k - 2)$-transitive for each natural number k less than $|X|$. (Recall that the alternating group is *by definition* a subgroup of the finitary symmetric group.)

(ii) The group $GL(V)$ has a 2-transitive action on the set $P_1(V)$ of all 1-subspaces of V.

This set is called the set of *projective points* of V; the set $P_2(V)$ of all 2-dimensional spaces is called the set of *projective lines*. Incidence between $P_1(V)$ and elements of $P_2(V)$ is the relation of containment of a 1-subspace in a 2-subspace of V. Then $PG(V) := (P_1(V), P_2(V))$ is an incidence system of points and lines with this axiom:

Linear Space Axiom. *Any two points are incident with a unique line.*

If $\dim V = 2$, $Gl(V)$ is doubly transitive on the set $P_1(V)$, the *projective line*. If the ground field for the vector space V is a finite field containing q elements, then there are $q + 1$ projective points being permuted. The group of actions induced—that is $GL(V)/K$ where K is the kernel of the action homomorphism—is the group $PGL(2, q)$, and it is 3-transitive on $P_1(V)$.

If $\dim V = 3$, it is true that any two distinct 2-subspaces of V meet in a 1-subspace. Thus we have the additional property:

Dual linear Axiom *Any two distinct lines meet at a point.*

Any system of points and lines, satisfying both the Linear Space Axiom and the Dual Linear Axiom is called a *projective plane*. Again, if the ground field of V is a finite field of q elements then we see that

$$|P_1(V)| = 1 + q + q^2 = |P_2(V)|$$

and each projective point lies on $q + 1$ projective lines, and, dually, each projective line is incident with exactly $q + 1$ projective points.

In the case $q = 2$ (which means that the field is $\mathbb{Z}/(2)$, integers mod 2), we get the system of seven points and seven lines which we have met as the second example in Sect. 4.3.5. We saw there that the full automorphism group of the projective plane of order 2 had order 168.

4.4.4 Permutation Characters

The results in this section only make sense for actions on a finite set X.

Suppose $f : G \to \text{Sym}(X)$ is a group action on a finite set X. Then $f(G)$ is a finite group of permutations of X. Without loss of generality we assume G itself is finite. (We do this, so that some sums taken over the elements of G are finite. But this is no drawback, for the assertions we wish to obtain can in general be applied to $f(G)$ as a faithfully acting finite permutation group.)

Associated with the finite action of G is the function

$$\pi : G \to \mathbb{Z}$$

which maps each element g of G to the number $\pi(g)$ of fixed points of g—that is, the number of letters that g leaves fixed. This function is called the *permutation character of the (finite) action* f, and, of course, is completely determined by f. The usefulness of this function is displayed in the following:

Theorem 4.4.7 (Burnside's Lemma)

(i) *The number of orbits with which G acts on X is the average number of letters fixed by the group elements:*

$$(1/|G|)\sum_{g \in G} \pi(g).$$

In particular, G is transitive, if and only if the average value of π over the group is 1.

(ii) *The number of orbits of G on $X^{(k)*}$ is*

$$(1/|G|)\sum_{g \in G} \pi(g)(\pi(g) - 1) \cdots (\pi(g) - k + 1).$$

(iii) *In particular, if G acts 2-transitively, then*

$$(1/|G|)\sum \pi(g)^2 = 2.$$

Also, if G is 3-transitive,

$$(1/|G|)\sum_{g \in G} (\pi(g))^3 = 5.$$

Proof (i) Count pairs $(x, g) \in X \times G$ with $x^g = x$. One way is to first choose $x \in X$ and then find $g \in G_x$. This count yields

$$\sum_{x \in X} |G_x| = \sum_{\text{orbits}} |G_x||O_x| = \sum_{\text{orbits}} |G| = |G| \cdot \text{no. of orbits.}$$

On the other hand we can count these pairs by choosing the group element g first, then $x \in X$ in $\pi(g)$ ways. Thus

$$|G| \cdot \text{no. of orbits} = \sum_{g \in G} \pi(g).$$

(ii) The number of ordered k-tuples of distinct letters left fixed by element g is $\pi(g) \cdot (\pi(g) - 1) \cdots (\pi(g) - k + 1)$ (since such a k-tuple must be selected from the fixed point set of g). We then simply apply part (i) with X replaced by $X^{(k)*}$.

(iii) If G is 2-transitive then the average value of $\pi(\pi - 1)$ is 1. But as such a group is also transitive, the average value of π is also 1. Thus the average value of π^2 is

$$\text{the average value of} (\pi(\pi - 1) + \pi) = 1 + 1 = 2.$$

Similarly, if G is 3-transitive,

$$\begin{aligned}
(1/|G|) \sum_{g \in G} (\pi(g))^3 &= (1/|G|) \sum_{g \in G} (\pi(g)(\pi(g) - 1)(\pi(g) - 2)) \\
&\quad + \sum_{g \in G} 3\pi(g)^2 - 2 \sum_{g \in G} \pi(g) \\
&= 1 + 2 \cdot 3 - 2 \cdot 1 = 5.
\end{aligned}$$

$\square$

The student interested in pursuing various topics brought up in Sects. 4.3 and 4.4 and the exercises below should consult *Permutation Groups* by Peter Cameron [9]. This excellent book is not only thoroughly up to date, but is crammed with fascinating side-topics.

4.5 Exercises

4.5.1 Exercises for Sects. 4.1–4.2

1. Show that in any finite group G of even order, there are an odd number of involutions. [Hint: Choose $t \in I$, where $I = inv(G)$, the set of all involutions in G. Then I partitions as $\{t\} + \Gamma(t) + \Delta(t)$, where $\Gamma(t) := C_G(t) \cap I - \{t\}$ and $\Delta(t) := t^G - C_G(t)$. Show why the last two sets have even cardinality.]
2. (Herstein) Show that if exactly one-half of the elements of a finite group are involutions, then G is a generalized dihedral group of order $2n$, n odd (recall the definition in Example 26(6) on p. 98 of a "generalized dihedral group"). [Hint: By Exercise (1) in this section, the number of involutions is an odd number equal to one-half the group order, and by Sylow's theorem all involutions are

conjugate, forcing $C_G(t) = \langle t \rangle$. The result follows from any Transfer Theorem such as Thompson's.[4]]

3. This has several parts:

(a) Show that if p is a prime, then any group of order p^2 is abelian. [Hint: It must have a non-trivial center, and the group mod its center cannot be a non-trivial cyclic group.]

(b) Let G be a finite group of order divisible by the prime p. Show that the number of subgroups of G or order p is congruent to one modulo p. [Hint: Make the same sort of decomposition of the class of groups of order p as was done for groups of order 2 in Exercise (1) in this section. You need to know something about the number of subgroups of order p in a group of order p^2.]

(c) If prime number p divides $|G| < \infty$, show that the number of elements of order p in G is congruent to -1 modulo p.

(d) (McKay's proof of parts (b) and (c) of this exercise.) Let G be a finite group of order divisible by the prime p. Let S be the set of p-tuples $(g_1, g_2, \ldots, g_p)$ of elements of G such that $\prod g_i = 1$.

i. Show that S is invariant under the action of σ, a cyclic shift which takes $(g_1, g_2, \ldots, g_p)$ to $(g_p, g_1, \ldots, g_{p-1})$.

ii. As $H = \langle \sigma \rangle \simeq Z_p$ acts on S, suppose there are a H-orbits of length 1 and b H-orbits of length p. Show that $|G|^{p-1} = |S| = a + bp$. As p divides the order of G conclude that p divides a.

iii. Show that $a \neq 0$. Conclude that a is the number of elements of order one or order p in G, and is a non-zero multiple of p.

iv. Let $P_1, \ldots, P_k$ be a full listing of the subgroups of order p in G. Show that $|\cup P_j| = a \equiv 0 \bmod p$, and deduce that $k \equiv 1 \bmod p$.

(e) Suppose p^k divides the order of G and P is a subgroup of order p^{k-1}. Show that p divides the index $[N_G(P) : P]$. [Hint: Set $N := N_G(P)$ and suppose, by way of contradiction, that p does not divide $[N : P]$. Then P is the unique subgroup of N of order p^{k-1} and so is characteristic in N. As p^k divides $|G|$, we see that p divides the index $[G : N]$. Now P acts on the left cosets of N in G by left multiplication, with N comprising a P-orbit of length one. Show that there are at least p such orbits of length one so that there is a left coset $bN \neq N$, with $PbN = bN$. Then $b^{-1}Pb \leq N$ forcing $b^{-1}Pb = P$ from the uniqueness of its order. But then $b \in N$, a contradiction.]

(f) Suppose the prime power p^k divides the order of the finite group G. Define a *p-chain of length k* to be a chain of subgroups of G,

$$P_1 < P_2 < \cdots < P_k,$$

---

[4]Almost certainly this is the elementary proof that Herstein had in mind rather than the proofs in several longer papers on this subject which have appeared in the AMA MONTHLY.

where P_j is a subgroup of G of order p^j, $j = 1, \ldots, k$. Show that the number of p-chains of G of length k is congruent to one modulo p. [Hint: Proceed by induction on k. The case $k = 1$ is part 4 (d) of this Exercise. One can assume the number m of p-chains of length $k - 1$ is congruent to 1 mod p. and these chains of length $k - 1$ partition the chains of length k. If c is a chain of length $k - 1$ with terminal member P_{k-1}, then any extension of c to a chain of length k is obtained by adjoining a group P_k of order p^k in the normalizer $N_G(P_{k-1})$. The number of such candidates is thus the number of subgroups of order p in $N_G(P_{k-1})/P_{k-1}$. Part 5 of this Exercise shows that p divides this group order, and so by part 4 (d), the number of Z_p's in the factor N/P is congruent to one mod p. Show that the conclusion is now forced.]

(g) From the previous result show that if p^k divides $|G|$, for $k > 0$, then the number of subgroups of order p^k is congruent to 1 mod p.

[Remark: Note that parts 3–7 of this Exercise did not require either the Sylow theorems of even the weaker Cauchy's Theorem that asserts that G contains an element of prime order p if p divides $|G|$. Indeed, these exercises essentially reprove them. Exercise (1) in this section and part 2 of this Exercise, on the other hand, required Cauchy's theorem in order to have Z_p's to discuss.]

4. Let G be a group of order $p^n q$, where p and q are primes and $n > 1$. Show that G is not a simple group. [Hint: We may assume that there are at least two p-sylow subgroups. Select a pair (P_1, P_2) of distinct p-Sylow subgroups so that their intersection $N = P_1 \cap P_2$ is as large as possible. Invoke Lemma 4.3.2, part (iii), to conclude that $N_G(N)$ contains distinct p-Sylow subgroups, and hence has order divisible by q. Choose $Q_1 \in \mathrm{Syl}_p(N_G(N))$. Then $G = P_1 Q$, and so

$$\langle M^G \rangle = \langle M^{P_1} \rangle.$$

The left side of this equation is a normal subgroup of G. But the right side of the presented equation is properly contained in P_1. Thus G is not simple unless $M = 1$. In the latter case, the q elements of $\mathrm{Syl}_p(G)$ pairwise intersect at the identity. Now count the number of elements of order dividing q.]

4.5.2 Exercises Involving Sects. 4.3 and 4.4

1. Show that any group of order fifteen is cyclic. [Hint: Use Sylow's Theorem to show that the group is the direct product of two cyclic p-sylow subgroups.]
2. Show that any group of order 30 has a characteristic cyclic subgroup of order 15.
3. Show that any group N which admits a group of automorphisms which is 2-transitive on its non-identity elements is an elementary abelian 2-group or a

cyclic group of order 3.

Extend this result by showing that if there is a group of automorphisms of N which is 3-transitive on the non-identity elements of N, then $|N| \leq 4$.

4. Show that if G faithfully acts as a k-fold transitive group on a set X, and G_x is a simple group, then (as k is at least 1) any proper normal subgroup of G is regular on X.

    Using a basic lemma of this chapter, show that if $1 \neq N$ is a normal subgroup of G, then G is the semidirect product NG_x (here $G_x \cap N = 1$ so G_x is a complement to N in G). Moreover, G_x acts $(k-1)$-fold transitively on the non-identity elements of N.

    Conclude that under these circumstances, G can be at most 4-fold transitive on X, and then only in the case that $|X| = 4$ and $G = \mathrm{Sym}(4)$.

5. Show that the group $\mathrm{Alt}(5)$ acts doubly transitively on its six 5-sylow subgroups.

6. Show that if N is a non-trivial normal subgroup of $G = \mathrm{Alt}(5)$, then either (i) N has order divisible by 30 or (ii) N normalizes and hence centralizes each 5-sylow subgroup. [Hint: Use the fact that G acts primitively on both 5 and 6 letters.]

7. Conclude that possibility (ii) of the previous exercise is impossible, by showing that a 5-sylow subgroup of $\mathrm{Alt}(5)$ is its own centralizer.

8. Using Exercises (5–7) in this section (the three preceding exercises), show in just a few steps, that $\mathrm{Alt}(5)$ is a simple group.

9. If G is a k-transitive group, and $G_{x_1,\ldots x_{k-1}}$ is simple, then, provided $k \geq 3$, and $|X| > 4$, G must be a simple group. So in this case, the property of simplicity is conferred from the simplicity of a subgroup. (This often happens for primitive groups of higher permutation rank, but the situation is not easy to generalize.) [Hint: Use Exercises (4–8) above.]

10. Provide a proof of the following result:

Lemma 4.5.1 *Suppose G is a group acting faithfully and primitively on a set X. Assume further that:*

(i) A is a normal subgroup of G_x, and
(ii) G is generated by the set of conjugates $A^G = \{A^g | g \in G\}$.

Then for any non-trivial normal subgroup N of G, we have $G = AN$.

[Hint: Use Corollary 4.4.2 and get $G = G_x N$ and then use (i) and (ii)].

11. Organize the last few exercises into a proof that all alternating groups $\mathrm{Alt}(n)$ are simple for $n \geq 5$.

12. (Witt's Trick) Suppose K is a k-transitive group acting faithfully on the set X, where $k \geq 2$.

    (a) Show that if $H := K_\alpha$, for some letter α in X, then $K = H + HtH$ for some involution t in K.

    (b) Suppose ∞ is not a letter in X, set $X' := X \cup \{\infty\}$, and set $S := \mathrm{Sym}(X')$. Suppose now that S contains an involution s such that

        i. s transposes ∞ and α,

ii. s normalizes H—i.e. $sHs = H$, and

iii. $o(st) = 3$, that is, $(st)^3 = 1$, while $st \neq 1$.

Then show that the subset $Y = K \cup KsK$ is a subgroup of S. [Hint: One need only show that Y is closed under multiplication, and this is equivalent to the assertion that $sKs \subset Y$. Decompose K; use the fact that $sH = Hs$ and that $sts = tst$.]

13. Show that in the previous Exercise Y is triply transitive on X'.

14. Suppose G is a simple group with an involution t such that the centralizer $C_G(t)$ is a dihedral group of order 8. Show the following:

   (a) The involution t is a central involution—that is, t is in the center of at least one 2-Sylow subgroup of G, or, equivalently, $|t^G|$ is an odd number. Thus any 2-Sylow subgroup of G is just D_8.

   (b) There are exactly two conjugacy classes of fours groups—say V_1^G and V_2^G, where V_1 and V_2 are the unique two fours subgroups of a 2-Sylow subgroup D of G.

   (c) There is just one conjugacy class of involutions in G.

   (d) Each fours group V_i is a *TI-group*—that is, a subgroup that intersects its distinct conjugates only at the identity group.

[Hints: This is basically an exercise in the use of Sylow's Theorem (Theorem 4.3.3), the Burnside Fusion Theorem (Theorem 4.3.7) and the Thompson Transfer Theorem (Theorem 4.3.8).

Part (a): Take $D = C_G(t) \simeq D_8$. Then its center is just $\langle t \rangle$, so $N_G(D) \leq C_G(t) = D$.

Part (b): Let V_i, $i = 1, 2$, be the two unique fours subgroups of $D \in \mathrm{Syl}(G)$. Since each V_i is normal in D, the Burnside Fusion theorem tells us that they could be conjugate in G only if they were already conjugate in $N_G(D) = D$, where it is patently untrue.

Part (c): Let $D = V_1 V_2 = C_g(t)$ where $V_1 \cap V_2 = \langle t \rangle$. Since $D \simeq D_8$, it also contains a normal cyclic group C of order 4. If no non-central involution of D is conjugate to the central involution t, then a contradiction to the simplicity of G is obtained from the Thompson Transfer Theorem, using C as the subgroup of index 2 in D. If $t^G \cap D \subseteq V_1$, the same argument, using V_1 in place of C, again yields a contradiction. Thus all involutions in D are conjugate to t.

Part (d): Clearly two conjugates of V_i could at best meet at a central involution, of which, by Part (c) there is only one class, t^G. In that case they would comprise two fours subgroups of the same 2-Sylow subgroup (the centralizer of their intersection), and hence could *not* be conjugate by Part (b).]

Remark A nice little geometry lurks in the above exercise. The points $\mathcal{P}$ are one class of fours groups—say V_1^G, and the lines $\mathcal{L}$ are the other class V_2^G. A line is declared to be *incident* with a point, if and only if the corresponding fours groups normalize each other—i.e. they lie in a common 2-Sylow subgroup. With the information above, the reader can easily prove that $N_G(V_i) \simeq \mathrm{Sym}(4)$, which tells us that each line is incident with just three points and each point lies on exactly three lines. Here are two examples:

(a) $|G| = 168$, $(\mathcal{P}, \mathcal{L})$ is the projective plane of order 2; 7 points, 7 lines.
(b) $|G| = 360$, $(\mathcal{P}, \mathcal{L})$ is the generalized quadrangle of order $(2, 2)$; 15 points, 15 lines.

Are these the only possibilities?

15. This exercise is really an optional student project. First show that if G if a 4-fold transitive group of permutations of the finite set X, then

$$(1/|G|)\sum_{g \in G} \pi(g)^4 = 15.$$

The object of the project would be to show that in general, if G is k-fold transitive on X then

$$(1/|G|)\sum_{g \in G} \pi(g)^k = B_k,$$

where B_k is the total number of partitions of a k-set (also known as *the k-th Bell number*). An analysis of the formulae involves one in expressing the number of sequences (with possible repetitions) of k elements from a set X in terms of ordered sequences with pair-wise distinct entries. The latter are expressed in terms of the so-called "falling factorials" $x(x-1)(x-2)\cdots(x-\ell+1)$ in the variable $x = |X|$. The coefficients involve Stirling numbers and the Bell number can be reached by known identities. But a more direct argument can be made by sorting the general k-tuples over X into those having various sorts of repetitions; this is where the partitions of k come in.

Chapter 5
Normal Structure of Groups

Abstract The Jordan-Hölder Theorem for Artinian Groups is a simple application of the poset-theoretic Jordan-Hölder Theorem expounded in Chap. 2. A discussion of commutator identities is exploited in defining the derived series and solvability as well as in defining the upper and lower central series and nilpotence. The Schur-Zassenhaus Theorem for finite groups ends the chapter. In the exercises, one will encounter the concept of normally-closed families of subgroups of a group G, which gives rise to several well-known characteristic subgroups, such as $\mathbf{O}_p(G)$, the *torsion subgroup*, and (when G is finite) the *Fitting subgroup*. Some further challenges appear in the exercises.

5.1 The Jordan-Hölder Theorem for Artinian Groups

5.1.1 Subnormality

As defined in Chap. 4, p. 99, a group is *simple* if and only if its only proper normal subgroup is the identity subgroup. (In particular, the identity group is not considered to be a simple group.)

A subgroup H is said to be *subnormal in G* if and only if there exists a *finite* chain of subgroups

$$H = N_0 \trianglelefteq N_1 \trianglelefteq \cdots \trianglelefteq N_k = G.$$

(Recall that this means each group N_i is normal in its successor N_{i+1}, but is not necessarily normal in the groups further up the chain.) We express the assertion that H is subnormal in G by writing

$$H \trianglelefteq \trianglelefteq G.$$

Let $\mathcal{SN}(G)$ be the collection of all subnormal subgroups of G. This can be viewed as a poset in two different ways: (1) First we can say that A is "less than or equal to B" if and only if A is a subgroup of B, that is, $\mathcal{SN}$ has the inherited structure of an *induced poset* of $\mathcal{S}(G)$, the poset of *all* subgroups of G with respect to the containment relation. (2) We also have the option of saying that A "is less than B" if and only if A is subnormal in B. The second way of defining a poset on $\mathcal{SN}$ would seem to be more conservative than the former. But actually the two notions are exactly the same, as can be concluded from the following Lemma.

Lemma 5.1.1 (Basic Lemma on Subnormality) *The following hold:*

(i) If $A \trianglelefteq \trianglelefteq B$ and $B \trianglelefteq \trianglelefteq G$, then $A \trianglelefteq \trianglelefteq G$.
(ii) If $A \trianglelefteq \trianglelefteq G$ and H is any subgroup of G, then $A \cap H \trianglelefteq \trianglelefteq H$.
(iii) If A is a subnormal subgroup of G, then A is subnormal in any subgroup that contains it.
(iv) If A and B are subnormal subgroups of G, then so is $A \cap B$.

Proof (i) This is immediate from the definition of subnormality: By assumption there are two finite ascending chains of subgroups, one running from A to B, and one running from B to G, each member of the chain is normal in its successor in the chain. Clearly concatenating the two chains produces a chain of subgroups from A to G, each member of the chain normal in the member immediately above it. So by definition, A is subnormal in G.

(ii) If $A \trianglelefteq \trianglelefteq G$ one has a finite subnormal chain $A = A_0 \trianglelefteq A_1 \trianglelefteq \cdots \trianglelefteq A_j := G$. Suppose H is any subgroup of G. It is then no trouble to see that $A \cap H \trianglelefteq A_1 \cap H \trianglelefteq A_2 \cap H \trianglelefteq \cdots A_j \cap H = G \cap H = H$, is a subnormal series from $A \cap H$ to H.

(iii) This assertion follows from Part (ii) where H contains A.

(iv) By Part (ii), $A \cap B$ is subnormal in B. Since B is subnormal in G, the conclusion follows by an application of Part (i). $\square$

At this point the student should be able to provide a proof that the two methods (1) and (2) of imposing a partial order on the subnormal subgroups lead to the same poset. We simply denote it $(\mathcal{SN}(G), \leq)$ (or simply $\mathcal{SN}$ if G is understood) since it *is* an induced poset of the poset of all subgroups of G. From Lemma 5.1.1, Part (iv), we see that

$\mathcal{SN}(G)$ *is a lower semilattice.*

Finally it follows from the second or third isomorphism theorem, that if

If A and B are distinct maximal normal subgroups of $G = \langle A, B \rangle = AB = BA$, then $G/A \simeq B/(A \cap B)$. Thus $\mathcal{SN}(G)$ is a semimodular lower semilattice and the mapping

$$\mu : \mathrm{Cov}(\mathcal{SN}) \rightarrow \text{simple groups}$$

(which records for every cover $X \trianglelefteq Y$ where X a maximal normal subgroup of Y, the isomorphism type of the simple group Y/X) is "semimodular" in the sense of Theorem 2.5.2, Sect. 2.5.3.

According to the general Jordan-Holder theory (advanced in Sect. 2.5.3, Theorem 2.5.2), μ can be extended to an *interval measure*

$$\mathcal{A}_{S\mathcal{N}} \to M.$$

from all *algebraic intervals* of $S\mathcal{N}$ into the additive monoid M of all multisets of isomorphism classes of simple groups—that is, the free additive semigroup of all formal non-negative integer linear combinations of isomorphism classes $[G]$ of simple groups:

$$\sum\nolimits_{\text{simp}} a_{[G]}[G], \quad \text{of finite support}.$$

If there is a finite unrefinable ascending chain in $S\mathcal{N}$ from 1 to G, then such a chain of subnormal subgroups is called a *composition series of* G. The multiset of the isomorphism classes of the simple groups A_{i+1}/A_i appearing in such a composition series $1 \trianglelefteq A_1 \trianglelefteq \cdots \trianglelefteq A_n = G$ is called the *multiset of composition factors*. The Theorem 2.5.2 states that these are the same multisets for any two composition series—the "measure" $\mu([1, G])$.

If you have been alert, you will notice that no assumption that the involved groups are finite has been made. The composition factors may be infinite simple groups.

But of course, in applying the Jordan-Hölder theorem, one is interested in pinning down a class of algebraic intervals. This is easy in the case of a finite group, for there, all intervals of $S\mathcal{N}(G)$ are algebraic. How does one approach this for infinite groups? The problem is that one can have an infinitely descending chain of subnormal subgroups, and also one can have an infinitely ascending chain of proper normal subgroups (just think of an abelian group with such an ascending chain).

However one can select certain subposets of $S\mathcal{N}$ which will provide us with many of these algebraic intervals: Suppose from any $S\mathcal{N}(G)$ we extract the set $S\mathcal{N}^*$ of all subnormal subgroups for which any downward chain of subgroups each normally containing its successor, must terminate in a finite number of steps. There are three other descriptions of this subcollection $S\mathcal{N}^*$, of subnormal subgroups: (i) they are the subnormal subgroups of G which belong to a finite unrefinable subnormal chain extended downward to the identity. (ii) They are the elements M of $S\mathcal{N}$ for which $[1, M]$ is an *algebraic* interval of $S\mathcal{N}$—see Chap. 2, Sect. 2.3.2. (iii) This is the special class of subgroups amenable to the Jordan-Holder Theorem—the subnormal subgroups of G which themselves possess a finite composition series. We call the elements of $S\mathcal{N}^*$ the *Artinian subgroups*. They are closed under the operation of taking pairwise meets in $S\mathcal{N}$ and so $S\mathcal{N}^*$ is clearly a lower semilattice, all of whose intervals are algebraic.

As a special case, we have:

Theorem 5.1.2 (The Jordan-Holder Theorem for general groups) *Let $S\mathcal{N}^*$ be the class of all Artinian subnormal subgroups of a group G. Then for any H in $S\mathcal{N}^*$, the list (with multiplicities) of all isomorphism classes of simple groups which appear* *as a cheif factor in any finite unrefinable subnormal chain of H is independent of the particular choice of that chain.*

Corollary 5.1.3 *Any two composition series (unrefinable subnormal series) of a finite group recreates through its chief factors the same list of isomorphism classes of finite simple groups, with mutiplicities respected.*

5.2 Commutators

Let G be a group. A *commutator* in G is an element of the form $x^{-1}y^{-1}xy$ (called a *commutator*) and is denoted $[x, y]$. A *triple commutator* $[x, y, z]$ is an element of shape $[[x, y], z]$, which the student may happily work out to be

$$[x, y]^{-1}z^{-1}[x, y]z = y^{-1}x^{-1}yxz^{-1}x^{-1}y^{-1}xyz.$$

If we write $u^x := x^{-1}ux$, the result of conjugation of u by x, then it is an exercise to verify the following identities, for elements x, y, and z in any group G:

$$[x, y] = [y, x]^{-1}. \tag{5.1}$$

$$[xy, z] = [x, z]^y[y, z] = [x, z][x, z, y][y, z]. \tag{5.2}$$

$$[x, yz] = [x, z][x, y]^z = [x, z][x, y][x, y, z]. \tag{5.3}$$

$$1 = [x, y^{-1}, z]^y[y, z^{-1}, x]^z[z, x^{-1}, y]^x. \tag{5.4}$$

$$[x, y, z][y, z, x][z, x, y] = [y, x][z, x][z, y]^x \times$$
$$[x, y][x, z]^y[y, z]^x[x, z][z, x]^y. \tag{5.5}$$

A commutator need not be a very typical element in a group. For example, in an abelian group, all commutators are equal to the identity.

If A, B and C are subgroups of a group G, then $[A, B]$ is the subgroup of G *generated* by all commutators of the form $a^{-1}b^{-1}ab = [a, b]$ as (a, b) ranges over $A \times B$. Similarly we write $[A, B, C] := [[A, B], C]$.

Theorem 5.2.1 (Basic Identities on Commutators of Subgroups) *Suppose A, B and C are subgroups of a group G.*

(i)   *A centralizes B if and only if $[A, B] = 1$.*
(ii) *$[A, B] = [B, A]$.*
(iii) *B always normalizes $[A, B]$.*
(iv) *A normalizes B if and only if $[A, B] \leq B$. In that case, $[A, B]$ is a normal subgroup of B.*

(v) If $f : G \to H$ is a group homomorphism, then $[f(A), f(B)] = f([A, B])$.
(vi) (The Three Subgroups Lemma of Phillip Hall) *If at least two of the three subgroups* $[A, B, C], [B, C, A]$ *and* $[C, A, B]$ *are trivial, then so is the third.*

Proof Part (i) A centralizes B if and only if every commutator of shape $[a, b], (a, b) \in A \times B$, is the identity. The result follows from this.

Part (ii) A subgroup generated by set X is also the subgroup generated by X^{-1}. Apply (5.1).

Part (iii) If $(a, b, b_1) \in A \times B \times B$, then it follows from (5.3) that

$$[a, b]^{b_1} = [a, b_1]^{-1}[a, bb_1] \in [A, B].$$

Part (iv) The following statements are clearly equivalent:

1. A normalizes B
2. $[a, b] \in B$ for all $(a, b) \in A \times B$
3. $[A, B] \le B$.

The normality assertion follows from part (iii).

Part (v) This is obvious.

Part (vi) Without loss of generality assume that $[A, B, C]$ and $[B, C, A]$ are the identity subgroup. Then all commutators $[a, b, c]$ and $[b, c, a]$ are the identity, no matter how the triplet (a, b, c) is chosen in $A \times B \times C$. Now by Eq. (5.4), any commutator $[c, a, b]$ is the identity, for any $(a, b, c) \in A \times B \times C$. This proves that $[C, A, B]$ has only the identity element for a generator. $\square$

Corollary 5.2.2 *Suppose again that A and B are subgroups of a group G.*

(i) If A *normalizes* B *and centralizes* $[A, B]$, *then* $A/C_A(B)$ *is abelian.*
   In particular, if $A \le \mathrm{Aut}(B)$ *and* A *centralizes* $[A, B] := \langle a^{-1}a^b|(a, b) \in A \times B \rangle$, *then* A *is an abelian group.*
(ii) If A *is a group of linear transformations of the vector space* V, *centralizing a subspace* W *as well as its factor space* V/W, *then* A *is an abelian group.*[1]

Proof All these results are versions of the following: By the Three Subgroups Lemma, $[A, B, A] = 1$ implies $[A, A, B] = 1$. That is the proof. $\square$

Of course anytime someone has some machinery, some mathematician wants to iterate it. We have defined the commutator of two subgroups in such manner that the result is also a subgroup. Thus inductively we may define a multiple commutator of subgroups in this way: Suppose $(A_1, A_2, \ldots)$ is a (possibly infinite) sequence of subgroups of a group G. Then for any natural number k, we define

$$[A_1, \ldots A_{k+1}] := [[A_1, \ldots A_k], A_{k+1}].$$

---

[1] This seems to be a frequently rediscovered result.

5.3 The Derived Series and Solvability

From the previous section, we see that the subgroup $[G, G]$ is a subgroup which

(i) is mapped into itself by any endomorphism $f : G \to G$ (a consequence of Theorem 5.2.1(V)). (This condition is called *fully invariant*. Since inner automorphisms are automorphisms which are in turn endomorphisms, "fully invariant"
implies "characteristic" implies "normal".)
(ii) yields an abelian factor, $G/[G, G]$ (let $f : G \to G/[G, G]$ and apply parts (i)
and (iii) of Theorem 5.2.1). Moreover, if G/N is abelian, then $[G, G] \leq N$, so
$[G, G]$ is the smallest normal subgroup of G, which possesses an abelian factor
(see Chap. 3).

The group $[G, G]$ is called the *commutator subgroup* or *derived subgroup* of G
and is often denoted G'.

Since we have formed the group $G' := [G, G]$, we may iterate this definition to
obtain the *second derived group*,

$$G'' := [[G, G], [G, G]].$$

Once on to a good thing, why not set $G' = G^{(1)}$ and $G'' = G^{(2)}$ and define

$$G^{(k+1)} := [G^{(k)}, G^{(k)}]$$

for all natural numbers k? The resulting sequence

$$G \geq G^{(1)} \geq G^{(2)} \geq, \cdots .$$

is called the *derived series of G*. Since each $G^{(k)}$ is characteristic in its successor
which in turn is characteristic in *its* successor, etc., we see the following:

Every $G^{(k)}$ is characteristic in G.

Note also that

$$G^{(k+l)} = (G^{(k)})^{(l)}. \tag{5.6}$$

A group G is said to be *solvable* if and only if the derived series

$$G \geq G^{(1)} \geq G^{(2)}, \ldots,$$

eventually becomes the identity subgroup after a finite number of steps.[2] The smallest
positive integer k such that $G^{(k)} = 1$ is called the *derived length* of G. Thus a group
is abelian if and only if it has derived length 1.

---

[2]In this case, for once, the name "solvable" is not accidental. The reason will be explained in the
Chap. 11 on Galois Theory.

Theorem 5.3.1 *(i) If $H \leq G$ and $A \leq G$, then $[A, H] \leq [A, G]$. In particular*

$$[H, H] \leq [G, H] \leq [G, G].$$

(ii) If $H \leq G$, then $H^{(k)} \leq G^{(k)}$. Thus, if G is solvable, so is H.
(iii) If $f : G \rightarrow H$ is a group homomorphism, then

$$f(G^{(k)}) = (f(G))^{(k)} \leq H^{(k)}.$$

(iv) If $N \trianglelefteq G$, then G is solvable if and only if both N and G/N are solvable.
(v) For a direct product,

$$(A \times B)^{(k)} = A^{(k)} \times B^{(k)}.$$

(vi) The class of all solvable groups is closed under taking finite direct products and taking subgroups and hence is closed under taking finite subdirect products. In particular, if G is a finite group, there is a normal subgroup N of G minimal among normal subgroups with respect to the property that G/N is solvable. (Clearly this normal subgroup is characteristic; it is called the solvable residual.*)*

Proof Part (i) follows from the containment,

$$\{[a, h] | (a, h) \in A \times H\} \subseteq \{[a, g] | (a, g) \in A \times G\}.$$

Part (ii) By Part (i), $H' \leq G'$. If we have $H^{(k-1)} \leq G^{(k-1)}$, then

$$H^{(k)} = [H^{(k-1)}, H^{(k-1)}] \leq [G^{(k-1)}, G^{(k-1)}] = G^{(k)},$$

utilizing $H^{(k-1)}$ and $G^{(k-1)}$ in the roles of H and G in the last sentence of Part (i). The result now follows by induction.

Part (iii) The equal sign is from Theorem 5.2.1, Part (v); the subgroup relation is from Part (i) above of this Theorem.

Part (iv) Suppose G is solvable. Then so are N and G/N by Parts (i) and (iii). Now assume, G/N and N are solvable. Then there exists natural numbers k and l, such that $(G/N)^{(k)} = 1$ and $N^{(l)} = 1$. Then by Part (iii), the former equation yields

$$G^{(k)}N/N = 1 \text{ or } G^{(k)} \leq N.$$

Then

$$G^{(k+l)} = (G^{(k)})^{(l)} \leq N^{(l)} \leq 1,$$

by Part (ii), so G is solvable.

Part (v) If the subgroups A and B of G centralize each other, it follows from Eqs. (5.2) or (5.3) that $[ab, a'b'] = [a, a'][b, b']$ for all $(a, b), (a', b') \in A \times B$. This shows $(A \times B)' \leq A' \times B'$. The reverse containment follows from Part (i). Thus the conclusion holds for $k = 1$. One need only iterate this result sufficiently many times to get the stated result for all k.

Part (vi) This is a simple application of the other parts. A subgroup of a solvable group is solvable because of the inequality of Part (ii). A finite direct product of solvable groups is solvable by Part (v). Thus, from the definition of *subdirect product*, a finite subdirect product of solvable groups is solvable. The last part follows from the remarks in Sect. 3.6.2. □

Remark Suppose $G_1, G_2, \ldots$ is a countably infinite sequence of finite groups for which G_k has derived length k (It is a fact that a group of derived length k exists for every positive integer k: perhaps one of those very rare applications of wreathed products.) Then the direct sum $\sum_{k \in \mathbb{N}} G_k = G_1 \oplus \cdots$ does not have a finite derived length and so is not solvable. Thus solvable groups are not closed under infinite direct products.

Recall that a *finite subnormal series* for G is a chain

$$G = N_0 \trianglerighteq N_1 \trianglerighteq N - 1 \cdots \trianglerighteq N_k = 1,$$

where each N_{j+1} is a normal subgroup of its predecessor, N_j, $j = 0, \ldots k - 1$, k is a natural number. (Of course what we have written is actually a *descending* subnormal series. We could just as well have written it backwards as an *ascending* series.) Recall also that a unrefinable subnormal series of finite length is called a *composition series*, and when such a thing exists, the factors N_j/N_{j+1} are simple groups called the *composition factors of G*. That the list of composition factors that appears is the same for all composition series, was the substance of the Jordan-Hölder Theorem for groups.

These notions make their reappearance in the following:

Corollary 5.3.2 (i) *A group G is solvable of derived length at most k if and only if it possesses a subnormal series*

$$G = N_0 \trianglerighteq N_1 \trianglerighteq N_1 \trianglerighteq \cdots \trianglerighteq N_k = 1.$$

such that each factor N_j/N_{j+1}, $j = 0, \ldots k - 1$, is abelian.
(ii) *A finite group is solvable if and only if it possesses a composition series with all composition factors cyclic of prime order (of course, the prime may depend on the particular factor taken).*

Proof Part (i) If G is solvable, the derived series is a subnormal series for which all factor groups formed by successive members of the series are abelian. Conversely, if the subnormal series is as given, then N_0/N_1 abelian implies $[G, G] = G' \leq N_1$. Inductively, if $G^{(j)} \leq N_j$, then N_j/N_{j+1} abelian implies $G^{(j+1)} \leq [N_j.N_j] \leq N_{j+1}$. So by induction, $G^{(j+1)} \leq N_{j+1}$ for all $j = 0, 1, \ldots k - 1$. Thus $G^{(k)} \leq N_k = 1$, and G is solvable.

Part (ii) If G possesses a composition series with all composition factors cyclic of prime order, G is solvable by Part (i). On the other hand, if G is finite and solvable it possesses a subnormal series

$$G = N_0 \trianglerighteq N_1 \trianglerighteq \cdots \trianglerighteq N_k = 1,$$

with each factor N_j/N_{j+1} a finite abelian group. But in any finite abelian group (where all subgroups are normal) we can form a composition series by choosing a maximal subgroup M_1 of A, a maximal subgroup M_2 of M_1, and so forth. Each factor will then be cyclic of prime order. Thus each interval $N_j \trianglerighteq N_{j+1}$ can be refined to a chain with successive factors of prime order. The result is a refinement of the original subnormal series to a composition series with all composition factors cyclic of prime order. □

There are many very interesting facts about solvable groups, especially finite ones, which unfortunately we do not have space to reveal. Two very interesting topics are these:

• Phillip Hall's theory of Sylow Systems. In a finite solvable group there is a representative S_i of $Syl_{p_i}(G)$, the p_i-Sylow subgroups of G such that $G = S_1 S_2 \cdots S_m$, and the S_i are permutable in pairs—that is $S_i S_j = S_j S_i$ where $p_1, \ldots p_m$ is the list of all primes dividing the order of G. Thus in any solvable group of order 120, for example, there must exist a subgroup of order 15. (This is not true, for example, of the non-solvable subgroup Sym(5) of that order.) A full account of Hall systems can be found in [22].

• The theory of formations. Basically, a *formation* is an isomorphism-closed class of finite solvable groups closed under finite subdirect products. Then each finite group G contains a characteristic subgroup $G_{\mathcal{F}}$ which is unique in being minimal among all normal subgroups N of G which yield a factor G/N belonging to $\mathcal{F}$. ($G_{\mathcal{F}}$ is called the *$\mathcal{F}$-residual of G*.) Whenever G is a finite group, let $D(G)$ be the intersection of all the maximal subgroups of G. This is a characteristic subgroup which we shall meet shortly, called the *Frattini subgroup* of G. A formation is said to be *saturated*, if for every solvable finite group G, $G/D(G) \in \mathcal{F}$ implies $G \in \mathcal{F}$. If $\mathcal{F}$ is saturated, then there exists a special class of subgroups, the $\mathcal{F}$-subgroups, subgroups in $\mathcal{F}$ described only by their embedding in G, which form a conjugacy class. Finite abelian groups comprise a formation but are not a saturated formation. A class of groups considered in the next section are the nilpotent groups and they *do* form a saturated formation. As a result one gets a theorem like this: "*Suppose X and Y are two nilpotent subgroups of a finite solvable group G, each of which is its own normalizer in G. Then X and Y are conjugate subgroups. Moreover, G must contain such groups.*"[3] A bit surprising. (The term 'nilpotent group' is defined in the following section.)

---

[3]This conjugacy class of subgroups was first announced in a paper by Roger Carter; they are called the class of "Carter subgroups" of G.

5.4 Central Series and Nilpotent Groups

5.4.1 The Upper and Lower Central Series

Let G be a fixed group. Set $\gamma_0(G) = G$, $\gamma_1(G) = [G, G]$, and inductively define

$$\gamma_{k+1}(G) := [\gamma_k(G), G],$$

for all positive integers k. Then the descending sequence of subgroups

$$G = \gamma_0(G) \geq \gamma_1(G) \geq \gamma_2(G) \geq \cdots,$$

is called the *lower central series* of the group G. Note that

$$\gamma_k(G) = [G, G, \ldots G] \text{ (with } k \text{ arguments)}.$$

Since any endomorphism $f : G \to G$, maps the arguments of the above multiple commutator into themselves, the endomorphism f maps each member of the lower central series into itself. A similar argument for homomorphisms allows us to state the following elementary result:

Lemma 5.4.1 *(i) Each member $\gamma_k(G)$ of the lower central series is a fully invariant subgroup of G.*

(ii) If $f : G \to H$ is a homomorphism of groups then

$$f(\gamma_k(G)) = \gamma_k(f(G)).$$

(iii) If H is a subgroup of G, then $\gamma_k(H) \leq \gamma_k(G)$.

Proof The proof is an easy exercise. □

A group G is said to be *nilpotent* if and only if its lower central series terminates at the identity subgroup in a finite number of steps. In this case G is said to *belong to nilpotence class k* if and only if k is the smallest positive integer such that $\gamma_k(G) = 1$. Thus abelian groups are the groups of nilpotence class 1.

The following is immediate from Lemma 5.4.1, and the definition of nilpotence.

Corollary 5.4.2 *(i) Every homomorphic image of a group of nilpotence class at most k is a group of nilpotence class at most k.*

(ii) Every subgroup of a group of nilpotence class at most k is also nilpotent of nilpotence class at most k.

(iii) Every finite direct product of nilpotent groups of nilpotence class at most k is itself nilpotent of nilpotence class at most k.

The beginning student should be able to supply immaculate proofs of these assertions by this time. [Follow the lead of the similar result for solvable groups in the previous section; do not hesitate to utilize the commutator identities in proving part (iii) of this theorem.]

Clearly, the definitions show that

$$[\gamma_k(G), \gamma_k(G)] \leq [\gamma_k(G), G] = \gamma_{k+1}(G),$$

so the factor groups $(\gamma_k(G))/(\gamma_{k+1}(G))$ are abelian. As a consequence,

Corollary 5.4.3 *Any nilpotent group is solvable.*

The converse fails drastically: Sym(3) is a solvable group of derived length 2. But is not nilpotent since its lower central series drops from Sym(3) to its subgroup $N \simeq Z_3$, and stabilizes at N for the rest of the series.

Now again fix G. Set $Z_0(G) := 1$, the identity subgroup; set $Z_1(G) = Z(G)$, the center of G (the characteristic subgroup of those elements of G which commute with every element of G); and inductively define $Z_k(G)$ to be that subgroup satisfying

$$Z_k(G)/Z_{k-1}(G) = Z(G/Z_{k-1}(G)).$$

That is, $Z_k(G)$ is the inverse image of the center of $G/Z_{k-1}(G)$ under the natural homomorphism $G \to G/Z_{k-1}(G)$. For example:

$$Z_2(G) = \{z \in G | [z, g] \in Z(G), \text{ for all } g \in G\}.$$

By definition $Z_k(G) \leq Z_{k+1}(G)$, so we obtain an ascending series of characteristic subgroups

$$1 = Z_0(G) \leq Z_1(G) \leq Z_2(G) \leq \cdots,$$

called the *upper central series of G*. Of course if G has a trivial center, this series does not even get off the ground. But, as we shall see below, it proceeds all the way to the top precisely when G is nilpotent.

Suppose now that G has nilpotence class k. This means $\gamma_k(G) = 1$ while $\gamma_{k-1}(G)$ is non-trivial. Then as

$$[\gamma_{k-1}(G), G] = \gamma_k(G) = 1,$$

G centralizes $\gamma_{k-1}(G)$, that is

$$\gamma_{k-1}(G) \leq Z_1(G).$$

Now assume for the purposes of induction, that

$$\gamma_{k-j}(G) \leq Z_j(G).$$

Then

$$[\gamma_{k-j-1}(G), G] \le \gamma_{k-j}(G) \le Z_j(G),$$

so

$$[\gamma_{k-j-1}, G] \le \{z \in G | [z, G] \le Z_j(G)\} := Z_{j+1}(G).$$

Thus, by the induction principle,

$$\gamma_{k-j}(G) \le Z_j(G), \text{ for all } 0 \le j \le k. \tag{5.7}$$

Putting $j = k$ in (5.7), we see that $Z_k(G) = \gamma_0(G) = G$. That is, *if the lower central series has length k, then the upper central series terminates at G in at most k steps.*

Now assume $G \ne 1$ has an upper central series terminating at G in exactly k steps—that is, $Z_k(G) = G$ while $Z_{k-1}(G)$ is a proper subgroup of G (as $G \ne 1$). Then, of course G/Z_{k-1} is abelian, and so $\gamma_1(G) = [G, G] \le Z_{k-1}(G)$. Now for the purpose of an induction argument, assume that

$$\gamma_j(G) \le Z_{k-j}(G).$$

Then

$$Z_{k-j}(G)/Z_{k-j-1}(G) = Z(G/Z_{k-j-1})$$

implies

$$[Z_{k-j}(G), G] \le Z_{k-j-1}(G),$$

so

$$\gamma_{j+1}(G) := [\gamma_j(G), G] \le [Z_{k-j}(G), G] \le Z_{k-(j+1)}(G).$$

Thus

$$\gamma_j(G) \le Z_{k-j}(G), \text{ for all } 0 \le j \le k. \tag{5.8}$$

Thus if $j = k$ in (5.8), $\gamma_k(G) \le Z_0(G) := 1$. Thus *if the upper central series terminates at G in exactly k steps, then the lower central series terminates at the identity subgroup in at most k steps.*

We can assemble these observations as follows:

Theorem 5.4.4 *The following assertions are equivalent:*

1. *G belongs to nilpotence class k. (By definition, its lower central series terminates at the identity in exactly k steps.)*
2. *The upper central series of G terminates at G in exactly k steps.*

There is an interesting local property of nilpotent groups:

Theorem 5.4.5 *If G is a nilpotent group, then any proper subgroup H of G is properly contained in its normalizer $N_G(H)$.*

Proof If $G = 1$, there is nothing to prove. Suppose H is a proper subgroup of G. Then there is a minimal positive integer j such that $Z_j(G)$ is not contained in H. Then there is an element z in $Z_j(G) - H$. Then $[z, H] \leq [z, G] \leq Z_{j-1}(G) \leq H$. Thus z normalizes H, but does not lie in it. $\square$

Corollary 5.4.6 *Every maximal subgroup of a nilpotent group is normal.*

Remark There are many groups with no maximal subgroups at all. Some of these are not even nilpotent. Thus one should not expect a converse to the above Corollary. This raises the question, though, whether the property of the conclusion of Theorem 5.4.5 characterizes nilpotent groups. Again one might expect that this is false for infinite groups. However, in the next subsection we shall learn that among finite groups it is indeed a characterizing property.

Here is a classical example of a nilpotent group: Suppose V is an n-dimensional vector space over a field F. A *chamber* is an ascending sequence of subspaces $S = (V_1, V_2, \ldots V_{n-1})$ where V_j has vector space dimension j. (This chain is unrefinable.) Now consider the group $B(S)$ of all linear transformations $T \in GL(V)$ such that T fixes each V_j and induces the identity transformation on each 1-dimensional factor-space V_j/V_{j-1}. Rendered as a group of matrices these are upper triangular matrices with "1's" on the diagonal. The reader may verify that the k-th member of the lower central series of this group consists of upper triangular matrices still having 1's on the diagonal, but having more and more upper subdiagonals all zero as k increases. Eventually one sees that $B(S)^{(n-1)}$ contains only the identity matrix.

5.4.2 Finite Nilpotent Groups

Let p be a prime number. As originally defined, a p-group is a group in which each element has p-power order. Suppose P is a finite p-group. Then by the Sylow Theorem, the order of G cannot be divisible by a prime distinct from p. Thus we observe

A finite group is a p-group if and only if the group has p-power order.

Now we have seen before (Lemma 4.3.2, part (ii)), that a non-trivial finite
p-group must possess a non-trivial center. Thus if $P \neq 1$ is a finite p-group, we
see that $Z(P) \neq 1$. Now if $Z(P) = P$, then P is abelian. Otherwise, $P/Z(P)$ has a
nontrivial center. Similarly, in general, if P is a finite p-group:

Either $Z_k(P) = P$ or $Z_k(P)$ properly lies in $Z_{k+1}(P)$.

As a result, a finite p-group possesses an upper central series of finite length,
and hence is a nilpotent group. Obviously, a finite direct product of finite p-groups
(where the prime p is allowed to depend upon the direct factor) is a finite nilpotent
group.

Now consider this

Lemma 5.4.7 *If G is a finite group and S is a p-Sylow subgroup of G, then
any subgroup H which contains the normalizer of S is itself self-normalizing—i.e.
$N_G(H) = H$ In particular $N_G(S)$ is self-normalizing.*

Proof Suppose x is an element of $N_G(H)$. Since P is a p-Sylow subgroup of G, it
is also a p-Sylow subgroup of $N_G(H)$ and H. Since $P^x \in \mathrm{Syl}_p(H)$, by Sylow's
theorem there exists an element $h \in H$ such that $P^x = P^h$. Then $xh^{-1} \in N_G(P) \leq
H$ whence $x \in H$. (Actually this could be done in one line, by exploiting the "Frattini
Argument" (Chap. 4; Corollary 4.3.4)). $\square$

Now we have

Theorem 5.4.8 *Let G be a finite group. Then the following statements are equivalent.*

 (i) *G is nilpotent.*
 (ii) *G is isomorphic to the direct product of its various Sylow subgroups.*
 (iii) *Every Sylow subgroup of G is a normal subgroup.*

Proof ((ii) implies (i)) Finite p-groups are nilpotent, as we have seen, and so too are
their finite direct products. Thus any finite group which is the direct product of its
Sylow subgroups is certainly nilpotent.

   ((ii) if and only (iii)) Now it is easy to see that a group is isomorphic to a direct product of its various p-Sylow subgroups, (p ranging over the prime divisors of the order
of G) if and only if each Sylow subgroup of G is normal. Being direct factors Sylow
subgroups of such a direct product are certainly normal. Conversely, if all p-Sylow
subgroups are normal, the direct product criterion of Chap. 3 (Lemma 3.5.1, Part (iii),
p. 96) is satisfied, upon considering the orders of the intersections $S_1 S_2 \cdots S_k \cap S_{k+1}$,
where $\{S_i\}$ is the full (one-element) collection of Sylow subgroups of G (one for each
prime).

   ((i) implies (iii)) So it suffices to show that in a finite nilpotent group, each p-sylow
subgroup P is a normal subgroup of G. If $N_G(P) = G$, there is nothing to prove. But
if $N_G(P)$ is a proper subgroup of G, then it is a proper self-normalizing subgroup by
Lemma 5.4.7. This completely contradicts Theorem 5.4.5 which asserts that every
proper subgroup of a nilpotent group is properly contained in its normalizer. $\square$

Fix a group G. A *non-generator* of G is an element x of G such that whenever x is a member of a generating set X—that is, when $x \in X$, and $\langle X \rangle = G$, then $X - \{x\}$ is also a generating set: $\langle X - \{x\} \rangle = G$. Thus x is just one of those superfluous elements that you never need. Now it is not easy to see from "first principles"—that is, *from the defining property alone*—that in a finite group, the product of a non-generator and a non-generator is again a non-generator. But we have this characterization of a non-generator:

Lemma 5.4.9 *Suppose G is a finite group. Let $D(G)$ be the intersection of all maximal subgroups of G. Then $D(G)$ is the set of all nongenerators of G.*

Proof Two statements are to be proven: (i) every non-generator lies in $D(G)$, and (ii) every element of $D(G)$ is a non-generator.

(i) Suppose x is a non-generator, and let M be any maximal subgroup of G. Then if x is not in M we must have $\langle \{x\} \cup M \rangle = G$, while $\langle M \rangle = M$, against the definition of "non-generator". Thus x lies in every maximal subgroup M and so $x \in D(G)$.

(ii) On the other hand assume that x is an arbitrary element of $D(G)$, and that X is a set of generators of G which counts x among its active participants. If $X - \{x\}$ does not generate G, then, as G is finite, there is a maximal subgroup M containing $H := \langle X - \{x\} \rangle$. But by assumption $x \in D(G) \leq M$, whence $\langle X \rangle \leq M$, a contradiction. Thus always it is the case that if $x \in X$, where X generates G, then also $X - \{x\}$ generates G. Thus x is a non-generator. $\square$

Remark This theorem really belongs to posets. Clearly finiteness can be replaced by the only distinctive property used in the proof: *that for every subset X of this poset, there is an upper bound—that is, an element of the filter P^X, which is either the 1-element of the poset or lies in a maximal member of the poset with the 1-element removed.* Its natural habitat is therefore lattices with the ascending chain condition. (We need upper semi-lattices so that we can take suprema; we need something like a lower-semilattice in order to define the Frattini element of the poset as the meet of all maximal elements. Finally the ascending chain condition makes sure that the maximal elements fully cover all subsets whose supremum is not "1". So at least lattices with the ascending chain condition (ACC) are more than sufficient for this result.)

Theorem 5.4.10 (Global non-generators) *If $X \cup D(G)$ generates a finite group G, then X generates G.*

Proof The argument is the same and has a similar poset-generalization. Suppose $\langle X \rangle$ is not G. Then it lies below a maximal subgroup M. By definition $D(G)$ must lie in M, and so $X \cup D(G)$ lies in M, which contradicts the hypothesis that $X \cup D(G)$ generates G. Thus we must abandon the foregoing supposition and conclude that $\langle X \rangle = G$. $\square$

The subgroup $D(G)$ is called the *Frattini subgroup of G*.[4] It is clearly a characteristic subgroup of G. In the case of finite groups, something special occurs which could be imitated only artificially in the context of posets:

Theorem 5.4.11 *In a finite group, the Frattini subgroup $D(G)$ is a characteristic nilpotent subgroup of G.*

Proof The "characteristic" part is clear, and has been remarked on earlier. Suppose $P \in \mathrm{Syl}_p(D(G))$. By the Frattini Argument, $G = N_G(P)D(G)$. It now follows from Theorem 5.4.10 that $N_G(P) = G$ and so in particular P is normal in $D(G)$. Thus every p-sylow subgroup of $D(G)$ is normal in it. Since it is finite, it is a direct product of its Sylow subgroups and so is nilpotent. $\square$

Corollary 5.4.12 (Wielandt) *For a finite group G the following are equivalent:*

  (i) *G is nilpotent.*
 (ii) *Every maximal subgroup of G is normal.*
(iii) *$G' \leq D(G)$.*

Proof If M is a maximal subgroup of the finite group G which also happens to be normal, then G/M is a simple group with no proper subgroups, and so must be a cyclic group of order p, for some prime p. Thus if all maximal subgroups are normal, the commutator subgroup lies in each maximal subgroup and so lies in the Frattini subgroup $D(G)$. On the other hand, if $G/D(G)$ is abelian, every subgroup above $D(G)$ is normal; so in particular, all maximal subgroups are normal. Thus Parts (ii) and (iii) are equivalent.

But if M is a maximal subgroup of a nilpotent group, M properly lies in its normalizer in G, and so $N_G(M) = G$. Thus the assertion of Part (i) implies Parts (ii) and (iii).

So it remains only to show that if every maximal subgroup is normal, then G is nilpotent. Since G is finite, we need only show that every Sylow subgroup is normal, under these hypotheses. Suppose P is a p-Sylow subgroup of G. Suppose $N_G(P) \neq G$. Then $N_G(P)$ lies in some maximal subgroup M of G. By Lemma 5.4.7, M is self-normalizing. But that contradicts our assumption that every maximal subgroup is normal. Thus $N_G(P) = G$ for every Sylow subgroup P. Now G is nilpotent by Theorem 5.4.8. $\square$

5.5 Coprime Action

Suppose G is a finite group possessing a normal abelian subgroup A, and let $F = G/A$ be the associated factor·group. Any *transversal* of A—that is, a system of coset representatives of A in G can be indexed by the elements of the factor group F. Any transversal $T := \{t_f | f \in F\}$ determines a function

---

[4] In some of the earlier literature one finds $D(G)$ written as "$\Phi(G)$". We do not understand fully the reason for the change.

$$\alpha_T : F \times F \to A$$

(called a *factor system*) by the rule that for all $(f, g) \in F$,

$$t_f \cdot t_g = t_{fg} \alpha_T(f, g).$$

In some sense, the factor system measures how far off the transversal is from being a group. Because of the associative law, one has

$$\alpha_T(fg, h)\alpha_T(f, g)^h = \alpha_T(f, gh)\alpha_T(g, h), \tag{5.9}$$

for all $f, g, h \in F$. (Notice that since A is abelian, conjugation by any element h of the coset Ax gives the same result $a^x = x^{-1}ax = h^{-1}ah$ which we have denoted by a^h.)

Now, given a transversal $T = \{t_f | f \in F\}$, any other transversal $S = \{s_f | f \in F\}$ is uniquely determined by a function $\beta : F \to A$, by the rule that

$$s_f = t_f \beta(f), f \in F.$$

Then the new factor system α_S for transversal S can be determined from the old one α_T by the equations

$$\alpha_S(f, g) = \alpha_T(f, g)[(\beta(fg))^{-1}\beta(f)^g \beta(g)] \text{ for all } f, g \in F. \tag{5.10}$$

The group G is said to *split over* A if and only if there is a subgroup H of G such that $G = HA$, $H \cap A = 1$—that is, if and only if some transversal H forms a subgroup. Such a subgroup H is called a *complement of A in G*. In that case, the factor system α_H for H is just the constant function at the identity element of A. Then, given G, A and transversal $T = \{t_f | f \in F\}$, we can conclude from Eq. (5.10) that G splits over A if and only if there exists a function $\beta : F \to A$ such that

$$\alpha_T(f, g) = \beta(fg)\beta(f)^{-g}\beta(g)^{-1}, \text{ for all } f, g \in F. \tag{5.11}$$

Now suppose A is a normal abelian p-subgroup of the finite group G and suppose P is a Sylow p-subgroup of G with the property that "P splits over A". That means that P (which necessarily contains A) contains a subgroup B such that $P = AB$ and $B \cap A = 1$. Let X be a transversal of P in G. Then $T := BX := \{bx | b \in B, x \in X\}$ is a transversal of A in G whose factor system α_T satisfies

$$\alpha_T(b, t) = 1 \text{ for all } (b, t) \in B \times T.$$

Now let m be the p-Sylow index $[G : P] = |X|$ in G. Since m is prime to $|A|$, there exists an integer k such that $mk \equiv 1 \bmod |A|$. Since A is abelian, the factors of any finite product of elements of A can be written in any order. So there is a function $\gamma : F \to A$ defined by

$$\gamma(f) := \prod_{x \in X} \alpha_T(x, f).$$

Now, multiplying each side of Eq. (5.9) as f ranges over F, one obtains

$$\gamma(h)\gamma(g)^h = \gamma(gh)\alpha_T(gh)^m.$$

Then, setting $\beta(g) := \gamma(g)^{-k}$ for all $g \in F$, we obtain Eq. (5.11). Thus G splits over A.

Noting that conversely, if G splits over A, so does P, we have in fact proved,

Theorem 5.5.1 (Gaschütz) *Suppose G is a finite group with a normal abelian p-subgroup A. Then G splits over A if and only if some p-Sylow subgroup of G splits over A.*

Next suppose A is a normal abelian subgroup of the finite group G and that H is a complement of A in G, so $G = HA$ and $H \cap A = 1$. Now suppose some second subgroup K was also a complement to A. Then there is a natural isomorphism $\mu : H \to K$ factoring through G/A, and a function $\beta : H \to A$ so that

$$\mu(h) = h\beta(h) \text{ for all } h \in H.$$

Expressed slightly differently, K is a transversal $\{k_h | h \in H\}$ whose elements can be indexed by H, and $k_h = h\beta(h)$ for all $h \in H$. Since K is a group, it is easy to see that for $h, g \in H$,

$$
\begin{aligned}
k_g \cdot k_h = h_{gh} &= (gh)\beta(gh) \\
&= g\beta(g) \cdot h\beta(h) \\
&= gh\beta(g)^h \beta(h).
\end{aligned}
$$

So

$$\beta(gh) = \beta(g)^h \beta(h), \text{ for all } g, h \in H. \tag{5.12}$$

Now assume $m = [G : A]$ is relatively prime to $|A|$, and select an integer n such that $mn \equiv 1 \bmod |A|$. Then as A is abelian, the product $\prod_H \beta(g)$, taken as g ranges over H, is a well-defined constant b, an element of A. Forming a similar product of the terms on both sides of Eq. (5.12) as g ranges over H, one now has

$$b = b^h \cdot \beta(h)^m,$$

so

$$\beta(h) = a \cdot a^{-h}, \text{ where } a = b^n, h \in H.$$

Then for every $h \in H$,

$$k_h = h\beta(h) = h \cdot a^{-h} \cdot a = a^{-1}ha.$$

Thus $K = a^{-1}Ha$ is a conjugate of H. So we have shown the following:

Lemma 5.5.2 *If A is a normal abelian subgroup of a group G of finite order, and the order of A is relatively prime to its index [G : A], then any two complements of A in G are conjugate.*

The reader should be able to use Gaschütz' Theorem and the preceeding Lemma to devise an induction proof of the following

Theorem 5.5.3 (The lower Schur-Zassenhaus Theorem) *Suppose N is a solvable normal subgroup of the finite group G whose order is relatively prime to its index [G : N]. Then the following statements are true:*

1. *There exists a complement H of N in G.*
2. *Any two complements of N in G are conjugate by an element of N.*

On the other hand, using little more than Sylow's theorem, a Frattini argument, and induction on group orders one can easily prove the following:

Theorem 5.5.4 (The upper Schur-Zassenhaus Theorem) *Suppose N is a normal subgroup of the finite group G, such that |N| and its index [G : N] are coprime. If G/N is solvable, then the following holds:*

1. *There exists a complement H of N in G.*
2. *Any two complements of N in G are conjugate by an element of N.*

Remark Both the lower and upper Schur-Zassenhaus theorems contain an assumption of solvability, either on N or G/N. However this condition can be dropped altogether, because of the famous Feit-Thompson theorem [18] which says that any finite group of odd order must be solvable. The hypothesis that $|N|$ and $[G : N]$ are coprime forces one of N or G/N to have odd order and so at least one is solvable. Of course the Feit-Thompson theorem is far beyond the sophistication of this relatively elementary course.

A somewhat over-looked application of the Schur-Zassenhaus theorem is the following:

Theorem 5.5.5 (Glauberman) *Let A be a group of automorphisms acting on a group G leaving invariant some coset Hx of a subgroup H. Then the following hold:*

1. *H is itself A-invariant.*
2. *If A and H have coprime orders (so that at least one is solvable), then A fixes an element of the coset Hx.*

Proof This is Exercise (10) in Sect. 5.6.1.

5.6 Exercises

5.6.1 Elementary Exercises

1. We say that a subgroup A of G is *subnormal in G* if and only if there is a finite ascending chain of subgroups:

$$A = N_0 < N_1 \cdots < N_m = G$$

from A to G, with each member of the chain normal in its successor—i. e. $N_i \unlhd N_{i+1}, i = 0, \ldots, m - 1$.

   (a) Using the Corollary to the Fundamental Theorem of Homomorphisms (Corollary 3.4.6), show that $\mathcal{SN}(G)$ is a semi-modular lower semi-lattice.
   (b) Use Theorem 2.4.2, part (ii), to conclude that in a group for which $\mathcal{SN}(G)$ has the descending condition—for example a finite group—the join of any collection of subnormal subgroups of a group is subnormal.

2. Suppose $\mathcal{N}$ is a family of normal subgroups of G. Let $\Sigma \mathcal{N}$ be defined to be the set of elements of G which are expressible as a *finite* product of elements each of which belongs to a subgroup in $\mathcal{N}$.

   (a) Show that $\Sigma \mathcal{N}$ is the subgroup of G generated by the subgroups in $\mathcal{N}$—i.e. $\Sigma \mathcal{N} = \langle \mathcal{N} \rangle := \langle \{N | N \in \mathcal{N}\} \rangle$.
   (b) We say that $\mathcal{N}$ is a *normally closed family*, if and only if for any non-empty subset $\mathcal{M} \subseteq \mathcal{N}$, $\langle \mathcal{M} \rangle \in \mathcal{N}$. A group is said to be a *torsion group* if and only if every element of G has finite order. (This does not mean that the group is finite: far from it: there are many infinite torsion groups.) Show that the family $\mathcal{T}(G)$ of all normal torsion subgroups of G, is a normally closed family. [HINT: When any element x is expressed as a finite product $x = a_1 \cdots a_k$ with $a_i \in N_i \in \mathcal{N}$, only a finite number of groups N_i are involved. So the proof comes down to the case $k = 2$.] Then the group

$$\text{Tor}(G) := \langle \mathcal{T}(G) \rangle \in \mathcal{T}(G)$$

   is a characteristic subgroup of G.

3. Fix a prime number p. A group H is said to be a *p-group* if and only if evey element of H has order a power of the prime p. Show that the collection of all normal p-subgroups of a (not necessarily finite) group G is normally closed. (In this case the unique maximal member of the collection of all normally closed p-subgroups of G is denoted $\mathbf{O}_p(G)$.)

4. Let G be any group, finite or infinite.

   (a) Show that any subnormal torsion subgroup of G lies in $\mathcal{T}(G)$.
   (b) Show that any subnormal p-subgroup of G lies in $\mathbf{O}_p(G)$.

5. Let $\mathbf{Q}$ denote the additive group of the rational numbers. The group $\mathbb{Z}_{p^\infty}$ is defined to be the subgroup of $\mathbf{Q}/\mathbb{Z}$ generated by the infinite set $\{1/p, 1/p^2, \ldots\}$. Show that this is an infinite p-group. Also show that it is *indecomposable*—that is, it is not the direct product of two of its proper subgroups.

6. Let $\mathcal{F}$ be a family of abstract groups which is closed under taking unrestricted subdirect products—that is, if H is a subgroup of the direct product $\prod_{\sigma \in I} X_\sigma$, with each $X_\sigma \in \mathcal{F}$, and the restriction of each canonical projection

$$\pi_\tau : \prod_{\sigma \in I} X_\sigma \to X_\tau$$

to H is an epimorphism, then $H \in \mathcal{F}$.

   (a) Show that there exists a unique normal subgroup $N_\mathcal{F}$ such that $N_\mathcal{F}$ is minimal with respect to being a normal subgroup which yields a factor $G/N_\mathcal{F} \in \mathcal{F}$[5]
   (b) Show that there is always a subgroup G' of G minimal with respect to these properties:
       (i) $G' \trianglelefteq G$,
       (ii) G/G' is abelian.
   (c) Explain why there is not always a subgroup N of G minimal with respect to being normal and having G/N a torsion group (or even a p-group). [Hint: Although torsion groups are closed under taking direct sums, they are not closed under taking direct products. This contrasts with Exercise (6) in this section where the family of groups was closed under *unrestricted* subdirect products.]
   (d) If G is finite, and $\mathcal{F}$ is closed under *finite subdirect products*—that is, the set of possibilities for $\mathcal{F}$ is enlarged because of the weaker requirement that it is enough that any subgroup H of a finite direct product of members of $\mathcal{F}$ which remains surjectve under the restrictions of the canonical projections, is still in $\mathcal{F}$—then there does exist a subgroup $N_\mathcal{F}$ which is minimal with respect to the two requirements: (i) $N_\mathcal{F} \trianglelefteq G$, and (ii) $G/N_\mathcal{F} \in \mathcal{F}$. (If π is any set of (non-negative) prime numbers, a group G is said to be a π-group

---

[5]Note that as a class of abstract groups $\mathcal{F}$ can either be viewed as either (i) a collection of groups closed under isomorphisms, or (ii) a disjoint union of isomorphism classes of groups. Under the former interpretation, the "membership sign", $\in$, as in "$G/N_\mathcal{F} \in \mathcal{F}$" is appropriate. But in the second case, "$\in$" need only be replaced by "an isomorphism with a member of $\mathcal{F}$". One must get used to this.

if and only every element of G has finite order divisible only by the primes in π. In particular, a π-group is a species of torsion group, and a p-group is a species of π-group.

Show that all π-groups are closed under finite subdirect products. [In this case we write

$$N_{\mathcal{F}} = O^{\pi}(G),\, O^{p}(G),\ \text{or}\ O^{p'}(G)$$

where $\pi = \{p\}$ or all primes except p, in the last two cases.]

7. Suppose $\mathcal{F}$ is a family of groups closed under normal products. Suppose G possesses a subgroup N that is maximal with respect to being both normal and being a member of $\mathcal{F}$ (this occurs when G is finite). Is it true that any subnormal subgroup of G which belongs to the family $\mathcal{F}$ lies in N? Prove it, or give a counterexample.

8. Suppose A and B are normal subgroups of a group G. Prove the following:

   (a) If A and B normalize subgroup C, then

   $$[AB, C] = [A, C][B, C].$$

   [Hint: Use the commutator identity (5.2) and the fact that B normalizes $[A, C]$ to show the containment of the left side in the right side.]

   (b) Show that the k-fold commutator $[AB, \ldots, AB]$ is a normal product

   $$\prod [X_1, X_2, \ldots, X_k],$$

   where the $X_i = A$ or B, and the product extends over all 2^k possibilities for the k-tuples $(X_1, \ldots, X_k)$. (Of course each factor in the normal product is a normal subgroup of G. But some of them occur more than once since the cases with $X_1 = A$ and $X_2 = B$, are duplicated in the cases with $X_1 = B$ and $X_2 = A$.)

   (c) Show that if A and B are nilpotent of class k and ℓ, respectively, then AB is nilpotent of class at most $k + \ell$. [Hint: Consider $\gamma_{k+\ell+1}(AB)$ and use the previous step.]

   (d) Prove that in a finite group, there exists a normal nilpotent subgroup which contains all other normal nilpotent subgroups of G. (This characteristic subgroup is called the *Fitting subgroup of G* and is denoted $F(G)$.)

9. Let G be a finite group. Show that if A and B are solvable normal subgroups of a group G, then their *normal product AB* is also solvable. [Hint: Use the commutator identities.] Conclude from this that if G is finite, then there exists a unique subgroup that is maximal with respect to being both normal and solvable. Clearly, it contains all other solvable normal subgroups of G.[6]

---

[6]This group is sometimes called the "solvable radical", but this usage, which intrudes on similar terminology from non-associative algebras, is far from uniform.

10. Prove Theorem 5.5.5. The finite group G contains a subgroup H, and A is a group of automorphisms of G which leaves the coset Hx invariant.

   (a) (i) Show that A leaves H invariant.
   (b) (ii) If A and H have coprime orders show that some element of Hx is fixed by the automorphism group A.

[Hint: For the first part, note that for every $a \in A$ $x^a = h_a x$ for some element $h_a \in H$. Compute $(hx)^a$ for arbitrary $h \in H$.
For the second part, note that in the semidirect product GA, A normalizes H by part (i) so $HA = AH$ is a subgroup of GA. Show that one can define a transitive action of HA on the elements of the coset Hx in which the elements of A act as they do as automorphisms, but the elements of H act by left multiplication. (One must show that the left-multiplication action of h^a is the composition of the actions of a^{-1}, of h and of a, in that order.) Let X be the subgroup of HA fixing a "letter"—say x—in Hx. Since H is transitive, $HA = HX$. Now X and A are two complements of H in the group. Apply the Schur Zassenhaus theorem.]

11. Here is an interesting simplicity criterion. Let G be a group acting faithfully and primitively on the set X, and let H be the stabilizer of some element of X. Assume

   (a) $G = G'$,
   (b) H contains a non-identity normal solvable subgroup A whose conjugates generate G.

Prove that G is simple.
[Hint: First note that $G = 1$ is not possible since $A \neq 1$. The imprimitive faithful action of G on X makes G and any non-trivial normal subgroup N of G transitive on X so $G = HN$. Then show AN is normal, and so contains all conjugates of A. The final contradiction that is around the corner must exploit the solvability of A.]

5.6.2 The Baer-Suzuki Theorem

The next few exercises spell out a subtle but well-known theorem known as the *Baer-Suzuki theorem*.[7] The proof considered here is actually due to Alperin and Lyons [2].

   In the following, G is a finite group. There is some new but totally natural notation due to Aschbacher. Suppose Σ is a collection of subgroups of G and H is a subgroup of G. Then the symbol $\Sigma \cap H$ denotes the collection of all subgroups in Σ which happened to be contained in the subgroup H. (Of course this collection could very well be empty.) If γ is any collection of subgroups of G, the symbol $N_G(\gamma)$ denotes

---

[7]Not to be confused with "Brauer-Suzuki Theorem".

the subgroup of G of all elements which under conjugation of subgroups leaves the collection γ invariant.

1. Suppose Ω is a G-invariant collection of p-subgroups of G. Let Q be any p-subgroup of G and let γ be any subset of $\Omega \cap Q$ for which $\gamma \subseteq N_G(\gamma)$. Show that either

   (a) $\gamma = \Omega \cap Q$ or
   (b) Then there is an element $X \in \Omega \cap Q$ which also lies in $N_G(\gamma)$.

[Hint: Without loss of generality, one may assume that $\gamma = \Omega \cap Q \cap N_G(\gamma)$. Then every element of $N_Q(N_Q(\gamma))$ leaves $\gamma = \Omega \cap N_Q(\gamma)$ invariant, so $N_Q(N_Q(\gamma)) = N_Q(\gamma)$. Since Q is a nilpotent group, γ is a Q-invariant collection of subgroups of Q and the conclusion follows easily.]

2. Prove the following:

Theorem 5.6.1 (The Baer-Suzuki Theorem) *Suppose X is some p-subgroup of the finite group G. Either $X \leq O_p(G)$ or $\langle X, X^g \rangle$ is not a p-subgroup for some conjugate $X^g \in X^G$.*

[HINT: Set $\Omega = X^G$, choose P a p-sylow subgroup of G which contains X, and set $\Delta := \Omega \cap P$. Since $\Delta = \Omega$ implies $\langle \Omega \rangle$ is a normal p-group, forcing the conclusion $X \in O_p(G)$, we may assume $\Omega - \Delta \neq \emptyset$.

Now show that for any subgroup $Y \in \Omega - \Delta$, $\langle Y, \Delta \rangle$ cannot be a p-group. [If R is a p-Sylow subgroup containing $\langle R, \Delta \rangle$, compare $|\Omega \cap R|$ and $|\Omega \cap P|$.]

Next consider the set $\mathcal{S}$ of all pairs (Y, γ) where $Y \in X^G - \Delta$, $\gamma \subseteq \Delta$ and $\langle Y, \gamma \rangle$ is a p-group. (We have just seen that this collection is non-empty.) Among these pairs, choose one, say $(Y, \gamma) \in \mathcal{S}$, so that $|\gamma|$ is maximal. Now if $\gamma = \emptyset$, $\langle X \rangle$ is not a p-group and we are done. Thus we may assume γ is non-empty.

Now set $Q := \langle Y, \gamma \rangle$, a p-group. Then, noting that $\langle Y, \Delta \cap Q \rangle$ is a p-group, the maximality of γ forces $\gamma = \Delta \cap Q$. Now $\bar{\gamma} := \cup_{X \in \gamma} X \subseteq P \subseteq N_G(\Delta)$ and $\bar{\gamma} \subseteq Q$ together imply

$$\bar{\gamma} \subseteq N_C(Q) \cap N_G(\Delta) \subseteq N_G(Q \cap \Delta) = N_G(\gamma).$$

By the previous exercise, either $\gamma = \Omega \cap Q$ or there is a subgroup $Y' \in \Omega \cap Q - \gamma$ with $Y' \subseteq N_G(\gamma)$. The former alternative is dead from the outset because of the existence of Y. Now Y' belongs to $X^G - \Delta$, and $\langle Y', \gamma \rangle$ is a p-subgroup, as it is a subgroup of Q. Thus (Y', γ) is also one of the extreme pairs in $\mathcal{S}$.

Similarly, since $\langle Y', \Delta \rangle$ is not a p-group, $\gamma \neq \Delta$ and so, applying the previous exercise once more with P in place of Q, there must exist a subgroup $Z \in \Delta - \gamma$, with $Z \in N_P(\gamma)$. Thus

$$\{Y', Z\} \cup \gamma \subseteq \Omega \cap N_G(\gamma).$$

Now $N_G(\gamma)$ is a proper subgroup of G (otherwise some element of X^G lies in the normal p-subgroup $\langle\gamma\rangle$). Now for the first time, we exploit induction on the group order to conclude that

$$\langle Y', \gamma \cup \{Z\}\rangle$$

is a p-group. Clearly this contradicts the maximality of $|\gamma|$, completing the proof.]

3. The proof just sketched is very suggestive of other generalizations, but there are limits. We say that a finite group is *p-nilpotent* if and only if $G/\mathbf{O}_{p'}(G)$ is a p-group.

   Show that the collection of finite p-nilpotent subgroups is closed under finite normal products. (In a finite group, there is then a unique maximal normal p-nilpotent subgroup, usually denoted $O_{p'p}(G)$.)

4. Is the following statement true? *Suppose X is a p-subgroup of the finite group G. Then either $X \in \mathbf{O}_{p'p}(G)$ or else there exists a conjugate $X^g \in X^G$ such that $\langle X, X^g\rangle$ is not p-nilpotent.* [Hint: Think about the fact that in any group, the group generated by two involutions is a dihedral group, and so is 2-nilpotent.]

Chapter 6
Generation in Groups

Abstract Here, the free group on set X is defined to be the automorphism group of a certain tree with labeled edge-directions. This approach evades some awkwardness in dealing with reduced words. The universal property that any group generated by a set of elements X is a homomorphic image of the free group on X, as well as the fact that a subgroup of a free group is free (possibly on many more generators) are easy consequences of this definition. The chapter concludes with a discussion of (k, l, m)-groups and the Brauer-Ree theorem.

6.1 Introduction

One may recall that for any subset X of a given group G, the *subgroup generated by* X, denoted $\langle X \rangle$, always has two descriptions:

1. It is the intersection of all subgroups of G which contain X;
2. It is the subset of all elements of G with are expressible as a finite product of elements which are either in X or are the inverse (in G) of an element of X.

The two notions are certainly equivalent, for the set of elements described in item 2, is closed under group multiplication and taking inverses and so, by the Subgroup Criterion (Lemma 3.2.3) is a subgroup of G containing X. On the other hand it must be a subset of every subgroup containing X. So this set is the same subgroup which was denoted $\langle X \rangle$.

But there is certainly a difference in the way the two notions *feel*. Our point of view in this chapter will certainly be in the spirit of the second of these two notions.

© Springer International Publishing Switzerland 2015

E. Shult and D. Surowski, *Algebra*, DOI 10.1007/978-3-319-19734-0_6

6.2 The Cayley Graph

6.2.1 Definition

Suppose we are given a group G and a subset X of G. X is said to be *a set of generators of G* if and only if $\langle X \rangle = G$. For any such set of generators X, we can obtain a directed graph $C(G, X)$ without multiple edges as follows:

1. The vertex set of $C := C(G, X)$ is the set of elements of G.
2. The directed edges are the ordered pairs of vertices: (g, gx), where $g \in G$ and $x \in X$. (It is our option to label such an edge by the symbol "x". Note that if x and y are distinct elements of X, then $gx \neq gy$ so each directed edge receives a unique label by this rule.)

The directed edge-labelled graph $C(G, X)$ is called the *Cayley graph of the generating set X of the group G*. Note that if the identity element 1 is a member of X, loops are possible. But of course 1, being a non-generator, can always be removed from X without disturbing any of our assumptions about X and G. If we do this, no loops will appear.

One now notices that *left* multiplication of all the vertices of C (that is, the elements of G) by a given element h of G induces an automorphism π_h of C (in fact one preserving the labels) since (hg, hgx) is also an edge labelled by "x".

Let us look at a very simple example. The set $X = \{a = (12), b = (23)\}$ is a set of generators of the group $G = \text{Sym}(3)$, the symmetric group on the set of letters $\{1, 2, 3\}$. The Cayley graph then has six vertices, and each vertex has two out-going edges labeled "a" and "b", respectively, and two in-going edges labeled "a" and "b" as well. Indeed, we could *coalesce* an outgoing and ingoing edge with the same label, as one *undirected* edge of that label. This only happens when the label indicates an *involution* of the group. In this case we get a hexagon with sides labelled "a" and "b" alternately.

6.2.2 Morphisms of Various Sorts of Graphs

Since we are talking about graphs, this may be a good time to discuss *homomorphisms of graphs*. But there are various sorts of graphs to discuss. The broad categories are

1. Simple graphs $\Gamma = (V, E)$ where V is the vertex set, and edges are just certain subsets of V of size two. (These graphs record all non-reflexive symmetric relations on a set V, for example the acquaintanceship relation among persons attending a large party.)
2. Undirected graphs with multiple edges. (This occurs when the edges themselves live an important life of their own—for example when they are the bridges connecting islands, airplane flights connecting cities, or doors connecting rooms.)

3. Simple graphs with directed edges. Here the edge set is a collection of *ordered pairs of vertices*. (This occurs when one is recording asymmetric relations, for example, which football team beat which in a football league, assuming each pair of teams plays each other at most once. Or the the comparison relation of elements in a poset. We even allow two relations between points, for example (a, b) and (b, a) may both be directed edges. This occurs in some cities where sites are connected by a network of both one-way and two-way streets.)
4. We could have any of the above, with labels attached to the edges. Cayley graphs, for example, can be viewed as directed labeled graphs.

A *graph homomorphism* $f : (V_1, E_1) \to (V_2, E_2)$ is a mapping, $f : V_1 \to V_2$ such that if $(a, b) \in E_1$ then either $f(a) = f(b)$ or $(f(a), f(b)) \in E_2$. If the two graphs are labeled, there are labelling mappings $\lambda_i : E_i \to L_i$ assigning to each (directed) edge, a label from a set of labels L_i, $i = 1, 2$. Then we require an auxilliary mapping $f' : L_1 \to L_2$, so that if $(a, b) \in E_1$ carries label x and if $(f(a), f(b))$ is an edge, then it too carries label $f'(x)$.

6.2.3 Group Morphisms Induce Morphisms of Cayley Graphs

Now suppose $f : G \to H$ is a surjective homomorphism of groups. Then it is easy to see that if X is a set of generators of G, then also $f(X)$ is a set of generators of $f(G) = H$. Now f is *a fortiori* a map from the vertex set of the Cayley graph $C(G, X)$ to the vertex set of the Cayley graph $C(H, f(X))$. For $(g, x) \in G \times X$, (g, gx) is a directed edge labelled x. Then clearly either $f(g) = f(gx)$ or else, $(f(g), f(gx)) = (f(g), f(g)f(x))$ is a directed edge of $C(H, f(X))$, labelled $f(x)$. Thus

Lemma 6.2.1 *If X is a set of generators of a group G and $f : G \to H$ is a surjective morphism of groups, then $f(X)$ is a set of generators of H and f induces a homomorphism of the Cayley graphs, $f : C(G, X) \to C(H, f(X))$ as labeled directed graphs.*[1]

6.3 Free Groups

6.3.1 Construction of Free Groups

The Cayley graphs of the previous section were defined by the *existence* of a group G (among other things). Now we would like to reverse this; we should like to start with a set X and obtain out of it a certain group which we shall denote $F(X)$.

---

[1]Note that the graph morphism has also denoted by the symbol f. This slight abuse of notation is motivated by the fact that f is indeed the mapping being applied to both vertices and labels of the domain Cayley graph.

We do this in stages.

First we give ourselves a fixed abstract set X. The *free monoid on X* is the collection $M(X)$, of "words" $x_1 x_2 \cdots x_k$ spelled with "letters" x_i chosen from the alphabet X.[2] The number k is allowed to be zero—that is, there is a unique "word" spelled with no letters at all—which we denote by the symbol ϕ and call the *empty word*. The number k is called the *length of the word* which, as we have just seen, can be zero.

The *concatenation* of two words $u = x_1 \cdots x_k$ and $w = y_1 y_2 \cdots y_\ell$, $x_i, y_j \in X$, is the word,

$$u * w := x_1 \cdots x_k y_1 \cdots y_\ell.$$

The word "bookkeeper" is the concatenation of "book" and "keeper"; 12345 is the concatenation $123 * 45$ or $12 * 345$, and so forth. Clearly the concatenation operation "$*$" is a binary operation on the set $W(X)$ of all words over the alphabet X, and it is associative (though very non-commutative). Moreover, for any word $w \in W(X)$, $\phi * w = w * \phi = w$. So the empty word ϕ is a two-sided identity with respect to the operation "$*$". Thus $M(X) = (W(X), *)$ is an associative semigroup with identity—that is, a monoid. Indeed, since it is completely determined by the abstract set X alone, it is called the *free monoid on (alphabet) X*.[3]

Now let $\sigma : X \to X$ be an *involutory fixed-point-free permutation of X*. This means σ is a bijection from X to itself such that

1. $\sigma^2 = 1_X$, the identity mapping on X, and
2. $\sigma(x) \neq x$ for all elements x in X.

We are going to construct a directed labelled graph $\Gamma = F(X, \sigma)$, called a *frame* which is completely determined by X and the involution σ.

Let $W^*(X)$ be the subset of $W(X)$ of those words $y_1 y_2 \cdots y_r$, r any non-negative integer, for which $y_i \neq \sigma(y_{i+1})$, for any $i = 1, \ldots, r - 1$. These are the words which never have a "factor" of the form $y\sigma(y)$. (Notice that these words contain no factors of the form $\sigma(y)y$ as well since, $\sigma(y)y = u\sigma(u)$ where $u = \sigma(y)$.) We call such words *reduced words*.

We now construct a labelled directed graph Γ whose vertices are these reduced words in $W^*(X)$. The set E of oriented labelled edges are of two types:

1. Ordered pairs $(w, wy) \in W^*(X) \times W^*(X)$ are to be directed edges labelled "y". (Note that the nature of the domain forces the reduced word w not to end in the letter $\sigma(y)$—otherwise wy would not be reduced.)
2. Similarly those ordered pairs of the form (wy, w) (under the same hypothesis that w not end in $\sigma(y)$) are directed edges labeled by "$\sigma(y)$".

The two types of directed edges are distinguished this way: for an edge of the first kind, the "head" of the directed edge is a word of length one more than it's "tail", while the reverse is the case for an edge of the second kind.

---

[2]Of course, to be formal about it, one can regard these as finite sequences $(x_1, \ldots, x_k)$, but the analogy with linguistic devices is helpful because the binary operation we use here is not natural for sequences.

[3]Monoids were introduced in Chap. 1.

Then the directed graph $\Gamma = (W^*(X), E)$ which we have defined and have called a *frame*, has these properties:

1. Let V_k be the set of words of $W^*(X)$ of length k. Then $V_0 = \{\phi\}$, and the vertex set partitions as $W^*(X) = V_0 + V_1 + V_2 + \ldots$.
2. Each vertex has all its outgoing edges labeled by X; it also has all its ingoing edges labeled by X. Each directed edge (a, b) that is labeled by $y \in X$, has its "transpose" (b, a) labelled by $\sigma(y)$.
3. For $k > 0$, each vertex in V_k receives just one ingoing edge from V_{k-1} and gives one outgoing edge to the same vertex in V_{k-1}. All further edges leaving or arriving at this vertex, are to or from vertices of V_{k+1}.
4. Given any ordered pair of vertices (w_1, w_2) (recall the w_i are reduced words) there is a unique directed path in Γ from vertex w_1 to vertex w_2. It follows that there are no circuits other than "backtracks"—that is circular directed walks of the form

$$(v_1, v_2, \ldots v_m, \ldots v_{2m+1})$$

where $v_j = v_{2m+1-j}$, (v_j, v_{j+1}) and (v_{j+1}, v_j) are all directed edges, $j = 1, \ldots m$.

As usual, the symbol $\mathrm{Aut}(\Gamma)$ denotes the automorphism group of the directed labeled graph Γ. At this point we define a monoid homomorphism

$$\mu : M(X) \to \mathrm{Aut}(\Gamma).$$

For each $y \in X$, and vertex w in $W^*(X)$ define

$$\mu(y)(w) := \begin{cases} yw & \text{if } w \text{ does not begin with letter } \sigma(y) \\ w' & \text{if } w \text{ has the form } \sigma(y)w'. \end{cases}$$

It is then quite clear that $\mu(y)$ is a label-preserving automorphism of Γ (this needs only be checked at vertices w of very short length where "cancellation" affects the terminal vertices).

Notice that by definition

$$\mu(\sigma(y)) = (\mu(y))^{-1},$$

the inverse of the automorphism induced by $\sigma(y)$.

Next, for any word $m = y_1 y_2 \cdots y_k$ in the monoid $M(X)$ we set

$$w^{\mu(m)} := \mu(m)(w) := \mu(y_k)(\mu(y_{k-1})(\cdots (\mu(y_1)(w)) \cdots) = w^{\mu(y_1)\cdots\mu(y_k)};$$

that is, $\mu(m)$ is the composition

$$\mu(y_k) \circ \cdots \circ \mu(y_1)$$

of automorphisms of Γ. Of course, since the automorphism group $\mathrm{Aut}(\Gamma)$ is assumed to act as right operators, we can express this composition as $\mu(y_1) \cdots \mu(y_k)$ where juxtaposition in this expression denotes multiplication in the group $\mathrm{Aut}(\Gamma)$.

Then, $\mu(m_1 m_2) = \mu(m_1)\mu(m_2)$ for all $(m_1, m_2) \in M(X) \times M(X)$ and $\mu(\phi) = 1_{W^*(X)}$. So μ is a monoid homomorphism

$$\mu : W(X) \to \mathrm{Aut}(\Gamma).$$

Note that μ is far from being 1-to-1. For each $y \in X$, $\mu(y\sigma(y)) = 1_\Gamma$, the identity automorphism on Γ. Thus all compound expressions derived from the empty word by successively inserting factors of the form $y\sigma(y)$—such as $ab\sigma(c)de\sigma(e)\sigma(d)cfg\sigma(g)\sigma(f)\sigma(b)\sigma(a)$–also induce the identity automorphism of Γ.

We now have

Lemma 6.3.1 *Let $G = Aut\ (\Gamma)$, the group of all label preserving automorphisms of the directed graph Γ.*

(i)   *The monoid homomorphism $\mu : M(X) \to G$ is onto.*

(ii) *G acts regularly on the vertices of Γ.*

(iii) *The group G is generated by the set $\mu(Y)$, where Y is any system of representatives of the σ-orbits on X.*

(iv) *Identities of the form $\mu(m) = 1_\Gamma$ occur if and only if the word $m \in M(X)$ is formed by a sequence of insertions of factors $y_i \sigma(y_i)$ starting with the empty word.*

(v)   *Γ is the Cayley graph of G with respect to the set of generators X.*

Proof Suppose g is an element of G fixing a vertex w in Γ. Then, as each (in- or out-) edge on w bears a unique label, every vertex next to w is fixed. Since every vertex u is connected to the vertex ϕ, the empty word in $W^*(X)$, by an "undirected path" of length equal to the length $\ell(u)$, the undirected version of Γ is connected. This forces the elements g in the first sentence of this paragraph to fix all vertices $W^*(X)$ of Γ.

But now it is clear that for any word $w \in W^*(X)$, regarded as a monoid element, $\mu(w)$ is an element of G taking vertex ϕ to w. Thus $\mu(M(X)) = G$, and 1. and 2. are proved. So is the statement that $\mu(X)$ is a set of generators of G in part (iii).

Suppose now, that $w = y_1 y_2 \cdots y_k \in M(X)$, satisfies $\mu(w) = 1_\Gamma$. Then successive premultiplications by y_k, then y_{k-1}, and so on, take the vertex ϕ on a journey along a directed path with successive arrows labelled $y_k, y_{k-1}, \ldots$; eventually returning to ϕ via a directed edge (y_1, ϕ). At each stage of this journey, the current image of ϕ is either taken to a vertex that is one length further or else one length closer to its original "home", vertex ϕ. So there is a position

$$v = \mu(y_j y_{j+1} \cdots y_k)(\phi)$$

in which a local maximum occurs in the distance from "home". This means that both $v^- = \mu(y_{j+1} \cdots y_k)(\phi)$ and $v^+ = \mu(y_{j-1} y_j \cdots y_k)(\phi)$ are one unit closer to the starting vertex ϕ than was v. But as pointed out in the construction of Γ, each vertex z at distance d from ϕ has only one out-edge and one in-edge to a vertex at distance $d - 1$ from ϕ, and the two closer vertices are the same vertex—say z_1—so that the two edges in and out of the vertex z are two orientations of an edge on the same pair of vertices $\{z, z_1\}$. It follows therefore that $v^- = v^+$ and that y_j, the label of directed edge (v^-, v), is the σ-image of the label y_{j-1} of the reverse directed edge (v, v^+). Thus

$$\begin{aligned}
\mu(w) &= \mu(y_1 y_2 \cdots y_{j-2} y_{j-1} y_j y_{j+1} \cdots y_k) \\
&= \mu(y_1 \cdots y_{j-1} y_{j+1} \cdots y_k) \\
&= 1_\Gamma
\end{aligned} \tag{6.1}$$

Now one applies induction on the length of $w' := y_1 \cdots y_{j-2} y_{j+1} \cdots y_k$, to complete the proof of (iii).

Part (iv) is obvious. $\square$

Now let Y be a system of representatives of the σ-orbits on X. Then we have

$$X = Y \cup \sigma(Y), \text{ and } Y \cap \sigma(Y) = \emptyset.$$

The group G above is canonically defined (one sees) by the set Y alone. To recover X one need only formally define X to be two disjoint copies of Y, and invoke any bijection $Y \to Y$ to define the involutory permutation $\sigma : X \to X$.

Moreover, we see from Part 3 of Lemma 6.3.1 that the only relations that can exist among the elements of G, are consequences of the relations $\mu(y) \circ \mu(\sigma(y)) = 1$, $y \in X$. Accordingly the group G is called the *free group on the set of generators Y* and is denoted $F(Y)$.[4]

Remark

1. (Uniqueness.) As remarked $F(Y)$ is uniquely defined by Y, so if there is a bijection $f : Y \to Z$, then there is an induced group isomorphism $F(Y) \simeq F(Z)$
2. For any frame $\Phi = F(Z, \zeta)$, $\text{Aut}(\Phi)$ is a free group $F(Z_0)$, for any system Z_0 of representatives of the ζ-orbits on Z.
3. (The inverse mapping on Γ.) Among the reduced words comprising the vertices of Γ, there is a well-defined involutory "inverse mapping"

$$\text{superscript -1} : W^*(X) \to W^*(X)$$

---

[4]Although this is the customary notation, unfortunately, it is not all that distinctive in mathematics. For example $F(X)$ can mean the field of rational functions in the indeterminates X over the field "F", as well as a myriad of meanings in Analysis and Topology.

which fixes only the empty word ϕ and takes the reduced word $w = y_1 y_2 \cdots y_k$ to the reduced word $w^{-1} := \sigma(y_k)\sigma(y_{k-1}) \cdots \sigma(y_1)$.

Thus the group-theoretic inverse mapping $\mu_w \to (\mu_w)^{-1}$ is extended to the vertices of Γ by the rule that $(\mu_w)^{-1} := \mu_{w^{-1}}$: that is, the inverse mapping of group elements induces a similar mapping on $W^*(X)$ via the bijection μ from the regular action of $G = \mathrm{Aut}(\Gamma)$ on the vertices of Γ.

Thus without ambiguity, we may write y^{-1} for $\sigma(y)$ and Y^{-1} for $\sigma(Y)$, so $Y \cap Y^{-1} = \emptyset$.

(It is this mapping on X which defines the graph Γ; there is no mention of a special system of representatives of σ-orbits. Thus if Y' is any other such system, then $F(Y) = F(Y')$ with "equals", rather than the weaker isomorphism sign.)

Corollary 6.3.2 *Any subgroup of a free group is also a free group.*

Proof Let $F(Y)$ be the free group on the set Y, let $X = Y \cup Y^{-1}$, and let Γ be the Cayley graph of $F(Y)$ with respect to the set of generators X. We used Γ to *define* $F(Y)$, for by the preceding lemma $F(Y)$ is the group of all label-and-direction-preserving automorphisms of Γ.

Let H be a non-trivial subgroup of $F(Y)$ and let Γ^H be the H-orbit on vertices of Γ which contains the vertex ϕ, the empty word. Now each vertex in Γ^H is a reduced word w and there is a unique directed path from ϕ to that vertex; moreover the sequence of labels from Y encountered along this directed path, are in fact the letters which in their order "spell" the reduced word w comprising this vertex. An *H-neighbor* of ϕ, is a vertex w of Γ^H such that along the unique directed path in Γ from ϕ to w, no intermediate vertex of Γ^H is encountered. Let Y^H denote the collection of all H-neighbors of ϕ. Note from the remarks above, and the fact that H is a subgroup of $F(X)$, that if the reduced word w is an H-neighbor of ϕ, then so is w^{-1}.

Now we construct a new labelled graph on the vertex set Γ^H. There is a directed edge $(x, y) \in \Gamma^H \times \Gamma^H$ if and only if $x = yw$ where yw is reduced and $w \in Y^H$. (This means $x = \mu(y)(w)$.) In a sense this graph can be extracted from the original graph Γ in the following way: If we think of Γ^H as embedded in the graph Γ, we are drawing a directed edge from x to y labelled w if an only if the reduced word formed by juxtaposing the labels of the successive edges along the unique directed path in Γ from x to y is w and w is in Y^H.

Now the graph Γ^H has as its vertices elements of $W^*(Y^H)$. There is a unique directed path in Γ^H connecting any two vertices. Also for every two vertices there are no directed edges or exactly two directed edges, (x, y) and (y, x) respectively labelled w and w^{-1} for a unique $w \in Y^H$. Thus the graph Γ^H has all of the properties that Γ had except that X has been replaced by Y^H. Thus as H is a regularly acting orientation- and label-preserving group of automorphisms of Γ^H, $H \simeq F(Y^H_+)$ where $Y^H = Y^H_- + Y^H_+$ is any partition of Y^H such that for each $w \in Y^H$, exactly one representative of $\{w, w^{-1}\}$ lies in Y^H_+. (Notice that we have used the inverse map on vertices of Γ in describing Γ^H.) Thus the graph Γ^H is a frame $F = F(Y^H, E)$, and H, being transitive, is by Lemma 6.3.1 the free group on Y^H_+. $\square$

Remark If X is finite with at least two elements, it is possible that the subgroup H is a free group on infinitely many generators (see Exercise (1) in Sect. 6.5.1).

6.3.2 The Universal Property

We begin with a fundamental theorem.

Theorem 6.3.3 *Suppose G is any group, and suppose X is a set of generators of G. Then there is a group epimorphism $F(X) \to G$, from the free group on the set X onto our given group G, taking X to X identically.*

Proof Here X is simply a set of generators of the group G, and so a subset of G. But a free group $F(X)$ can be constructed on any set, despite whatever previous roles the set X had in the context of G. One constructs the formal set $Y = X + X^{-1} = $ two disjoint copies of X, with the notational convention that the exponential operator -1 denotes both a bijection $X \to X^{-1}$ and its inverse $X^{-1} \to X$ so that $(x^{-1})^{-1} = x$. Recall that any arbitrary element of the free group $F(X)$ was a *reduced word in the alphabet Y*—that is, an element of $W^*(Y)$, consisting of those words which do *not* have the form $w_1 y y^{-1} w_2$ for any $y \in Y$. (Recall that these were the vertices of the Cayley graph $\Gamma = \Gamma(Y)$ which we constructed in the previous subsection.)

Now there is a mapping f which we may think of as our "amnesia-recovering morphism". For each word $w = u_1 \cdots y_k \in W^*(Y)$, $f(w)$ is the element of G we obtain when we "remember" that each y_i is one of the generators of G or the inverse of one. Clearly $f(w_1 w_2) = f(w_1) f(w_2)$, since this is how elements multiply when we remember that they are group elements. Then f is onto because every element of G can be expressed as a finite product of elements of X and their inverses. There is really nothing more to prove. $\square$

One should bear in mind that the epimorphism of the theorem completely depends on the pair (G, X). If we replace X by another set of generators, Y—possibly of smaller cardinality—one gets an entirely different epimorphism: $F(Y) \to G$.

6.3.3 Generators and Relations

Now consider the same situation as in the previous subsection: A group G contains a set of generators X, and there is an epimorphism $f := f_{G,X} : F(X) \to G$, taking

X to X "identically". But there was on the one hand, a directed labelled graph Γ (the "frame") that came from the construction in Lemma 6.3.1. This was seen to be the Cayley graph $C(F(X), X)$. Since $X \cup X^{-1}$ is also a set of generators of G, one may form the Cayley graph $C(G, X \cup X^{-1})$. Then by Lemma 6.2.1, there is a label-preserving vertex-surjective homomorphism of directed labelled graphs

$$f : \Gamma \to C(G, X \cup X^{-1}),$$

from the frame $\Gamma = F(X \cup X^{-1}, -1)$ to the Cayley graph.

Where Γ had no circuitous directed walks other than backtracks, the Cayley graph $C(G, X \cup X^{-1})$ may now have directed circuits. In fact it must have them if f is not an isomorphism. For suppose $f(w_1) = f(w_2)$. If $w_1 \neq w_2$, there is in Γ a *unique directed walk from* w_1 to w_2. Its image is then a directed circuit in $C(G, X \cup X^{-1})$.

Since G acts regularly on its Cayley graph by left multiplication, every directed circular walk in $C(G, X U X^{-1})$ can be translated to one that begins and ends at the identity element 1 of G. Ignoring the peripheral attached backtracks, we can take such a walk to be a circuit

$$(1, x_1, x_1 x_2, x_1 x_2 x_3, ..., x_1 x_2 \cdots x_k = 1), x_i \in X \cup X^{-1},$$

where $x_1 x_2 \cdots x_k$ is *reduced* in the sense that for $i = 1, \ldots, k - 1$, $x_i \neq x_{i+1}^{-1}$. The equation

$$x_1 \cdots x_k = 1$$

where a reduced word in the generators is asserted to be 1, is called a *relation*. In a set of relations, the words that are *being* asserted to be the identity are called *relators*. Obviously, the distinction is only of grammatical significance since the relators determine the relations and vice versa.

Let $\mathcal{R}$ be a set of relators in the set of generators X; precisely, this means that $\mathcal{R}$ is a set of reduced words in $W^*(X \cup X^{-1})$. Then the normal closure of $\mathcal{R}$ in the free group $F(X)$, is the subgroup $\langle \mathcal{R}^{F(X)} \rangle_{F(X)}$, generated by the set of all conjugates in $F(X)$ of all relators in $\mathcal{R}$. Clearly, this is the intersection of all normal subgroups of $F(X)$ which contain $\mathcal{R}$. A group G is said to be *presented by the generators X and relations $\mathcal{R}$*, if and only if

$$G \simeq F(X)/\langle \mathcal{R}^{F(X)} \rangle_{F(X)}.$$

This is expressed by writing

$$G = \langle X | \mathcal{R} \rangle.$$

This is the "free-est" group which is generated by a set X and satisfying the relations $\mathcal{R}$ in the very real sense that any other group which satisfies these relations is a homomorphic image of the presented group. Stated more precisely we have

Theorem 6.3.4 *Let* $\langle X|\mathcal{R}\rangle$ *be a presented group, and let* $\phi : X \to H$ *be a function, where* H *is a group with identity element* e. *Then* ϕ *extends uniquely to a homomorphism* $\langle X|\mathcal{R}\rangle \to H$ *if and only if*

$$\phi(x_1)\phi(x_1)\cdots\phi(x_r) = e$$

whenever $x_1 x_2 \cdots x_r \in \mathcal{R}$.

In writing out a presentation, the collection $\mathcal{R}$ of words declared to be equal to the identity element can be replaced by equations that are equivalent to such declarations. For example, if $\mathcal{R} = \{xyx, x^2yx\}$ one can replace $\mathcal{R}$ by equations $\{yx^2 = 1, x^2y = x^{-1}\}$, or $\{xy = x^{-1}, x^3y = x^{-1}\}$. Let us look at some examples:

Example 29 (The infinite dihedral group D_∞) $G = \langle x, y | x^2 = y^2 = 1 \rangle$. As you can see we allow some latitude in the notation. The set of generators here is $X = \{x, y\}$; but we ignore the extra wavy brackets in displaying the presentation. The set of relators is $\mathcal{R} = \{x^2, y^2\}$, but we have actually written out the explicit relations.

This is a group generated by two involutions x and y. Nothing is hypothesized at all about the order of the product xy, so it generates an infinite cyclic subgroup $N = \langle xy \rangle$. Now setting $u := xy$, the generator of N, we see that $yx = u^{-1}$ and that $x(xy)x = yx$ so $xux = u^{-1}$. Similarly $yuy = u^{-1}$. Moreover, $uy = x$ shows $Ny = Nx$. Thus

$$xN = Nx = Ny = yN$$

and so N is a normal cyclic subgroup of index 2 in G. One of the standard models of this group is given by its action on the integers.

Example 30 (The dihedral group of order 2n, denoted D_{2n}) Here $G = \langle x, y | x^2 = y^2 = (xy)^n = 1 \rangle$. Everything is just as above except that the element $u = xy$ now has order n. This means the subgroup $N = \langle u \rangle$ (which is still normal) is cyclic of order n, so $G = N + xN$ has order $2n$. Note that this group is a homomorphic image of the group of Example 29. When $n = 2$, the group is an elementary abelian group of order 4—the so-called *Klein fours group*.

Example 31 Consider the following fairly simple presentation: $G = \langle x, y | xy = y^2x, yx = x^2y \rangle$. It follows that

$$y^{-1}xy = y^{-1}y^2x = yx = x^2y = xxy,$$

so that $y^{-1} = x$ (after right multiplying the extreme members of the above equation by $(xy)^{-1}$). But then

$$e = xy = y^2x = y(yx) = y,$$

so $y = e$, implying $x = e$. In other words, the relations imposed on the generating elements of G are so destructive that the group defined is actually the trivial group.

The general question of calculating the order of a group presented by generators and relations, is not only difficult, but in certain instances can be shown to be an impossible task. (This is a consequence of Boone's theorem on the the the so-called *word problem in group theory*.)

Example 32 (*Polyhedral or (k, l, m)-groups*) Here k, l, and m are integers greater than one. The presentation is

$$G := \langle x, y | x^k = y^l = (xy)^m = 1 \rangle.$$

The case that $k = l = 2$ is the dihedral group of order $2m$ of Example 30, above. We consider a few more cases:

(1) THE CASE $(k = 2, m = 2)$:   Here $u = xy$ has order 2. But then x and u are involutions, and the product $xu = y$ has order l. So G is a homomorphic image of the dihedral group of order $2l$, namely:

$$\langle x, u | x^2 = u^2 = (xu)^l \rangle \simeq D_{2l}.$$

It is not difficult to see that this homomorphism is an isomorphism.

(2) THE CASE $(k = 2, l = 3, m = 3)$:   Here the relations are

$$x^2 = y^3 = (xy)^3 = 1.$$

There is a group satisfying the relations. The alternating group on four letters contains a permutation $x = (12)(34)$ of order 2, an element $y = (123)$ of order 3 for which the product $u = (12)(34)(123) = (134)$ has order three, and these elements generate Alt(4). Thus there is an epimorphism of the presented $(2, 3, 3)$-group onto Alt(4). But the reader can check that the relations imply

$$x(yxy^2) = y^2xy.$$

The left side is the product of two involutions, while the right side is an involution. It follows that the involutions x and $x_1 = yxy^2$ commute. But now the equations just presented show that

$$x^{y^2} = x_1, \text{ and } x^y = xx_1.$$

Thus the abelian fours group $K := \langle x, x_1 \rangle$ is normalized by y, so the presented $(2, 3, 3)$-group G has coset decomposition

$$G = K + yK + y^2K,$$

and so has order 12.
It follows that G is isomorphic to Alt(4).

(3) THE CASE $(k = 2, l = 3, m = 4)$:   Here $x^2 = y^3 = (xy)^4 = 1$, so $xyxy = y^{-1}xy^{-1}x$ and $yxyx = xy^{-1}xy-1$. Now conjugation by x inverts $y^{-1}y^x$ since

$$(y^{-1}y^x) = (y^{-1})^x \cdot y = [y^{-1} \cdot ((y^{-1})^x)^{-1}]^{-1} = (y^{-1}y^x)^{-1}.$$

But

$$y^{-1}y^x = y^2 \cdot xyx = y(yxyx) = y(xy^{-1}xy^{-1}) = (xy^{-1})^{-1} \cdot y^{-1} \cdot (xy^{-1})$$

and is a conjugate of y^{-1} and so has order 3. But yx is conjugate to xy, so yx has order four. Summarizing,

(i) y has order 3,
(ii) $yxyx$ has order 2, and
(iii) $y \cdot yxyx = y^2xyx$ has order 3.

Thus,

$$H := \langle y, yxyx \rangle$$

is a $(2, 3, 3)$-group, and so by the previous case, has order at most 12. But obviously

$$G = \langle x, y \rangle = \langle x, H \rangle = H + Hx,$$

so $|G| = 24$. But in the case of $\mathrm{Sym}(4)$, setting $x = (12)$ and $y := (134)$, we see that there is an epimorphism

$$G \to \mathrm{Sym}(4).$$

The orders force[5] this to be an isomorphism, whence

$$\text{any } (2, 3, 4)\text{-group is } \mathrm{Sym}(4). \tag{6.2}$$

(4) THE CASE $(k, l, m) = (2, 3, 5)$:   So here goes. We certainly have

$$G = \langle x, y | x^2 = y^3 = (xy)^5 = 1 \rangle.$$

Again set $z := xy$. Then $y = xz$ has order three.
Then for x and z, we have the relations

$$x^2 = z^5 = 1. \tag{6.3}$$
$$(xz)^3 = 1. \tag{6.4}$$

---

[5]Of course this could also be worked out by showing that the Cayley graph must complete itself in a finite number of vertices. But the reasons seem equally *ad hoc* when done this way.

But the latter implies

$$xzx = z^{-1}xz^{-1} \tag{6.5}$$

$$xz^{-1}x = zxz. \tag{6.6}$$

Now set $t = zxz^{-1}$, an involution. Then

$$(ty)^3 = (zxz^{-1} \cdot xz)^3 = (z(zxz)z)^3 = (z^2xz^2)^3 \tag{6.7}$$

$$= z^2xz^4xz^4xz^2 = z^2x(xzx)xz^2 = z^5 = 1. \tag{6.8}$$

Thus $H = \langle t, y \rangle$ is a $(2, 3, 3)$-subgroup of order divisible by 6, and so is Alt(4). It contains a normal subgroup K with involutions $t, r := t^y$ and t^{y^2}, and setting $Y = \langle y \rangle$, one has $H = Y + YrY$.

Next let s be the involution $z^{-1}(t^y)z$, which, upon writing t and y in terms of x and z becomes

$$s = z^{-1}xz^3xz.$$

Now

$$sy = z^{-1}xz^3xz \cdot xz = (xz)^{-1}(z^3xz^2)(xz)$$
$$= (xz)^{-1}z^{-1}(z^{-1}xz)z \cdot xz$$

is a conjugate of x and so has order 2. Thus s inverts y.
Also

$$sr = st^y = (z^{-1}xz^3xz)(xz^3xz)$$
$$= (xz)^{-1}z^3xzxz^3(xz)$$
$$= (xz)^{-1}z^3(z^{-1}xz^{-1})z^3(xz)$$
$$= (xz)^{-1}(z^2xz^2)(xz)$$

is conjugate to z^2xz^2 which has order 3 by Eq. 6.7. Consider now the collection

$$M = KHK = H + HsH = H + Hs + Hst + Hst^y + Hst^{y^2},$$

a set of 60 elements closed under taking inverses. We show that M is closed under multiplication. Using the three results

$$M = H + HsH$$
$$H = Y + YrY$$
$$sY = Ys$$

we have $MM \subseteq M$ if and only if

$$srs \in M.$$

But this is true since $(sr)^3 = 1$ implies

$$srs = rsr \in HsH \subseteq M.$$

Finally, to get $G = M$ we must show that at least two of the three generators x, y or z lies in M. Now M already contains y, and

$$s = z^{-1}xz^3xz = (xz)^{-1}z^3(xz) = y^{-1}z^3y,$$

and so M contains z^3 and $z = (z^3)^2$. Now, as $|G| = 60$ and G contains a subgroup of index 5, there is a homomorphism $\rho : G \to \mathrm{Sym}(5)$ whose image has order at least $30 = \mathrm{lcm}\{2, 3, 5\}$. Because of the known normal strucure of $\mathrm{Sym}(5)$ obtained in the previous chapter, the only possibility is that ρ is a monomorphism and that $G \simeq \mathrm{Alt}(5)$.

Again, this could have been proved using the Cayley graph. Neither proof is particularly enlightening.[6]

Example 33 (*Coxeter Groups*) This class covers all groups generated by involutions whose only further relations are the declared orders of the pairwise products of these involutions. First we fix abstract set X. A *Coxeter Matrix M over X* is a function $M : X \times X \to \mathbb{N} \cup \{\infty\}$ which associates with each ordered pair $(x, y) \in X \times X$, a natural number $m(x, y)$ or the symbol "∞" so that (1) $m(x, x) = 1$ and (2) that $m(x, y) = m(y, x)$—that is, the matrix is "symmetric".[7]

The *coxeter group W(M)* is then defined to be the presented group

$$W(M) := \langle X | \mathcal{R} \rangle.$$

where $\mathcal{R} = \{(xy)^{m(x,y)} | (x, y) \in X \times X\}$. Thus $W(M)$ is generated by involutions (because of the 1's on the 'diagonal' of M). If X is non-empty, the group is always non-trivial. In 1934, H M. S. Coxeter [6] classified the M for which $W(M)$ is a finite group. These finite Coxeter groups have an uncanny way of appearing all over a large part of important mathematics: Lie Groups, Algebraic Groups, and Buildings, to name a few areas.

---

[6] The real situation is basically topological. It is a fact that a (k, l, m)-group is finite if and only if

$$1/k + 1/l + 1/l > 1.$$

The natural proof of this is in terms of manifolds and curvature. Far from discouraging anyone, such a connection shows the great and mysterious "interconnectedness" of mathematics!

[7] Of course if X is finite, it can be totally ordered so that the data provided by M can be encoded in an ordinary matrix. We just continue to say "matrix" in the infinite case—perhaps out of love of analogy.

Remark As remarked on p. 174, there are many instances in which the choice of relations, $\mathcal{R}$ may very well imply that the presented group has order only 1. For example if $G := \langle x | x^3 = x^{232} = 1 \rangle$ then G has order 1. In fact, if we were to select at random an arbitrary set of relations on a set X of generators, the odds are that the presented group has order 1. From this point of view, it seems improbable that a presented group is *not* the trivial group. But this is illusory, since any non-trivial group bears *some* set of relations, and so the presented group with these relations is also non-trivial in these cases. So most of the time, the non-trivial-ness of a presented group is proved by finding an object on which the presented group can act in a non-trivial way.[8] Thus a Coxeter group $W(M)$ is known not to be trivial for the reason that it can be faithfully represented as an orthogonal group over the real field, preserving a bilinear form $B := B_M : V \times V \to \mathbb{R}$ defined by M. Thus Coxeter Groups are non-trivially acting groups, in which each generating involution acts as a reflection on some real vector space with some sort of inner product. That is the trivial part of the result. The *real* result is that the group acts *faithfully* in this way. This means that *there is no product of these reflections equal to the identity transformation unless it is already a consequence of the relations given in M itself.* This is a remarkable theorem.

6.4 The Brauer-Ree Theorem

If G acts on a finite set Ω, and X is any subset of G, we let $\nu(X)$ be defined by the equation

$$\nu(X) := \sum\nolimits_{\Gamma_i = \langle X \rangle\text{-orbit}} (|\Gamma_i| - 1),$$

where the sum ranges over the $\langle X \rangle$-orbits Γ_i induced on Ω. Thus if 1 is the identity element, $\nu(1) = 0$, and if t is an involution with exactly k orbits of length 2, then $\nu(t) = k$.

Theorem 6.4.1 (Brauer-Ree) *Suppose $G = \langle x_1, \ldots, x_m \rangle$ acts on Ω and $x_1 \cdots x_m = 1$. Then*

$$\sum\nolimits_{i=1}^{m} \nu(x_i) \geq 2\nu(G).$$

Proof Let $\Omega = \Omega_1 + \cdots + \Omega_k$ be a decomposition of Ω into G-orbits, and for any subset X of G, set

$$\nu_i(X) = \sum\nolimits_{\Gamma_{ij} = \langle X \rangle\text{-orbits in } \Omega_i} (|\Gamma_{ij}| - 1),$$

---

[8]It is the perfect analogue of Locke's notion of "substance": something on which to hang properties.

so $\nu(X) = \sum \nu_i(X)$. By induction, if $k > 1$, $\sum \nu_i(x_j) \geq 2(|\Omega_i| - 1)$, so

$$\sum_j \nu(x_j) = \sum_{i,j} \nu_i(x_j) \geq \sum_i 2(|\Omega_i| - 1) = 2\nu(G).$$

Thus we may assume $k = 1$, so G is transitive.

Now each x_i induces a collection of disjoint cycles on Ω. We propose to replace each such cycle of length d, by the product of $d - 1$ transpositions $t_1 \cdots t_{d-1}$. Then the contribution of that particular cycle of x_i to $\nu(x_i)$ is $\sum \nu(t_i) = d - 1$ and x_i itself is the product of the transpositions which were used to make up each of its constituent cycles. We denote this collection of transpositions $T(x_i)$ so

$$\nu(x_i) = \sum_{t \in T(x_i)} \nu(t) \text{ and } x_i = \prod_{t \in T(x_i)} t,$$

in an appropriate order. Since $x_1 \cdots x_m = 1$, we have

$$\prod_i \prod_{t \in T(x_i)} t = 1,$$

or, letting T be the union of the $T(x_i)$, we may say $\prod_{t \in T} t = 1$, with the transpositions in an appropriate order. But as i ranges over $1, \ldots m$, the cycles cover all points of Ω, since G is transitive. It follows that the graph $\mathcal{G}(T); = (\Omega, T)$, where the transpositions are regarded as edges, is connected. It follows that

$$\langle T \rangle = \mathrm{Sym}(\Omega).$$

But $\mathrm{Sym}(\Omega)$ is transitive and so $\nu(\mathrm{Sym}(\Omega)) = \nu(G)$. Since $\prod_T t = 1$, $\langle T \rangle = \mathrm{Sym}(\Omega)$, and $\sum_T \nu(t) = \sum_i \nu(x_i)$, it suffices to prove the asserted inequality for the case that $G = \mathrm{Sym}(\Omega)$ and the x_i are transpositions.

So now, $G = \mathrm{Sym}(\Omega)$, $x_1 \cdots x_m = 1$, where the x_i are a generating set of transpositions. Among the x_i we can select a minimal generating set $\{s_1, \ldots, s_k\}$ taken in the order in which they are encountered in the product $x_1 \cdots x_m$. We can move these s_i to the left in the product $\prod x_i$ at the expense of conjugating the elements they pass over, to obtain

$$s_1 s_2 \cdots s_k y_1 y_2 \cdots y_l = 1, k + l = m,$$

where the y_j's are conjugates of the x_i's. Now by the Feit-Lyndon-Scott Theorem (Theorem 4.2.5, p. 112), the product $s_1 \cdots s_k$ is an n-cycle (where $n = |\Omega|$). Thus $k \geq n - 1$. But then $y_1 \cdots y_l$ is the inverse n-cycle, and so $l \geq n - 1$, also. But $k + l = m = \sum \nu(x_i)$ and $n - 1 = \nu(G)$, so the theorem is proved. $\square$

6.5 Exercises

6.5.1 Exercises for Sect. 6.3

1. Show that if $|X| > 1$, then it is possible to have $|X|$ finite, while $|Y^H|$ is infinite. That is, a free group on a finite number of generators can have a subgroup which is a free group on an infinite number of generators. [Hint: Show that one may select members of Y^H first, and choose them with unbounded length.

2. This is a project. Is it possible for a group to be a free group on more than one cardinality of generator? The question comes down to asking whether $F(X) \simeq F(Z)$ implies there is a bijection $X \to Z$. Clearly the graph-automorphism definition of a free group is the way to go.

3. Another project: What can one say about the automorphism group of the free group $F(X)$? Use the remarks above to show that it is larger than $\text{Sym}(X)$. [Hint: Consider the free group on one generator].

4. Suppose $G = \langle x, y | \mathcal{R} \rangle$. Show that $G = \{e\}$, the identity group, if $\mathcal{R}$ implies either of the following relations:

   (a) $yx = y^2 x, xy^3 = y^2 x,$
   (b) $xy^2 = y^3 x, x^2 y = yx^3.$

5. Prove that

   $$\langle x, y | x^4 = e, y^2 = x^2, yxy^{-1} = yx^3 \rangle \simeq \langle r, s, t | r^2 = s^2 = t^2 = rst \rangle.$$

6. Prove that the group

   $$G = \langle x, y | x^2 = y^3 = (xy)^3 = e \rangle$$

   has order at most 12. Conclude that $G \simeq Alt(4)$.

7. Prove that

   (a) $|\langle x, y | x^2 = y^3 = (xy)^4 = e \rangle| = 24$
   (b) $|\langle x, y | x^2 = y^3 = (xy)^5 = e \rangle| = 60$

8. Let $Q_{2^{a+1}} := \langle x, y | x^{2^a} = e, y^2 = x^{2^{a-1}}, yxy^{-1} = x^{-1} \rangle$. Show that $|Q_{2^{a+1}}| = 2^{a+1}$. When $a > 1$, the group $Q_{2^{a+1}}$ is called the *generalized quaternion group*. It contains a unique involution. Can you prove this?

9. Suppose G is a (k, l, m)-group. Using only the directed Cayley graph $C(G, X)$ where X is the set of two elements $\{x, y\}$ of orders 2 and 4, Show that

   (a) any $(2, 4, 4)$-group has order 24 and so is $\text{Sym}(4)$.
   (b) any $(2, 3, 5)$-group has order 60.

10. The following chain of exercises was inspired by a problem in a recent book on Group Theory. That book asks the student to show that any simple group of order 360 is the alternating group on six letters. As a "Hint", it asks the reader

to consider the index of a 5-Sylow normalizer. This would be a useful hint if there were just six 5-Sylow subgroups. Unfortunately, there are 36 of them, and the proof does not appear to be a simple consequence of group actions and the Sylow theorems. However the statement can be proved as a useful application of (k, l, m)-groups.

In the following G is a simple group of order 360. The reader is asked to prove that G is isomorphic to the alternating group on six letters. It will suffice to prove that it has a subgroup of index 6. In this series of exercises, the student is asked to assume

(H) *There is no subgroup of index 6 in G.*

and show that this leads to a contradiction.

(1) Show that G contains 36 5-Sylow subgroups, and so contains 144 elements of order 5. [Hint: Use Sylow's theorem, the simplicity of G, and (H).]
(2) The 3-Sylow subgroups are abelian, and each intersects its other conjugates only at the identity element. [Hint: If x were an element of order three contained in two distinct 3-Sylow subgroups, then the number d of 3-Sylow subgroups contained in $C_G(x)$ is 4 or 40 and the latter choice makes $|C_G(x)|$ large enough to conflict with (H). So $T = Syl_3(C_G(x))$ has 4 elements acted on transitively by conjugation, with one of the 3-Sylows inducing a 3-cycle. If F is the kernel of the action of $C_G(x)$ on T, then $C_G(x)/F$ is isomorphic to $Alt(4)$ or $Sym(4)$. The latter choice forces $C_G(x)$ to have index 5. Show that $C_G(x)$ now contains a normal fours group K. Since K is normalized by a 2-Sylow subgroup containing it, $N_G(K)$ now has index at most 5 against (H) and simplicity.]
(3) The collection $\Omega = Syl_3(G)$ has 10 members, G acts doubly transitively on Ω and G contains 80 elements of order 3 or 9. [Hint: If $|\Omega| = 40$, there are 320 non-identity elements of 3-power order and this conflicts with (1).]
(4) Prove the following:
  i. Every involution t fixes exactly two letters of Ω.
  ii. If t normalizes 3-Sylow subgroup P, then t centralizes only the identity element of P.
  iii. The 2-Sylow subgroup of $N_G(P)$ is cyclic of order 4.
  iv. P is elementary abelian.
  [Hint: Since t must act on the elements of $\Omega - \{P\}$ exactly as it does on the elements of P, t must fix 2 or 4 letters of Ω. But it must also fix an odd number of letters (why?). Now that $t^{-1}at = a^{-1}$ for all $a \in P$, t is the unique involution in a 2-Sylow subgroup of $N_G(P)$ (explain). The other parts follow.]
(5) There are no elements of orders 6, 10 or 15.
(6) Fix an involution t and let $\mathcal{A}$ be the full collection of 40 subgroups of order 3 in G. (Of course these are partitioned in 10 sets of four each, by the 3-Sylow subgroups.) By Step (5), we know that if y is an element of order 3, then the

product ty has order 2, 3, 4 or 5. From our results on (k, l, m)-groups, we have a corresponding partition of $\mathcal{A}$ with these components:

$$\mathcal{A}_2 := \{A \in \mathcal{A} | \langle t, A \rangle = \text{Sym}(3)\} \qquad\qquad (6.9)$$

$$\mathcal{A}_3 := \{A \in \mathcal{A} | \langle t, A \rangle = \text{Alt}(4)\} \qquad\qquad (6.10)$$

$$\mathcal{A}_4 := \{A \in \mathcal{A} | \langle t, A \rangle = \text{Sym}(4)\} \qquad\qquad (6.11)$$

$$\mathcal{A}_5 := \{A \in \mathcal{A} | \langle t, A \rangle = \text{Alt}(5)\}, \qquad\qquad (6.12)$$

Prove the following:

(i) $|\mathcal{A}_2| = 8$

(ii) $|\mathcal{A}_3| = 8$:

(iii) $|\mathcal{A}_4| = 8$

[Hint: For (i), count Z_3's inverted by t. For (ii), let $\mathcal{K}(t)$ be the set of Klein fours groups K which contain t. (There are just two.) Each such member is normalized by four members of $\mathcal{A}$, and, of course, these four belong to $\mathcal{A}_3$. Similarly for (iii), let $\mathcal{K}'(t)$ be the fours groups which are normalized by t but do not contain t. There is a bijection between $\mathcal{K}'(t)$ and the set of involutions of $C_G(t) - \{t\}$. If $K \in \mathcal{K}'(t)$, then t normalizes two of the elements of $\mathcal{A}$ in $N_G(K) \simeq \text{Sym}(4)$, but generates the whole $N_G(K)$ with the other two.]

(7) Conclude that $\mathcal{A}_5$ is not empty.

6.5.2 Exercises for Sect. 6.4

1. Use the Brauer-Ree Theorem to show that Alt(7) is not a $(2, 3, 7)$-group. [Hint: Use the action of $G = \text{Alt}(7)$ on the 35 3-subsets of the seven letters. (We will call these 3-subsets "triplets".) With respect to this action show that an involution fixes exactly seven triplets, and so has ν-value 14. Any element of order 3 contributes $\nu = 22$, and that an element of order 7 fixes no triplet, acts in five 7-cycles and contributes $\nu = 30$.]

6.6 A Word to the Student

6.6.1 The Classification of Finite Simple Groups

The Jordan-Holder theorem associates with each finite group, a multiset of finite simple groups—the so-called *chief factors*. What are these groups? Many many theorems are known which, under the induction hypothesis reduce to the case that a minimal counter-example to the theorem, say G, is a finite simple group. If one had a list of these groups, then one need only run through the list, verifying that each of these groups either satisfies the conjectured conclusion or belongs to some other class of groups forbidden by the hypothesis. This is how the classification of maximal subgroups of the finite classical groups and the classification of flag-transitive linear spaces proceeded [13].

After decades of intensive effort, the classification of finite simple groups was finally realized in 2004 when the last steps, classifying the so-called "quasi-thin groups" was published (M. Aschbacher and S. Smith) [2–4].

Here is the over-all conclusion:

Theorem 6.6.1 *Suppose G is a finite simple group. Then G belongs to one of the following categories:*

1. *G is cyclic of prime order.*
2. *G is an alternating group on at leas five letters*
3. *G belongs to one of* 16 *families of groups defined as automorphism groups of objects defined by buildings.*
4. *G is one of a strange list of* 26 *so-called "sporadic groups".*

The sporadic groups have no apparent algebraic reason for existing, yet they do exist. If anything should spur the sense of mystery about our logical universe it is this! (The student may consult a complete listing of these groups in the books of Aschbacher [1], Gorenstein [9] or Greiss [10].)

There have been many estimates of the number of pages that a convincing proof of this grand theorem would require. For some authors it is 10,000 for others 30,000.

Like similar endeavors in other branches of mathematics, a great deal of the difficulty is encountered when the degrees of freedom are small. For finite groups, this occurs for groups with "small" 2-rank; that is, groups of odd order and groups whose 2-Sylow subgroup has a cyclic subgroup of index 2. Here, the theory of group characters, both ordinary, exceptional and modular, plays a central role.

Group characters are certain complex-valued functions on a finite group which are indicators of ways to represent a finite group as a multiplicative group of matrices (a "*representation*" of the group). This beautiful subject involves a vast interplay of ring theory, linear algebra and algebraic number theory. No book on general algebra could begin to do justice to this topic. Accordingly, the interested student is advised to seek out books that are entirely devoted to representation theory and/or character theory. The classic text for representation theory is *Representation Theory of Finite Groups and Associative Algebras* by Curtis and I. Reiner [7]. Excellent books on the theory of group characters are by M. Isaacs [12], W. Feit [8], M. J. Collins [5] and L. C. Grove [11], but there are many others.

References

1. Aschacher M (1993) Finite group theory. Cambridge studies in advanced mathematics, vol 10. Cambridge University Press, Cambridge
2. Aschbacher M, Stephen S (2004) The structure of strong quasithin K-groups. Mathematical surveys and monographs, vol 112. American Mathematical Society, Providence

3. Aschbacher M, Smith S (2004) The classication of quasithin groups II, main theorems: the classication of QTKE-groups. Mathematical surveys and monographs, vol 172. American Mathematical Society, Providence
4. Aschbacher M, Lyons R, Smith S, Solomon R (2004) The classication of nite simple groups of characeristic 2 type. Mathematical surveys and monographs, vol 112. American Mathematical Society, Providence
5. Collins MJ (1990) Representations and characters of finite groups. Cambridge studies in advanced mathematics, vol 22. Cambridge University Press, Cambridge
6. Coxeter HMS (1934) Discrete groups generated by reections. Ann Math 35:588–621
7. Curtis CW, Reiner I (1962) Representation theory of finite groups and associative algebras. Pure and applied mathematics, vol IX. Interscience Publishers, New York, pp xiv+685
8. Feit W (1967) Characters of finite groups. W. A. Benjamin Inc., New York
9. Gorenstein D (1968) Finite groups. Harper & Row, New York
10. Griess R (1998) Twelve sporadic groups. Springer monographs in mathematics. Springer, Berlin
11. Grove LC (1997) Groups and characters. Pure and applied mathematics interscience series. Wiley, New York
12. Isaacs IM (1994) Character theory of finite groups. Dover Publications Inc., New York
13. Kleidman P, Liebeck M (1990) The subgroup structure of the finite classical groups. London mathematical society lecture note series, vol 129. Cambridge University Press, Cambridge

Chapter 7
Elementary Properties of Rings

Abstract Among the most basic concepts concerning rings are the poset of ideals (left, right and 2-sided), possible ring homomorphisms, and the group of units of the ring. Many examples of rings are presented—for example the monoid rings (which include group rings and polynomial rings of various kinds), matrix rings, quaternions, algebraic integers etc. This menagerie of rings provides a playground in which the student can explore the basic concepts (ideals, units, etc.) *in vivo*.

7.1 Elementary Facts About Rings

7.1.1 Introduction

In the next several parts of this course we shall take up the following topics and their applications:

1. The Galois Theory.
2. The Arithmetic of Integral Domains.
3. Semisimple Rings.
4. Multilinear Algebra.

All of these topics require some basic facility in the language of rings and modules.

For the purposes of this survey course, all rings possess a multiplicative identity. This is not true of all treatments of the theory of rings. But virtually every major application of rings is one which uses the presence of the identity element (for example, this is always part of the axiomatics of Integral Domains, and Field Theory). So in order to even have a language to discuss these topics, we must touch base on a few elementary definitions and constructions.

7.1.2 Definitions

Recall from Sect. 1.2 that a monoid is a set M with an associative binary operation which admits a two-sided identity element. Thus if $(M, \cdot)$ is a monoid, there exists

© Springer International Publishing Switzerland 2015

E. Shult and D. Surowski, *Algebra*, DOI 10.1007/978-3-319-19734-0_7

an element "1" with the property, that $1 \cdot m = m \cdot 1 = m$ for all $m \in M$. Notice that Proposition 1.2.1 tells us that such a two-sided identity is unique in a monoid. Indeed the "1" is also the unique left identity element as well as the unique right identity element.

A *ring* is a set R endowed with two distinct binary operations, called *(ring) addition* and *(ring) multiplication*—which we denote by "$+$" and by juxtaposition or "$\cdot$", respectively—such that

Addition laws: $(R, +)$ is an abelian group.
Multiplicative Rules: $(R, \cdot)$ is a monoid.
Distributive Laws: For all $a, b, c \in R$,

$$a(b + c) = ab + ac \tag{7.1}$$
$$(a + b)c = ac + bc \tag{7.2}$$

Just to make sure we know what is being asserted here, for any elements a, b and c of R:

1. $a + (b + c) = (a + b) + c$
2. $a + b = b + a$
3. There exists a "zero element" (which we denote by 0_R, or when the context is clear, simply by 0) with the property that $0_R + a = a$ for all ring elements a.
4. Given $a \in R$, there exists a unique element $(-a)$ in R such that $a + (-a) = 0_R$.
5. $a(bc) = (ab)c$.
6. There exists a multiplicative "identity element" (which we denote by 1_R or just by 1 when no confusion is likely to arise) with the property that $1_R a = a1_R = a$, for all ring elements $a \in R$.
7. Then there are the distributive laws which we have just stated—that is, if $a, b, c \in R$, then $a(b + c) = ab + ac$ and $(a + b)c = ac + bc$.

As we know from the group theory, the zero element 0_R is unique in R. Also given element x in R, there is exactly one additive inverse $-x$, and that $-(-x) = x$. The reader should be warned that in the axioms for a ring, the zero element may not be distinct from the multiplicative identity 1_R (although this is required for integral domains as seen below). If they are the same, it is easy to see that the ring contains just one element, the zero element.

Lemma 7.1.1 *For any elements a, b in R:*

1. *$(-a)b = -(ab) = a(-b)$, and*
2. *$0 \cdot b = b \cdot 0 = 0$.*

Some notation: Suppose X and Y are subsets of the set R of elements of a ring. We write

$$X + Y := \{x + y \mid x \in X, y \in Y\} \tag{7.3}$$

$$XY := \{xy | x \in X, y \in Y\}, \tag{7.4}$$
$$-X := \{-x | x \in X\} \tag{7.5}$$

Species of Rings

- A ring is said to be *commutative* if and only *multiplication* of ring elements is commutative. A consequence of the distributive laws in a commutative ring is the famous "binomial theorem" familiar to every student of algebra:

Theorem 7.1.2 (The Binomial Theorem) *If a and b are elements of a commutative ring R, then for every positive integer n,*

$$(a + b)^n = \sum_{k=0}^{n} [n!/(n-k)!k!]a^k b^{n-k}.$$

- A *division ring is a ring in which* $(R^* := R - \{0\}, \cdot)$ *is a group.*
- A *commutative division ring is called a* field.
- A *commutative ring in which* $(R^*, \cdot)$ *is a monoid—i.e. the non-zero elements are closed under multiplication and possesses an identity—is called* an integral domain.

There are many other species of rings—Artinian, Noetherian, Semiprime, Primitive, etc. which we shall meet later, but there is no immediate need for them at this point.

There are many examples of rings that should be familiar to the reader from previous courses. The most common examples are the following:

$\mathbb{Z}$, the *integers*, an integral domain,
$\mathbb{Q}$, the field of *rational numbers*,
$\mathbb{R}$, the field of *real numbers*,
$\mathbb{C}$, the field of *complex numbers*,
$\mathbb{Z}/(n)$, the ring of integral residue classes mod n,
$K[x]$, the ring of polynomials with coefficients from the field K.
$M_n(R)$, the ring of n-by-n matrices with entries in a commutative ring R, under ordinary matrix addition and matrix multiplication.

More examples are developed in Sects. 7.3 and 7.4. In particular the reader will revisit polynomial rings as part of a general construction.

Finally, if R is a ring with multiplicative identity element 1, then a *subring* of R is a subset S of R such that is itself is a ring under the addition and multiplication operations of R. Thus

1. S is a subgroup of $(R, +)$—i.e. $S = -S$ and $S + S = S$. In particular $0_R \in S$.
2. $SS \subseteq S$—i.e. S is closed under multiplication.
3. S contains a multiplicative identity element e_S (which may or may not be the identity element of R).

Thus the ring of integers, $\mathbb{Z}$, is a subring of any of the three fields $\mathbb{Q}$, $\mathbb{R}$ and $\mathbb{C}$, of rational, real and complex numbers, respectively. But the set of even numbers is *not* a subring of the ring of integers.

If R is an arbitrary ring and B is any subset of R, the *centralizer of B in R* will denote the set

$$C_R(B) := \{r \in R | rb = br \text{ for all } b \in B\}.$$

We leave the reader to verify that the centralizer $C_R(B)$ is a subring of R (Exercise (6) in Sect. 7.5.1). In the case that $B = R$, the centralizer of B is called the *center of R*.[1]

7.1.3 Units in Rings

Suppose R is a ring with multiplicative identity element 1_R. We say that an element x in R has a *right inverse* if and only if there exists an element x' in R such that $xx' = 1_R$. Similarly, x has a *left inverse* if and only if there exists an element x'' in R such that $x''x = 1_R$. (If R is not an integral domain these right and left inverses might not be unique.) An element x in R is called a *unit* if and only if it possesses a *two-sided inverse*, that is, an element that is both a right and a left inverse in R. The element 1_R, of course, is a unit, so we always have at least one of these. We observe

Lemma 7.1.3 *1. An element x has a right inverse if and only if $xR = R$. Similarly, x has a left inverse if and only if $Rx = R$.*
  2. If both a right inverse and a left inverse of an element x exist, then these inverses are equal, and represent the unique (two-sided) inverse of x.
  3. The set of units of any ring form a group under multiplication.

Proof Part 1. If x has a right inverse xR contains $xx' = 1_R$, so xR contains R. The converse, that $xR = R$ implies that $xx' = 1_R$, for some element x' in R, is transparent. The left inverse story follows upon looking at this argument through a mirror.

Part 2. Suppose x_l and x_r are respectively left and right inverses of an element x. Then

$$x_r = (1_R)x_r = (x_l x)x_r = x_l(xx_r) = x_l(1_R) = x_l.$$

Thus $x^{-1} := x_l$ is a two sided inverse of x and is unique, since it is equal to *any* right inverse of x.

---

[1] Sometimes the literature calls this the *centrum of R*.

Part 3. Suppose x and y are units of the ring R with inverses equal to x^{-1} and y^{-1}, respectively. Then xy has $y^{-1}x^{-1}$ as both a left and a right inverse. Thus xy is a unit. Thus the set $U(R)$ of all units of R is closed under an associative multiplication, has an identity element with respect to which every one of its elements has a 2-sided inverse. Thus it is a group. $\square$

A ring in which every non-zero element is a unit must be a division ring and conversely.

Example 34 Some examples of groups of units:

1. In the ring of integers, the group of units is $\{\pm 1\}$, forming a cyclic group of order 2.
2. The integral domain $\mathbb{Z} \oplus \mathbb{Z}i$ of all complex numbers of the form $a + bi$ with a, b integers and $i = \sqrt{-1}$, is called the ring of *Gaussian integers*. The units of this ring are $\{\pm 1, \pm i\}$ forming the cyclic group of order four under multiplication.
3. The complex numbers of the form $a + b(e^{2i\pi/3})$, a and b integers, form an integral domain called *the Eisenstein numbers*. The units of this ring are $\{1, z, z^2, z^3 = -1, z^4, z^5\}$, where $z = -e^{2i\pi/3}$, forming a cyclic group of order 6.
4. The number $\lambda := 2 + \sqrt{5}$ is a unit in the domain $D := \{a + b\sqrt{5} \mid a, b \in \mathbb{Z}\}$ since $(2 + \sqrt{5})(-2 + \sqrt{5}) = 1$. Now λ is a real number, and since $1 < \lambda$, we see that the successive powers of λ form a monotone increasing sequence, and so we may conclude that the group $U(D)$ of units for this domain is an infinite group.
5. In the ring $R := \mathbb{Z}/\mathbb{Z}n$ of residue classes of integers mod n, under addition and multiplication of these residue classes, the units are the residue classes of the form $w + \mathbb{Z}n$ where w is relatively prime to n (of course, if we wish, we may take $0 \le w \le n-1$.) They thus form a group of order $\phi(n)$, where ϕ denotes the *Euler phi-function*. Thus the units of $R = \mathbb{Z}/8\mathbb{Z}$ are $\{1 + 8\mathbb{Z}, 3 + 8\mathbb{Z}, 5 + 8\mathbb{Z}, 7 + 8\mathbb{Z}\}$, and form a group isomorphic to the *Klein fours group* rather than the cyclic group of order four.[2]
6. In the ring of n-by-n matrices with entries from a field F, the units are the invertible matrices and they form the general linear group, $GL(n, F)$, under multiplication.

7.2 Homomorphisms

Let R and S be rings. A mapping $f : R \to S$ is called a *ring homomorphism* if and only if

1. For all elements a and b of R, $f(a + b) = f(a) + f(b)$.
2. For all elements a and b of R, $f(ab) = f(a)f(b)$.

---

[2] See p. 75 for the general definition of "fours group".

A basic consequence of Part 1 of this definition is that f induces a homomorphism of the underlying additive group, so, from our group theory, we know that $f(-a) = -f(a)$, and $f(0_R) = 0_S$. In fact any polynomial expression formed from a finite set of elements using only the operations of addition, multiplication and additive inverses gets mapped by f to a similar expression with all the original elements replaced by their f-images. Thus:

$$f([(a+b)c^2 + (-c)bc]a) = [(f(a)+f(b))(f(c))^2 + (-f(c))(f(b)f(c))]f(a).$$

As was the case with groups, the composition $g \circ f : R \to T$ of two ring homomorphisms $f : R \to S$ and $g : S \to T$, must also be a ring homomorphism. The homomorphism is *onto* or is an *epimorphism* or a *surjection*, is *injective*, or is an *isomorphism*, if and only if these descriptions apply to the induced group homomorphism of the underlying additive groups.

An *automorphism* of a ring R is an isomorphism $\alpha : R \to R$. The composition $\beta \circ \alpha$ of automorphisms is itself an automorphism, and the collection of all automorphisms of R form a group $\mathrm{Aut}(R)$ under the operation of composition of automorphisms.

An *antiautomorphism* of a ring R is an automorphism α of the additive group $(R, +)$ such that for any elements a and b of R,

$$\alpha(ab) = \alpha(b)\alpha(a).$$

Note that since the multiplicative identity element of a ring R is unique (see the first paragraph of Sect. 7.1.2, p. 186), both automorphisms and antiautomorphisms must leave the identity element fixed. If α is an automorphism, the set $C_R(\alpha)$ of elements left fixed by α form a subring of R containing its multiplicative identity element. However, if α is an antiautomorphism, the fixed point set, while closed under addition, need not be closed under multiplication. But, if by chance a subring S of R is fixed elementwise by the antiautomorphism α then S must be a commutative ring. In fact, for any commutative ring R, the identity mapping on R is an antiautomorphism (as well as an automorphism).

Antiautomorphisms and automorphisms of a ring R can be composed with these results:

antiautomorphism $\circ$ antiautomorphism $=$ automorphism

antiautomorphism $\circ$ automorphism $=$ antiautomorphism

automorphism $\circ$ antiautomorphism $=$ antiautomorphism

Example 35 The complex conjugation mapping that sends $\alpha = a + bi$ to $\bar{\alpha} = a - bi$, where a and b are real numbers, is certainly a familiar example of a field automorphism. Now let F be any field, let n be a positive integer, and let σ be any automorphism of the field F, and let $M_n(F)$ be the ring of all $n \times n$ matrices having entries in the field F. The operations are entry-wise addition of matrices, and ordinary matrix multiplication. IF $X \in M_n(F)$, the expression $X = (x_{ij})$ means that x_{ij} is the entry in the ith row and jth column of X.

(a) The mapping $m(\sigma) : S : M_n(F) \to M_n(F)$ which replaces each matrix entry by its σ-image, is transparently an automorphism of the ring $M_n(F)$.

(b) The *transpose mapping*, $T : M_n(F) \to M_n(F)$ replaces each entry a_{ij} by a_{ji} for any matrix $A = (a_{ij}) \in M_n(F)$. Suppose $A = (a_{ij})$ and $B = (b_{ij})$ are elements of $M_n(F)$. Then the (i, j)-entry of the matrix AB is the "dot" product of the ith row of A with the jth column of B—that is, $\sum_k a_{ik} b_{k,j}$. On the other hand, the (ji)th entry of $B^T A^T$ is the dot product of the jth row of B^T (which is j-column of B) and the ith column of A^T (which is the ith row of A—the (i, j)th entry of AB. Thus this product is the transpose of the matrix AB, and we may write

$$B^T A^T = (AB)^T.$$

Since T respects addition, we see that the transpose mapping is an antiautomorphism of the ring $M_n(F)$.

(c) One can then compose the two mappings to obtain the antiautomorphism

$$S \circ T = T \circ S : M_n(F) \to M_n(F),$$

which we call the σ-*transpose mapping*.

7.2.1 Ideals and Factor Rings

Again let R be a ring. A *left ideal* is a subgroup $(I, +)$ of the additive group $(R, +)$, such that $RI \subseteq I$. Similarly, a subgroup $J \leq (R, +)$ is a **right ideal** if $JR \subseteq J$. Finally, a *(two-sided) ideal* is a left ideal which is also a right ideal. The parentheses are to indicate that if we use the word "ideal" without adornment, a two-sided ideal is intended. A subset X of R is a two sided ideal if and only if

$$X + X = X \tag{7.6}$$
$$-X = X \tag{7.7}$$
$$RX + XR \subseteq X. \tag{7.8}$$

Note that while a 2-sided ideal $I \subseteq R$ is closed under both addition and multiplication, it cannot be regarded as a *subring* of R unless it possesses its own multiplicative identity element.

If I is a two sided ideal in R, then, under addition, it is a (normal) subgroup of the commutative additive group $(R, +)$. Thus we may form the factor group $R/I := (R, +)/(I, +)$ whose elements are the additive cosets $x + I$ in R.

Not only is R/I an additive group, but there is a natural multiplication on this set, induced from that fact that all products formed from two such additive cosets, $x + I$ and $y + I$, lie in a unique further additive coset. This is evident from

$$(x + I)(y + I) \subseteq xy + xI + Iy + II \subseteq xy + I + I + I = xy + I. \qquad (7.9)$$

With respect to addition and multiplication, R/I forms a ring which we call the *factor ring*. Moreover, by Eq. (7.9), the homomorphism of additive groups $R \rightarrow R/I$ is a ring epimorphism.

7.2.2 The Fundamental Theorems of Ring Homomorphisms

Suppose $f : R \rightarrow S$ is a homomorphism of rings. Then the *kernel of the homomorphism* f is the set

$$\ker f := \{x \in R | f(x) = 0_S\}.$$

We have

Theorem 7.2.1 (The Fundamental Theorem of Ring Homomorphisms)

1. *The kernel of any ring homomorphism is a two-sided ideal.*
2. *If $f : R \rightarrow S$ is a homomorphism of rings, then there is a natural isomorphism between the factor ring $R/\ker f$ and the image ring $f(R)$, which takes each coset $x + \ker f$ to $f(x)$.*

Proof Part 1. Suppose $f : R \rightarrow S$ is a ring homomorphism and that x is an element of $\ker f$. Then as $\ker f$ is the kernel of the group homomorphism of the underlying additive groups of the two rings, $\ker f$ is a subgroup of the additive group of R. Moreover, for any element r in R, $f(rx) = f(r)f(x) = f(r)0_S = 0_S$. Thus $R(\ker f) + (\ker f)R \subseteq \ker f$, and so $\ker f$ is a two-sided ideal.

Part 2. The image ring $f(R)$ is a subring of S formed from all elements of S which are of the form $f(x)$ for at least one element x in R. We wish to define a mapping $\phi : R/(\ker f) \rightarrow f(R)$ which will take the additive coset $x + \ker f$ to the image element $f(x)$. We must show that this mapping is well-defined. Now if $x + \ker f = y + \ker f$, then $x + (-y) \in \ker f$, so $0_S = f(x + (-y)) = f(x) + (-f(y))$, whence $f(x) = f(y)$. So changing the coset representative does not change the image, and so the mapping is well-defined. (In fact the ϕ-image of $x + \ker f$ is found simply by applying f to the entire coset, that is, $f(x + \ker f) = f(x)$.)

The mapping ϕ is easily seen to be a ring homomorphism since

$$\phi((x + \ker f) + (y + \ker f)) = \phi(x + y + \ker f) = f(x + y)$$
$$= f(x) + f(y) = \phi(x + \ker f) + \phi(y + \ker f)$$

and

$$\phi((x + \ker f)(y + \ker f)) = \phi(xy + \ker f) = f(xy)$$
$$= f(x)f(y) = \phi(x + \ker f) \cdot \phi(y + \ker f).$$

Clearly ϕ is onto since by definition, $f(x) = \phi(x + \ker f)$ for any image element $f(x)$. If $\phi(x + \ker f) = \phi(y + \ker f)$ then $f(x) = f(y)$ so $f(x + (-y)) = 0_S$, and this gives $x + \ker f = y + \ker f$. Thus ϕ is an injective mapping.

We have assembled all the requirements for ϕ to be an isomorphism. $\square$

Before leaving this subsection, there is an important class of ring homomorphisms which play a role in the theory of R-algebras.[3]

Let A and B be two rings that contain a common subring R lying in the center of both A and B. A ring homomorphism $f : A \to B$ for which $f(ra) = rf(a) = f(a)r$ for all ring elements $a \in A$ and $r \in R$ is called an R-*homomorphism*. (We shall generalize this notion to R-modules in the next chapter.)

7.2.3 The Poset of Ideals

We wish to make certain statements that hold for the set of all left ideals of R as well as for the set of all right ideals and all (two-sided) ideals of R. We do this by asserting "X holds for (left, right) ideals with property Y". This means the statement is asserted three times: once about ideals with property Y, again about left ideals with property Y, and again about right ideals with property Y. It does save a little space, and helps point out a uniformity among the theorems. But on the other hand it makes statements difficult to scan, and sometimes renders the statement a little less indelible in the student's memory. So we will keep this practice at a minimum.

Lemma 7.2.2 *Suppose A and B are (left, right) ideals in the ring R. Then*

1. $A + B := \{a + b | (a, b) \in a \times B\}$ *is itself a (left, right) ideal and is the unique such ideal minimal with respect to containing both A and B.*
2. $A \cap B$ *is a (left, right) ideal which is the maximal such ideal contained in both A and B.*
3. *The set of all (left, right) ideals of R form a partially ordered set—that is a poset— with respect to the containment relation. In view of the preceding two conclusions, this poset is a lattice.*

Theorem 7.2.3 (The Correspondence Theorem for Rings) *Suppose $f : R \to S$ is an epimorphism of rings—that is, $S = f(R)$. Then there is a one-to-one correspondence*

---

[3] R-algebras are certain rings that are also R-modules in a way that is compatible with ring multiplication. Since we have not yet defined R-modules, we cannot introduce R-algebras at this point. Nor do we need to do so. R-homomorphisms between rings can still be defined at this point and we do require this concept for describing a universal property of polynomial rings on p. 207.

between the left (right) ideals of S and the left (right) ideals of R which contain ker f.[4]
Under this correspondence, 2-sided ideals are matched only with 2-sided ideals.

Proof The statement of the theorem is to be read as two statements, one about
isomorphic posets of left ideals and one about isomorphic posets of right ideals. We
prove the theorem only for left ideals. The proof for right ideals is virtually a mirror
reflection of the proof we give here.

If J is a left ideal of the ring S, set $f^{-1}(J) := \{r \in R | f(r) \in J\}$. Clearly if a
and b are elements of $f^{-1}(J)$, and r is an arbitrary element of R, then

$$f(ra) = f(r)f(a) \in J$$
$$f(-a) = -f(a) \in J$$
$$f(a+b) = f(a) + f(b) \in J$$

So ra, $-a$, and $a + b$ all belong to $f^{-1}(J)$. Thus $f^{-1}(J)$ is a left ideal of R
containing $f^{-1}(0_S) = \ker f$. If also J is a right ideal of S, we have $JS \subseteq J$. Then
$f(f^{-1}(J)R) \subseteq Jf(R) = JR \subseteq J$, whence $f^{-1}(J)R \subseteq f^{-1}(J)$. Thus $f^{-1}(J)$ is
a right ideal if and only if J is.

Now if I is a left ideal of R, then so is $f(I)$ a left ideal of $f(R) = S$. But if
I contains ker f, then $f^{-1}(f(I)) = I$, because, for any element $x \in f^{-1}(f(I))$,
there is an element $i \in I$ such that $f(x) = f(i)$. But then $x + (-i) \in \ker f \subseteq I$,
whence $x \in I$.

We thus see that the operator f^{-1} induces a mapping of the poset of left ideals of
S onto the set of left ideals of R containing ker f. It remains only to show that this
(containment-preserving) operator is one-to-one. Suppose $I = f^{-1}(J_1) = f^{-1}(J_2)$.
We claim $J_1 \subseteq J_2$. If not, there is an element $y \in J_1 - J_2$. Since f is an epimorphism,
y has the form $y = f(x)$, for some element x in R. Then x is in $f^{-1}(J_1)$ but not
$f^{-1}(J_2)$, a contradiction. Thus we have $J_1 \subseteq J_2$. By symmetry, $J_2 \subseteq J_1$ and so
$J_1 = J_2$. This completes the proof. □

A SECOND PROOF We can take advantage of the fact that the kernel of the surjective
ring homomorphism $f : R \to S$ is also the kernel of the homomorphism $f^+ :$
$(R, +) \to (S, +)$ of the underlying additive groups and exploit the fact that the
Correspondence Theorem for Groups already presents us with a poset isomorphism

$$\mathcal{A}(R : \ker f) \to \mathcal{A}(S : 0),$$

where $\mathcal{A}(M : N)$ denotes the poset of all subgroups of M which contain the subgroup
N. The isomorphism is given by sending subgroup A of $\mathcal{A}(R : \ker f)$ to the subgroup
$f(A)$ of S.

We need only show that if X is a subgroup of $\mathcal{A}(R : S)$. then

1. $RX \subseteq X$ if and only if $Sf(X) \subseteq f(X)$.
2. $XR \subseteq X$ if and only if $f(X)S \subseteq f(X)$.

---

[4]The latter poset is the principle filter generated by $ker f$ in the poset of left (right) ideals of R.

Both of these are immediate consequences of the fact that f is a ring homomorphism, and $S = f(R)$. $\square$

As in the case of groups, when we say *maximal* (left, right) ideal, we mean a maximal member in the poset of all *proper* (left, right) ideals. Thus the ring R itself, though an ideal, is not a maximal ideal.

Lemma 7.2.4 *Every (left, right) ideal lies in a maximal (left, right) ideal.*

Proof Let P be the full poset of all proper ideals, all proper left ideals or all proper right ideals of the ring R. The proof is essentially the same for these three cases. Let

$$C := J_\sigma | \sigma \in I\}$$

be an a totally ordered subset (that is, an ascending chain) in the poset P. Let U be the set-theoretic union of the J_σ. Then, from the total ordering, any ring element in U belongs to some J_σ. Thus any (left, right) multiple of that element by an element of R belongs to U. Thus it is easy to see that RUR, RU, or UR is a subset of U in the three respective cases of P. Similarly, $-U = U$ and using the total ordering of the chain, $U + U \subseteq U$. Thus U is an (left, right) ideal in R. If $U = R$, then 1_R would belong to some J_σ, whence RJ_σ, $J_\sigma R$ and $RJ_\sigma R$ would all coincide with R, against $J_\sigma \in P$. Thus U, being a proper ideal, is a member of P and so is an upper bound in P to the ascending chain C.

Since any ascending chain in P has an upper bound in P, it follows from Zorn's Lemma (see p. 26) that any element of P is contained in a maximal member of P. $\square$

A *simple ring* is a ring with exactly two (two-sided) ideals: $0 := \{0_R\}$ and R. (Note that the "zero-ring" in which $1_R = 0_R$ is not considered a simple ring here. Simple rings are non-trivial). Since every non-zero element of a division ring D is a unit, 0 is a maximal ideal of D. Thus any division ring is a simple ring. The really classical example of a simple ring, however, is the full matrix ring over a division ring, which (when the division ring is a field) appears among the examples in Sect. 7.4.

Lemma 7.2.5 *1. An ideal A of R is maximal in the poset of two-sided ideals if and only if R/A is a simple ring.*
2. An ideal A of a commutative ring R is maximal if and only if R/A is a field.

Proof Part 1, is a direct consequence of the Correspondence Theorem 7.2.3 of the previous subsection.

Part 2. Set $\bar{R} := R/A$. If $\bar{R}$ is a field, it is a simple ring, and A is maximal by Part 1. If A is a maximal ideal, the commutative ring $\bar{R}$ has only 0 for a proper ideal. Thus for any non-zero element $\bar{x}$ of $\bar{R}$, $\bar{x}\bar{R} = \bar{R}\bar{x} = \bar{R}$. Thus any non-zero element of $\bar{R}$ is a unit. Thus $\bar{R}$ is a commutative division ring, and hence is a field. The proof is complete. $\square$

Suppose now that A and B are ideals in a ring R. In a once-only departure from our standard notational convention—special for (two-sided) ideals—we let the symbol AB denote the *additive subgroup of* $(R, +)$ *generated by the set of elements* $\{ab \mid (a, b) \in A \times B\}$ (the latter would ordinarily have been written as "AB" in our standard notation). That is, for ideals A and B, AB is the set of all finite sums of the form

$$\sum a_j b_j, \quad a_j \in A, \ b_j \in B.$$

If we multiply such an expression on the left by a ring element r, then by the left distributive laws, each a_j is replaced by $r a_j$ which is in A since A is a left ideal, yielding thus another such expression. Similarly, right multiplication of such an element by any element of R yields another such element. Thus AB is an ideal.[5]

Such an ideal AB clearly lies in $A \cap B$, but it might be even smaller. Indeed, if B lies in the *right annihilator of* A, that is the set of elements x of R such that $Ax = 0$, then $AB = 0$, while possibly $A \cap B$ is a non-zero ideal (with multiplication table all zeroes, of course).

This leads us to still another construction of new ideals from old: Suppose A and B are right ideals in the ring R with $B \leq A$. The *right residual quotient* is the set

$$(B : A) := \{r \in R \mid Ar \subseteq B\}.$$

Similarly, if B and A are left ideals of R, with $B \leq A$, the same symbol $(B : A)$ will denote the *left residual quotient*, the set of ring elements r such that $rA \subseteq B$.[6]

Lemma 7.2.6 *If $B \leq A$ is a chain of right (left) ideals of the ring R, then the right (left) residual quotient $(B : A)$ is a two-sided ideal of R.*

Proof We prove here only the right-hand version. Set $W := (B : A)$. Clearly $W + W \subseteq W$, $W = -W$, and $WR \subseteq W$, since B is a right ideal. It remains to show $RW \subseteq W$. Since A is also a right ideal,

$$A(RW) = (AR)W \subseteq AW \subseteq B,$$

so by definition, $RW \subseteq W$. $\square$

A *prime ideal* of the ring R is an ideal—say P—such that whenever two ideals A and B have the property that $AB \subseteq P$, then either $A \subseteq P$, or $B \subseteq P$. (Note: here AB is the product of two ideals as defined above. But of course, since P is an additive group, AB (as a product of two ideals) lies in P if and only if AB (the old set product) lies in P—so the multiple notation is not critical here.)

---

[5]Indeed, one can see that it would be a two-sided ideal even if A were only a left ideal and B were a right ideal. It is unfortunate to risk this duplication of notation: AB as ideal product, or set product, but one feels bound to hold to the standard notation of the literature—when there *is* a standard—even if there are small overlaps in that standard.

[6]The common notation $(B : A)$ should not cause confusion. One knows which concept is intended, from the fact that one is either talking about right ideals A and B or left ideals, or, in the case that the ideals are two-sided, by the explicit use of the word "right" or "left" in front of "residual quotient".

A rather familiar example of a prime ideal would be the set of all multiples of a prime number—say 137, to be specific—in the ring of integers. This set is both a prime ideal and a maximal ideal. The ideal {0} in the ring of integers would be an example of a prime ideal which is not a maximal ideal.

We have the following analog of Lemma 7.2.5 for prime ideals in a commutative ring:

Lemma 7.2.7 *Let R be a commutative ring, and let P be an ideal of R. Then the following conditions are equivalent:*

(i) P is a prime ideal;
(ii) R/P is an integral domain—a commutative ring whose non-zero elements are closed under multiplication.

Proof (i) implies (ii). Suppose P is a prime ideal. Suppose $x + P$ and $y + P$ are two cosets of P (that is, elements of R/P) for which $(x + P)(y + P) \subseteq P$. Then $xy \in P$ and so $(xR)(yR) \subseteq P$. But since R is commutative, xR and yR are ideals of R whose product lies in the prime ideal P. It follows that one of these ideals is already in P forcing one of the cosets $x + P$ or $y + P$ to be the zero coset $0 + P = P$, contrary to assumption. Thus the non-zero elements of the factor ring (the non-zero additive cosets of P) are closed under multiplication. That makes R/P and integral domain.

(ii) implies (i). Now suppose P is an ideal of the commutative ring R for which R/P is an integral domain. We must show that P is a prime ideal. Suppose A and B are ideals of R with the property that $AB \subseteq P$. Then $(A + P)/P$ and $(B + P)/P$ are two ideals of A/P whose product ideal $(A + P)(B + P)/P = (AB + P)/P$ is the zero element of R/P. Since R/P is an integral domain, it is not possible that both ideals, $(A + P)/P$ and $(B + P)/P$ contain non-zero cosets. Thus, one of these ideals of R/P is already the zero coset. That forces $A \subseteq P$ or $B \subseteq P$. Since, for any two ideals, A and B, $AB \subseteq P$ implies $A \subseteq P$ or $B \subseteq P$, P must be a prime ideal. $\square$

Theorem 7.2.8 *In any ring, a maximal ideal is a prime ideal.*

Proof Suppose M is a maximal ideal in R and A and B are ideals such that $AB \subseteq M$. Assume A is not contained in M. Then $A + M = R$, by the maximality of M, but $RB = AB + MB \subseteq M$, so B belongs to the two-sided ideal $(M : R)$. Since M is a left ideal, M also belongs to this right residual quotient. Thus we see that $M + B$ lies in this right residual quotient $(M : R)$. But the latter cannot be R since 1_R does not belong to this residual quotient. Thus $M + B$ is a proper ideal of R and maximality of M as a two-sided ideal forces $B \subseteq M$.

Thus either $A \subseteq M$ or (as just argued) or $B \subseteq M$. Thus M is a prime ideal.

7.3 Monoid Rings and Polynomial Rings

7.3.1 Some Classic Monoids

Recall that a monoid is a set with an associative operation (say, multiplication) and an identity element with respect to that operation.

Example 36 (FREQUENTLY ENCOUNTERED MONOIDS)

1. Of course any group is a monoid.
2. The positive integers $\mathbb{Z}^+$ under multiplication. The integer 1 is the multiplicative identity.
3. The set $\mathbb{N}$ of natural numbers (non-negative integers) under addition. Of course, 0 is the identity relative to this operation.
4. $\mathcal{M}(X)$, the multisets over a set X. The elements of this monoid consist of all mappings $f : X \to \mathbb{N}$ which achieve a non-zero value at only finitely many elements of X. The addition of functions f and g yields the function $f + g$ whose value at x is defined to be $f(x) + g(x)$, the sum of two natural numbers (see p. xiii).
   When $X = \{x_1, \ldots, x_n\}$ is finite, there are three common ways to represent the monoid of multisets.

   (a) We can think of a multiset as a sequence of n natural numbers. In this way, the mapping $f : X \to \mathbb{N}$ is represented by the sequence

   $$(f(x_1), f(x_2), \ldots, f(x_n)).$$

   Addition is performed entry-wise, that is, if $\{a_i\}$ and $\{b_i\}$ are two sequences of natural numbers of length n, then

   $$(a_1, a_2, \ldots, a_n) + (b_1, b_2, \ldots, b_n) = (a_1 + b_1, a_2 + b_2, \ldots, a_n + b_n).$$

   (b) We can regard a multiset f as an inventory of the number of objects of each type from a list of types, X. We can then represent this inventory for f as a linear combination $\sum_i f(x_i)x_i$, where two such linear combinations are regarded as distinct if they differ at any coefficient. Addition is then performed by simply merging the two inventories—in effect adding together the linear combinations coefficient-wise. For example if $X = \{a, b, o\}$, where $a =$ apples, $b =$ bananas, and $o =$ oranges, then $(a+2b)+(b+2o) = a + 3b + 2o$—that is, an apple and two bananas added to a banana and two oranges is one apple, three bananas and two oranges. Thus one *can* add apples and oranges in this system![7]

---

[7]See p. 51 to realize that the positive-integer linear combinations simply encode the mappings of finite support that are elements of the monoid of multisets defined there. We used just such a

(c) Finally one can represent a multiset over $X = \{x_1, \ldots, x_n\}$ as a *monomial in commuting indeterminates*. Thus the mutiset f is represented as a product $\prod x_i^{f(x_i)}$, a monomial in the commuting indeterminates $\{x_i\}$. (Terms with a zero exponent are represented by 1 or are omitted.) For example, if $n = 4$, then $x_1^2 x_3^2 x_4^3 = x_4^3 x_3^2 x_2^0 x_1^2$. Multiplication here, is the usual multiplication of monoids one learns in freshman algebra, where exponents of a common "variable" can be added. We denote this monoid by the symbol $\mathcal{M}^*(X)$, to emphasize that the operation is multiplication. Of course we have a monoid isomorphism $\mathcal{M}(X) \to \mathcal{M}^*(X)$ in this case.[8] This isomorphism takes a sequence $(a_1, \ldots, a_n)$ of natural numbers to the monomial $x_1^{a_1} \cdots x_n^{a_n}$.

5. **The Free Monoid over X.** This was the monoid whose elements are the words of finite length over the alphabet X. These are sequences (possibly empty) of juxtaposed elements of X, $x_1 x_2 \cdots x_m$ where $m \in N$ and the "letters" x_i are elements of X, as well as the empty word which we denote by ϕ. The associative operation is *concatenation* of words, the act of writing one word right after another. The empty word is an identity element. We utilized this monoid in constructing free groups in Chap. 6, p. 166.

   In particular, the free monoid over the single letter alphabet $X = \{x\}$, consists of $\phi, x, xx, xxx, \ldots$ which we denote as $1, x, x^2, x^3, \ldots$, etc. Clearly this monoid is isomorphic to the monoid N of non-negative integers under addition (the second item in this list).

6. Let L be a lower semillattice with a "one" element 1. Thus, L is a partially ordered set $(P, \leq)$ with the property that the poset ideal generated by any finite subset of P possesses a unique maximal element. Thus if $X = \{x_1, \ldots, x_n\} \subseteq P$, then the poset ideal $I := \{y \in P \mid y \leq x_i, \text{ for } i = 1, \ldots, n\}$ contains a unique maximal element called the "meet" of X and denoted $\bigwedge X = x_1 \wedge \cdots \wedge x_n$. When applied to pairs of elements, "$\wedge$" becomes an associative binary operation on P. In this way $(P, \wedge)$ forms a monoid $M(L)$, whose multiplicative identity is the element 1. Similarly one derives a monoid $M(P, \vee)$ from an upper semilattice P with a zero element, using the "join" operation. One just applies the construction of the previous paragraph to its dual semilattice.

7.3.2 General Monoid Rings

Let R be any ring, and let M be a monoid. The *monoid ring RM* is the set of all functions $f : M \to R$ such that $f(m) \neq 0$ for only finitely many elements $m \in M$.

(Footnote 7 continued)
multiset monoid over the set of isomorphism classes of simple groups in proving the Jordan-Hölder Theorem for groups.
[8] The need for a multiplicative representation (like $\mathcal{M}^*(X)$) rather than an additive one (like $\mathcal{M}(X)$) occurs when a monoid is to be embedded into the multiplicative semigroup of elements of a ring. Without it there would be a confusion of the two additive operations.

We make RM into a ring as follows. First of all, addition is defined by "pointwise addition" of functions, i.e., if f, $g \in RM$, then we define

$$(f + g)(m) = f(m) + g(m) \in R, \ m \in M.$$

Clearly, this operation gives RM the structure of an additive abelian group. Next, multiplication is defined by what is called *convolution* of functions. That is, if f, $g \in RM$, then we define

$$(f * g)(m) = \sum_{m'm''=m} f(m')g(m'') \in R, \ m \in M. \tag{7.10}$$

The summation on the right hand side of Eq. (7.10) is taken over all pairs $(m', m'') \in M \times M$ such that $m'm'' = m$. Note that since f and g are non-zero at only finitely many elements of the domain M, the sum on the right side is a sum of finitely many non-zero terms, and so is a well-defined element of R.

If 1_M is the identity of the monoid M and if 1 is the identity of R, then the function $e : M \to R$ defined by setting

$$e(m) = \begin{cases} 1 & \text{if } m = 1_M, \\ 0 & \text{if } m \neq 1_M \end{cases}$$

is the multiplicative identity in RM. Indeed, if $f \in RM$, and if $m \in M$, then

$$(e * f)(m) = \sum_{m'm''=m} e(m')f(m'')$$
$$= f(m)$$

and so $e * f = f$. Similarly, $f * e = f$, proving that e is the multiplicative identity of RM.

Next, we establish the associative law for multiplication. Thus, let $f, g, h \in RM$, and let $m \in M$. Then, through repeated use of the distributive law among elements of R, we have:

$$(f * (g * h))(m) = \sum_{m'm''=m} f(m')(g * h)(m'')$$
$$= \sum_{m'm''=m} f(m') \left(\sum_{n'n''=m''} g(n')h(n'') \right)$$
$$= \sum_{m'm''=m} \left(\sum_{n'n''=m''} f(m')g(n')h(n'') \right)$$
$$= \sum_{m'n'n''=m} f(m')g(n')h(n'')$$

$$= \sum_{k'm''=m} \left(\sum_{m'n'=k'} f(m')g(n') \right) h(n'')$$

$$= \sum_{k'm''=m} (f * g)(k')h(m'')$$

$$= ((f * g) * h)(m).$$

The distributive laws in RM are much easier to verify, and so we conclude that $(RM, +, *)$ is indeed a ring.

One might ask: "Why all this business about mappings $M \to R$? Why not just say the elements of RM are just R-linear combinations of elements of M such as

$$\sum_{m \in M} a_m m$$

with only finitely many of the a_m non-zero?" The answer is that we can—but only with the following:

Proviso: *linear combinations of elements of M which display different coefficients are to be regarded as distinct elements of RM.*[9]

The correspondence between R-valued functions on the monoid M and finite R-linear sums as above is given by the mapping

$$f \to \sum_{m \in M} f(m)m \in RM.$$

Since the "Proviso" given above is in force, this correspondence is bijective and is easily seen to be a ring isomorphism.

Writing the elements of RM as R-linear combinations of elements of M sometimes makes calculation easier. Thus if $\alpha = \sum a_m m$, and $\beta = \sum b_n n$ are elements of RM (so only finitely many of the a_m and b_m are non-zero), then

$$\alpha\beta = \sum_{m \in M} \left(\sum_{kn=m} a_k b_n \right) m.$$

---

[9]In mathematics the phrase "linear combinations of a set of objects X" is used, even when those objects themselves satisfy linear relations among themselves. So we can't just say a monoid ring RM is the collection of R-linear combinations of the elements of M until there is some insurance that the elements of M *do not already satisfy some R-linear dependence relation*. That is why one uses the formalism of mappings $M \to R$. Two mappings differ if and only if they differ in value at some argument in M.

Note that since R may not be a commutative ring, the a_k's must always appear to the left in the products $a_k b_n$ in the formula presented above.

7.3.3 Group Rings

A first example is the so-called *group ring*. Namely, let G be a multiplicative group (regarded, of course, as a monoid), let R be a ring, and call the resulting monoid ring RG the *R-group ring of G*. Thus, elements of RG are the finite sums $\sum_{g \in G} a_g g$ where each $a_g \in R$. (The sums are finite because only finitely many of the coefficients a_g are non-zero.) Note that the group ring is commutative if and only if R is a commutative ring and G is an abelian group.[10]

7.3.4 Möbius Algebras

Let M be the monoid $M(P, \vee)$ of part 6 of Example 36. Here $(P, \leq)$ is an upper semilatice, with a "zero" element. Thus for any two elements a and b of P, there is an element $a \vee b$ which is the unique minimal element in the filter $\{y \in P \,|\, y \geq a, y \geq b\}$ generated by a and b. Then for any ring R, we may form the monoid ring RM. Since M is commutative in this case, the monoid ring is commutative if the ring R is commutative.

In the case that P is finite, Theorem 2.4.2 forces the semilattice P to be a lattice. In that particular case, if R is a field F, then the monoid ring FM is called a *Möbius Algebra of P over F* and is denoted $A_V(P)$. There are many interesting features of this algebra, that can be used to give short proofs for the many identities involving the Möbius function of a poset. The reader is referred to the classic paper of Curtis Greene [1].

7.3.5 Polynomial Rings

The Ring $R[x]$

If M is the free monoid on the singleton set $\{x\}$, then we write $R[x] := RM$ and call this the *polynomial ring* over R in the *indeterminate x*. At this point the reader, who almost certainly has had previous experience with abstract algebra, will have

---

[10]Modules over group rings are the source of the vast and important subject known as "Representation Theory". Unfortunately it is beyond the capacity of a book such as this one on basic higher algebra to do justice to this beautiful subject.

encountered polynomials, but might not recognize them in the present guise. Indeed, polynomials are most typically represented as elements (sometimes referred to as "formal sums") of the form

$$a_0 + a_1 x + a_2 x^2 + \cdots + a_n x^n,$$

where the "coefficients" $a_0, a_1, \ldots, a_n$ come from the ring R, and polynomials with different coefficients are regarded as different polynomials (the Proviso). If f is written as above, and if $a_n \neq 0$, then we say that f has *degree n*.

We mention in passing that polynomials are frequently written in "functional form:"

$$f(x) = \sum_{i=0}^{n} a_i x^i.$$

This is arguably misleading, as such a notation suggests that x is a "variable," ranging over some set. This is not the case, of course, as x is the fixed generator of the free monoid on a singleton set and therefore does not vary at all. Yet, we shall retain this traditional notation as we shall show below—through the so-called "evaluation homomorphism"—that the functional notation above is not unreasonable.

Note that if R is a ring, and if we regard $R \subseteq R[x]$ as above, then we see that R and the element x commute. Suppose now that R and S are rings, that $R \subseteq S$ and that $s \in S$ is an element of S that commutes with every element of R. The subring of S generated by R and s is denoted by $R[s]$. A moment's thought reveals that every element of $R[s]$ can be written as a polynomial in s with coefficients in R; that is, if $\alpha \in R[s]$, then we can write α as

$$\alpha = \sum_{i=0}^{n} a_i s^i,$$

for some nonnegative integer n, and where the coefficients $a_i \in R$. In this situation, we can define the *evaluation homomorphism* $E_s : R[x] \to R[s] \subseteq S$ by setting

$$E_s(f(x)) = f(s) := \sum_{i=0}^{n} a_i s^i,$$

where $f(x) = \sum_{i=0}^{n} a_i x^i$. That is to say, the evaluation homomorphism (at s) simply "plugs in" the ring element s for the indeterminate x.

The evaluation homomorphism is, as asserted, a homomorphism of rings. To prove this we need only show that E_s preserves addition and multiplication. Let $f(x), g(x) \in R[x]$, let $f(x)$ have degree k and let $g(x)$ have degree m. Setting $n = \max\{k, m\}$ we may write $f(x) = \sum_{i=0}^{n} a_i x^i$, $g(x) = \sum_{i=0}^{n} b_i x^i \in R[x]$. Then

$$E_s(f(x) + g(x)) = \sum_{i=0}^{n} (a_i + b_i)s^i$$

$$= \sum_{i=0}^{n} a_i s^i + \sum_{i=0}^{n} b_i s^i$$

$$= f(s) + g(s) = E_s(f(x)) + E_s(g(x)).$$

Next, since

$$f(x)g(x) = \sum_{l=0}^{k+m} \sum_{i=0}^{l} a_i b_{l-i} x^l,$$

we conclude that

$$E_s(f(x)g(x)) = \sum_{l=0}^{k+m} \sum_{i=0}^{l} a_i b_{l-i} s^l$$

$$= \left(\sum_{i=0}^{k} a_i s^i \right) \left(\sum_{j=0}^{m} b_j s^j \right)$$

$$= f(s)g(s) = E_s(f(x)) E_s(g(x)).$$

Therefore, $E_s : R[x] \to S$ is a ring homomorphism whose image is obviously the subring $R[s] \subseteq S$.

The Polynomial Ring $R\{X\}$ in Non-commuting Indeterminates

Let X be a set and let R be a fixed ring. The *free monoid on* X was defined on p. 166 and denoted $M(X)$. It consists of sequences $(x_1, \ldots, x_n)$, $n \in \mathbf{N}$, of elements x_i of X which are symbolically codified as "words" $x_1 x_2 \cdots x_n$. (When n is the natural number zero, then we obtain the empty word which is denoted ϕ.) The monoid operation is concatenation (the act of writing one word w_2 right after another w_1 to obtain the word $w_1 w_2$).

The resulting monoid ring $RM(X)$ consists of all mappings $f : M(X) \to R$ which assume a value different from $0 \in R$ at only finitely many words of $M(X)$. Addition is pointwise and multiplication is the convolution defined on p. 200. Then, as described there, $RM(X)$ is a ring with respect to addition and multiplication. This ring is called the *polynomial ring in non-commuting indeterminates* X and is denoted $R\{X\}$.[11]

Suppose S is a ring containing the ring R as a subring, and let B be a subset of the centralizer in S of R—i.e. $B \subseteq C_S(R)$. Suppose $\alpha : X \to B$ is some mapping of sets. We can then extend α to a function $\alpha : M(X) \to C_S(R)$, by setting

----

[11] This notation is not completely standard.

$\alpha(w) = \alpha(x_1)\alpha(x_2) \cdots \alpha(x_n)$ exactly when w is the word $x_1 x_2 \cdots x_n$, and setting $\alpha(\phi) = 1_R$, the multiplicative identity element of the ring R as well as $C_S(R)$.

For each function $f : M(X) \to R$ which is an element of the monoid ring $RM(X)$, define

$$E_\alpha^*(f) := \sum_{w \in M(X)} f(w)\alpha(w). \tag{7.11}$$

Since $f(w)$ can be non-zero for only finitely many words w of the free monoid $M(X)$, the quantity on the right side of Eq. (7.11) is a well-defined element of S. [Of course we can rewrite all this in a way that more closely resembles conventional polynomial notation. Following the discussion on p. 202 we can represent the function $f \in RM(X)$ in polynomial notation as

$$p_f := \sum_{w \in \text{Supp } f} f(w) \cdot w$$

where Supp f is the "support" of f—the finite set of words w for which $f(w) \neq 0 \in R$. Then

$$E_\alpha^*(f) = E_\alpha^*(p_f) := \sum_{w \in \text{Supp } f} f(w)\alpha(w),$$

a finite sum of elements of $RC_S(R) \subseteq S$.]

In Exercise (6) in Sect. 7.5.3 the student is asked to verify that $E_\alpha^* : RM(X) = R\{X\} \to S$ is a ring homomorphism, and is the unique ring homomorphism from $RM(X)$ to S that extends the mapping $\alpha : X \to B$.

The Polynomial Ring $R[X]$ over a Set of Commuting Indeterminates

We have defined a multiplicative version of the monoid of multisets over X, namely $\mathcal{M}^*(X)$ (see Example 36, part 4(c)). The elements of this monoid are the multiplicative identity element 1 and all finite monomials products $\prod_{i=1}^{n} x_i^{a_i}$, where $\{x_1, \ldots, x_n\}$ may range over any finite subset of X. There is a natural monoid homomorphism

$$\deg : \mathcal{M}^*(X) \to (\mathbb{N}, +)$$

which is defined by

$$\prod_{i=1}^{n} x_i^{a_i} \mapsto \sum_{i=1}^{n} a_i.$$

The image deg m is called the *degree* of m, for $m \in \mathcal{M}^*(X)$.

The monoid ring $R\mathcal{M}^*(X))$, is denoted $R[X]$ in this case. As remarked, its elements can be uniquely described as formal sums $\sum a_m m$ where $m \in \mathcal{M}^*(X)$, $a_m \in R$, and only finitely many of the coefficients a_m are non-zero. The adjective "formal" is understood to mean that two such sums are distinct if and only they exhibit some difference in their coefficients (our Proviso from p. 201). These sums are called *polynomials*. Since $\mathcal{M}^*(X)$ is a commutative monoid, the polynomial ring $R[X]$ is a commutative ring if and only if R is commutative.

A polynomial is said to be *homogeneous* if and only if all of its non-zero monomial summands possess the same degree, or if it is the zero polynomial. This common degree of non-zero summands of a homogeneous polynomial h is called the *degree* of that polynomial, and is again denoted deg h.

Any polynomial $p = \sum a_m m$ can be expressed as a sum of homogeneous polynomials $p = \sum_i h_i$ where h_i has degree $i \in \mathbb{N}$, simply by grouping the monomial summands according to their degree. Clearly, the uniqueness of the coefficients a_m which define p tells us that decomposition of an arbitrary polynomial as a sum of homogeneous polynomials is unique. The *degree of a non-zero polynomial* $p = \sum_{i=0}^{d} h_i \in R[X]$ is defined to be the highest value $d = \deg h_i$ achieved for a non-zero homogeneous summand in p. By a further extension of notation, it is denoted deg p. The convention is that the zero polynomial (where all $a_m = 0$) possesses every possible degree in $\mathbb{N}$.

Recall from p. 187 that an integral domain is a commutative ring in which the non-zero elements are closed under multiplication.

Lemma 7.3.1 *If D is an integral domain, then so is the polynomial ring $D[X]$.*

Proof The proof proceeds in three steps.

(Step 1)   *If D is an integral domain and $X = \{x\}$, then $D[x]$ is an integral domain.*

For any non-zero polynomial $p = \sum a_i x^i$ (a finite sum), the *lead coefficient* is the coefficient of the highest power of x appearing among the non-zero terms of the sum. The distributive law alone informs us that the lead coefficient of a product of two non-zero polynomials is in fact the product of the lead coefficients of each polynomial, and this is non-zero since D is an integral domain. Thus the product of two non-zero polynomials cannot be zero, and so $D[x]$ is an integral domain.

(Step 2)   *If X is a finite set, and D is an integral domain, then $D[X]$ is also an integral domain.*

We may suppose $X = \{x_1, \ldots, x_n\}$ and proceed by induction on n. The case $n = 1$ is Step 1. Suppose the assertion of Step 2 were true for $n - 1$. Then $D' := D[x_1, \ldots, x_{n-1}]$ is an integral domain. Then $D[X] = D'[x_n]$ is an integral domain by applying Step 1 with D' in the role of D.

(Step 3)   *If X is an infinite set and D is an integral domain, then $D[X]$ is an integral domain.*

Consider any two non-zero polynomials p and q in $D[X]$. Each is a finite D-linear combination of monomials and so both p and q lie in a subring $D[X']$ where X' is a finite subset of X. But $D[X']$ is itself an integral domain by Step 2, and so the product pq cannot be zero.

The proof is complete. $\square$

Theorem 7.3.2 (Degrees in polynomial domains) *If D is an integral domain, then $D[X]$ possesses the following properties:*

1. *Let D_i denote the set of homogeneous polynomials of degree i.* (Note that this set includes the zero polynomial and is closed under addition) *The degree of the product of two non-zero homogeneous polynomials is the sum of their degrees. In general, the domain $D[X]$ possesses a direct decomposition as additive groups (actually D-modules)*

$$D[X] = D_0 \oplus D_1 \oplus D_2 \oplus \cdots \oplus D_n \oplus \cdots$$

   where $D_i D_j \subseteq D_{i+j}$.
2. *For arbitrary non-zero polynomials p and q (whether homogeneous or not) the degree of their product is the sum of their degrees. The group of units consists of the units of D (embedded as polynomials of degree zero in $D[X]$).*
3. *An element of $D[X]$ is said to be irreducible if and only if it cannot be expressed as a product of two non-units.* (Notice that this definition does not permit an irreducible element to be a unit.) *If $D = F$ is a field, then every non-zero non-unit of $F[X]$ is a product of finitely many irreducible polynomials.*[12]

Proof Part 1. Let h_1 and h_2 be non-zero homogeneous polynomials of degrees d_1 and d_2, respectively. Then all formal monomial terms delivered to the product $h_1 h_2$ by use of the distributive law, possess the same degree $d_1 + d_2$. But can all the coefficients of these monomials be zero? Such an event is impossible since, by Lemma 7.3.1, $D[X]$ is an integral domain. The decomposition of $D[X]$ as a direct sum of the D_i follows from the uniqueness of writing any polynomial as a sum of homogeneous polynomials.

Part 2. The degree of any polynomial is the highest degree of one of its homogeneous summands. That the degrees add for products of homogeneous polynomials, now implies the same for arbitrary polynomials. It follows that no polynomial of positive degree can be a unit of $D[X]$, and so all units are found in $D \simeq D_0$, and are thus units of D itself.

Part 3. Let $D = F$ be a field. Suppose $p \in D[X]$ were not a finite product of irreducible elements. Then p is itself not irreducible, and so can be written as a product $p_1 p_2$ of two non-units. Suppose one of these factors—say p_1, had degree zero. Then p_1 would be a non-zero element of F. But in that case, p_1 would be a unit of $F[X]$ contrary to its being irreducible. Thus p_1 and p_2 have positive degrees, and by Part 2, these degrees must be less than the degree of p. An easy induction on degrees reveals that both of the p_i are finite products of irreducible elements, and therefore $p = p_1 p_2$ is also such a product. $\square$

Clearly the polynomial ring $R[x]$ studied in an earlier part of this section is just the special case of $R[X]$ where $X = \{x\}$. For $R[X]$ there is also a version of the evaluation homomorphism E_s of $R[x]$. The reader is asked to recall the definition of R-homomorphism which was introduced on p. 193.

---

[12] A much stronger statement about factorization in $D[X]$ is investigated in Chap. 10.

Theorem 7.3.3 (The Universal Property of Polynomial Rings) *Suppose B is a commutative ring containing R as a subring. Let $X = \{x_1, \dots, x_n\}$, and let $\phi : X \to B$ be any function. Then there exists an R-homomorphism*

$$\hat{\phi} : R[X] \to B$$

which extends ϕ.

Proof For each monomial $m = \prod_{i=1}^{n} x_i^{a_i}$ set $\hat{\phi}(m) = \prod_{i=1}^{n} \phi(x_i)^{a_i}$, which is a unique element of B. For each polynomial $p = \sum a_m m$, where m is a monomial in $\mathcal{M}^*(X)$, we set

$$\hat{\phi}(p); = \sum_m a_m \hat{\phi}(m). \tag{7.12}$$

In other words, to effect $\hat{\phi}$, one merely substitutes $\phi(x_i)$ for x_i, leaving all coefficients the same. Note that when m is the unique monomial in the x_i of degree zero (i.e. all the exponents a_i are zero) them $m = 1$, the multiplicative identity element of $R[X]$. Then by definition

$$\hat{\phi}(1) = \hat{\phi}(x_1^0 x_2^0 \cdots x_n^0) = \prod_{i=1}^{n} \phi(x_i)^0 = 1_B,$$

the multiplicative identity element of the ring B.

For all polynomials $p, p_1, p_2 \in R[X]$, and ring elements $r \in R$, one has

$$\hat{\phi}(rp) = r\hat{\phi}(p)$$
$$\hat{\phi}(p_1 + p_2) = \hat{\phi}(p_1) + \hat{\phi}(p_2)$$
$$\hat{\phi}(p_1 p_2) = \hat{\phi}(p_1)\hat{\phi}(p_2).$$

The first equation follows from the definition of $\hat{\phi}$ given in Eq. (7.12). The next two equations are natural consequences of $\hat{\phi}$ being a substitution transformation.

Since B is commutative, it follows that $\hat{\phi} : R \to B$ is an R-homomorphism as defined on p. 193. $\square$

7.3.6 Algebraic Varieties: An Application of the Polynomial Rings $F[X]$

This brief section is a side issue. Its purpose is to illustrate the interplay between polynomials in many commuting indeterminates and other parts of Algebra. The evaluation homomorphism plays a key role in this interplay.

Let n be a positive integer and let V be the familiar n-dimensional vector space $F^{(n)}$ of all n-tuples over a field F. Since the field F is commutative, the ring $F[X] := F[x_1, x_2, \ldots, x_n]$ of all polynomials in n (commuting) indeterminates $\{x_1, \ldots, x_n\}$ having coefficients in the field F, is a commutative ring.

Recall from Theorem 7.3.3 that if f is a mapping taking x_i to $a_i \in F$, $i = 1, \ldots, n$, we obtain an F-homomorphism

$$\hat{f} : F[X] \to F.$$

Since $\hat{f}$ is defined by substituting a_i for x_i, it makes sense to denote the image of polynomial p under this mapping by the symbol $p(a_1, \ldots, a_n)$.

Notice that for every vector $v = (a_1, \ldots, a_n) \in V$, there exists a unique mapping $\alpha(v) : X \to F$ taking x_i to a_i and so, by Theorem 7.3.3, there exists an F-homomorphism of rings: $\hat{\alpha}(v) : F[X] \to F$. Thus each polynomial p defines a *polynomial function* $e(p) : V \to F$ taking v to $\hat{\alpha}(v)(p)$—that is, it maps vector $v = (a_i, \ldots, a_n)$ to the field element $p(a_1, \ldots, a_n)$. (Harking back to our discussion in one variable, p is a polynomial, not a function, but with the aid of the evaluation mappings, it now *determines* a polynomial function $e(p) : V \to F$.)

The vectors v such that $e(p)(v) = 0 \in F$ are called the *zeroes of polynomial p*. Collections of vectors in V that are the common zeroes of some set of polynomials is called an *affine variety*.

Suppose we begin with an ideal J in $F[X]$. Then the *variety of zeroes of J* is defined to be the subset

$$\mathcal{V}(J) := \{(a_1, \ldots, a_n) \in V \mid p(a_1, \ldots, a_n) = 0 \text{ for all } p \in J\}.$$

In other words, $\mathcal{V}(J)$ is the set of vectors which are zeroes of each polynomial $p(x_1, \ldots, x_n)$ in J. Clearly, if we have ideals $I \subseteq J \subseteq F[X]$, then $\mathcal{V}(I) \supseteq \mathcal{V}(J)$, and so we have an order-reversing mapping

$$\mathcal{V} : \text{the poset of ideals of } F[X] \to \text{ the poset of subsets of } V.$$

Conversely, for every subset $X \subseteq V$, we may consider the collection

$$\mathcal{I}(X) := \{p(x_1, \ldots, x_n) \in F[X] \mid p(a_1, \ldots, a_n) = 0 \text{ for all } (a_1, \ldots, a_n) \in X\}.$$

Clearly $\mathcal{I}(X)$ is an ideal of $F[X]$ for any subset $X \subseteq V$ and that if $Y \subseteq X \subseteq V$, then $\mathcal{I}(Y) \supseteq \mathcal{I}(X)$. Therefore we have an order-reversing mapping of these posets in the opposite direction:

$$\mathcal{I} : \text{poset of subsets of } V \to \text{poset of ideals of } F[X].$$

Moreover, one has the *monotone relations*

$$\bar{J} := \mathcal{I}(\mathcal{V}(J)) \geq J$$
$$\bar{X} := \mathcal{V}(\mathcal{I}(X)) \leq X$$

for every ideal $J \subseteq F[X]$ and every subset $X \subseteq V$. Thus the pair of mappings forms a *Galois connection* in the sense of Chap. 2, and the two "bar" operations, ideal $J \rightarrow \bar{J}$, and subset $X \rightarrow \bar{X}$, are closure operators (idempotent order preserving maps on a poset). Their images—the "closed" objects—are a special subposet which defines a topology on the respective ambient sets. If we start with an ideal J, its closure $\bar{J}$ turns out to be the ideal

$$\sqrt{J} := \{x \in F[X] | x^n \in J, \text{ for some natural number } n\}.$$

This is the content of the famous "zeroes Theorem" of David Hilbert.[13] An analogue for varieties would be a description of the closure $\bar{X}$ of X in terms of the original subset X. These image sets $\bar{X}$ are called *(affine) algebraic varieties*. That there is not a uniform description of affine varieties in terms of V alone should be expected. The world of "V" allows only such a primitive language to describe things that we cannot say words like "nilpotent", "ideal", etc. That is why the Galois connection is useful. The mystery of these "topology-inducing" closed sets can be pulled back to a better understood algebraic world, the poset of ideals.

Is the new topology on V really new? Of course, it is not ordained in heaven that any vector space actually has a topology which is more interesting than the discrete one. In the case of finite vectors spaces, one cannot expect anything better, and indeed every subset of a finite vector space is an affine algebraic variety. Also, if F is the field of complex numbers, the two topologies coincide, although that takes a non-trivial argument. But there are cases where this topology on algebraic varieties differs from other standard topologies on V.

Every time this happens, one has a new opportunity to do analysis another way, and to deduce new theorems.

7.4 Other Examples and Constructions of Rings

7.4.1 Examples

Example 37 CONSTRUCTIONS OF FIELDS Perhaps the three most familiar fields are the rational numbers $\mathbb{Q}$, the real numbers $\mathbb{R}$, and the complex numbers $\mathbb{C}$. In addition

---

[13]Note that $\sqrt{J}$, as defined, is not necessarily an ideal when F is replaced by a non-commutative division ring.

we have the field of integers modulo a prime. Later, when we begin the Galois theory and the theory of fields, we shall meet the following ways to construct a field:

1. Forming the factor ring D/M where M is a maximal ideal in an integral domain D.
2. Forming the ring of quotients of an integral domain.
3. Forming the topological completion of a field.

The first of these methods is already validated in Theorem 7.2.5. However, making use of this theorem requires that we have sufficient machinery to determine that a given ideal is maximal. Using the Division Algorithm and the Euclidean Trick (Lemmas 1.1.1 and 1.1.2, respectively), we can flesh this out in the ring $\mathbb{Z}$ of integers, as follows.

First of all, let I be an ideal in $\mathbb{Z}$. We shall show that I is a *principal ideal*, that is, $I = \mathbb{Z}n := \{mn \mid m \in \mathbb{Z}\}$ for some integer n. If $I = 0$, then it is already clear that $I = \mathbb{Z}0$. Thus, we assume that $I \neq 0$. Since $x \in I$ implies that $-x \in I$, we may conclude that $I \cap \mathbb{Z}^+ \neq \emptyset$..

Thus, we let n be the least positive integer in I; we claim that $I = \mathbb{Z}n$. Suppose $x \in I$. By the Division Algorithm there are integers q and r with $x = qn + r$, $0 \leq r < n$. Since $r = x - qn \in I$, we infer that $r = 0$, by our minimal choice of n. Thus $x = qn$ is a multiple of n. Since x was an arbitrary member of I, one has $I = \mathbb{Z}n$.

Next, we shall show that the ideal $I = \mathbb{Z}p$ is a maximal ideal if and only if p is prime. To this end, let p be prime and let I be an ideal of $\mathbb{Z}$ containing $\mathbb{Z}p$. If $\mathbb{Z}p \subsetneq I$, then there exists an element x in the set $I - \mathbb{Z}p$. Thus, we see that $p \nmid x$. We can then apply the Euclidean Trick to obtain integers s and t with $sp + tx = 1$. But as $p, x \in I$, we find that $1 \in I$, forcing $I = \mathbb{Z}$. Therefore, when p is prime, the ideal $\mathbb{Z}p$ is maximal. The converse is even easier; we leave the details to the reader.

The complex numbers $\mathbb{C}$ is also an example of this sort. The integral domain $\mathbb{R}[x]$ also possesses an analogue of the Euclidean algorithm utilizing the degree of a polynomial to compare the remainder r with a in the standard equation $a = qb + r$ produced by the algorithm. As a result, an argument similar to that of the previous paragraph employing the Euclidean trick shows that the principal ideal $\mathbb{R}[x]p(x)$ of all multiples of the polynomial $p(x)$, is a maximal ideal whenever $p(x)$ is a polynomial that is prime in the sense that it cannot be factored into two non-units in $\mathbb{R}[x]$—that is, $p(x)$ is an *irreducible* polynomial in $\mathbb{R}[x]$. Now the polynomial $x^2 + 1$ cannot factor into two polynomials of $\mathbb{R}[x]$ of degree one, so it is irreducible and the ideal $((x^2 + 1)) := \mathbb{R}[x](x^2 + 1)$ is therefore maximal. The factor ring $F = \mathbb{R}[x]/((x^2 + 1))$ is therefore a field. Each element of the field is a coset with the unique form $a + bx + ((x^2 + 1))$, $a, b \in \mathbb{R}$. Setting $\mathbf{i} := x + ((x^2 + 1))$ and $\mathbf{1} = 1 + ((x^2 + 1))$ we see that

(i) $\mathbf{1}$ is the multiplicative identity of F,
(ii) $\mathbf{i}^2 = -\mathbf{1}$, and that
(iii) every element of the field F has a unique expression as $a\mathbf{1} + b\mathbf{i}$ for real numbers a and b.

One now sees that F conforms to the usual presentation of the field $\mathbb{C}$ of complex numbers.

The field $\mathbb{Q}$ on the other hand is the ring of quotients of the ring of integers $\mathbb{Z}$, while $\mathbb{R}$ is the topological completion of the order topology on $\mathbb{Q}$ (its elements are the familiar Dedekind cuts). Thus each of the most commonly used fields of mathematics, $\mathbb{Q}$, $\mathbb{R}$, and $\mathbb{C}$, exemplify one of the three methods of forming a field listed at the beginning of this example.

Example 38 RINGS OF FUNCTIONS. Let X be a set and let R be a ring. Set $R^X :=$ {functions $X \to R$}. For $f, g \in R^X$, we define their *point-wise* sum and product by setting

$$(f + g)(x) := f(x) + g(x), \ (fg)(x) := f(x)g(x),$$

where the sum and product on the right-hand sides of the equations above are computed in the ring R. The additive identity is the function mapping X identically to $0 \in R$; the multiplicative identity maps X identically to $1 \in R$. One easily checks the remaining ring properties in R^X.

Example 39 DIRECT PRODUCT OF RINGS. Suppose $\{R_\sigma\}_{\sigma \in I}$ is a family of rings indexed by I. Let 1_σ be the multiplicative identity element of R_σ. Now we can can form the direct product of the additive groups $(R_\sigma, +)$, $\sigma \in I$:

$$P := \prod_{\sigma \in I}(R_\sigma, +).$$

Recall that the elements of the product are functions $f : I \to \bigcup_{\sigma \in I} M_\sigma$ (disjoint union), with the property that $f(\sigma) \in R_\sigma$ for each $\sigma \in I$. We can then convert this additive group into a ring by defining the multiplication of two elements f and g of P by the rule that fg is the function $I \to \bigcup_{\sigma \in I} R_\sigma$ defined by

$$(fg)(\sigma) := f(\sigma)g(\sigma), \ \sigma \in I,$$

where the juxtaposition on the right indicates multiplication in the ring R_σ. (This sort of multiplication is often called "coordinate-wise multiplication." Note also that this example generalizes Example 38 above.) It is an easy exercise to show that this multiplication is associative and is both left and right distributive with respect to addition in P. Finally, one notes that the function 1_P whose value at σ is 1_σ, the multiplicative identity of the ring R_σ, $\sigma \in I$, is the multiplicative identity in P; so P is indeed a ring.

Now that we understand a direct product of rings, is there a corresponding direct sum of rings? The additive group $(S, +)$ of such a ring should be a direct sum $\oplus_I (R_\sigma, +)$ of additive groups of ring. But in that case, the multiplicative identity element 1_S of the ring S must be uniquely expressible as a function which vanishes on all but a finite subset F of indices in I. But if multiplication is the "coordinate-wise multiplication" of its ambient direct product, we have a problem when I is infinite. For in that case, there is an indexing element $\sigma \in I - F$, and there exists a function p_σ which vanishes at all indices distinct from σ, but assumes the value 1_σ at σ. Then we should have $1_S \cdot p_\sigma = 0 \neq p_\sigma$ contradicting the assumption that 1_S was the identity element. Since our definition of ring requires the existence of a multiplicative identity we see that any infinite direct sum of rings,

$$\oplus_I R_\sigma = \oplus_F R_\sigma,$$

although possessing most of the properties of a ring, is not actually a ring itself. This is an expense one pays in maintaining a multiplicative identity.

Example 40 OPPOSITE RINGS. Suppose we have a ring R with a given multiplication table. We can define a new multiplication (let's call it "∘") by transposing this multiplication table. That is, $a \circ b := ba$ for all a and b in R. The new operation is still distributive with respect to addition with both distributive laws simply transposed, and it is easy for the student to verify that $(R, +, \circ)$ is a ring. (Remember, we need a multiplicative identity, and the associative law should at least be checked.) We call this ring the *opposite ring* of R and denote it by the symbol $\text{Opp}(R)$.

One may now observe that any antiautomorhism of R is essentially an isormorphism $R \to \text{Opp}(R)$.

Example 41 ENDOMORPHISM RINGS. Let $(A, +)$ be an additive abelian group and let $\text{End}(A) := \text{Hom}(A, A)$ denote the set of all group homomorphisms $A \to A$. We give $\text{End}(A)$ the structure of a ring as follows. First of all, addition is defined point-wise, i.e., $(f + g)(a) := f(a) + g(a)$, $f, g \in \text{End}(A)$, $a \in A$. The reader should have no difficulty in showing that $\text{End}(A)$ is an abelian group relative to the above addition, with identity being the mapping that sends every element of A to the additive identity $0 \in A$.

The multiplication in $\text{End}(A)$ is just function composition $fg := f \circ g : A \to A$, which is always associative. The identity function 1_A which takes every element $a \in A$ to itself is clearly the multiplicative identity. Next, note that if $f, g, h \in \text{End}(A)$, and if $a \in A$, then

$$f(g + h)(a) = f(g(a) + h(a)) = fg(a) + fh(a) = (fg + fh)(a),$$

Similarly, one has the right distributive law, and so it follows that $\text{End}(A)$ is a ring relative to the above-defined addition and multiplication.

In case A is a right vector space over the field F, we may not only consider $\text{End}(A)$ as above (emphasizing only the structure of A as an additive abelian group), but we may also consider the set

$$\text{End}_F(A) := \{F\text{-linear transformations } f : A \to A\}.$$

Since the point-wise sum of F-linear transformations is again a F-linear transformation, as is the composition of two F-linear transformations, we conclude that $\text{End}_F(A)$ is, in fact, a subring of $\text{End}(A)$.

When the F-vector space A has finite dimension n, then by choosing a basis, every linear transformation of A into itself can be rendered uniquely as an $n \times n$ matrix with coefficients in F. Then addition and composition of linear transformations is duplicated as matrix addition and multiplication. In this guise, $\mathrm{End}_F(A)$ is seen to be isomorphic to the *full matrix ring over F*—the ring $\mathrm{M}_n(F)$ of all $n \times n$ matrices over F, with respect to matrix addition and multiplication.

Now suppose $\sigma \in \mathrm{Aut}(F)$, the groups of automorphisms of F. Then σ determines a bijection $m(\sigma) : M_n(F) \to M_n(F)$ which takes each matrix $A = (a_{ij})$ to $A^{m(\sigma)} := ((a_{ij})^{\sigma})$. Recall that for matrices $A = (a_{ij})$ and $B = (b_{ij})$, the (i, j)-entry of AB is the sum $\sum_k a_{ik} b_{kj}$. Observing that applying the field automorphism σ to this sum is the same as applying it to each a_{ik} and b_{kj} one sees that

$$(AB)^{m(\sigma)} = A^{m(\sigma)} B^{m(\sigma)}.$$

Similarly $m(\sigma)$ preserves sums of matrices and so $^{m(\sigma)}$ is an automorphism of the matrix ring $M_n(F)$.

Next define the *transpose mapping* $T : M_n(F) \to M_n(F)$ which takes each matrix $A = (a_{ij})$ to its *transpose* $A^T := (a_{ji})$—in other words, the entries of the matrix are reflected across its main diagonal. Now if $A = (a_{ij})$ and $B = (b_{ij})$ are matrices of $M_n(F)$ it is straightforward to check that both $(AB)^T$ and $B^T A^T$ possess the same (i, j)th entry, namely $\sum_k a_{jk} b_{ki}$ (see Example 35, p. 190). Since it is clear from the definition of T that it preserves addition, one now sees that the transpose mapping is an *antiautomorphism* of the ring $M_n(F)$.

We finally ask the reader to observe that the antiautomorphism T and the automorphism $m(\sigma)$ are commuting mappings on $M_n(F)$. Thus (from the data given at the beginning of Sect. 7.2) the composition of these two mappings in any order is one and the same antiautomorphism of the ring $M_n(F)$.

The ring $M_n(F)$ is a simple ring for every positive integer n and field F. We will not prove this until much later; it is only offered here as a "sociological" statement about the population of mathematical objects which are known to exist. It would be sadistic to propose definitions to students which turn out to be inhabited by nothing—providing yet one more description of the empty set. So there is some value in pointing out (without proof) that certain things exist.

Example 42 QUATERNIONS. As in the previous Example 37, we consider the field $\mathbb{C}$ of complex numbers with each complex number uniquely represented in the form $a + bi$, $a, b \in \mathbb{R}$, where $i^2 = -1 \in \mathbb{R}$. *Complex conjugation* is the bijection $\mathbb{C} \to \mathbb{C}$ which maps each complex number $\alpha = a + bi$, $a, b \in \mathbb{R}$ to its complex conjugate $\overline{\alpha} = a - bi$. One can easily check that for any complex numbers α and β

$$\overline{\alpha + \beta} = \overline{\alpha} + \overline{\beta} \tag{7.13}$$

$$\overline{\alpha\beta} = \overline{\alpha}\overline{\beta} \tag{7.14}$$

so that complex conjugation is an *automorphism* of the field $\mathbb{C}$.

It is useful to define the "norm" $N(\alpha)$ of a complex number as the real number $\alpha\overline{\alpha}$. Thus, if $\alpha = a + bi$, then $N(\alpha) = a^2 + b^2$, which is always a non-negative real number, and is zero only if $a = b = 0 = \alpha$. Moreover, since complex conjugation is an automorphism of the complex number field, we have

$$N(\alpha\beta) = N(\alpha)N(\beta) \text{ for all } \alpha, \beta \in \mathbb{R}.$$

Next let $\mathbb{H}$ be the collection of all 2-by-2 matrices of the form

$$h(\alpha, \beta) := \begin{pmatrix} \alpha & \beta \\ -\overline{\beta} & \overline{\alpha} \end{pmatrix}.$$

where α and β are complex numbers. Clearly by Eq. (7.13), $\mathbb{H}$ is an additive subgroup of the full group $M_2(\mathbb{C})$ of all 2-by-2 matrices with entries in $\mathbb{C}$ under matrix addition. But $\mathbb{H}$ is actually a subring of the matrix ring $M_2(\mathbb{C})$, since

$$\begin{pmatrix} \alpha & \beta \\ -\overline{\beta} & \overline{\alpha} \end{pmatrix}\begin{pmatrix} \gamma & \delta \\ -\overline{\delta} & \overline{\gamma} \end{pmatrix} = \begin{pmatrix} \alpha\gamma - \beta\overline{\delta} & \alpha\delta + \beta\overline{\gamma} \\ -\overline{\beta}\gamma - \overline{\alpha}\overline{\delta} & -\overline{\beta}\delta + \overline{\alpha}\overline{\gamma} \end{pmatrix}$$

by ordinary matrix mutiplication. Thus, using Eq. (7.14), we have

$$h(\alpha, \beta)h(\gamma, \delta) = h(\alpha\gamma - \beta\overline{\delta}, \alpha\delta + \beta\overline{\gamma}). \tag{7.15}$$

Clearly the 2-by-2 identity matrix I_2 is the multiplicative identity of the ring $\mathbb{H}$.

Notice that the complex-transpose mapping defined by

$$\begin{pmatrix} \alpha & \beta \\ -\overline{\beta} & \overline{\alpha} \end{pmatrix}^{\tau} = \begin{pmatrix} \overline{\alpha} & -\beta \\ -\overline{\beta} & \alpha \end{pmatrix}$$

induces an antiautomorphism $\tau : \mathbb{H} \rightarrow \mathbb{H}$ (see Example 35, p. 190). Thus $h(\alpha, \beta)^{\tau} = h(\overline{\alpha}, -\beta)$.

Next one may notice that

$$D(h(\alpha, \beta)) := h(\alpha, \beta)h(\alpha, \beta)^{\tau} = (N(\alpha) + N(\beta))I_2, \tag{7.16}$$

the sums of the norms of two complex numbers times the 2×2 identity matrix. Since such norms are non-negative real numbers, the sum can be zero only if each summand norm is zero, and this forces $\alpha = \beta = 0 \in \mathbb{C}$ so that $h(\alpha, \beta)$ is the 2-by-2 zero matrix, the zero element of the ring $\mathbb{H}$. Thus if $h(\alpha, \beta) \neq 0 \in \mathbb{H}$, it is an invertible matrix of $M_2(\mathbb{C})$, whose inverse is given by

$$h(\alpha, \beta)^{-1} = h(d^{-1}\alpha, d^{-1}(-\beta)), \text{ where } d = (N(\alpha) + N(\beta)),$$

and this inverse clearly belongs to $\mathbb{H}$. Thus we see that every non-zero element of $\mathbb{H}$ is a unit in $\mathbb{H}$. Thus $\mathbb{H}$ is a division ring. The reader may check that it is not a commutative ring, and so the ring $\mathbb{H}$—called "*the division ring of real quaternions*"— is our first example of a division ring which is not a field. (The student is encouraged to peruse Exercise (7) in Sect. 7.5.4.)

Example 43 THE DIRICHLET ALGEBRA AND MÖBIUS INVERSION. This example comes from Number Theory. Let $P = (\mathbb{Z}^+, |)$ denote the divisor poset on the set of positive integers. Thus one writes $a|b$ if integer a divides integer b, $a, b, \in \mathbb{Z}^+$.

Let $A := \text{Hom}(\mathbb{Z}^+, \mathbb{C})$ denote the set of all functions from the positive integers into the complex numbers, viewed as a vector space over $\mathbb{C}$. A binary operation "$*$" is defined on A in the following way: if f and g are functions, then $f * g$ is defined by the formula

$$(f * g)(n) := \sum_{d_1,d_2;d_1d_2=n} f(d_1)g(d_2).$$

It is easily seen that "$*$" is distributive with respect to vector addition and that

$$[(f * g) * h](n) = \sum_{d_1,d_2,d_3;d_1d_2d_3=n} f(d_1g(d_2)h(d_3)$$
$$= [f * (g * h)]$$

so the associative law holds. Define the function ϵ by

$$\epsilon(1) = 1 \text{ and } \epsilon(n) = 0, \text{ if } n > 1.$$

Then $\epsilon * f = f * \epsilon = f$ for all $f \in A$. Thus $A = (A, +.*)$ is a commutative ring (actually a $\mathbb{C}$-algebra) whose multiplication $*$ is called "Dirichlet multiplication" (see [23], p. 23).

The *zeta function* is the constant function ζ such that

$$\zeta(n) = 1, \text{ for all } n \in \mathbb{Z}^+.$$

The zeta function is a unit in the ring A, and its inverse, μ is called the *Möbius function*. The definition of the Möbius function depends on the fact that every positive integer n is uniquely expressible (up to the order of the factors) as a product of positive prime numbers, $n = p_1^{a_1} \cdots p_r^{a_r}$.[14] (In the case $n = 1$ all exponents a_i are zero and the product is empty.) Then $\mu(n)$ is given by:

$$\mu(n) = 1, \text{ if } n = 1$$
$$\mu(n) = 0 \text{ if some } a_i \geq 2,$$
$$\mu(n) = (-1)^r (\text{ when each } a_i = 1).$$

---

[14] A consequence of the Jordan-Hölder Theorem.

Lemma 7.4.1 *If $n > 1$*

$$\sum_{d|n} \mu(d) = 0,$$

where the sum is over all positive integer divisors of n.

Proof Suppose $n = p_1^{a_1} \cdots p_r^{a_r}$, where the p_i are positive primes. Then

$$\sum_{d|n} \mu(d) = \sum_{(e_1,\ldots,e_r)} \mu(p_1^{e_1} \cdots p_r^{e_r}) \text{ where the } e_i \text{ are 0 or 1}$$

$$= 1 - r + \binom{r}{2} + \cdots + (-1)^r$$

$$= (1-1)^r = 0.$$

$\square$

Lemma 7.4.2 $\mu * \zeta = \zeta * \mu = \epsilon$.

Proof By the definition of Dirichlet multiplication, $(\mu * \zeta)(1) = \zeta(1) \cdot \mu(1) = 1$. If $n > 1$,

$$(\mu * \zeta)(n) = \sum_{d|n} \mu(d) = 0$$

by Lemma 7.4.1. $\square$

Theorem 7.4.3 (MÖBIUS INVERSION) *If the functions f and g are related by the equation*

$$g(n) = \sum_{d|n} f(d) \text{ for all } n, \tag{7.17}$$

then

$$f(n) = \sum_{d|n} g(d)\mu(n/d) = \sum_{d|n} \mu(d)g(n/d) \tag{7.18}$$

for all positive integers n.

Proof Equation (7.17) asserts the $g = \zeta * f$. Since ζ is a unit, $f = \zeta^{-1} * g = \mu * g$. $\square$

Remark In the 1930s the group-theorist Phillip Hall recognized that the theory of the Dirichlet Algebra and its Mobius function could be generalized to more general partially ordered sets. This idea was fully expounded by Gian-Carlo Rota using an arbitrary locally finite poset to replace the integer divisor poset $P = (\mathbb{Z}^+, |)$ (see [2]). Here the analogue of the Dirichlet Algebra is defined over an arbitrary commutative ring and is call the *Incidence Algebra* in this context.[15] The zeta function and

---

[15]The incidence algebra in general is not commutative. The commutativity of the Dirichlet algebra was a consequence of the fact that the poset of divisors of any integer n is self-dual.

its inverse, the Mobius function, are still defined for the incidence algebras. Beyond this, Rota's theory exposed a labyrinth of unexpected identities involving the Möbius function. More on this interesting subject can be harvested from [1] and [2].

7.4.2 Integral Domains

Let D be any integral domain. Recall from p. 187 that this term asserts that D is a commutative non-zero ring such that if $a, b \in D$ are both nonzero elements, then $ab \neq 0$. In other words, $D^* := D \backslash \{0\}$ is closed under multiplication (and hence, from the other ring axioms, D^* is a multiplicative monoid).

The Cancellation Law

A very useful consequence of the definition of integral domain is the following:

Lemma 7.4.4 (The cancellation law) *Let D be an integral domain. Suppose for elements a, b, c in D, one has $ab = ac$. If a is non-zero, then $b = c$.*

Proof Indeed, $ab = ac$ implies that $a(b - c) = 0$. Since a is non-zero, and non-zero elements are closed under multiplication, one must conclude that $b - c = 0$, forcing $b = c$. $\square$

Domains Which Are Not Like the Integers

Obviously, the example *par excellence* of an integral domain is the ring of integers. But perhaps that is also why it may be the least representative example.

The integers certainly enjoy every property that defines an integral domain.

But there are many further properties of the integers which are not shared by many integral domains. For example, we already showed in Example 37 above that every ideal I of the ring of integers is a *principal ideal*. Such an integral domain is called a *principal ideal domain*. Other elementary examples of principal ideal domains include

$\mathbb{Z}[i] = \{a + bi \in \mathbb{C} \mid a, b \in \mathbb{Z}\}$ (the Gaussian integers);
$\mathbb{Z}[\omega] := \{a + b\omega \in \mathbb{C} \mid a, b \in \mathbb{Z}\}$ (the Eisenstein numbers)
(Here, $\omega = (-1 + \sqrt{-3})/2$, a root of $x^2 + x = 1$.)
$\mathbb{Z}[\sqrt{-2}] := \{a + b\sqrt{-2} \in \mathbb{C} \mid a, b \in \mathbb{Z}\}$.
$F[x]$, where F is any field.

We shall postpone till Chap. 9 the proofs that the above integral domains are actually principal ideal domains.

One the other hand, we shall give an example of a perfectly respectable integral domain which is not a principal ideal domain. This is the domain

$$D := \{a + b\sqrt{-5} \mid a, b \in \mathbb{Z}\}.$$

Now consider the ideal $I = 3D + (2 + \sqrt{-5})D$. We shall show that D is not a principal ideal. We start by showing that $I \neq D$. Were I equal to D, then we would have $1 \in I$, forcing the existence of elements $x = a + b\sqrt{-5}$, $y = c + d\sqrt{-5} \in D$, $a, b, c, d \in \mathbb{Z}$ with

$$1 = 3x + (2 + \sqrt{-5})y = (3a + 2c - 5d) + (3b + 2d + c)\sqrt{-5}. \qquad (7.19)$$

Thus

$$3b + c + 2d = 0 \qquad (7.20)$$
$$3a + 2c - 5d = 1. \qquad (7.21)$$

Then $c = -3b - 2d$, and substitution for c in Eq. (7.21) yields

$$3a - 6b - 9d = 1.$$

But this is impossible, as 1 is not an integer multiple of 3. Therefore, I is a proper ideal in D.

Now, by way if contradiction, assume I is a principle ideal, so that $I = \zeta D$, for some element $\zeta \in D$. Notice that complex conjugation induces an automorphism of D taking any element $z = a + b\sqrt{-5}D$ with $a, b \in \mathbb{Z}$ to $\bar{z} := a - b\sqrt{-5}$. Then the *norm* of z, which is defined by $N(z) := z\bar{z} = a + 5b^2$ is a mapping $N : D \to \mathbb{Z}$ which preserves multiplication. Precisely,

$$N(yz) = yz\bar{y}\bar{z} = yz\bar{y}\bar{z} = (y\bar{y})(z\bar{z}) = N(y)N(z).$$

Thus if $I = \zeta D$, then the norm of every element of the ideal I must be a multiple of $N(\zeta)$. Now $I = 3D + (2 + \sqrt{-5})D$ contains these two elements:

$$3 - (2 + \sqrt{-5}) = 1 - \sqrt{-5}$$
$$3 - 2(2 + \sqrt{-5}) = -1 - 2\sqrt{-5}$$

of norms 6 and 21, respectively. It follows that $\zeta = e + f\sqrt{-5}$ ($e, f \in \mathbb{Z}$), must have norm 1, or 3. If $N(\zeta) = e^2 + 5f^2 = 1$, then $\zeta = \pm 1$. But that would force $I = D$ which we have already shown is impossible. Thus, $N(\zeta) = 3$, which is also impossible, since 3 is not of the form $n^2 + 5m^2$ for any integers m and n. We conclude that I cannot be a principle ideal.

As a matter of fact we have $9 = 3.3 = (2+\sqrt{-5})(2-\sqrt{-5})$, a factorization into irreducible elements (elements that are not a product of two or more non-units) in two distinct ways. So the ring D does not have the property of unique factorization into primes that is enjoyed by the integers.

The Characteristic of an Integral Domain

Suppose now that D is an integral domain. We define the *characteristic* of D as the smallest natural number n such that

$$nx = \underbrace{x + x + \cdots x}_{n \text{ terms}} = 0$$

for *every* element x in D. This already says that $o(x)$ divides n for every $x \in D$, where $o(x)$ denotes the order of x in the additive abelian group $(D, +)$. Note that if $x \neq 0$ then

$$\underbrace{x + x + \cdots + x}_{o(1) \text{ terms}} = x(\underbrace{1 + 1 + \cdots + 1}_{o(1) \text{ terms}}) = 0,$$

from which we conclude that $o(x)$ divides $o(1)$. Conversely, from

$$x(\underbrace{1 + 1 + \cdots + 1}_{o(x) \text{ terms}}) = \underbrace{x + x + \cdots + x}_{o(x) \text{ terms}} = 0,$$

together with the fact that D is an integral domain, we see that (since $x \neq 0$) $1 + 1 + \cdots + 1 = 0$ ($o(x)$ terms), and so it follows also that $o(1)|o(x)$. Thus $o(x) = o(1)$ for every $x \in D$.

Note finally that if $o(1)$ is not a prime and is written as $o(1) = km$ for natural numbers m and n, then $0 = km1 = (k1)(m1)$ forcing $k1 = 0$ or $m1 = 0$, contradicting that all nonzero elements of D possess a constant additive order, as established in the previous paragraph.

Therefore, $o(1)$ must, in fact, be prime.

We have therefore shown that

Lemma 7.4.5 *For any integral domain D, one of the following two alternatives hold:*

(i)   *No nonzero element of D has finite additive order, or*
(ii)   *There exists a prime number p such that $px = 0$ for every element x of D. Moreover, for every integer n, such that $0 < n < p$, and every nonzero element $x \in D$, $nx \neq 0$.*

In the former case (i) we say that D *has characteristic zero*, while in the latter case (ii) we say that D *has characteristic p*.

It should be clear that all subrings of an integral domain of characteristic c ($c = 0$ or $c = p$, for some prime p) are integral domains of the same characteristic c. In particular, if D is any subring of the complex numbers $\mathbb{C}$, then it is an integral domain of characteristic 0. Thus, the Gaussian Integers, the Eisenstein numbers, or any other subring of the reals or complex numbers, must be an integral domain of characteristic 0. The field $\mathbb{Z}/p\mathbb{Z}$, where p is a prime number, clearly has positive characteristic p.

The following lemma shows that integral domains of non-zero characteristic p may possess certain endomorphisms. It will be crucial for our analysis of finite fields and inseparable extensions in Chap. 11.

Lemma 7.4.6 *Suppose D is an integral domain of positive characteristic p, a prime number. The pth power mapping $D \to D$, which maps each element $r \in D$ to its pth power r^p, is an endomorphism of D.*

Proof Since any integral domain D is commutative, one has $(xy)^p = x^p y^p$ for any $x, y \in D$—that is the pth power mapping preserves multiplication. It remains to show that for any $x, y \in D$,

$$(x + y)^p = x^p + y^p. \tag{7.22}$$

Since we are in a commutative ring, the "binomial theorem" (Theorem 7.1.2) holds: thus:

$$(x + y)^p = \sum_{k=0}^{k=p} x^k y^{p-k} \binom{p}{k}.$$

Using the fact that p is a prime number, one can show that each combinatorial number

$$\binom{p}{k} = p!/(k!(p-k)!)$$

is a multiple of p whenever $0 < k < p$.[16] The result then follows. $\square$

Finite Integral Domains

We shall conclude this subsection with a result that is not only interesting in its own right, but whose proof involves one of the most important "counting arguments" in all of mathematics. This is the so-called "Pigeon Hole Principle," which says simply that any injective mapping of a finite set into itself must also be surjective.

Theorem 7.4.7 *Let D be a finite integral domain. Then D is a field.*

Proof Let d be a nonzero element of D; we shall show that d has a multiplicative inverse. Note that right multiplication by d defines an injective mapping $D \to D$.

---

[16]In fact this argument is a special case of the same argument used in Wielandt's proof of the Sylow Theorems (Theorem 4.3.3, p. 117).

For if $ad = bd$ then $a = b$ by the Cancellation Law (see Lemma 7.4.4). By the Pigeon Hole Principle, this mapping must be surjective and so there must exist an element $c \in D$ such that $cd = 1$. This says that c is the multiplicative inverse of d and the proof is complete. $\square$

7.5 Exercises

7.5.1 Warmup Exercises for Sect. 7.1

1. Give a formal proof of Lemma 7.1.1 using only the axioms of a ring and known consequences (e.g. relevant Lemmata in Chap. 3) of the fact that $(R, +)$ is a commutative group.
2. Do the even integers form a ring? Explain your answer.
3. Let K be one of the familiar fields, $\mathbb{Q}, \mathbb{R}$, or $\mathbb{C}$, or let K be the ring of integers $\mathbb{Z}$. Let $M_2(K)$ be the collection of all 2-by-2 matrices with entries in K. Define addition and multiplication by the standard formulae

$$\begin{pmatrix} a_1 & b_1 \\ c_1 & d_1 \end{pmatrix} + \begin{pmatrix} a_2 & b_2 \\ c_2 & d_2 \end{pmatrix} = \begin{pmatrix} a_1 + a_2 & b_1 + b_2 \\ c_1 + c_2 & d_1 + d_2 \end{pmatrix}$$

$$\begin{pmatrix} a_1 & b_1 \\ c_1 & d_1 \end{pmatrix} \cdot \begin{pmatrix} a_2 & b_2 \\ c_2 & d_2 \end{pmatrix} = \begin{pmatrix} a_1 a_2 + b_1 c_2 & a_1 b_2 + b_1 d_2 \\ c_1 a_2 + d_1 c_2 & c_1 b_2 + d_1 d_2 \end{pmatrix}.$$

   (a) Show that $M_2(K)$ is a ring with respect to matrix addition and multiplication.
   (b) Explicitly demonstrate that $M_2(K)$ is not a commutative ring.
   (c) Show that the subset

$$T_2(K) = \left\{ \begin{pmatrix} a & b \\ 0 & d \end{pmatrix} \mid a, b, d \in K \right\}$$

   is a subring.
   (d) Is the set

$$T_2(K) = \left\{ \begin{pmatrix} 0 & b \\ 0 & 0 \end{pmatrix} \mid d \in K \right\}$$

   a subring of $M_2(K)$?
   (e) In the three cases where K is a field, is $M_2(K)$ a division ring?
   (f) Now suppose $K = \mathbb{Z}$, the ring of integers.
      (i) Let

$$A = \left\{ \begin{pmatrix} a & b \\ c & d \end{pmatrix} \mid a, b, c, d \in \mathbb{Z}, c \text{ is even} \right\}.$$

Is A a subring of $M_2(\mathbb{Z})$?

(ii) Let

$$A = \left\{ \begin{pmatrix} a & b \\ -b & a \end{pmatrix} | a, b \in \mathbb{Z}, \right\}.$$

Is B a subring of $M_2(\mathbb{Z})$?

4. Consider the subset S of the field $\mathbb{Q}$ of rational numbers consisting of all fractions a/b, where the non-zero integer b is not divisible by 5. Is S a subring of $\mathbb{Q}$? Explain your answer.

5. Let R be the ring $\mathbb{Z}/(12)$ of integer residues classes modulo 12. List the units of R and write out a multiplication table for the multiplicative group that they form. Can you identify this multiplicative group as one that you have met before?

6. Let B be any non-empty subset of a ring R. Show explicitly that the centralizer $C_R(B)$ is a subring of R. [Hint: Use the three-point criterion of p. 193.]

7. (The Binomial Theorem for Commutative Rings) Let R be any commutative ring, let x and y be two elements of R, and let n be a positive integer. Show that

$$(x + y)^n = \sum_{k=0}^{k=p} x^k y^{n-k} \binom{n}{k},$$

where as usual

$$\binom{n}{k} := n!/(k!(n-k)!),$$

denotes the combinatorial number. [Hint: Use induction on n and the usual recursion identities for the combinatorial numbers.]

7.5.2 Exercises for Sect. 7.2

1. Suppose $\alpha : R \rightarrow S$ is a surjective homomorphism of rings. Suppose T is a subring of S. Is the set

$$\alpha^{-1}(T) := \{r \in R | \alpha(r) \in T\}$$

a subring of R? (Be aware that subrings must contain their own multiplicative identity element.)

2. Let R be a commutative ring and let P be an ideal of R. Prove that P is a prime ideal if and only if whenever $a, b \in R$ with $ab \in P$ then $a \in P$ or $b \in P$.

3. Prove Lemma 7.2.7.

4. In a commutative ring, a non-zero element a is said to be a *zero divisor* if and only if there exists a non-zero element b such that $ab = 0$. Thus an integral

domain could be characterized as a non-zero commutative ring with no zero divisors.

Show that if R is a commutative non-zero ring with only finitely many zero divisors, then R is either an integral domain or is a finite ring. [Hint: For any zero divisor a, consider the surjective homomorphism $\phi : (R, +) \to (aR, +)$ of additive abelian groups, defined by $r \mapsto ar$, for all $r \in R$. Then note that every non-zero element of aR and every non-zero element of ker ϕ is a zero divisor. Apply the group isomorphism $R/\ker \phi \simeq (aR, +)$ to deduce that ker ϕ has finite index as a subgroup of $(R, +)$. (Explain why the homomorphism ϕ need not be a ring homomorphism.)]

5. Let R be a ring and let I, J be ideals of R. We say that I, J are *relatively prime* (or are *comaximal*) if $I + J = R$. Prove that if I, J are relatively prime ideals of R, then $IJ + JI = I \cap J$.

6. (a) Prove the *Chinese Remainder Theorem*: Let R be a ring and let I, J be relatively prime ideals of R. Then the ring homomorphism $R \to R/I \times R/J$ given by $r \mapsto (r + I, r + J)$ determines an isomorphism

$$R/(I \cap J) \cong R/I \times R/J.$$

   (b) More generally, if $I_1, I_2, \ldots, I_k \subseteq R$ are pairwise relatively prime, then the ring homomorphism $R \to R/I_1 \times R/I_2 \times \cdots \times R/I_r$, $r \mapsto (r + I_1, r + I_2, \ldots, r + I_r)$ determines an isomorphism

$$R/(I_1 \cap I_2 \cap \cdots \cap I_k) \cong R/I_1 \times R/I_2 \times \cdots \times R/I_k.$$

   [Hint: For the homomorphism in part (a), the kernel is easy to identify. The issue is showing its surjectivity. For this use $I + J = R$ to show that $0 \times (R/J)$ and $(R/I) \times 0$ are both contained in the image of the homomorphism. Part (b), cries out for an induction proof.]

7. Let R be a ring and let $I \subseteq R$ be an ideal. Show that if $I \subseteq P_1 \cup P_2 \cup \cdots \cup P_r$, where $P_1, P_2, \ldots P_r$ are prime ideals, then $I \subseteq P_j$ for some index j. [Hint: for each $i = 1, 2, \ldots, r$, let $x_i \in I - (P_1 \cup P_2 \cup \cdots \cup P_{i-1} \cup P_{i+1} \cup \cdots \cup P_r)$. Show that $x_1 + x_2 x_3 \cdots x_r \notin P_1 \cup \cdots \cup P_r$.]

8. Recall the definition of residual quotient, $(B : A) := \{r \in R | Ar \subseteq B\}$. Let R be a ring and let I, J and K be right ideals of R. Prove that $((I : J) : K) = (I : JK)$. (Here, following the custom of the ring-theorists, the symbol JK denotes the additive group generated by the set product JK. Of course it is a right ideal of R.)

9. Let R be a commutative ring and let I be an ideal of R. Define the *radical* of I to be the set

$$\sqrt{I} = \{r \in R | r^m \in I \text{ for some positive integer } m\}.$$

Show that $\sqrt{I}$ is an ideal of R containing I.

10. Let R be a commutative ring, and let $Q \subseteq R$ be an ideal. We say that Q is *primary* if $ab \in Q$ and $a \notin Q$ implies that $b^n \in Q$ for some positive integer n. Prove the following for the primary ideal $Q \subseteq R$:

   (a) $P := \sqrt{Q}$ is a prime ideal containing Q. In fact P is the smallest prime ideal containing Q. (In this case we call Q a *P-primary* ideal.)
   (b) If Q is a P-primary ideal, $ab \in Q$, and $a \notin P$, then $b \in Q$.
   (c) If Q is a P-primary ideal and I, J are ideals of R with $IJ \subseteq Q$, $I \not\subseteq P$, then $J \subseteq Q$.
   (d) If Q is a P-primary ideal and if I is an ideal $I \not\subseteq P$, then $(Q : I) = Q$.

11. Suppose that P and Q are ideals of the commutative ring R satisfying the following:

   (a) $P \supseteq Q$.
   (b) If $x \in P$ then for some positive integer n, $x^n \in Q$.
   (c) If $ab \in Q$ and $a \notin P$, then $b \in Q$.

   Prove that Q is a P-primary ideal.
12. Assume that $Q_1, Q_2, \ldots, Q_r$ are all P-primary ideals of the commutative ring R. Show that $Q_1 \cap Q_2 \cap \ldots \cap Q_r$ is a P-primary ideal.
13. Let R be a commutative ring and let $Q \subseteq R$ be an ideal. Prove that Q is a primary ideal if and only if the only zero divisors of R/Q are nilpotent elements. (An element r of a ring is called *nilpotent* if $r^n = 0$ for some positive integer n.)
14. Consider the ideal $I = n\mathbb{Z}[x] + x\mathbb{Z}[x] \subseteq \mathbb{Z}[x]$, where $n \in \mathbb{Z}$ is a fixed integer. Prove that I is a maximal ideal of $\mathbb{Z}[x]$ if and only if n is a prime.
15. If R is a commutative ring and $x \in \bigcap\{M | M$ is a maximal ideal$\}$, show that $1 + x \in \mathcal{U}(R)$, the group of units of R.

7.5.3 Exercises for Sect. 7.3

1. Let R be a ring and let M be a monoid with the following property:

   (F) *Each element m possesses only finitely many distinct factorizations $m_1 m_2$ in M.*

   Define the *completion*, $(RM)^*$ of the monoid ring RM to be the set of *all* functions $f : M \to R$ with point-wise addition and convolution multiplication. (The point is that we are no longer limited to functions of finite support.) Show that one still obtains a ring and that the usual monoid ring RM occurs as a subring of $(RM)^*$.
2. The completion of the polynomial ring $R[x]$ is usually denoted $R[[x]]$. Note that $R[[x]]$ can be viewed as the ring of "power series" in the indeterminate x. Show that while the only units of $R[x]$ are in fact the units of R, $R[[x]]$ has far more units. In particular, note that

$$(1-x)^{-1} = 1 + x + x^2 + \cdots = \sum_{i=0}^{\infty} x^i.$$

Show that, in fact, $f(x) = \sum_{i=0}^{\infty} a_i x^i$ is a unit of $R[[x]]$ if and only if a_0 is a unit of R. [Hint: Show that if $f(x) \cdot g(x) = 1$, where $f(x) = \sum a_i x^i$, $g(x) = \sum b_i x^i$, $a_i, b_i \in R$, then $a_0 b_0 = 1$ so a_0 is a unit.

On the other hand if a_0 is a unit in R, one can set $b_0 = a_0^{-1}$, and show inductively that there exist solutions b_i to

$$\sum_{i=0}^{n} a_{n-i} b_i = 0$$

in which each b_i is a $\mathbb{Z}$-linear combination of monomial words in a_0^{-1} and $\{a_1, a_2, \ldots, a_n\}$.]

3. Let R be a ring, and let M be the free commutative monoid on the two-element set $\{x, y\}$. Thus $M = \{x^i y^j \mid i, j \in \mathbb{N}\}$, where, as usual, we employ the convention that when the natural number zero appears as an exponent, one reads this as the monoid identity element. (Note that M can also be identified with the monoid of multisets over $\{x, y\}$) The monoid algebra RM is usually denoted by $R[x, y] :=$ RM and is called a *polynomial ring* in two commuting indeterminates. Elements can be written as finite sums of the form

$$f(x, y) = \sum_{i,j \geq 0} a_{ij} x^i y^j, \ a_{ij} \in R.$$

On the other hand, the polynomial ring construction can clearly be iterated, giving the polynomial ring $R[x][y] := (R[x])[y]$. Show that these rings are isomorphic.

4. (a) Suppose M and N are monoids with identity elements 1_M and 1_N (the monoid operation in both cases is multiplication and is denoted by juxtaposition of its arguments). A *monoid homomorphism* from M to N is a mapping $\phi : M \to N$ with the property that $\phi(1_M) = 1_N$ and $\phi(m_1 m_2) = \phi(m_1)\phi(m_2)$ for all elements $m_1, m_2 \in M$. Fix a ring R and form the monoid rings RM and RN. Consider the mapping $\rho(\phi) : RM \to RN$ which is defined by

$$\sum_{m \in M} r_m \cdot m \mapsto \sum_{m \in M} r_m \phi(m),$$

where $r_m \in R$ and as usual, the elements r_m are non-zero for only finitely many m's. Show that $\rho(\phi)$ is a ring homomorphism. Show that $\rho(\phi)$ is one-to-one (onto) if and only if ϕ is one-to-one (onto). In particular, if ϕ is an automorphism of M, then $\rho(\phi)$ is an automorphism of RM.

(b) Similarly we may define an *antiautomorphism of the monoid M* to be a bijection $\tau : M \to M$ such that $\tau(m_1 m_2) = \tau(m_2)\tau(m_1)$. Prove that τ must fix the identity element 1_M. In addition, let α be an antiautomorphism of the ring R. Let the mapping $\rho(\alpha, \tau) : RM \to RM$ be defined by

$$\sum\nolimits_{m \in Suppf} f(m)m \mapsto \sum\nolimits_{m \in Suppf} f(m)^\sigma m^\tau.$$

Show that $\rho(\sigma, \tau)$ is an antiautomorphism of the monoid ring RM.

(c) Let rev $: M(X) \to M(X)$ be the antiautomorphism of the free monoid on X which rewrites each word with the letters appearing in reverse order. Thus $\mathrm{rev}(x_1 x_2 \cdots x_n) = x_n x_{n-1} \cdots x_2 x_1$. Show that if R is a commutative ring, then the mapping $\beta : R\{X\} \to R\{X\}$ defined by

$$\sum\nolimits_{w \in M(X)} r_w w \to \sum\nolimits_{w \in W} r_w \, \mathrm{rev}(w),$$

is an antiautomorphism of the ring $R\{X\}$. (Here as usual, the r_w are ring elements which are non-zero for only finitely many words w.)

(d) Let β be the antiautomorphism of the preceding part of this exercise. Show that the subset $C_{R\{X\}}(\beta)$ of polynomials left fixed by β are the R-linear combinations of the palindromic words, and, if $|X| > 1$, that this set is not closed under multiplication.

5. Suppose X is a finite set, and $R[X]$ is the ring of polynomials whose commuting indeterminates are the elements of X. We suppose B is a ring which is either the zero ring or contains R in its center. Suppose $\phi : X \to B$ is any mapping into the commutative ring B. As in Theorem 7.3.3, we define the mapping $\hat{\phi} : R[X] \to B$ which maps every polynomial $\sum_{m \in M} a_m m$ (with only finitely many of the coefficients a_m in R nonzero) to the element $\sum_m a_m \phi(m)$. This mapping simply substitutes $\phi(x_i)$ for x_i in each polynomial in $R[X]$ and $x_i \in X$ Show that $\hat{\phi}$ possesses the following properties:

$$\hat{\phi}(b_1 + b_2) = \hat{\phi}(b_1) + \hat{\phi}(b_2)$$
$$\hat{\phi}(b_1 b_2) = \hat{\phi}(b_1)\hat{\phi}(b_2)$$
$$\hat{\phi}(rb) = r\hat{\phi}(b)$$

for any $b.b_1, b_2$ in B and any r in R.

6. Let $\alpha : X \to B \subseteq C_S(R)$ be a mapping from a set X into a set B which is contained in the centralizer, in ring S, of a subring R.

(a) Show that the corresponding evaluation mapping

$$E_\alpha^* : R\{X\} \to S$$

is a ring homomorphism. (See p. 204 for definitions.)

(b) (Uniqueness) If we regard X as the set of words of length one in the free monoid on X, show that E_α^* is the unique ring homomorphism $h : R\{X\} \to S$ which extends the given mapping $\alpha : X \to B$. [Show that $h(p) = E_\alpha^*(p)$ for each polynomial p.]

7. Let $X = \{x_1, \ldots, x_n\}$ be any non-empty finite set. Recall that $M(X)$ denotes the free monoid on the set X—that is, the set of all "words" of finite length in the alphabet X with the concatenation operation. Also one may recall that $\mathcal{M}(X)$ is the additive monoid of multisets of X—that is, sequences $(a_1, \ldots, a_n)$ of non-negative integers. Define the "letter inventory mapping" $\kappa : M(X) \to \mathcal{M}(X)$ where, for a word w,

$$\kappa(w) = (a_1, \ldots, a_n)$$

if and only if, for each i, the letter x_i occurs exactly a_i times in the word w. Show that κ is a monoid homomorphism (as defined in the first part of Exercise 4). From the same exercise, conclude that for any commutative ring R, the mapping κ extends to an R-homomorphism of rings $R\{X\} \to R[X]$.

8. Let R be a subring of ring S and let B be a set of pairwise commuting elements of $C_S(R)$. Suppose $\alpha : X \to B$ is a mapping of set X into B.

(a) Show that the mapping $E_\alpha : R[X] \to S$ defined on p. 204 is a homomorphism of rings.

(b) (Uniqueness) Show that if X is regarded as a subset of $R[X]$ (namely as the set of those monomials which are words of length one), then for any ring homomorphism $\nu : R[X] \to S$ which extends $\alpha : X \to B$, one must have $\nu = E_\alpha$.

9. Again let $\alpha : X \to B$ be a mapping of set X into B, a collection of pairwise commuting elements of the centralizer in ring S of its subring R. Let κ be the monoid homomorphism of Exercise 7 and let

$$\rho(\kappa) : R\{X\} = RM(X) \to R\mathcal{M}(X) = R[X]$$

be its extension to the monoid rings as in Exercise 4. Show that

$$E_\alpha^* = E_\alpha \circ \rho(\kappa).$$

[Hint: Use the uniqueness of E_α^*.]

10. Let n be a positive integer. Suppose $A = (a_{ij})$ is an $n \times n$ matrix with entries drawn from the ring R. For $j \in \{1, \ldots, n\}$, let e_j be the n-tuple of $R^{(n)}$ whose jth entry is $1 \in R$, the multiplicative identity of R, and all of whose remaining entries are equal to $0 \in R$. The matrix A is said to be R-invertible if and only if each e_j is an R-linear combination of the rows of A. Let $X = \{x_1, \ldots, x_n\}$, and let

$$\sigma(A) : R[X] \to R[X]$$

which replaces each polynomial $p(x_1, \ldots, x_n)$ by the polynomial

$$p\left(\sum a_{1i}x_i, \sum a_{2i}x_i, \ldots, \sum a_{ni}x_i\right),$$

the parameter i in each sum ranging from 1 to n. Show that if A is R-invertible, then $\sigma(A)$ is an automorphism of $R[X]$. [Hint: Use the evaluation homomorphism E_α for an appropriate triplet (α, B, S). Then show that it is onto.]

11. Let G be a finite group, let F be a field, and let $\chi : G \to F^\times$ be a homomorphism of G into the multiplicative group $F^\times := F\backslash\{0\}$. Define the element $\epsilon_\chi = \sum_{g\in G} \chi(g^{-1})g \in FG$ and show that $\epsilon_\chi^2 = |G|\epsilon_\chi$.

12. Same assumptions as the above exercise. Show that the element ϵ_χ, commutes with every element of FG by showing that for all $g \in G$, $g\epsilon_\chi = \chi(g)\epsilon_\chi$.

7.5.4 Exercises for Sect. 7.4

1. Let $\{R_\sigma\}_{\sigma\in I}$ be a family of rings and let $P = \prod_{\sigma\in I} R_\sigma$ be the direct product of these rings. If K is any subset of the index set I, set $P_K := \{f \in P \mid f(\sigma) = 0 \text{ for all } \sigma \notin K\}$. Show that P_K forms a (two-sided) ideal of P.

2. Let X be a set and let R be a ring. In Example 38 we defined the ring of functions R^X. Prove that

$$R^X \cong \prod_{x\in X} R_x,$$

where $R_x \simeq R$ for each $x \in X$.

3. Prove that any non-zero subring of an integral domain contains the multiplicative identity element of the domain. [Hint: use Lemma 7.4.4.]

4. Prove that the polynomial ring $\mathbb{Z}[x]$ is not a principal ideal domain by showing that the ideal $I := 2\mathbb{Z}[x] + x\mathbb{Z}[x]$ is not a principal ideal.

5. Let d be a positive integer congruent to -1 modulo 4 and consider the integral domain

$$D := \{a + b\sqrt{-d}\mid a, b \in \mathbb{Z}\}$$

whose operations are inherited as a subring of the complex number field $\mathbb{C}$. Show that the ideal $I := 2D + (1 + \sqrt{-d})D$ is not a principal ideal.

6. Suppose $\omega = e^{2i\pi/3} = (-1 - \sqrt{-3})/2$, so ω is a complex number satisfying the identity: $\omega^2 = -\omega - 1$. On p. 189 and again on p. 218 we introduced the *Eisenstein numbers*, the integral domain $\mathbb{Z}[\omega]$. Show that the mapping $\mathbb{Z}[\omega] \to \mathbb{Z}[\omega]$ defined by

$$a + b\omega \mapsto a + b\omega^2, a, b \in \mathbb{Z}$$

is an automorphism of this ring.

7. Let $\tau : M_2(\mathbb{C}) \to M_2(\mathbb{C})$ be the composition of the transpose antiautomorphism, T, and the automorphism $m(c)$ induced by complex conjugation ($c : \mathbb{C} \to \mathbb{C}$) of the matrix entries. (See Example 41 for notation and the fact that $\tau = c \circ T = T \circ c$.) Let $\mathbb{H} = \{h(\alpha, \beta) | (\alpha, \beta) \in \mathbb{C} \times \mathbb{C}\}$ be the ring of real quaternions viewed as a subring of M_2). (See p. 215.)

   (a) Show that as an antiautomorphism of $M_2(\mathbb{C})$, that τ^2 is the identity automorphism—that is, τ is an involution.
   (b) Show that $\mathbb{H}$ is invariant under this antiautomorphism.
   (c) Establish that $h(\alpha, \beta)^\tau = h(\alpha^c, -\beta)$ and conclude that the centralizer in $\mathbb{H}$ of the antiautomorphism τ (that is, its fixed points) is the center of $\mathbb{H}$, the set $\mathbb{R}\mathbf{I}_2 = \{h(r, 0) | r \in \mathbb{R}\}$, the real multiples of the 2-by-2 identity matrix.
   (d) Show that for any $h \in \mathbb{H}$, $D(h) := hh^\tau$ is a "real" element—that is an element of the center $Z(\mathbb{H})$. [Hint: Show that hh^τ is fixed by τ.]
   (e) Compute that $D(h(\alpha, \beta)) = N(\alpha) + N(\beta) = det(h(\alpha, \beta))$.
   (f) Show that $D(h_1 h_2) = D(h_1)D(h_2)$, for all $h_1, h_2 \in \mathbb{H}$. [Hint: Use the fact that τ is an antiautomorphism of $\mathbb{H}$ and that $D(\mathbb{H})$ is in the center of $\mathbb{H}$.]
   (g) Let
   $$\mathbb{H}|_{\mathbb{Z}} := \{h(\alpha, \beta) | \alpha, \beta \text{ Gaussian integers}\}.$$

   Show that $\mathbb{H}|_{\mathbb{Z}}$ is a subring of $\mathbb{H}$ which is also invariant under the antiautomorphism τ. [Hint: Just check the rules for multiplying and applying τ for the quaternionic numbers $h(\alpha, \beta)$.]
   (h) Using the previous two steps, prove that the set of integers which are the sum of four perfect squares, is closed under multiplication. (This set turns out to include all natural numbers.)

References

1. Greene C (1982) The Möbius function of a partially ordered set. In: Rival I (ed) Appeared in ordered sets. D. Reidel Publishing Company, Dordrecht, pp 555–581
2. Rota G-C (1964) On the foundations of combinatorial theory I. Theory of Mobius functions Zeitschrift für Wahrscheinlichkeitstheorie, Band 2, Heft 4. Springer, Berlin, pp 340–368

Chapter 8
Elementary Properties of Modules

Abstract One way to study rings is through R-modules, which (in some possibly non-faithful way) serve to "represent" the ring. But such a view is restrictive. With R-modules one enjoys an increased generality. Any property possessed by an R-module can conceivably apply to the ring R itself, as a module over itself. Also, any universal property gains a greater strength, when one enlarges the ambient realm in which the property is stated—in this case from rings R, to their R-modules. This chapter still sticks to basics: homomorphisms, submodules, direct sums and products and free R-modules. Beyond that, chain conditions on the poset of submodules can be seen to have important consequences in two areas: endomorphisms of modules, and their generation. The former yields the Krull-Remak-Schmidt Theorem concerning the uniqueness of direct decompositions into indecomposable submodules, while, for the latter, the ascending chain condition is connected with finite generation via Noether's Theorem. (The existence of rings of integral elements is derived from this theorem.) The last sections of the chapter introduce exact sequences, projective and injective modules, and mapping properties of $\mathrm{Hom}(M, N)$—a hint of the category-theoretic point of view to be unfolded at the beginning of Chap. 13.

8.1 Basic Theory

8.1.1 Modules over Rings

Fix a ring R. A *left module over R* or *left R-module* is an additive group $M = (M, +)$, which admits a composition (or *scalar multiplication*)

$$R \times M \to M$$

(which we denote as a juxtaposition of arguments, as we do for multiplication in R) such that for all elements $a, b \in R$, $m, n \in M$,

$$a(m + n) = am + an \tag{8.1}$$

$$(a + b)m = am + bm \tag{8.2}$$

© Springer International Publishing Switzerland 2015

E. Shult and D. Surowski, *Algebra*, DOI 10.1007/978-3-319-19734-0_8

$$(ab)m = a(bm) \tag{8.3}$$

$$1 \cdot m = m, \tag{8.4}$$

where 1 denotes the multiplicative identity element of R. It follows from $0_R + 0_R = 0_R$ that for any element m of M, $0_R m + 0_R m = 0_R m$. Thus $0_R m = 0_M$ for any $m \in M$. Here, of course, 0_R denotes the additive identity of the additive group $(R, +)$ while 0_M denotes the additive identity of the additive group $(M, +)$.

A *right R-module* is an additive group $M = (M, +)$ admitting a right composition $M \times R \to M$, such that, subject to the same conventions of denoting this composition by juxtaposition, we have for all $a, b \in R$, $m, n \in M$,

$$(m + n)a = ma + na \tag{8.5}$$

$$m(a + b) = ma + mb \tag{8.6}$$

$$m(ab) = (ma)b \tag{8.7}$$

$$m1 = m. \tag{8.8}$$

Again we can easily deduce the identities

$$m0_R = 0_M \tag{8.9}$$

$$m(-a) = -(ma) \tag{8.10}$$

where 0_R and 0_M are the additive identity elements of the groups $(R, +)$ and $(M, +)$, and where m is an arbitrary element of M.

Notice that for a left R-module, M, the set $\{r \in R | rM = 0\}$ is a 2-sided ideal of R. It is called the *left annihilator* of M. Similarly, for a right R-module M the set $\{r \in R | Mr = 0\}$ is also a 2-sided ideal, called the *right annihilator of M*. In either case (left module or right), the module is said to be *faithful* if and only if the corresponding annihilating ideal is the 0-ideal. (See Exercise (3) in Sect. 8.5.1.)

8.1.2 The Connection Between R-Modules and Endomorphism Rings of Additive Groups

Consider first a left R-module M. For each element $r \in R$, the mapping $\rho(r)$: $(M, +) \to (M, +)$ defined by $m \mapsto rm$, is an endomorphism of the additive group $(M, +)$. (This is a consequence of Eq. (8.1).) As remarked in Example 41 of Chap. 7, the set $\text{End}(M)$ of all endomorphisms of the additive group $(M, +)$ forms a ring under point-wise addition of endomorphisms and composition of endomorphisms. Now for any elements r, s in the ring R and $m \in M$, Eq. (8.3) yields

$$\rho(rs)m = (rs)m = r(sm) = \rho(r)[\rho(s)(m)] = (\rho(r) \circ \rho(s))(m).$$

Thus

$$\rho(rs) = \rho(r) \circ \rho(s). \tag{8.11}$$

Similarly, Eq. (8.2) yields

$$\rho(r+s) = \rho(r) + \rho(s). \tag{8.12}$$

These equations tell us that $\rho : R \to \mathrm{End}(M)$ is a homomorphism of rings.

Notice that $\ker \rho = \{r \in R | rM = 0\}$, is the left annihilator of M as defined in the previous subsection.

Conversely, if we are given such a ring homomorphism $\rho : R \to \mathrm{End}(M)$, where M is an additive group, then M acquires the structure of a left R-module by setting $rm := \rho(r)(m)$, for all $r \in R$, $m \in M$.

In the same vein, a right R-module structure on the abelian group $(M, +)$ is equivalent to specifying a ring homomorphism $R \to \mathrm{Opp}(R)(\mathrm{End}(M))$ Here $\mathrm{Opp}(R)$ is the additive group of all endomorphisms of M endowed with a multiplication $\circ_o$ where, for any endomorphisms α and β,

$$\alpha \circ_o \beta = \beta \circ \alpha,$$

the endomorphism obtained by first applying α and then applying β in that chronological order. (See Example 40.)

Remark At times one might wish to convert a left R-module to a right S-module, or the reverse. We have the following constructions:

Suppose R and S are rings whose multiplicative identity elements are denoted e_R and e_S, respectively.

1. *Let M be a left R-module. Suppose $\psi : S \to \mathrm{Opp}(R)$ is a ring homomorphism for which $\psi(e_S) = e_R$. For each $m \in M$, and $s \in S$ define*

$$ms := \psi(s)m.$$

Since $\psi(s) \in R$, and M is a left R-module, the right side is a well-defined element of M. With multiplication $M \times S \to M$ in this way, it is easy to show that M becomes a right S module.

2. *Similarly if M is a right S module, then, from a ring homomorphism $\phi : R \to \mathrm{Opp}(S)$ for which $\phi(e_R) = e_S$, one may convert M into a left R-module.*

Proof of the appropriate module axioms is requested in Exercises in Sect. 8.5.2.

8.1.3 Bimodules

Finally, there is a third type of module, an *(R,S)-bimodule*. Here M is a left R-module and at the same time a right S-module, and the left and right compositions are connected by the law

$$r(ms) = (rm)s, \text{ for all } (r, m, s) \in R \times M \times S. \tag{8.13}$$

Although this resembles a sort of "associative law", what it really says is that every endomorphism of M induced by right multiplication by an element of ring S commutes with every endomorphism induced by left multiplication by an element of ring R. We shall consider bimodules in more detail later, especially in our discussions of tensor products (see Sect. 13.3).

Here are some examples of bimodules.

Example 1 Let T be a ring and let R and S be subrings of T, each containing the multiplicative identity element e_T of T. Then $(T, +)$ is an additive group, which admits a composition $T \times S \to T$. It follows directly from the ring axioms, that with respect to this composition, T becomes a right S module. We denote this module by the symbol T_S when we wish to view T in this way.

Also, $(T, +)$ admits a composition $R \times T \to T$ with respect to which T becomes a *left* R-module which we denote by $_RT$.

Finally $(T, +)$ admits a composition

$$S \times T \times R \to T$$

inherited directly from ring multiplication. With respect to this composition T becomes an (R, S)-bimodule, which we denote by the symbol $_RT_S$. Notice that here, the special law for a bimodule (Eq. 8.13) really *is* the associative law of T (restricted to certain triples, of course). (Note that in this example, T can be replaced by any 2-sided ideal I, to form a (R, S)-bimodule $_RI_S$.)

Example 2 If R is a commutative ring, and if M is a right R-module, we may define a left R-module structure on M by defining $r \cdot m := mr$, $r \in R$, $m \in M$. Note that the commutativity of R guarantees the validity of condition (8.3) since

$$r(ms) = r(sm) = (rs)m = m(rs)(mr)s = (rm)s. \tag{8.14}$$

Thus we see that M is also an (R, R)-bimodule. Bimodules constructed in this way have a role to play in defining multiple tensor products and the tensor algebra (pp. 481–493), and for this reason we give them a special name: *symmetric bimodules*.

Example 3 Here is a simple variation on the preceding example. Let σ be an automorphism of the commutative ring R. Then a right R-module M can be converted into an (R, R)-bimodule by declaring that $mr = r^\sigma m$ for all $(m, r) \in M \times R$. (One simply emulates the sequence of Eq. (8.14) applying σ or its inverse when ring elements pass to the left or right over module elements.)

Remark In these two examples, the commutativity of R is more or less forced. For suppose M is an arbitrary right R-module and σ is automorphism of the additive group $(R, +)$. If we attempt to convert M into an (R, R)-bimodule by the rule $mr = r^\sigma m$ for all $(m, r) \in M \times R$, three conclusions are immediate: (i) σ is easily seen to be an antiautomorphism of R (that is, $(rs)^\sigma = s^\sigma r^\sigma$), (ii) σ takes the right annihilator of M_R, that is, the two-sided ideal $I = \mathrm{Ann}_R(M_R) := \{r \in R | Mr = 0\}$, to the left annihilator $I^\sigma = \mathrm{Ann}_R({}_R M)$, and (iii) that the factor ring R/I is commutative. Thus if $I = I^\sigma$, we have recovered Example 3 above with R replaced by $R' := R/I$.

8.1.4 Submodules

Fix a (right) R-module M. A *submodule* is an additive subgroup of $(M, +)$ that is closed under all scalar multiplications by elements of R.

It is easy to see that a subset N is a submodule of the right R-module M, if and only if

  (i) $N + N = N$
  (ii) $-N = N$, and
  (iii) $NR \subseteq N$.

It should be remarked that the last containment is actually the equality $NR = N$, since right multiplication by the identity element 1_R induces the identity endomorphism of N.

Suppose now $\mathcal{N} = \{N_\sigma | \sigma \in I\}$ is a family of submodules of the right R-module M. The intersection

$$\bigcap_{\sigma \in I} N_\sigma$$

of all these submodules is clearly a submodule of M. This is the unique supremum of the poset of all submodules contained in all of the submodules in $\mathcal{N}$.

Now let X be an arbitrary subset of the right R-module M. The *submodule generated by* X is defined to be the intersection of all submodules of M which contain the subset X and is denoted $\langle X \rangle_R$. From the previous paragraph, this intersection is the unique smallest submodule in the poset of all submodules of M which contain X.

This object has another description. Let $\bar{X}$ be the set of all elements of M which are finite (right) R-linear combinations of elements from X—that is elements of the form $x_1 r_1 + \cdots + x_n r_n$ where n is any natural number, the x_i belong to X and the r_j belong to the ring R. Then $\bar{X}$ is a subgroup (recall that $x(-1_R) = -x$) and $\bar{X}R = \bar{X}$, so $\bar{X}$ is a submodule. On the other hand, it must lie in any submodule which contains X and so must coincide with $\langle X \rangle_R$.

In the case that $X = \{x_1, \ldots, x_n\}$ is a finite set,

$$\bar{X} = x_1 R + \cdots + x_n R.$$

A right R-module is said to be *finitely generated* if and only if it is generated by a finite set. Thus right R-module N is finitely generated if and only if

$$N = x_1 R + \cdots + x_n R$$

for some finite subset $\{x_1, \ldots x_n\}$ of its elements.

Now once again let $\mathcal{N} = \{N_\sigma | \sigma \in I\}$ be a family of submodules of the right R-module M. The *sum over $\mathcal{N}$* is the submodule of M generated by the set-theoretic union $\cup\{N | N \in \mathcal{N}\}$ of all the submodules in the family $\mathcal{N}$ and is denoted $\sum_{\sigma \in I} N_\sigma$. It is therefore the set of all finite R-linear combinations of elements which lie in at least one of the submodules in $\mathcal{N}$. But of course, as each submodule is invariant under right multiplication by elements of R, this simply consists of all finite sums of elements lying in at least one module of the family. In any event, this is the unique member of the set of all submodules of M which contain every member of $\mathcal{N}$.

We summarize this information in

Lemma 8.1.1 *Let R be a ring, and let M be a left R-module. The poset of all submodules of M (with containment as the partial order) is a lattice with unrestricted meets and joins, given respectively by the intersection and sum operations on arbitrary families of submodules.*

Obviously, a corresponding lemma holds for the poset of submodules of a given *left* module over some ring R.

8.1.5   Factor Modules

Now suppose N is a submodule of the right R-module M. Since N is a subgroup of the additive group $(M, +)$, we may form the factor group M/N whose elements are the additive cosets $x + N$, $x \in M$ of N. Next, define an R-scalar multiplication $M/N \times R \to M/N$ by setting $(x + N)r := xr + N$. That this is well defined is an immediate consequence of the fact that $Nr \subseteq N$ for each $r \in R$. Showing the remaining requisite properties (8.5)–(8.8) is equally routine, completing the definition of the quotient right R-module M/N. Again, this construction can be duplicated for left R-modules: For each submodule N of a left R-module M, we obtain a left R-module M/N.

8.1.6 The Fundamental Homomorphism Theorems for R-Modules

Suppose M and N are right R-modules (for the same ring R). An *R-module homomorphism* $M \to N$ is a homomorphism $f : (M, +) \to (N, +)$ between the underlying additive groups, such that

$$f(m)r = f(mr), \qquad (8.15)$$

for all $m \in M$ and all $r \in R$. In other words, these are group homomorphisms which commute with the right multiplications by each of the elements of the ring R.

It follows from the definition that the composition $\beta \circ \alpha$ of two module homomorphisms $\alpha : A \to B$ and $\beta : B \to C$, is an R-homomorphism $A \to C$.

Before going further, we will find it convenient to adopt a common visual-linguistic device for discussing compositions of homomorphisms. Suppose $\alpha : A \to B$, $\beta : B \to C$ and $\gamma : A \to C$ are homomorphisms. This data is assembled in the following diagram:

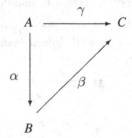

We say that the "diagram commutes" if and only if $\gamma = \beta \circ \alpha$, the composition of α and β.

We use similar language for square diagrams:

$$
\begin{array}{ccc}
A & \xrightarrow{\gamma} & C \\
\alpha \downarrow & & \downarrow \delta \\
C & \xrightarrow{\beta} & D
\end{array}
$$

Again, such a diagram "commutes" if and only if $\delta \circ \gamma = \beta \circ \alpha$.

The adjectives which accrue to an R-homomorphism are exactly those which are applicable to it as a homomorphism of additive groups: Thus the R-homomorphism f is

- an *R-epimorphism*
- an *R-monomorphism*
- an *R-isomorphism*

if and only if f is an epimorphism, monomorphism, or isomorphism, respectively, of the underlying additive groups.

We define the *kernel of an R-homomorphism* $f : M \to N$ among right R-modules, simply to be the kernel of f as a homomorphism of additive groups. Explicitly, ker f is the set of all elements of M which are mapped to the additive identity 0_N of the additive group N. But the equation $0_N R = 0_N$ shows that if x lies in ker f, then the entire set xR must be contained in ker f since $f(xR) = f(x)R = 0_N R = 0_N$. Thus the kernel of an R-homomorphism must be an R-submodule. At the same time, the image, $f(M) \subseteq N$ of the homomorphism is easily checked to be a submodule of N. Note finally that if $K \subseteq M$ is a submodule of M, then there exists a surjective homomorphism $\pi : M \to M/K$, defined by setting $\pi(m) := m + K \in M/K$, $m \in M$. This is called the *projection homomorphism* of M *onto* M/K.

The student should be able to restate the obvious analogues of the above definitions for left R-modules as well.

Now we can state

Theorem 8.1.2 (The Fundamental Theorem of Module Homomorphisms) *Suppose $f : M \to N$ is a homomorphism of right (left) R-modules.*

(i) *If $K \subseteq$ ker f is a submodule of M, then the mapping $\bar{f} : M/K \to N$ defined by setting $\bar{f}(m + K) := f(m)$ is a well-defined module homomorphism making the diagram below commute:*

$$
\begin{array}{ccc}
M & \xrightarrow{\;f\;} & N \\[4pt]
{\scriptstyle \pi}\big\downarrow & \nearrow{\scriptstyle \bar{f}} & \\[4pt]
M/K & &
\end{array}
$$

(ii) *If $K =$ ker f, then the homomorphism $\bar{f} : M/K \to f(M)$ is an isomorphism.*

Proof We have already seen that the mapping $\bar{f} : M/K \to N$ is a well-defined homomorphism of abelian groups. However, as

$$\bar{f}((x + K)r) = f(xr) = f(x)r = (\bar{f}(x + K))r,$$

we infer that $\bar{f}$ is also an homomorphism of R-modules, proving part (i). Note that if $K =$ ker f, then we again recall that the induced homomorphism $\bar{f} : M/K \to f(M)$ is an isomorphism of abelian groups. From part (i) $\bar{f}$ is an R-module homomorphism, proving part (ii). $\square$

Theorem 8.1.3 (i) (The Correspondence Theorem for R-modules.) *If $f : M \to N$ is an R-epimorphism of right (left) R-modules, then there is a poset isomorphism between $\mathcal{S}(M,$ ker $f)$, the poset of all submodules of M which contain ker f and the poset $\mathcal{S}(N, 0_N)$ of all submodules of N.*

(ii) (The Composition Theorem for R-modules.) *Suppose $A \leq B \leq M$ is a chain in the poset of submodules of a right (left) R-module M. Then there is an R-isomorphism*

$$M/A \to (M/B)/(B/A)$$

Proof If $N' \in \mathcal{S}(N, 0_N)$, then $f^{-1}(N') \in \mathcal{S}(M, \ker f)$ and $ff^{-1}(N') = N'$. On the other hand, if $M' \in \mathcal{S}(M, \ker f)$ and if $M'' = f^{-1}f(M'')$, then $f^{-1}f(M'') \supseteq M''$. However, if there exists an element $m \in f^{-1}f(M'')\backslash M''$, then $f(m) = f(m'')$ for some $m'' \in M''$. But then, $m - m'' \in \ker f$, forcing $m - m'' \in M''$ and so $m \in M''$, a contradiction. Therefore the homomorphism f establishes a bijection $\mathcal{S}(M, \ker f) \to \mathcal{S}(N, 0_N)$, proving part (i).

For part (ii), note that the projection homomorphism $p : M \to M/A$ contains the submodule B in its kernel. Therefore, the induced mapping $M/B \xrightarrow{\bar{p}} M/A$ is a well-defined surjective R-module homomorphism. However, since it is clear that $\ker \bar{p} = A/B$, we may apply Theorem 8.1.2 to infer that $(M/B)/(A/B) \cong M/A$, proving part (ii). $\square$

Theorem 8.1.4 (The Modularity Theorem for R-modules.) *Suppose A and B are submodules of the right (left) R-module M. Then there is an R-isomorphism*

$$(A + B)/A \to B/(B \cap A).$$

Proof We have the composite homomorphism

$$B \hookrightarrow A + B \xrightarrow{\pi} (A + B)/A,$$

whose kernel is obviously $B \cap A$. Since every element of $(A + B)/A$ is of the form $(a + b) + A = b + A$, $a \in A$, $b \in B$, we infer that the above homomorphism is also surjective. Now apply Theorem 8.1.3, part (ii). $\square$

8.1.7 The Jordan-Hölder Theorem for R-Modules

An R-module is said to be *irreducible* if and only if its only submodules are itself and the zero submodule—that is, it has no proper submodules.

Let $\mathrm{Irr}(R)$ denote the collection of all isomorphism classes of *non-zero* irreducible R modules. Suppose M is a given R-module, and let $P(M)$ be the lattice of all its submodules. Recall that an *interval [A,B]* in $P(M)$ is the set $\{C \in P(M) | A \leq C \leq B\}$ (this is considered to be undefined if A is not a submodule of B). Therefore, in this context, a *cover* is an interval $[A, B]$ such that A is a maximal proper submodule of B. Finally, recall that a non-empty interval $[A, B]$ is said to be *algebraic* if and only if there exists a finite unrefinable chain of submodules from A to B.

Because of the Correspondence Theorem for R-modules (Theorem 8.1.3), we see that (A, B) is a cover if and only if B/A is a non-trivial irreducible R-module. Thus we have a well-defined mapping

$$\mu : \mathrm{Cov}(P(M)) \to \mathrm{Irr}(R),$$

from the set of all covers in the lattice $P(M)$ to the collection of all non-zero irreducible R-modules. If A_1 and A_2 are distinct proper submodules of a submodule B, with each A_i maximal in B, then $A_1 + A_2 = B$, and by the Modularity Theorem for R-modules (Theorem 8.1.4), $A_2/(A_1 \cap A_2)$ (being isomorphic to B/A_1) is an irreducible R-module. Thus if $\{[A_1, B], [A_2, B]\} \in \mathrm{Cov}(P(M))$ and $A_1 \neq A_2$, then $[A_1 \cap A_2, A_i]$, $i = 1, 2$ are also covers. This is the statement that

- *The poset $P(M)$ of all submodules of the R-module M, is a semi-modular lattice.*

At this point the Jordan-Hölder Theorem (Theorem 2.5.2 of Chap. 2) for semi-modular lower semilattices, implies the following:

Theorem 8.1.5 (Jordan-Hölder Theorem for R-modules.) *Let M be an arbitrary R-module, and let $P(M)$ be its poset of submodules.*

(i) *Then the mapping*
$$\mu : Cov(P(M)) \to Irr(R),$$

extends to an interval measure

$$\mu : Alg(P(M)) \to \mathcal{M}(Irr(R)),$$

from the set of all algebraic intervals of $P(M)$ to the monoid of multisets over the collection of all isomorphism classes of non-trivial irreducible R-modules. Specifically, the value of this measure μ at any algebraic interval $[A, B]$, is the multiset which inventories the collection of irreducible modules A_{i+1}/A_i which appear for any finite unrefinable chain $A = A_0 < A_1 < \cdots < A_k = B$ according to their isomorphism classes. The multiset of isomorphism classes of irreducible modules is the same for any such chain.

(ii) *Suppose M itself possesses a composition series—that is, there exists a finite unrefinable properly ascending chain of submodules beginning at the zero submodule and terminating at M. The composition factors are the finitely many non-trivial irreducible submodules formed from the factors between successive members of the chain. Then the collection of irreducible modules so obtained is the same (up to isomorphism) with the same multiplicities, no matter what composition series is used. In particular all composition series of M have the same length.*

8.1.8 Direct Sums and Products of Modules

Let I be any index set, and let $\{M_\sigma | \sigma \in I\}$ be a family of right R-modules indexed by I. Recall that the direct product of the additive groups $(M_\sigma, +)$ is the collection of all functions

$$f : I \to \bigcup_{\sigma \in I} M_\sigma \quad \text{(disjoint union)}$$

subject to the condition that $f(\sigma) \in M_\sigma$ for each $\sigma \in I$. This had the structure of an additive group $\prod_{\sigma \in I} M_\sigma$ with addition defined "coordinatewise," i.e., $(f + g)(\sigma') = f(\sigma') + g(\sigma')$, $f, g \in \prod M_\sigma$, $\sigma' \in I$. Here, since each M_σ is an R-module, $\prod M_\sigma$ can be endowed with the structure of a right R-module, as follows. If $f \in \prod M_\sigma$, and $r \in R$, we define $fr \in \prod M_\sigma$ to be the function defined by $fr(\sigma) := f(\sigma)r$. The right R-module so defined is called *the direct product* over the family $\{M_\sigma \mid \sigma \in I\}$ of R-modules.

The *direct sum* of the family $\{M_\sigma \mid \sigma \in I\}$ of R-modules is the submodule of $\prod M_\sigma$ consisting of all functions $f \in \prod M_\sigma$ such that $f(\sigma) \neq 0_{M_\sigma}$ for only finitely many $\sigma \in I$. (This is easily verified to be a submodule of $\prod M_\sigma$.)

The direct product $\prod M_\sigma$ comes equipped with a family of projection homomorphisms $\pi_{\sigma'} : \prod M_\sigma \to M_{\sigma'}$, $\sigma' \in I$, defined by setting $\pi_{\sigma'}(f) = f(\sigma') \in M_{\sigma'}$. Likewise the direct sum comes equipped with a family of coordinate injections $\mu_{\sigma'} : M_{\sigma'} \to \prod M_\sigma$, defined by the rule $\mu_{\sigma'}(m_{\sigma'}) := f_{\sigma'} \in \prod M_\sigma$, where

$$f_{\sigma'}(\sigma) = \begin{cases} m_{\sigma'} & \text{if } \sigma = \sigma' \\ 0 & \text{if } \sigma \neq \sigma'. \end{cases}$$

The direct sum of the R-modules M_σ, $\sigma \in I$ is denoted $\bigoplus_{\sigma \in I} M_\sigma$; again, note that this is a submodule of the direct product $\prod_{\sigma \in I} M_\sigma$. These projection and injection homomorphisms satisfy certain "universal conditions;" which are expounded in Sect. 8.4.1.

We note that each $\mu_{\sigma'}(M_{\sigma'}) \cong M_{\sigma'}$; furthermore it is routine to verify that $\bigoplus M_\sigma = \sum \mu_\sigma(M_\sigma) \subseteq \prod M_\sigma$.

As was the case for abelian groups, when I is a finite set, say, $I = \{1, \ldots, n\}$, the notions of direct product and direct sum coincide, and we write

$$\prod_{i=1}^{n} M_i = \bigoplus_{i=1}^{n} M_i = M_1 \oplus M_2 \oplus \cdots \oplus M_n.$$

Let M be a right R-module and assume that $\{M_\sigma \mid \sigma \in I\}$ is a family of submodules with $M = \sum_{\sigma \in I} M_\sigma$. We have a natural surjective homomorphism $p : \bigoplus M_\sigma \to \sum M_\sigma$ via

$$p(f) := \sum_{\sigma \in I} f(\sigma) \in \sum_{\sigma \in I} M_\sigma. \tag{8.16}$$

Since $f \in \bigoplus M_\sigma$ implies that $f(\sigma) \neq 0$ for only finitely many $\sigma \in I$, we see that the sum in Eq. (8.16) makes sense (being a finite sum). Again, the verification that p is a surjective R-module homomorphism is entirely routine. When p is injective (and hence is an isomorphism), we say that $M = \sum M_\sigma$ is an *internal direct sum* of the submodules M_σ, $\sigma \in I$; when there is no danger of confusion, we shall write $\sum M_\sigma = \bigoplus M_\sigma$.

Theorem 8.1.6 (How To Recognize Direct Sums Internally) *Let M be a right R-module and let $\{M_\sigma \mid \sigma \in I\}$ be a family of submodules. Then $M = \bigoplus M_\sigma$ if and only if*

(i) $M = \displaystyle\sum_{\sigma \in I} M_\sigma$, *and*

(ii) *for each $\sigma \in I$, $M_\sigma \cap \displaystyle\sum_{\nu \neq \sigma} M_\nu = 0$.*

Proof Assuming conditions (i) and (ii) above, it suffices to show that the homomorphism $p : \bigoplus M_\sigma \to M = \sum M_\sigma$ defined above is injective. Thus, let $f \in \bigoplus M_\sigma$, and assume that $p(f) = \sum f(\sigma) = 0$. But then, for any $\sigma \in I$,

$$f(\sigma) = -\sum_{\nu \neq \sigma} f(\nu) \in M_\sigma \cap \sum_{\nu \neq \sigma} M_\nu = 0;$$

since $\sigma \in I$ was arbitrary, we conclude that $f = 0$, proving that $p : \bigoplus M_\sigma \to M$ is injective.

Conversely, assume that $p : \bigoplus M_\sigma \to M$ is an isomorphism. Then, as $p(\bigoplus M_\sigma) = \sum M_\sigma$, we conclude already that $M = \sum M_\sigma$. Next, if $m \in M_\sigma \cap \sum_{\nu \neq \sigma} M_\nu$, we have elements $m_{\sigma'} \in M_{\sigma'}$, $\sigma' \in I$ satisfying $m_\sigma = -\sum_{\nu \neq \sigma} m_\nu$. But then, if we define $f \in \bigoplus M_\sigma$ by $f(\sigma') := m_{\sigma'}$, $\sigma' \in I$, we conclude that $f \in \ker p$, a contradiction. $\square$

Hypothesis (ii) of the preceding Theorem 8.1.6 can be relaxed when the collection of potential summands is finite.

Corollary 8.1.7 (Recognizing finite internal direct sums) *Suppose $\{M_1, \ldots, M_n\}$ is a collection of submodules of the right R-module M and suppose they "span" M—that is, $M = \sum_{i=1}^{n} M_i$. Then $M \simeq M_1 \oplus \cdots \oplus M_n$ if and only if, for each i, $1 \leq i \leq n - 1$, one has*

$$M_{i+1} \cap \sum_{j=1}^{i} M_j = 0.$$

The proof is left as Exercise (4) in Sect. 8.5.1.

8.1.9 Free Modules

Let M be a right R-module, and let $\mathcal{B}$ be a subset of M. An R- *linear combination* of elements of $\mathcal{B}$ is any expression of the form $\sum_{b \in \mathcal{B}} br_b \in M$, where each $r_b \in R$ and where only finitely many terms $br_b \neq 0$. We say that $\mathcal{B}$ *spans* M if each element of M can be expressed as an R-linearly combination of elements in $\mathcal{B}$. In this case, it is clear that $M = \sum_{b \in \mathcal{B}} bR$.

The subset $\mathcal{B} \subseteq M$ is said to be R- *linearly independent* if the linear combination $\sum br_b = 0$ if and only if each "scalar" $r_b = 0$. Finally, we say that $\mathcal{B}$ is a *basis* of M if and only if $\mathcal{B}$ is R-linearly independent and spans M.

Lemma 8.1.8 *The following statements about a right R-module are equivalent:*

(i) *M has a basis.*
(ii) *There exists a subset $\mathcal{B}$ of M such that each element of M is uniquely expressible as an R-linear combination of elements of $\mathcal{B}$.*
(iii) *M is a direct sum of submodules isomorphic to R_R.*

Proof (i) implies (ii). By (i) M contains a basis $\mathcal{B}$, an R-linearly independent spanning set. If $\sum_{\mathcal{B}} b_i r_i$ and $\sum_{\mathcal{B}} b_i s_i$ were two R-linear combinations for the same element $m \in M$, then their difference $\sum_{\mathcal{B}} b_i (r_i - s_i)$ would be an R-linear combination representing the module element 0. By the R-linear independence of $\mathcal{B}$, one has $r_i = s_i$ for all i, so the R-linear combinations for m are the same. Thus (ii) holds.

(ii) implies (i) Since we can always write $0 = \sum_{\mathcal{B}} b_i \cdot 0$ the uniqueness of the R-linear combinations of elements of $\mathcal{B}$ representing 0, establishes that $\mathcal{B}$ is an R-linearly independent. Since $\mathcal{B}$ is presented as a spanning set, it is in fact a basis.

(iii) implies (ii). Suppose M is a direct sum of modules isomorphic to R_R. Then for some set B, we can regard M as the module of all functions $f : B \to R$ where $f(b) \neq 0$ for only finitely many $b \in B$, under point-wise addition and right multiplications by elements of R. For each such f let $\mathrm{supp}(f) = \{b \in B | f(b) \neq 0\}$, a finite set. Let $\mathcal{B}$ be the collection of functions $\{\epsilon_b | b \in B\}$ where $\epsilon_b(y) = 1_R$ if $y = b$ and is 0_R otherwise (here 0_R and 1_R are the additive and multiplicative identity elements of the ring R). Then for any function f we may write $f = \sum_{b \in \mathrm{supp}(f)} \epsilon(b) f(b)$. In case $f = 0 \in M$, each $f(b) = 0$. It follows that every element of the direct sum has a unique expression as an R-linear combination of the functions in $\mathcal{B}$. Thus (ii) holds.

(i) Implies (iii) Now suppose the R-module M has a basis B. Suppose $b \in B$. Then the uniqueness of expression means that for any $r, s \in R$, $br = bs$ implies $r = s$. Thus the map which sends br to r is a well-defined isomorphism $bR \to R_R$ of right R-modules. The directness of the sum $M = \oplus_{b \in B} bR$ follows from the uniqueness of expression. Thus (iii) holds.

The proof is complete. $\square$

We hasten to warn the reader that a given R-module need not have a basis (or even a nontrivial linearly independent set). One need only consider the case that M is not a faithful module. At the other extreme, if R is any ring, the corresponding right R-module $M = R_R$ has the basis $\{1\}$, where 1 is the multiplicative identity element of R. (Note that, in fact, if $u \in R$ is any unit (=invertible element) of R, then $\{u\}$ is a basis of R.)

A *free (left) right R-module* is simply an R-module having a basis—or equivalently the other two properties of Lemma 8.1.8.

The fundamental universal property of free modules is the following.

Theorem 8.1.9 (The Universal Property of Free Modules) *Let F be a free R-module with basis $\mathcal{B} \subseteq F$. Let M be an arbitrary R-module, and let m_b, $b \in \mathcal{B}$ be arbitrary elements of M. Then there exists a unique R-module homomorphism $\phi : F \to M$ such that $\phi(b) = m_b$, $b \in \mathcal{B}$.*

Proof The desired homomorphism $\phi : F \to M$ will be defined in stages. First set $\phi(b) = m_b$, for all $b \in \mathcal{B}$. Next we set

$$\phi\left(\sum_{b \in \mathcal{B}} br_b\right) = \sum_{b \in \mathcal{B}} m_b r_b \in M. \tag{8.17}$$

Note that Eq. 8.17 produces a well-defined mapping $F \to M$, since any element of F has a *unique* representation as an R-linear combination of the basis elements in $\mathcal{B}$.

Next, it must be shown that this mapping obeys the fundamental properties of an R-homomorphism. Consider elements $x, x' \in F$, and if $r, r' \in R$. Since $\mathcal{B}$ is a basis of F, there are unique representations

$$x = \sum_{b \in \mathcal{B}} br_b, \quad x' = \sum_{b \in \mathcal{B}} br'_b$$

for suitable scalars $r_b, r'_b \in R$. Therefore,

$$\begin{aligned}
\phi(xr + x'r') &= \phi\left(\sum_{b \in \mathcal{B}} br_b r + \sum_{b \in \mathcal{B}} br'_b r'\right) \\
&= \phi\left(\sum_{b \in \mathcal{B}} b(r_b r + r'_b r')\right) \\
&= \sum_{b \in \mathcal{B}} m_b(r_b r + r'_b r') \\
&= \sum_{b \in \mathcal{B}} m_b r_b r + \sum_{b \in \mathcal{B}} m_b r'_b r' \\
&= \phi(x)r + \phi(x')r',
\end{aligned}$$

proving that ϕ is, indeed, a homomorphism of right R-modules. This proves the result. $\square$

The following result is very nearly obvious, but important nonetheless.

Theorem 8.1.10 *Let M be an R-module. Then there exists a free R-module F and a surjective homomorphism $\epsilon : F \to M$.*

Proof Form the direct sum $F = \bigoplus_{m \in M} R_m$, where each $R_m \simeq R_R$. From the above, F is a free R-module having basis $\{e_m \mid m \in M\}$ where the elements e_m are defined, as usual, by requiring that

$$e_{m'}(m) = \begin{cases} 1 & \text{if } m' = m \\ 0 & \text{if } m' \neq m. \end{cases}$$

Also, using Theorem 8.1.9, we see that there is a (unique) R-module homomorphism $\epsilon : F \to M$ with $\epsilon(e_m) = m$, $m \in M$. Obviously, ϵ is surjective, completing the proof. $\square$

8.1.10 Vector Spaces

Let D be a division ring. A right D-module M is called a *right vector space over D*. Similarly, a left D-module is a *left vector space over D*.

The student should be able to convert everything that we define for a right vector space to the analogous notion for a left vector space. There is thus no need to go through the discussion twice, once for left vector spaces and once again for right vector spaces—so we will stick to right vector spaces for this discussion.

The elements of the vector space are called *vectors*. We say that a vector v *linearly depends* on a set $\{x_1, \ldots, x_n\}$ of vectors if there exist elements $\alpha_1, \ldots, \alpha_n \in D$, such that

$$v = x_1\alpha_1 + \cdots + x_n\alpha_n.$$

Equivalently, v linearly depends on $\{x_1, \ldots, x_n\}$ if and only if v is contained in the (right) submodule generated by $x_1, \ldots, x_n$. Clearly from this definition, the following properties hold:

Reflexive Property. x_1 linearly depends on $\{x_1, \ldots, x_n\}$.
Transitivity. If v linearly depends on $\{x_1, \ldots, x_n\}$ and each x_i linearly depends on $\{y_1, \ldots, y_m\}$, then v linearly depends on $\{y_1, \ldots, y_m\}$.
The Exchange Condition. If vector v linearly depends on $\{x_1, \ldots, x_n\}$ but not on $\{x_1, \ldots, x_{n-1}\}$, then x_n linearly depends on $\{x_1, \ldots, x_{n-1}, v\}$.

Thus *linear dependence* is a dependence relation as defined in Sect. 2.6.2. It follows from Theorems 2.6.2 and 2.6.3 that maximal linearly independent sets exist and span V. Such a set is then a basis of V, and so it follows immediately that a vector space over a division ring D is a free D-module. Next, by Theorem 2.6.4, any two maximal linearly independent sets in the vector space V have the same cardinality. Therefore this cardinality is an invariant of V and is called *the dimension of V over D* and is denoted $\dim_D(V)$. Thus in particular, if V is a finitely generated D-module, then it is finite dimensional.

The above discussion motivates the following general question. If M is a free R-module over the arbitrary ring R, must it be true that any two bases of M have the same cardinality? The answer is no, in general, but is known to be true, for example, if the right R-module R_R is Noetherian. (See Sect. 8.2 for the definition of Noetherian. For the proof of a basis in this case, see [1], *An Introduction to Homological Algebra*, Academic Press, 1979, Theorem 4.9, p. 111 [33].) If the ring R is commutative, then we have the following affirmative result:

Theorem 8.1.11 *If M is a free module over the commutative ring R, then any two bases of M have the same cardinality.*

Proof We may, using Zorn's lemma, extract a maximal ideal $J \subseteq R$; thus $F := R/J$ is a field. Furthermore, one has that M/MJ is an R/J-module, i.e., is an F-vector space. Let $B \subseteq M$ be an R-basis of M; as we have just seen that the dimension of a vector space is well defined, it suffices to show that the set $\overline{B} := \{b + MJ \mid b \in B\}$ is an F-basis of M/MJ. First, it is obvious that $\overline{B}$ spans M/MJ. Next, an F-linear dependence relation $\sum_{b_i \in B}(b_i + MJ)(r_i + J) = 0$ translates into a relation of the form $\sum b_i r_i \in MJ$. Therefore, there exist elements $s_i \in J$ such that $\sum b_i r_i = \sum b_i s_i$; but as the elements b_i are R-linearly independent, we infer that each $r_i = s_i$, i.e., that each $r_i \in J$. Therefore, each $r_i + J = 0 \in F$, proving that $\overline{B}$ is F-linearly independent.

This shows that the cardinality of any basis of the free-module M is equal to the dimension of the F-vector space M/MJ. $\square$

As a result of Theorem 8.1.11, we see that we may unambiguously define the *rank of a free module* over a commutative ring to be the cardinality of any basis.

8.2 Consequences of Chain Conditions on Modules

8.2.1 Noetherian and Artinian Modules

Let M be a right R-module. We are interested in the poset $P(M) := (P, \leq)$, of all submodules of M, partially ordered by inclusion. We already mentioned in Sect. 8.1.7 that $P(M)$ is a semimodular lower semi-lattice for the express purpose of activating the general Jordan-Hölder theory. Actually, the fundamental homomorphism theorems of Sect. 8.1.6 imply a much stronger property:

Theorem 8.2.1 *The poset $P(M) = (P, \leq)$ of all submodules of a right R-module is a modular lattice with a minimal element 0 (the lattice "zero") and a unique maximal element, M (the lattice "one").*

Recall from Sect. 2.5.4 that a lattice L is modular if and only if

(M) $a \geq b$ implies $a \wedge (b \vee c) = b \vee (a \wedge c)$ for all a, b, c in L.

This condition is manifestly equivalent to its dual:

(M*) $a \leq b$ implies $a \vee (b \wedge c) = b \wedge (a \vee c)$ for all a, b, c in L.

In the poset $P(M)$ the condition (M) reads:

If A, B, and C are submodules with B contained in A, then

$$A \cap (B + C) = B + (A \cap C).$$

This is an easy argument, achieved by showing that an arbitrary element of one side is an element of the set described by the other side.

We say that the right R-module M *is right Noetherian* if and only if the poset $P(M)$ satisfies the ascending chain condition (ACC) (See Sect. 2.3.3). Recall from Lemma 2.3.4 that this is equivalent to saying that any nonempty collection $\mathcal{C}$ of submodules of M has a maximal member—that is, a module not properly contained in any other module in $\mathcal{C}$. This formulation will be extremely useful in the sequel. Similarly, we say that a right R-module M is *Artinian* if and only if the poset $P(M)$ satisfies the descending chain condition (DCC). By Lemma 2.3.5 this is equivalent to the assertion that any nonempty collection of submodules of M must contain a minimal member.

Of course, there are left-module versions of these concepts: leading to "left-Noetherian", and "left-Artinian" modules. When it is clear that one is speaking in the context of (left) right-R-modules, to say that a module is Noetherian is understood to mean that it is (left) right Noetherian. Similarly, "Artinian" means "(left) right Artinian" when speaking of (left) right R-modules.

Finally, the chain conditions enjoy certain hereditary properties:

Lemma 8.2.2 (i) *Suppose N is a submodule of the right R-module M. Then M is Noetherian (or Artinian), if and only if both M/N and N are Noetherian (Artinian).*

(ii) *Suppose $M = \sum_{i=1}^{n} N_i$, a finite sum of its submodules. Then M is Noetherian (Artinian) if and only if each N_i is Noetherian (Artinian).*

(iii) *Let $\{A_i | i = 1, \ldots, n\}$ be a finite family of submodules of M. Then $M/(A_1 \cap \ldots \cap A_n)$ is Noetherian (Artinian) if and only if each factor module M/A_i is Noetherian (Artinian).*

Proof These results are consequences of three facts:

1. The poset $P(M) = (P, \leq)$ of all submodules of a right R-module M is a modular lattice with a minimal element 0 (the lattice "zero") and a unique maximal element, M (the lattice "one") (Theorem 8.2.1).

2. Purely lattice-theoretic results concerning chain conditions in a modular lattice given in Sect. 2.5.4—in particular, Lemmas 2.5.6 and 2.5.7.
3. The Correspondence Theorem for Modules (Theorem 8.1.3, part (i)) which produces a isomorphism between the principle filter $P(M)^N$ and the poset $P(M/N)$ of submodules of the factor module M/N.

In Exercises (2–4) in Sect. 8.5.2, respectively, the reader is asked to supply formal proofs of the three parts of the Lemma using the facts listed above. $\square$

We say that the module M has a *finite composition series* if and only if there is a finite unrefinable chain of submodules in $P(M)$ proceeding from 0 to M. Now since $P(M)$ is a modular lattice, so are each of the intervals $[A, B] := \{X \in P(M)| A \leq X \leq B\}$. Moreover, from the modular property, we have that any interval $[A, A + B]$ is poset isomorphic to $[A \cap B, B]$. We then have:

Lemma 8.2.3 *The following three conditions for an R-module M are equivalent:*

(i)   *M has a finite composition series.*
(ii) *M is both Noetherian and Artinian.*
(iii) *Every unrefinable chain of submodules if finite.*

Proof Note first that (i) implies (iii) by Theorem 2.5.5. Next, as any chain of submodules is contained in an unrefinable chain by Theorem 2.3.1, we see immediately that (iii) implies (ii). Finally, assume condition (ii). Then as M is Artinian, M contains a minimal submodule, say M_1. Again, as M is Artinian, the set of submodules of M properly containing M_1 has a minimal member, say M_2. We continue in this fashion to obtain an unrefinable chain $0 \subsetneq M_1 \subsetneq M_2 \subsetneq \ldots$. Finally, since M is Noetherian, this unrefinable chain must stabilize, necessarily at M, which therefore produces a finite composition series for M. $\square$

8.2.2 Effects of the Chain Conditions on Endomorphisms

The Basic Results

Let $f : M \to M$ be an endomorphism of the right R-module M. As usual, we regard f as operating from the left. Set f^0 to be the identity mapping on M, $f^1 := f$ and inductively define $f^n := f \circ f^{n-1}$, for all positive integers n. Then f^n is always an R-endomorphism, and so its image $M_n := f^n(M)$ and its kernel $K_n := \ker(f^n)$ are submodules. We then have two series of submodules, one ascending and one descending:

$$0 = K_0 \leq K_1 \leq K_2 \leq \ldots$$
$$M = M_0 \geq M_1 \geq M_2 \geq \ldots$$

Lemma 8.2.4 *Let f be an endomorphism of the right R-module M and let $K_n :=$ $\ker(f^n)$ and $M_n := f^n(M)$ as above. The following statements hold:*

(i) $M_n = M_{n+1}$ *implies* $M = M_n + K_n$.

(ii) $K_n = K_{n+1}$ *implies* $M_n \cap K_n = 0$.

(iii) If M is Noetherian, then for all sufficiently large n, $M_n \cap K_n = 0$.

(iv) If M is Artinian, then $M = M_n + K_n$ for all sufficiently large n.

(v) If M is Noetherian and f is surjective, then f is an automorphism of M.

(vi) If M is Artinian and f is injective, then again f is an automorphism of M.

Proof (i) Since $M_n = f(M_n)$ we have $M_n = M_{2n}$. Suppose x is an arbitrary element of M. Then $f^n(x) = f^{2n}(y)$, for some element y. But we can write $x = f^n(y) + (x - f^n(y))$. Now the latter summand is an element of K_n while the former summand belongs to M_n. Thus $x \in M_n + K_n$, proving (i).

For (ii), suppose $K_n = K_{n+1}$, so that $K_n = K_{2n}$. Choose $x \in K_n \cap M_n$. Then $x = f^n(y)$ for some element y, and so

$$0 = f^n(x) = f^{2n}(y).$$

But as $K_{2n} = K_n$, $f^{2n}(y) = 0$ implies $f^n(y) = 0 = x$. Thus $K_n \cap M_n = 0$.

(iii) If M is Noetherian, the ascending series $K_0 \le K_1 \le \ldots$ must stabilize at some point, so there exists a natural number n such that $K_n = K_{n+1}$. By part (ii), $M_n \cap K_n = 0$, and so (iii) holds.

(iv) If M is Artinian, the descending series $M_0 \ge M_1 \ge M_2 \ldots$ eventually stabilizes and so we have $M_n = M_{n+1}$ for sufficiently large n. Now apply part (i).

(v) Note that if f is surjective, $M = M_0 = M_n$ for all n. But M is Noetherian, so by part (iii), $M_n \cap K_n = 0$ for some n. Thus $K_n = 0$. But since the K_i are ascending, $K_1 \le K_n = 0$, which makes f injective. Hence f is a bijective endomorphism, that is, an automorphism, whence (v).

(vi) Finally, if f is injective, $0 = K_n$ for all n; since M is Artinian, $M_n = M_{n+1}$ for sufficiently large n. Apply part (iv) to conclude that $M = M_n + K_n$, i.e., that $M = M_n$. Since $M_1 \ge M_n$ we have, a fortiori, that $M = M_1$ proving that f is surjective. Hence f is a bijection once more, proving (vi). $\square$

Corollary 8.2.5 *Suppose M is both Artinian and Noetherian and $f : M \to M$ is an endomorphism. Then for some natural number n,*

$$M = M_n \oplus K_n.$$

Schur's and Fitting's Lemmas

Schur's Lemma and Fitting's lemma both address the structure of the endomorphism ring $\operatorname{End}_R(M)$ of an R-module M. First of all, recall from Sect. 8.1.7 that an R-module $M \neq 0$ is said to be *irreducible* if M has no nontrivial proper R-submodules.

Lemma 8.2.6 (Schur's Lemma.) *If M is an irreducible R-module, then $\text{End}_R(M)$ is a division ring.*

Proof If $0 \neq \phi \in \text{End}_R(M)$, then as $\ker \phi$ and $\text{im } \phi$ are both submodules of M, we conclude that $\phi : M \to M$ must be bijective, and hence possesses an inverse, i.e., $\text{End}_R(M)$ is a division ring. $\square$

Next, a module M is said to be *indecomposable* if and only if it is not the direct sum of two non-trivial submodules. Fitting's Lemma generalizes Schur's lemma, as below:

Theorem 8.2.7 (Fitting's Lemma.) *Suppose $M \neq 0$ is an indecomposable right R-module having a finite composition series. Then the following hold:*

(i) *For any R-endomorphism $f : M \to M$, f is either invertible (that is, f is a unit in the endomorphism ring $S := \text{End}_R(M)$) or f is nilpotent (which means that $f^n = 0$, the trivial endomorphism, for some natural number n).*

(ii) *The endomorphism ring $S = \text{End}_R(M)$ possesses a unique maximal right ideal which is also the unique maximal left ideal, and which consists of all nilpotent elements of S.*

Proof (i) First of all, by Lemma 8.2.3, M has a finite composition series if and only if it is both Artinian and Noetherian. It follows from Corollary 8.2.5 that for sufficiently large n, $M = M_n \oplus K_n$. Since M is indecomposable, either $M_n = 0$ or $K_n = 0$. In the former case f is nilpotent. In the latter case f is injective, and it is also surjective since $M = M_n$. Thus f is an automorphism of M and so possesses a two-sided inverse in S.

(ii) It remains to show that the nilpotent elements of S form a two-sided ideal, and to prove the uniqueness assertions about this ideal.

Suppose f is nilpotent. Since $M \neq 0$, $K_1 = \ker f \neq 0$, and so for any endomorphism $s \in S$, neither sf nor fs is injective. So, by the dichotomy imposed by Part (i), both sf and fs must be nilpotent.

Next, assume that f and g are nilpotent endomorphisms while $h := f + g$ is not. Then by the dichotomy, h possesses a 2-sided inverse h^{-1}. Then

$$id_M = hh^{-1} = fh^{-1} + gh^{-1},$$

where id_M denotes the identity mapping $M \to M$. Since $s = id_M - t$, it follows that the two endomorphisms, $s := fh^{-1}$ and $t := gh^{-1}$, commute and, by the previous paragraph, are nilpotent. Then there is a natural number m large enough to force $s^m = t^m = 0$. Then

$$id_M = (id_M)^{2m} = (s + t)^{2m} = \sum_{j=0}^{2m} \binom{2m}{j} s^{2m-j} t^j.$$

Each monomial $s^{2m-j} t^j = 0$ since at least one of the two exponents exceeds $m - 1$. Therefore, $id_M = 0$, an absurdity as $M \neq 0$. Thus h must also be nilpotent, and so the set nil(S) of all nilpotent elements of S is closed under addition. From the conclusion of the previous paragraph this set of nilpotent elements nil(S) forms a two-sided ideal.

That this ideal is the unique maximal member of the poset of right ideals of S follows from the fact that $S - \text{nil}(S)$ consists entirely of units. A similar statement holds for the poset of left ideals of S. The proof is complete. $\square$

The Krull-Remak-Schmidt Theorem

The Krull-Remak-Schmidt Theorem asserts that any Artinian right R-module can be expressed as a direct sum of indecomposable submodules, and that this decomposition is unique up to the isomorphism type and order of the summands. We begin with a lemma that will be used in the uniqueness aspect.

Lemma 8.2.8 Let $f : A = A_1 \oplus A_2 \to B = B_1 \oplus B_2$ be an isomorphism of Artinian modules, and define homomorphisms $\alpha : A_1 \to B_1$ and $\beta : A_1 \to B_2$ by the equation

$$f(a_1, 0) = (\alpha(a_1), \beta(a_1))$$

for all $a_1 \in A_1$. If α is an isomorphism, then there exists an isomorphism

$$g : A_2 \to B_2.$$

Proof Clearly $\alpha = \pi_1 \circ f \circ \mu_1$ and $\beta = \pi_2 \circ f \circ \mu_1$, where π_j is the canonical coordinate projection $B \to B_j$, $j = 1, 2$, and μ_1 is the canonical embedding $A_1 \to A_1 \oplus A_2$ sending a_1 to $(a_1, 0)$.

Suppose first that $\beta = 0$. In this case, the composition $\pi_2 \circ f : A \to B_2$ is surjective and has kernel precisely $A_1 \oplus 0 \leq A$ and hence induces an isomorphism $\overline{\pi_2 \circ f} : A_2 \cong A/(A_1 \oplus 0) \overset{\cong}{\to} B_2$.

So our proof will be complete if we can construct from f another isomorphism $h : A \to B$ such that $h(a_1, 0) = (\alpha(a_1), 0)$ for all $a_1 \in A_1$. We define h in the following way: if $f(a_1, a_2) = (b_1, b_2)$, set

$$h(a_1, a_2) := (b_1, b_2 - \beta\alpha^{-1}(b_1)).$$

Clearly h is an R-homomorphism, and is easily checked to be injective. Now $h \circ f^{-1} : B \to B$ is an endomorphism of an Artinian module. Since h is injective, so is $h \circ f^{-1}$. Then by Lemma 8.2.4, part (vi), $h \circ f^{-1}$ is surjective which clearly implies that h is surjective. Since $h(a_1, 0) = (\alpha(a_1), 0)$, the proof is complete. $\square$

The next lemma addresses the existence aspect of the Remak-Krull-Schmidt Theorem:

Lemma 8.2.9 Suppose A is an Artinian right R-module. Then A is a direct sum of finitely many indecomposable modules.

Proof Let $\mathcal{C}$ be the family of submodules $C \subseteq A$ for which the statement of the lemma is false. We shall assume, by way of contradiction, that $\mathcal{C} \neq \emptyset$. Since A is Artinian, $\mathcal{C}$ has a minimal member, say $C \in \mathcal{C}$. Clearly, C is not itself indecomposable, and so must itself decompose: $C = C' \oplus C''$, for proper nontrivial submodules $0 < C', C'' < C$. By minimality of $C \in \mathcal{C}$, we infer that both C' and C'' must decompose into direct sums of indecomposable submodules. But then so does $C = C' \oplus C''$, a contradiction. Thus, $\mathcal{C} = \emptyset$ and the proof is complete. $\square$

The above preparations now lead to:

Theorem 8.2.10 (The Krull-Remak-Schmidt Theorem) *If A is an Artinian and Noetherian right R-module, then A decomposes into a direct sum of finitely many indecomposable right R-modules. This decomposition is unique up to the isomorphism type and order of the summands.*[1]

Proof By Lemma 8.2.9 we know that A admits such a decomposition. As for the uniqueness, assume that A, A' are isomorphic right R-modules admitting decompositions

$$A = A_1 \oplus A_2 \oplus \cdots \oplus A_m, \quad A' = A'_1 \oplus A'_2 \oplus \cdots \oplus A'_n,$$

where the submodules $A_i \subseteq A$, $A'_j \subseteq A'$ are indecomposable submodules. We may assume that $m \leq n$ and shall argue by induction on m that $m = n$ and that (possibly after rearrangement) $A_i \cong_R A'_i$, $i = 1, 2, \ldots, n$.

Assume that $f : A \to A'$ is the hypothesized isomorphism, and denote by $\pi_i : A \to A_i$, $\pi'_j : A' \to A'_j$, $\mu_i : A_i \to A$, $\mu'_j : A'_j \to A'$ the canonical projection and coordinate injection homomorphisms. We now define the R-module homomorphisms

$$\alpha_i := \pi'_1 \circ f \circ \mu_i : A_i \to A'_1,$$
$$\alpha'_i := \pi_i \circ f^{-1} \circ \mu'_1 : A'_1 \to A_i,$$

$i = 1, 2, \ldots, m$. One then checks that

$$\sum_{i=1}^{m} \alpha_i \circ \alpha'_i$$

is the identity mapping on the indecomposable module B_1. By Fitting's Lemma (Theorem 8.2.7) not all the $\alpha_i \circ \alpha'_i$ can be nilpotent. Reindexing if necessary, we may assume $\alpha_1 \circ \alpha'_1$ is not nilpotent. Then it is in fact an automorphism of A'_1, and so $\alpha_1 : A_1 \to A'_1$ is an isomorphism. Furthermore, we have

---

[1] Here is a precise rendering of the uniqueness statement: Let $A = A_1 \oplus \cdots \oplus A_m$ be isomorphic to $B = B_1 \oplus \cdots \oplus B_n$ where the A_i and B_j are indecomposable. Then $m = n$ and there is a permutation ρ of the index set $I = \{1, \ldots, n\}$ such that A_i is isomorphic to $B_{\rho(i)}$ for all $i \in I$.

$$f(a_1, 0, \ldots, 0) = (\alpha_1(a_1), \ldots) \in A'.$$

Applying Lemma 8.2.8, there is an isomorphism

$$A_2 \oplus \cdots \oplus A_m \cong A'_2 \oplus \cdots \oplus A'_n.$$

Now we apply the induction hypothesis to deduce that $m = n$, and that for some permutation σ of $J = \{2, \ldots, n\}$, A_i is isomorphic to $A'_{\sigma(i)}$, $i \in J$. The result follows. $\square$

8.2.3 Noetherian Modules and Finite Generation

We can come to the point at once:

Theorem 8.2.11 *Suppose M is a right R-module, for some ring R. Then M is right Noetherian if and only if every right submodule of M is finitely generated.*

Proof Assume that M is a Noetherian right R-module, and let $N \subseteq M$ be a submodule. Let $\mathcal{N}$ be the family of finitely-generated submodules of N; since M is Noetherian we know that $\mathcal{N}$ must have a maximal member N_0. If $N_0 \subsetneq N$, then there exists an element $n \in N_0 \backslash N$. But then $N_0 + nR$ is clearly finitely generated; since $N_0 \subsetneq N_0 + nR \subseteq N$, we have an immediate contradiction. Therefore, N must be finitely generated.

Conversely, assume that every submodule of M is finitely generated, and let $M_1 \subseteq M_2 \subseteq \cdots$ be a chain of submodules of M. Then it is clear that $M_\infty := \bigcup_{i=1}^{\infty} M_i$ is a submodule of M, hence must itself be finitely generated, say $M_\infty = x_1 R + x_2 R \cdots + x_n R$. But since $x_1, \ldots, x_n \in \bigcup M_i$, we infer that there must exist some integer m with $x_1, \ldots, x_n \in M_m$, which guarantees that $M_m = M_{m+1} = \ldots$. Since the ascending chain $M_1 \subseteq M_2 \subseteq \cdots$ was arbitrary, M is by definition Noetherian. $\square$

8.2.4 E. Noether's Theorem

The reader shall soon discover that many properties of a ring are properties which are lifted (in name) from the R-modules R_R or $_R R$. We say that a ring R is a *right Noetherian ring* if and only if R_R is a Noetherian right R-module. In view of Theorem 8.2.11 we are saying that a ring R is right Noetherian if and only *every right ideal is finitely generated*. In an entirely analogous fashion, we may define the concept of a *left Noetherian ring*.

Note that there is a huge difference between saying, on the one hand, that a module is finitely generated, and on the other hand, that *every* submodule is finitely generated.

Indeed if R is a ring, then the right R-module R_R is certainly finitely generated (viz., by the identity $1 \in R$). However, if R_R is not right Noetherian, then there will exist right ideals of R that are not finitely generated.

However, if the ring R is right Noetherian, then this difference evaporates: this is the content of Noether's Theorem.

Theorem 8.2.12 (E. Noether's Theorem) *Let R be a right Noetherian ring, and let M be a right R-module. Then M is finitely generated if and only if M is right Noetherian.*

Proof Obviously, if M is Noetherian, it is finitely generated. Conversely, we may write $M = x_1 R + x_2 R + \cdots + x_n R$ and each submodule $x_i R$, being a homomorphic image of R_R, is a Noetherian module. Therefore, M is Noetherian by virtue of Exercise (2) in Sect. 8.5.2. $\square$

Although the above theorem reveals a nice consequence of Noetherian rings, the applicability of the theorem appears limited until we know that such rings are rather abundant. One very important class of Noetherian rings are the principal ideal domains. These are the integral domains such that every ideal is generated by a single element, and in Chap. 10, we shall have quite a bit more to say about modules over such rings. The basic prototype here is the ring $\mathbb{Z}$ of integers. That this is a principal ideal domain can be shown very easily, as follows. Let $0 \neq I \subseteq \mathbb{Z}$ be an ideal and let d be the least positive integer contained in I. If $x \in I$, then by the "division algorithm" (Lemma 1.1.1) there exist integers q and r with $x = qd + r$ and where $0 \le r < d$. Since $r = x - qd \in I$, we conclude that by the minimality of d, $r = 0$, i.e., x must be a multiple of d. Therefore,

Lemma 8.2.13 *The ring of integers $\mathbb{Z}$ is a principal ideal domain.*

Note that since an integral domain is by definition commutative, it does not matter whether we are speaking of right ideals or left ideals here.

This argument used only certain properties shared by all Euclidean Domains (see Sect. 9.3). All of them are principal ideal domains by a proof almost identical to the above. But principal ideal domains are a larger class of rings, and most of the important theorems (about their modules and the unique factorization of their elements into primes) encompass basically the analysis of finite-dimensional linear transformations and the elementary arithmetic of the integers. So this subject shall be taken up in much greater detail in a separate chapter (see Chap. 10). Nonetheless, as we shall see in the next subsection, just knowing that $\mathbb{Z}$ is a Noetherian ring, yields the important fact that "algebraic integers" form a ring.

8.2.5 Algebraic Integers and Integral Elements

Let D be a Noetherian integral domain and let K be a field containing D. An element $\alpha \in K$ is said to be *integral over D* if and only if α is the zero of a polynomial in

$D[x]$ having lead coefficient 1. In this case, of course, some power of α is a D-linear combination of lower powers, say

$$\alpha^n = a_{n-1}\alpha^{n-1} + \cdots + a_2\alpha^2 + a_1\alpha + a_0,$$

where n is a positive integer and the coefficients a_i lie in D.

In the particular case that $D = \mathbb{Z}$ and K is a subfield of the complex numbers, then any element $\alpha \in K$ that is integral over $\mathbb{Z}$ is called an *algebraic integer*. Thus $\sqrt{3}$ is an algebraic integer (as it is a zero of the polynomial $x^2 - 3 \in \mathbb{Z}[x]$), but one can show, however, that $\frac{1}{2}\sqrt{3}$ is not. The complex number $\omega = (1 - \sqrt{-3})/2$ is an algebraic integer because $w^2 + w + 1 = 0$.

Their is a simple criterion for an element of K to be integral over D.

Lemma 8.2.14 *Let K be a field containing the Noetherian integral domain D. Then an element α of K is integral over D if and only if the subring $D[\alpha]$ is a finitely generated D-module.*

Proof First suppose α is integral over D. We must show that $D[\alpha]$ is a finitely generated D-module. By definition, there exists a positive integer n such that α^n is a D-linear combination of the elements of $X := \{1, \alpha, \alpha^2, \ldots, \alpha^{n-1}\}$. Let $M = \langle X \rangle_D$ be the submodule of $D[\alpha]$ generated by X. Now αX contains α^n together with lower powers of α all of which are D-linear combinations of elements of X. Put more succinctly, $M\alpha \subseteq M$. It now follows that all positive integer powers of α lie in M, and so $M = D[\alpha]$. Since M is generated by the finite set X, $D[\alpha]$ is indeed finitely generated.

Now suppose $D[\alpha]$ is a finitely generated D-module. We must show that α is integral over D—that is, α is a zero of a monic polynomial in $D[x]$. We now apply Noether's Theorem (Theorem 8.2.12) to infer that, in fact, $D[\alpha]$ is a Noetherian D-module. But it contains the following chain of submodules:

$$D \leq D + \alpha D \leq D + \alpha D + \alpha^2 D \leq \ldots$$

and the Noetherian condition forces this chain to terminate, say at

$$D + \alpha D + \cdots + \alpha^{n-1} D. \tag{8.18}$$

Then $\alpha^n D$ must lie in the module presented in (8.18) and so the element α^n is a D-linear combination of lower powers of α. Of course that makes α integral over D. $\square$

The main thrust of this subsection is the following:

Theorem 8.2.15 *Let K be a field containing the Noetherian integral domain D and let $\mathcal{O}$ denote the collection of all elements of K that are integral over D. Then $\mathcal{O}$ is a subring of K.*

Proof Using the identity $(-\alpha)^j = (-1)^j \alpha^j$, it is easy to see that if α is the zero of a monic polynomial in $D[x]$, then so is $-\alpha$. Thus $-\mathcal{O} = \mathcal{O}$.

It remains to show that the set $\mathcal{O}$ is closed under the addition and multiplication operations of K. Let α and β be any elements of $\mathcal{O}$. Then there exist positive integers n and m such that the subring $D[\alpha]$ is generated (as a D-module) by powers α^i, $i < n$ and similarly the D-module $D[\beta]$ is generated by powers β^j, $j < m$. Thus every monomial expression $\alpha^k \beta^\ell$ in the subring $D[\alpha, \beta]$ is a D-linear combination of the finite set of monomial elements $\{\alpha^i \beta^j | i < n, j < m\}$. It follows that the D-module $D[\alpha, \beta]$ is finitely generated. Now since D is a Noetherian ring, Theorem 8.2.12 implies that $D[\alpha, \beta]$ is Noetherian.

Now $D[\alpha + \beta]$ and $D[\alpha\beta]$ are submodules of the Noetherian module $D[\alpha, \beta]$ and so are themselves Noetherian by Theorem 8.2.11. But then, invoking once again the fact that D is a Noetherian ring, we see that $D[\alpha + \beta]$ and $D[\alpha\beta]$ are both finitely generated D-modules, and so are in $\mathcal{O}$ by Lemma 8.2.14. Thus $\mathcal{O}$ is indeed a subset of the field K which is closed under addition and multiplication in K. $\square$

Since the algebraic integers are simply the elements of the complex number field $\mathbb{C}$ which are integral over the subring $\mathbb{Z}$ of integers, we immediately have the following:

Corollary 8.2.16 *The algebraic integers form a ring.*

In the opinion of the authors, the spirit of abstract algebra is immortalized in this maxim:

• *a good definition is better than many lines of proof.*

Perhaps nothing illustrates this principle better than the proof of Theorem 8.2.15 just given. In the straightforward way of proving this theorem one is obligated to produce *monic* polynomials with coefficients in D for which $\alpha + \beta$ and $\alpha\beta$ are roots; not an easy task. Yet the above proof was not hard. That was because we had exactly the right definitions!

8.3 The Hilbert Basis Theorem

This section gives us a second important application of Theorem 8.2.12.

In this section we consider polynomial rings of the form $R[x_1, x_2, \ldots, x_n]$, where $x_1, x_2, \ldots, x_n$ are *commuting* indeterminates. In the language of Sect. 7.3, these are the monoid rings RM where M is the free commutative monoid on the set $\{x_1, x_2, \ldots, x_n\}$. However, and in contrast to most textbooks, we shall not require that the coefficient ring R be commutative. We recall that by Exercise (3) in Sect. 7.5.3, the iterated polynomial ring $R[x][y]$ can be identified with $R[x, y]$. Since R is not assumed to be commutative, we see *a fortiori* that the polynomial ring $R[x]$ may not be commutative. However, we hasten to remind the reader that the indeterminate x commutes with every element of R. As a result of this, we infer immediately that $R[x]$ inherits a natural right R-module structure.

Theorem 8.3.1 (The Hilbert Basis Theorem) *If R is right Noetherian, then $R[x]$ is also right Noetherian.*

Proof It suffices to show that if A is a right ideal of $R[x]$, then A is finitely generated as a right $R[x]$-module. Let A^+ be the collection of all leading coefficients of all polynomials lying in A; that is

$$A^+ := \left\{ c_m \mid \text{there exists } f(x) = \sum_{j=0}^{m} c_j x^j \in A, \ \deg f(x) = m \right\} \subseteq R.$$

Since A is a right ideal in the polynomial ring $R[x]$, A^+ is a right ideal in R. Since R is right Noetherian, we know that A^+ must be finitely generated as a right R-module, so we may write

$$A^+ = a_1 R + a_2 R + \cdots + a_n R$$

for suitable elements $a_1, a_2 \ldots, a_n \in A^+$. From the definition of A^+, there exist polynomials $p_i(x)$ in A having lead coefficients a_i, $i = 1, 2, \ldots, n$. Let $n_i := \deg p_i(x)$ and set $m := \max\{n_i\}$. Now let $R_m[x]$ be the polynomials of degree at most m in $R[x]$ and set

$$A_m = A \cap R_m[x].$$

We claim that

$$A = A_m + p_1(x)R[x] + p_2(x)R[x] + \cdots + p_n(x)R[x]. \tag{8.19}$$

Let $f(x)$ be a polynomial in A of degree $N > m$ and write

$$f(x) = \alpha x^N + \text{ terms involving lower powers of } x.$$

Then, as α is the leading coefficient of $f(x)$ we have $\alpha \in A^+$, and so

$$\alpha = a_1 r_1 + a_2 r_2 + \cdots a_n r_n$$

for suitable $r_1, r_2, \ldots, r_n \in R$. Then

$$f_1(x) := f(x) - x^{N-n_1} p_1(x) r_1 - x^{N-n_2} p_2(x) r_2 - \cdots - x^{N-n_n} p_n(x) r_n$$

is a polynomial in A which has degree less than N. In turn, if the degree of $f_1(x)$ exceeds m this procedure can be repeated. By so doing, we eventually obtain a polynomial $h(x)$ in A of degree at most m, which is equal to the original $f(x)$ minus an $R[x]$-linear combination of $\{p_1(x), p_2(x), \ldots, p_n(x)\}$. But since each $p_i(x)$ and $f(x)$ all lie in A, we see that $h(x) \in A_m$. Thus

$$f(x) \in A_m + p_1(x)R[x] + \cdots p_n(x)R[x].$$

Since $f(x)$ was an arbitrary member of A, Eq. (8.19) holds.

In view of Eq. (8.19), it remains only to show that $A_m R[x]$ is a finitely generated $R[x]$-module, and this is true if A_m is a finitely generated R-module. But A_m is an R-submodule of $R_m[x]$ which is certainly a finitely generated R-module (for $\{1, x, \ldots x^m\}$ is a set of generators). By Noether's Theorem (Theorem 8.2.12) $R_m[x]$ is Noetherian and hence so is A_m. Thus, A_m is a finitely-generated right R-module, and the proof is complete. $\square$

Corollary 8.3.2 *If R is right Noetherian, then so is $R[x_1, \ldots, x_n]$.*

Remark As expected, there is an obvious left version of the above result. Thus, if R is left Noetherian, than $R[x]$ is also a left Noetherian ring.

8.4 Mapping Properties of Modules

In this section we shall consider the "calculus of homomorphisms" of modules—the idea that constructions can be characterized by mapping properties. Such an approach, leads to the important notions of exact sequences, projective and injective modules, and other notions which feed the rich lore of homological algebra.

8.4.1 The Universal Mapping Properties of the Direct Sum and Direct Product

Characterizing the Direct Sum by a Universal Mapping Property

Recall that the formal direct sum over a family $\{M_\sigma | \sigma \in I\}$ of right R-modules is the set of all mappings $f : I \to \cup_\sigma M\sigma$, with $f(\sigma) \in M_\sigma$, and $f(\sigma)$ non-zero for only finitely many indices σ. We think of the indices as indicators of coordinates, and the value $f(\sigma)$ as the σ-coordinate of the element f. This is an R-module $\oplus_\sigma M_\sigma$ under the usual coordinate-wise addition and application of right operators from R.

There is at hand a canonical collection of injective R-homomorphisms: $\kappa_\sigma : M_\sigma \to \oplus_\sigma M_\sigma$ defined by setting

$$(\kappa_\sigma(a_\sigma))(\tau) = \begin{cases} a_\sigma & \text{if } \sigma = \tau \\ 0 & \text{if } \sigma \neq \tau \end{cases}$$

where a_σ is an arbitrary element of M_σ. That is, κ_σ takes an element a of M_σ to the unique element of the direct sum whose coordinate at σ is a, and whose other coordinates are all zero.

Theorem 8.4.1 *Suppose $M = \oplus_{\sigma \in I} M_\sigma$ is the direct sum of modules $\{M_\sigma | \sigma \in I\}$ with canonical homomorphisms $\kappa_\sigma : M_\sigma \to M$. Then, for every module B, and* *family of homomorphisms $\phi_\sigma : M_\sigma \to B$, there exists a unique homomorphism $\phi : M \to B$ such that $\phi \circ \kappa_\sigma = \phi_\sigma$ for all $\sigma \in I$. In fact, this property characterizes the direct sum up to isomorphism.*

Proof We define $\phi : M \to B$ by the equation

$$\phi(a) := \sum_\sigma \phi_\sigma(a(\sigma)).$$

Since $a(\sigma) \in M_\sigma$, by definition, and only finitely many of these summands are non-zero, the sum describes a unique element of B. It is elementary to verify that ϕ, as we have defined it, is an R-homomorphism. Moreover, if a_τ is an arbitrary element of M_τ, then

$$\phi \circ \kappa_\tau(a_\tau) = \phi(\kappa_\tau(a_\tau)) = \sum_{\sigma \in I} \phi_\sigma(\kappa_\tau(a_\tau(\sigma))) = \phi_\tau a_\tau.$$

Thus we have $\phi \circ \kappa_\tau = \phi_\tau$, for all τ.

Now suppose there were a second homomorphism $\lambda : M \to B$, with the property that $\lambda \circ \kappa_\tau = \phi_\tau$, for all $\tau \in I$. Then

$$\lambda(a) = \sum_{\sigma \in I} \lambda(\kappa_\sigma(a(\sigma))) = \sum_{\sigma \in I} \phi_\sigma(a(\sigma)) = \phi(a).$$

Thus $\lambda = \phi$. So the homomorphism ϕ which we have produced is unique.

Now assume that the R-module N, and the collection of morphisms $\lambda_\sigma : M_\sigma \to N$ satisfy the universal mapping conditions of the last part of the theorem.

First apply this property to M when $B = N$ and $\phi_\sigma = \lambda_\sigma$. Then there is a homomorphism $\lambda : M \to N$ with $\lambda \circ \kappa_\sigma = \lambda_\sigma$.

Similarly, there is a homomorphism $\kappa : N \to M$ for which $\kappa \circ \lambda_\sigma = \kappa_\sigma$. It follows that

$$\kappa \circ \kappa_\sigma = \kappa \circ \lambda_\sigma = \kappa_\sigma = 1_M \circ \kappa_\sigma,$$

where 1_M is the identity morphism on M. So by the uniqueness of the homomorphism, one may conclude that $\kappa \circ \lambda = 1_M$. Similarly, reversing the roles, $\lambda \circ \kappa = 1_N$. Therefore κ is an isomorphism. The proof is complete. $\square$

We have seen from our earlier definition of the direct sum $M = \oplus_{\sigma \in I} M_\sigma$ that there are projection mappings $\pi_\sigma : M \to M_\sigma$ which, for any element $m \in M$, reads off its σ-coordinate, $m(\sigma)$. One can see that these projection mappings are related to the canonical mappings κ_σ by the equations

$$\pi_i \circ \kappa_i = 1 \text{ (the identity mapping on } M_i) \tag{8.20}$$

$$\pi_i \circ \kappa_j = 0 \text{ if } i \neq j. \tag{8.21}$$

We remark that the existence of the projection mappings π_i with the compositions just listed could have been deduced directly from the universal mapping property which characterizes the direct sum. Let δ_{ij} be the morphism $M_j \to M_i$ which is the zero map if $i \neq j$, but is the identity mapping on M_i if $i = j$. If we apply the universal property with M_i and δ_{ij} in the respective roles of B and ϕ_i, we obtain the desired morphism $\phi = \pi_i : M \to M_i$ with $\pi_i \circ \kappa_j = \delta_{ij}$.

Characterizing the Direct Product by a Universal Mapping Property

We can now characterize the direct product of modules by a universal property which is the complete dual of that for the direct sum.

Theorem 8.4.2 *Suppose* $M = \prod_{\sigma \in I} M_\sigma$ *is the direct product of the modules* $\{M_\sigma | \sigma \in I\}$ *with canonical projections* $\pi_\sigma : M \to M_\sigma$. *Then, for every module* B, *and family of homomorphisms* $\phi_i : B \to M_i$ *there exists a unique homomorphism* $\phi : B \to M$ *such that* $\pi_i \circ \phi = \phi_i$, *for each* $i \in I$. *This property characterizes the direct product up to isomorphism.*

Proof The canonical projection mapping π_i, recall, reads off the i-th coordinate of each element of M. Thus, if $m \in M$, m is a mapping $I \to \cup M_\sigma$ with $m(i) \in M_i$, and $\pi_i(m) := m(i)$.

Now define the mapping $\phi : B \to M$ by the rule:

$$\phi(b)(i) := \phi_i(b),$$

for all $(i, b) \in I \times B$. Then certainly ϕ is an R-homomorphism. Moreover, one checks that

$$(\pi_i \circ \phi)(b) = \pi_i(\phi(b)) = (\phi(b))(i) = \phi_i(b).$$

Thus $\pi_i \circ \phi = \phi_i$.

Now if $\psi : B \to M$ were an R-homomorphism which also satisfied $\pi_i \circ \psi = \phi_i$, for all $i \in I$, then we should have

$$(\psi(b))(i) = \pi_i(\psi(b)) = (\pi_i \circ \psi)(b) = \phi_i(b) = (\phi(b))(i),$$

at each i, so $\psi(b) = \phi(b)$ at each $b \in B$—i.e. $\psi = \phi$ as mappings.

Now suppose N were some R-module equipped with morphisms $\pi'_i : N \to M_i$, $i \in I$ satisfying this mapping property: whenever there is a module B with morphisms $\phi_i : B \to M_i$, then there is a morphism $\phi_B : B \to N$ such that $\pi'_i \circ \phi_B = \phi_i$.

First we use the property for M. Setting $B = N$ and $\phi_i = \pi'_i$, we obtain (in the role of ϕ) a morphism $\pi' : N \to M$ such that $\pi_i \circ \pi' = \pi'_i$.

Similarly, using the same property for N (with M and π_i in the roles of B and the ϕ_i), we obtain a homomorphism $\pi : M \to N$ such that $\pi'_i \circ \pi' = \pi_i$. It follows that

$$\pi_i \circ \pi' \circ \pi = \pi'_i \circ \pi = \pi_i = \pi_i \circ 1_M.$$

By the uniqueness, one has $\pi' \circ \pi = 1_M$. Similarly $\pi \circ \pi' = 1_N$. Therefore π is an isomorphism.

As a final remark, the universal mapping property can be used to show the existence of mappings $\kappa_i : M_i \to \prod_{i \in I} M_i$ such that $\pi_i \circ \kappa_j$ is the identity map on M_i if $i = j$, and is the zero mapping $M_i \to M_j$ if $i \neq j$. $\square$

8.4.2 $\mathrm{Hom}_R(M, N)$

The Basic Definition

We shall continue to assume that R is an arbitrary ring with identity element 1_R. If M, N are right R-modules, then we may form the abelian group $\mathrm{Hom}_R(M, N)$ consisting of R-module homomorphisms $M \to N$. Indeed, the abelian group structure on $\mathrm{Hom}_R(M, N)$ is simply given by pointwise addition: if $\phi, \theta \in \mathrm{Hom}_R(M, N)$, then $(\phi + \theta)(m) := \phi(m) + \theta(m) \in N$.

$\mathrm{Hom}_R(M, N)$ Is a Left R-Module

We remark here that $\mathrm{Hom}_R(M, N)$ possesses a natural structure of a left R-module: For each ring element $r \in R$ and homomorphism $\phi \in \mathrm{Hom}_R(M, N)$, the mapping $r \cdot \phi$ is defined by

$$(r \cdot \phi)(m) := \phi(mr) = \phi(m)r$$

Notice, by this definition, that for $r, s \in R$, and $m \in M$,

$$(r \cdot (s \cdot \phi))(m) = (s \cdot \phi)(mr) = (\phi(mr))s \tag{8.22}$$
$$= (\phi(m)r)s = \phi(m)(rs) = ((rs) \cdot \phi)(m). \tag{8.23}$$

In the preceding equations left multiplication of maps by ring elements was indicated by a "dot". The sole purpose of this device was to clearly distinguish the newly-defined left ring multiplications of maps from right ring multiplication of the module elements in the equations. However, from this point onward, the "dot" operation will be denoted by juxtaposition as we have been doing for left modules generally.

Clearly $r\phi \in \mathrm{Hom}_R(M, N)$, and the reader can easily check that the required equations of mappings hold:

$$r(\phi_1 + \phi_2) = r\phi_1 + r\phi_2,$$
$$(r + s)\phi_1 = r\phi_1 + s\phi_1, \text{ and}$$
$$(rs)\phi_1 = r(s\phi_1),$$

where $r, s \in R$ and $\phi_i \in \text{Hom}_R(M, N)$, $i = 1, 2$. In particular, when R is a commutative ring, one may regard $\text{Hom}_R(M, N)$ as a *right* R-module simply by passing scalars from one side to another. (For more about making $\text{Hom}_R(M, N)$ into a module, see the Exercise Sect. 8.5.3.)

In the special case that $N = R_R$, and M is any right R-module, we obtain a left R-module $M^* := \text{Hom}_R(M, R_R)$ called the *dual module of M*. We shall have use for this latter, especially when R is a field.

Basic Mapping Properties of $\text{Hom}_R(M, N)$

We note that Hom_R has the following mapping properties. If $\alpha : M_1 \to M_2$ and $\beta : N_1 \to N_2$ are homomorphisms of right R-modules, then we have an induced homomorphism of abelian groups,

$$\text{Hom}(\alpha, \beta) : \text{Hom}_R(M_2, N_1) \to \text{Hom}_R(M_1, N_2), \quad \phi \mapsto \beta \circ \phi \circ \alpha. \qquad (8.24)$$

Let id_M denote the identity mapping $M \to M$.[2]

The reader should have no difficulty in verifying that $\text{Hom}(id_M, \alpha)$ really is a homomorphism of abelian groups. In particular, if we fix a right R-module M, then the induced homomorphism of abelian groups are:

$$\text{Hom}(id_M, \beta) : \text{Hom}_R(M, N_1) \to \text{Hom}_R(M, N_2), \quad \phi \mapsto \beta \circ \phi. \qquad (8.25)$$

In a language that will be developed later in this book, this says that the assignment $\text{Hom}_R(M, \bullet)$ is a "functor" from the "category" of right R-modules to the "category" of abelian groups. (See Sect. 13.2 for a full exposition.) Likewise, if N is a fixed right R-module, and if $\alpha : M_2 \to M_1$ is an R-module homomorphism, then we have an induced homomorphism of abelian groups

$$\text{Hom}(\alpha, id_N) : \text{Hom}_R(M_1, N) \to \text{Hom}_R(M_2, N), \quad \phi \mapsto \phi \circ \alpha.$$

This gives rise to another "functor," $\text{Hom}_R(\bullet, N)$, from the category of right R-modules to the category of abelian groups, but is "contravariant" in the sense that the directions of the homomorphisms get reversed:

$$M_1 \overset{\phi}{\leftarrow} M_2 \text{ gets mapped to } \text{Hom}_R(M_1, N) \overset{\beta}{\to} \text{Hom}_R(M_2, N).$$

---

[2]The notation "id_M" serves to distinguish the identity mapping on the set M from the multiplicative identity element 1_M of any ring or monoid which unfortunately happens to be named "M".

8.4.3 Exact Sequences

Definition

Let $\phi : M \to N$ be a homomorphism of right R-modules, and set $M'' := \phi(M)$, $M' = \ker \phi$. Then we may build a sequence of homomorphisms

$$M' \xrightarrow{\mu} M \xrightarrow{\epsilon} M'',$$

where $\mu : M' \hookrightarrow M$ is just the inclusion homomorphism and where $\epsilon(m) = \phi(m) \in M''$ for all $m \in M$. Furthermore, by definition, we notice the seemingly innocuous fact that $\ker \epsilon = \operatorname{im} \mu$.

More generally, suppose that we have a sequence of right R-modules and R-module homomorphisms:

$$\cdots \to M_{i-1} \xrightarrow{\phi_{i-1}} M_i \xrightarrow{\phi_i} M_{i+1} \to \cdots .$$

We say that the above sequence *exact at M_i* if $\operatorname{im} \phi_{i-1} = \ker \phi_i$. The sequence above is said to be *exact* if it is exact at every M_i. Note that the homomorphism $\mu : M \to N$ is injective if and only if the sequence $0 \to M \xrightarrow{\mu} N$ is exact. Likewise $\epsilon : M \to N$ is surjective if and only if the sequence $M \xrightarrow{\epsilon} N \to 0$ is exact. Of particular interest are the *short exact sequences*. These are exact sequences of the form

$$0 \to M' \xrightarrow{\mu} M \xrightarrow{\epsilon} M'' \to 0.$$

Note that in this case Theorem 8.1.2 simply says that $M'' \cong M/\mu(M')$.

The Left-Exactness of Hom

The "hom functors" come very close to preserving short exact sequences, as follows.

Theorem 8.4.3 *Let $0 \to N' \xrightarrow{\mu} N \xrightarrow{\epsilon} N''$ be an exact sequence of right R-modules, and let M be a fixed R-module. Then the following sequence of abelian groups is also exact:*

$$0 \to \operatorname{Hom}_R(M, N') \xrightarrow{\operatorname{Hom}(1_M, \mu)} \operatorname{Hom}_R(M, N) \xrightarrow{\operatorname{Hom}(1_M, \epsilon)} \operatorname{Hom}_R(M, N'').$$

Likewise, if $M' \xrightarrow{\mu} M \xrightarrow{\epsilon} M'' \to 0$ is an exact sequence of right R-modules, and if N is a fixed R-module, then the following sequence of abelian groups is exact:

$$0 \to \operatorname{Hom}_R(M'', N) \xrightarrow{\operatorname{Hom}(\mu, 1_N)} \operatorname{Hom}_R(M, N) \xrightarrow{\operatorname{Hom}(\epsilon, 1_N)} \operatorname{Hom}_R(M', N).$$

Proof Assume that $0 \to N' \xrightarrow{\mu} N \xrightarrow{\epsilon} N''$ is exact and that M is a fixed R-module. First of all, note that if $\phi' \in \mathrm{Hom}_R(M, N')$, then since μ is injective, we see that $0 = \mathrm{Hom}(M, \mu)(\phi') = \mu \circ \phi'$ clearly implies that $\phi' = 0$, and so $\mathrm{Hom}(M, \mu) : \mathrm{Hom}_R(M, N') \to \mathrm{Hom}_R(M, N)$ is injective. Next, since $\mathrm{Hom}(M, \epsilon) \circ \mathrm{Hom}(M, \mu) = \mathrm{Hom}(M, \mu \circ \epsilon) = \mathrm{Hom}(M, 0) = 0$, we have $\mathrm{im}\,\mathrm{Hom}(M, \mu) \subseteq \ker \mathrm{Hom}(M, \epsilon)$. Finally, suppose that $\phi \in \ker \mathrm{Hom}(M, \epsilon)$. Then $\epsilon \circ \phi = 0 : N \to M''$ which says that for each $n \in N$, $\phi(n) \in \ker \epsilon = \mathrm{im}\,\mu$, i.e. $\phi(n) \in \mathrm{im}\,\mu$ for each $n \in N$. But since $\mu : M' \to M$ is injective, we may write $\phi(n) = \mu \circ \phi'(n)$ for some uniquely defined element $\phi'(n) \in M'$. Finally, since $\phi' : N \to M'$ is clearly a homomorphism, we are finished as we have proved that $\mathrm{Hom}(M, \mu)(\phi') = \phi$. The second statement in the above theorem is proved similarly. $\square$

If the sequences in Theorem 8.4.3 were completed to short exact sequences, it would be natural to wonder whether the resulting sequences would be exact. This is not the case as the following simple counterexample shows. Consider the short exact sequence of $\mathbb{Z}$-modules (i.e., abelian groups):

$$0 \to \mathbb{Z} \xrightarrow{\mu_2} \mathbb{Z} \xrightarrow{\epsilon} \mathbb{Z}/\mathbb{Z}2 \to 0,$$

where $\mu_2 : \mathbb{Z} \to \mathbb{Z}$ is multiplication by 2: $\mu_2(m) = 2m$, and where ϵ is the canonical quotient mapping. The sequence below

$$0 \to \mathrm{Hom}_{\mathbb{Z}}(\mathbb{Z}/\mathbb{Z}2, \mathbb{Z}) \xrightarrow{\mathrm{Hom}(1,\mu_2)} \mathrm{Hom}_{\mathbb{Z}}(\mathbb{Z}/\mathbb{Z}2, \mathbb{Z}) \xrightarrow{\mathrm{Hom}(1,\epsilon)} \mathrm{Hom}_{\mathbb{Z}}(\mathbb{Z}/\mathbb{Z}2, \mathbb{Z}/\mathbb{Z}2) \to 0$$

is not exact as $\mathrm{Hom}_{\mathbb{Z}}(\mathbb{Z}/\mathbb{Z}2, \mathbb{Z}) = 0$, whereas $\mathrm{Hom}_{\mathbb{Z}}(\mathbb{Z}/\mathbb{Z}2, \mathbb{Z}/\mathbb{Z}2) \cong \mathbb{Z}/\mathbb{Z}2$, and so $\mathrm{Hom}(\mathbb{Z}/\mathbb{Z}2, \epsilon) : \mathrm{Hom}_{\mathbb{Z}}(\mathbb{Z}/\mathbb{Z}2, \mathbb{Z}) \to \mathrm{Hom}_{\mathbb{Z}}(\mathbb{Z}/\mathbb{Z}2, \mathbb{Z}/\mathbb{Z}2)$ certainly cannot be surjective.

Thus, we see that the "hom functors" do not quite preserve exactness. The extent to which short exact sequences fail to map to short exact sequences—and the remedy for this deficiency is the subject of *homological algebra*. (In a subsequent chapter, we shall encounter another "functor" (namely that derived from the tensor product) from right R-modules to abelian groups which also fails to preserve short exact sequences).

Split Exact Sequences

Note that given any right R-modules M', M'', we can always build the following rather trivial short exact sequence:

$$0 \to M' \xrightarrow{\mu_1} M' \oplus M'' \xrightarrow{\pi_2} M'' \to 0,$$

where μ_1 is inclusion into the first summand and π_2 is projection onto the second summand. Such a short exact sequence is called a *split short exact sequence*. More generally, we say that the short exact sequence $0 \to M' \xrightarrow{\mu} M \xrightarrow{\epsilon} M'' \to 0$ *splits* if and only if there is an isomorphism $\alpha : M \to M' \oplus M''$ making the following ladder diagram commute[3]:

$$
\begin{array}{ccccccccc}
0 & \longrightarrow & M' & \xrightarrow{\ \mu\ } & M & \xrightarrow{\ \epsilon\ } & M'' & \longrightarrow & 0 \\
& & \downarrow{\scriptstyle 1_{M'}} & & \downarrow{\scriptstyle \alpha} & & \downarrow{\scriptstyle 1_{M''}} & & \\
0 & \longrightarrow & M' & \xrightarrow{\ \mu_1\ } & M' \oplus M'' & \xrightarrow{\ \pi_2\ } & M'' & \longrightarrow & 0
\end{array}
$$

In turn, the following lemma provides simple conditions for a short exact sequence to split:

Lemma 8.4.4 *Let*

$$0 \to M' \xrightarrow{\mu} M \xrightarrow{\epsilon} M'' \to 0 \qquad\qquad (*)$$

be a short exact sequence of right R-modules. Then the following conditions are equivalent:

(i) There exists an R-module homomorphism $\sigma : M'' \to M$ such that $\epsilon \circ \sigma = 1_{M''}$.
(ii) There exists an R-module homomorphism $\rho : M \to M'$ such that $\rho \circ \mu = 1_{M'}$.

The short exact sequence (*) splits.

Proof Clearly condition (iii) implies both (i) and (ii); we shall be content to prove that (i) implies (iii), leaving the remaining implication to the reader. Note first that since $\epsilon \circ \sigma = 1_{M''}$, it follows immediately that $\sigma : M'' \to M$ is injective. In particular, $\sigma : M'' \to \sigma(M'')$ is an isomorphism. Clearly it suffices to show that $M = \mu(M') \oplus \sigma(M'')$ (internal direct sum). If $m \in M$ then $\epsilon(m - \sigma(\epsilon(m))) = \epsilon(m) - (\epsilon \circ \sigma)(\epsilon(m)) = \epsilon(m) - \epsilon(m) = 0$, which says that $m - \sigma(\epsilon(m)) \in \ker \epsilon = \operatorname{im} \mu$. Therefore, there exists $m' \in M'$ with $m - \sigma(\epsilon(m)) = \mu(m')$, forcing $m = \mu(m') + \sigma(\epsilon(m)) \in \mu(M') + \sigma(M'')$. Finally, if for some $m' \in M'$, $m'' \in M''$, so that $\mu(m') = \sigma(m'')$, then $m'' = \epsilon(\sigma(m'')) = \epsilon(\mu(m')) = 0$, which implies that $\mu(M') \cap \sigma(M'') = 0$, proving that $M = \mu(M') \oplus \sigma(M'')$, as required. $\square$

8.4.4 Projective and Injective Modules

Let R be a ring and let P be an R-module. We say that P is *projective* if every diagram of the form

$$
\begin{array}{ccccc}
& & P & & \\
& & \downarrow{\scriptstyle \phi''} & & \\
M & \xrightarrow{\ \epsilon\ } & M'' & \longrightarrow & 0 \qquad \text{(exact row)}
\end{array}
$$

---

[3]Recall the definition of "commutative diagram" on p. 237.

can be embedded in a commutative diagram of the form

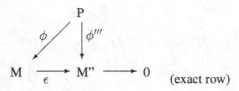

$$M \xrightarrow{\epsilon} M" \longrightarrow 0 \qquad \text{(exact row)}$$

Note first that any free right R-module is projective. Indeed, assume that F is a free right R-module with basis $\{f_\alpha \mid \alpha \in \mathcal{A}\}$ and that we have a diagram of the form

$$
\begin{array}{c}
F \\
\downarrow \phi'' \\
M \xrightarrow{\epsilon} M'' \longrightarrow 0 \qquad \text{(exact row)}
\end{array}
$$

For each $m''_\alpha := \phi''(f_\alpha) \in M''$, $\alpha \in \mathcal{A}$, select an element $m_\alpha \in M$ with $\epsilon(m_\alpha) = m''_\alpha$, $\alpha \in \mathcal{A}$. As F is free with basis $\{f_\alpha \mid \alpha \in \mathcal{A}\}$, there exists a (unique) homomorphism $\phi : F \to M$ with $\phi(f_\alpha) = m_\alpha$, $\alpha \in \mathcal{A}$. This gives the desired "lifting" of the homomorphism ϕ'' :

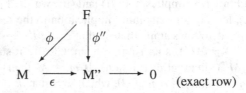

$$M \xrightarrow{\epsilon} M" \longrightarrow 0 \qquad \text{(exact row)}$$

We have the following simple characterization of projective modules.

Theorem 8.4.5 *The following conditions are equivalent for the R-module P.*

 (i) P is projective.
 (ii) Every short exact sequence $0 \to M' \to M \to P \to 0$ splits.
(iii) P is a direct summand of a free R-module.

Proof That (i) implies (ii) follows immediately from Lemma 8.4.4. Now assume condition (ii) and let $\epsilon : F \to P$ be a surjective R-module homomorphism, where F is a free right R-module. Let $K = \ker \epsilon$ and consider the commutative diagram

$$
\begin{array}{c}
P \\
\sigma \nearrow \quad \downarrow 1_P \\
0 \longrightarrow K \xrightarrow{\mu} F \xrightarrow{\epsilon} P \longrightarrow 0
\end{array}
$$

where $\mu : K \hookrightarrow F$, and where the homomorphism $\sigma : P \to F$ exists by the projectivity of P. Apply Lemma 8.4.4 to infer that $F \cong \mu(K) \oplus \sigma(P)$; since $\sigma(P) \cong P$. Thus (iii) holds.

Finally, we prove that (iii) implies (i). Thus, let F be a free right R-module which admits a direct sum decomposition $F = N \oplus P$, where $N \subseteq F$ is a submodule. From the diagram

$$
\begin{array}{c}
P \\
\downarrow \phi' \\
M \xrightarrow{\ \epsilon\ } M'' \longrightarrow 0
\end{array}
$$

we obtain the following diagram.

$$
\begin{array}{ccc}
F = N \oplus P & \underset{\pi}{\overset{\mu}{\rightleftarrows}} & P \\
\theta \downarrow & & \downarrow \phi \\
M & \xrightarrow{\ \epsilon\ } & M'' \longrightarrow 0
\end{array}
$$

Here $\pi : N \oplus P \to P$ is the projection of the sum onto the right coordinate, and $\mu : P \to N \oplus P$ is the injection onto $0 \oplus P$ so that $\pi \circ \mu = 1_P$. The homomorphism $\theta : F \to M$ satisfying $\epsilon \circ \theta = \phi \circ \pi_2$ exists since, as already observed above, free modules are projective. However, setting $\bar{\phi} := \theta \circ \mu$ gives the desired lifting of ϕ, and so it follows that P is projective. $\square$

Dual to the notion of a projective module is that of an *injective* module. A right R-module I is said to be *injective* if every diagram of the form

$$
\begin{array}{c}
I \\
\uparrow \theta' \\
0 \longrightarrow M' \xrightarrow{\ \mu\ } M \quad \text{(exact row)}
\end{array}
$$

can be embedded in a commutative diagram of the form

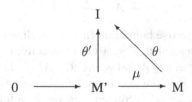

In order to obtain a characterization of injective modules we need a concept dual to that of a free module. In preparation for this, however, we first need the concept of a *divisible* abelian group. Such a group D satisfies the property that for every $d \in D$ and for every $0 \neq n \in \mathbb{Z}$, there is some $c \in D$ such that $nc = d$.

Example 44 1. The most obvious example of a divisible group is probably the additive group $(\mathbb{Q}, +)$ of rational numbers.
2. A moment's thought should reveal that if F is any field of characteristic 0, then $(F, +)$ is a divisible group.
3. Note that any homomorphic image of a divisible group is divisible. Of paramount importance is the divisible group $\mathbb{Q}/\mathbb{Z}$.
4. Let p be a prime, and let $\mathbb{Z}(p^\infty)$ be the subgroup of $\mathbb{Q}/\mathbb{Z}$ consisting of elements of p-power order. Then $\mathbb{Z}(p^\infty)$ is a divisible group. (You should check this. Note that there is a direct sum decomposition $\mathbb{Q}/\mathbb{Z} = \bigoplus \mathbb{Z}(p^\infty)$, where the direct sum is over all primes p).
5. Note that the direct product of any number of divisible groups is also divisible.

The importance of divisible groups is displayed by the following.

Theorem 8.4.6 *Let D be an abelian group. Then D is divisible if and only if D is injective.*

Proof Assume first that D is injective. Let $d \in D$, and let $0 \neq n \in \mathbb{Z}$. We form the diagram

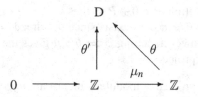

where $\mu_n : \mathbb{Z} \to \mathbb{Z}$ is multiplication by n, $\theta'(m) = md$, $m \in \mathbb{Z}$, and where θ is the extension of θ' guaranteed by the injectivity of D. If $\theta(1) = c$ then $nc = n\theta(1) = \theta(n) = \theta(\mu_n(1)) = \theta'(1) = d$, proving that D is divisible.

Conversely, assume that D is divisible and that we are given a diagram of the form

$$D$$
$$\theta' \uparrow$$
$$0 \longrightarrow A' \xrightarrow{\mu} A$$

where $\mu : A' \to A$ is an injective homomorphism of abelian groups. We shall, for the sake of convenience, assume that $A' \leq A$ via the injective homomorphism μ. Let $\mathcal{P} = \{(A'', \theta'')\}$ where $A' \leq A'' \leq A$ and where $\theta'' : A'' \to D$ is a homomorphism with $\theta''|_{A'} = \theta'$. We make $\mathcal{P}$ into a poset via the relation $(A''_1, \theta''_1) \preccurlyeq (A''_2, \theta''_2)$ if and only if $A''_1 \leq A''_2$ and $\theta''_2|_{A''_1} = \theta''_1$. Since $(A', \theta') \in \mathcal{P}$, it follows that $\mathcal{P} \neq \emptyset$. If $\{(A''_\alpha, \theta''_\alpha)\}_{\alpha \in \mathcal{A}}$ is a chain in $\mathcal{P}$, we may set $A'' = \bigcup_{\alpha \in \mathcal{A}} A''_\alpha$, and define $\theta'' : A'' \to D$ by setting $\theta''(a'') = \theta''_\alpha(a'')$ where $a'' \in A''_\alpha$. Since $\{(A''_\alpha, \theta''_\alpha)\}_{\alpha \in \mathcal{A}}$ is a chain, it follows that $\theta''(a'')$ doesn't depend upon the particular choice of index $\alpha \in \mathcal{A}$ with $a'' \in A_\alpha''$. Therefore, we see that the chain $\{(A_\alpha'', \theta_\alpha'')\}_{\alpha \in \mathcal{A}}$ has (A'', θ'') as an upper bound. This allows us to apply Zorn's lemma and infer the existence of a maximal element $(A_0'', \theta_0'') \in \mathcal{P}$. Clearly the proof is complete once we show that $A_0'' = A$.

Assume, by way of contradiction, that there exists an element $a \in A \backslash A_0''$, and let m be the order of the element $a + A_0''$ in the additive group A/A_0''. We divide the argument into the two cases according to $m = \infty$ and $m < \infty$.

$m = \infty$. In this case it follows easily that $\mathbb{Z}\langle a \rangle$ is an infinite cyclic group and that $\mathbb{Z}\langle a \rangle \cap A_0'' = 0$. As a result, $A'' := \mathbb{Z}\langle a \rangle + A_0'' = \mathbb{Z}\langle a \rangle \oplus A_0''$ and an extension of θ_0'' can be obtained simply by setting $\theta''(ka + a_0'') = \theta_0''$, $k \in \mathbb{Z}$, $a_0'' \in A_0''$. This is a well-defined homomorphism $A'' \to D$ extending θ_0''. But that contradicts the maximality of (A_0'', θ_0'') in the poset $\mathcal{P}$.

$m < \infty$. Here, we have that $ma \in A_0''$ and so $\theta_0''(ma) \in D$. Since D is divisible, there exists an element $c \in D$ with $mc = \theta_0''(ma)$. Now define $\theta'' : A'' := \mathbb{Z}\langle a \rangle + A_0'' \to D$ by setting $\theta''(ka + a_0'') = kc + \theta_0''(a_0'') \in D$. We need to show that θ'' is well defined. Thus, assume that k, l are integers and that $a_0'', b_0'' \in A_0''$ are such that $ka + a_0'' = la + b_0''$. Then $(k-l)a = b_0'' - a_0'' \in A_0''$ and so $k - l = rm$ for some integer r. Therefore

$$kc - lc = rmc$$
$$= r\theta_0''(ma)$$
$$= \theta_0''(rma)$$
$$= \theta_0''(b_0'' - a_0'')$$
$$= \theta_0''(b_0'') - \theta_0''(a_0''),$$

which says that

$$kc + \theta_0''(a_0'') = lc + \theta_0''(b_0'')$$

and so $\theta'' : A'' \to D$ is well defined. As clearly θ'' is a homomorphism extending θ_0'', we have once again violated the maximality of $(A_0'', \theta_0'') \in \mathcal{P}$, and the proof is complete. $\square$

Let R be a ring, and let A be an abelian group. Define $M := \mathrm{Hom}_\mathbb{Z}(R, A)$; thus M is certainly an abelian group under point-wise operations. Give M the structure of a right R-module via

$$(f \cdot r)(s) = f(rs), \ r, s \in R, \ f \in M.$$

It is easy to check that the above recipe gives $\mathrm{Hom}_\mathbb{Z}(R, A)$ the structure of a right R-module. (See Exercise (1) in Sect. 8.5.3.)

The importance of the above construction is found in the following.

Proposition 8.4.7 *Let R be a ring and let D be a divisible abelian group. Then the right R-module $\mathrm{Hom}_\mathbb{Z}(R, D)$ is an injective R-module.*

Proof Suppose that we are given the usual diagram

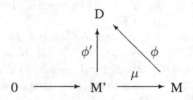

$$\text{Hom}_{\mathbb{Z}}(R, D)$$
$$\theta' \uparrow$$
$$0 \longrightarrow \quad M' \quad \xrightarrow{\ \mu\ } M \quad \text{(exact row)}$$

The homomorphism $\theta' : M' \to \text{Hom}_{\mathbb{Z}}(R, D)$ induces the homomorphism $\phi' : M' \to D$ by setting $\phi'(m') = \theta'(m')(1)$, for all $m' \in M'$. Since D is divisible, it is injective and so we may form the commutative diagram:

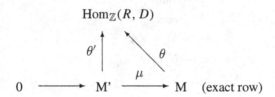

$$D$$
$$\phi' \uparrow \quad \nwarrow \phi$$
$$0 \longrightarrow M' \xrightarrow{\ \mu\ } M$$

In turn, we now define $\theta : M \to \text{Hom}_{\mathbb{Z}}(R, D)$ by setting $\theta(m)(r) := \phi(mr) \in D$. It is routine to check that θ is a homomorphism of right R-modules and that the following diagram commutes:

$$\text{Hom}_{\mathbb{Z}}(R, D)$$
$$\theta' \uparrow \quad \nwarrow \theta$$
$$0 \longrightarrow M' \xrightarrow{\ \mu\ } M \quad \text{(exact row)}$$

Thus $\text{Hom}_{\mathbb{Z}}$ is an injective R-module. $\square$

Recall that any free right R-module is the direct sum of a number of copies of R_R, and that *any* R-module is a homomorphic image of a free module. We now define a *cofree* R-module to be the direct *product* (not *sum*!) of any number of copies of the injective module $\text{Hom}_{\mathbb{Z}}(R, \mathbb{Q}/\mathbb{Z})$. Note that by Exercise (12) in Sect. 8.5.3, a cofree module is injective. We now have

Proposition 8.4.8 *Let M be an R-module. Then M can be embedded in a cofree R-module.*

Proof Let $0 \neq m \in M$ and pick a nonzero element $\alpha(m) \in \mathbb{Q}/\mathbb{Z}$ whose order $o(m)$ in the additive group $\mathbb{Q}/\mathbb{Z}$ is equal to that of m in the additive group M. (For, example, one could choose $\alpha(M) = (1/o(m)) + \mathbb{Z}$.) This gives an additive group homomorphism $\alpha : \mathbb{Z}\langle m \rangle \to \mathbb{Q}/\mathbb{Z}$ given by $\alpha(km) = k\alpha(m)$, for each integer k. As $\mathbb{Q}/\mathbb{Z}$ is a divisible abelian group, there exists a homomorphism of abelian groups $\alpha_m : M \to \mathbb{Q}/\mathbb{Z}$ extending $\alpha : \mathbb{Z}\langle m \rangle \to \mathbb{Q}/\mathbb{Z}$. In turn, this homomorphism gives a homomorphism $\beta_m : M \to \mathrm{Hom}_{\mathbb{Z}}(R, \mathbb{Q}/\mathbb{Z})$ defined by setting $\beta_m(m')(r) = \alpha_m(m'r) \in \mathbb{Q}/\mathbb{Z}$. It is routine to verify that β_m is indeed, a homomorphism of right R-modules.

Next, we form the cofree module $C := \prod_{0 \neq m \in M} \mathrm{Hom}_{\mathbb{Z}}(R, \mathbb{Q}/\mathbb{Z})_m$, where each $\mathrm{Hom}_{\mathbb{Z}}(R, \mathbb{Q}/\mathbb{Z})_m = \mathrm{Hom}_{\mathbb{Z}}(R, \mathbb{Q}/\mathbb{Z})$. Collectively, the homomorphisms β_m, $0 \neq m \in M$, give a homomorphism $\mu : M \to C$ via

$$\mu(m')(m) = \beta_m(m') \in \mathrm{Hom}_{\mathbb{Z}}(R, \mathbb{Q}/\mathbb{Z}).$$

It remains only to show that μ is injective. Suppose m were a non-zero element in ker μ. Then $\mu(m)(m') = 0$ for all $0 \neq m' \in M$. But this says in particular that $0 = \beta_m(m) \in \mathrm{Hom}_{\mathbb{Z}}(R, \mathbb{Q}/\mathbb{Z})$, forcing $\alpha_m(m) = \beta_m(m)(1) = 0$. That assertion, however, produces a contradiction since $\alpha_m(m) = \alpha(m)$ was always chosen to have the same additive order $o(m)$ of the non-zero element m in $(M, +)$. $\square$

Finally we have the analogue of Theorem 8.4.5, above; the proof is entirely dual to that of Theorem 8.4.5.

Theorem 8.4.9 *The following conditions are equivalent for the R-module I.*

(i) I is injective.
(ii) Every short exact sequence $0 \to I \to M \to M'' \to 0$ splits.
(iii) I is a direct summand of a cofree R-module.

8.5 Exercises

8.5.1 Exercises for Sect. 8.1

1. Show that any abelian group is a $\mathbb{Z}$-module. [Hint: Write the group operation as addition, and define the multiplication by integers.]
2. (Right modules from left modules). In the remark on p. 233, a recipe is given for converting a left R module into a right S-module, by the formula $ms := \psi(s)m$, where $m \in M$, a left R-module, $s \in S$ and where $\psi : S \to \mathrm{Opp}(R)$ is a ring homomorphism mapping the multiplicative identity element of S to that of R. Show that all the axioms of a right S-module are satisfied by this rule.
3. (Annihilators of modules as two-sided ideals.) Suppose that M is a right R-module.

   (a) Show that $\mathrm{Ann}(M) := \{r \in R | Mr = 0_M\}$ is a two-sided ideal in R. (It is called the *annihilator of M*.)
   (b) Suppose $A \leq B \leq M$ is a chain of submodules of the right R-module M. Show that

   $$(A : B) := \{r \in R | Br \subseteq A\}$$

is a 2-sided ideal of R. [Hint: $(A : B)$ is the annihilator of the submodule B/A. Use the previous step. Note that if A and B are right ideals of R (that is, submodules of R_R), then $\text{Ann}(B/A)$ is just the *residual quotient* $(A : B)$ defined on p. 196.]

4. Prove Corollary 8.1.7. [Hint: Establish the criterion of Lemma 8.1.6.]
5. (Finite generation)

   (a) Let $(\mathbb{Q}, +)$ be the additive group of the field of rational numbers. Being an abelian group, this group is a $\mathbb{Z}$-module by Exercise (1) in Sect. 8.5.1. Show that $(\mathbb{Q}, +)$ is not finitely generated as a $\mathbb{Z}$-module. [Hint: Pay attention to denominators.]

   (b) Suppose M is a finitely generated right R-module. Show that for any generating set X for M, there is a finite subset X_0 of X which also generates M. [Hint: By assumption there exists a finite generating set F. Write each member of F as a finite R-linear combination of elements of X.]

6. Let M_α, M'_α, $\alpha \in \mathcal{A}$ be right R-modules, and let $\phi_\alpha : M_\alpha \to M'_\alpha$, $\alpha \in \mathcal{A}$ be R-module homomorphisms. Prove that there is a unique homomorphism $\prod \phi_\alpha : \prod M_\alpha \to \prod M'_\alpha$ making each diagram below commute:

$$
\begin{array}{ccc}
\prod_{\alpha \in \mathcal{A}} M_\alpha & \xrightarrow{\ \prod \phi_\alpha\ } & \prod_{\alpha \in \mathcal{A}} M'_\alpha \\
\pi_\beta \downarrow & & \pi'_\beta \downarrow \\
M_\beta & \xrightarrow{\ \phi_\beta\ } & M'_\beta
\end{array}
$$

7. Let M_α, M'_α, $\alpha \in \mathcal{A}$ be right R-modules, and let $\phi_\alpha : M_\alpha \to M'_\alpha$, $\alpha \in \mathcal{A}$ be R-module homomorphisms. Prove that there is a unique homomorphism $\oplus \phi_\alpha : \oplus M_\alpha \to \oplus M'_\alpha$ making each diagram below commute:

$$
\begin{array}{ccc}
M_\beta & \xrightarrow{\ \pi_\beta\ } & M'_\beta \\
\mu_\beta \downarrow & & \mu'_\beta \downarrow \\
\bigoplus_{\alpha \in \mathcal{A}} M_\alpha & \xrightarrow{\ \oplus \phi_\alpha\ } & \bigoplus_{\alpha \in \mathcal{A}} M'_\alpha
\end{array}
$$

8. Prove, or provide a counterexample to the following statement. Any submodule of a free module must be free.
9. Prove that the direct sum of free R-modules is also free.

8.5.2 Exercises for Sects. 8.2 and 8.3

1. Show that any submodule or factor module of a Noetheriean (Artinian) module is Noetherian (Artinian). [Hint: This is Sect. 2.3, Corollary 2.3.7.]
2. Suppose $M = N_1 + \cdots + N_k$, a sum of finitely many submodules N_i. Show that if each submodule N_i is Noetherian (Artinian) then M is Noetherian (Artinian). [Hint: This is Sect. 2.5.4, Lemma 2.5.7, part (i).]
3. Show that if N is a submodule of M then M is Noetherian (Artinian) if and only if both N and M/N are Noetherian (Artinian). [Hint: Isn't this just the content of Lemma 2.5.6 of Sect. 2.5.4?]
4. Show that if M/A_i is a Noetherian (Artinian) right module for a finite collection of submodules $A_1, \ldots A_n$, then so is $M/(A_1 \cap \cdots \cap A_n)$. [Hint: Refer to Sect. 2.5.4, Lemma 2.5.7, part (ii).]
5. Let R be a ring. Suppose the R-module R_R is Noetherian (Artinian)—that is, the ring R is *right Noetherian (right Artinian)*. Show that any finitely generated right R-module is Noetherian (Artinian). [Hint: Such a right module M is a homomorphic image of a free module $F = x_1 R \oplus \cdots \oplus x_k R$ over a finite number of generators. Then apply other parts of this exercise.]
6. Suppose the right R-module M satisfies the ascending chain condition on submodules. Then every generating set X of M contains a finite subset that generates M.
7. Let F be a field and let $V = F^{(n)}$, the vector space of n-tuples of elements of F, where n is a positive integer. Recall that an n-tuple $\alpha := (a_1, \ldots, a_n)$ is a *zero of the polynomial* $p(x_1, \ldots, x_n)$ if and only $p(a_1, \ldots, a_n) = 0$ (more precisely, the polynomial p is in the kernel of the evaluation homomorphism $e_\alpha : F[x_1, \ldots, x_n] \to F$ (see p. 209)).

   (a) Show that if $\{p_1, p_2, \ldots\}$ is an infinite set of polynomials in the polynomial ring $F[x_1, \ldots, x_n]$, and V_i is the full set of zeros of the polynomial p_i, then the set of common zeros of the p_i—that is the intersection $\cap V_i$—is actually the intersection of finitely many of the V_i.

   (b) Show that for any subset X of V there exists a finite set $S_X := \{p_1, \ldots p_k\}$ polynomials such that *any* polynomial which vanishes on X is an F-linear combination of those in the finite set F.

   (c) Recall that a *variety* is the set of common zeros in V of some collection of polynomials in $F[x_1, \ldots, x_n]$. The varieties in V form a poset under the inclusion relation (see p. 209). Show that the poset of varieties of V possesses the descending chain condition. [Hint: All three parts exploit the Hilbert Basis Theorem.]

8.5.3 Exercises for Sect. 8.4

1. Let M be a left R-module and let A be an abelian group. Show that the right R-scalar multiplication on $\mathrm{Hom}_{\mathbb{Z}}(M, A)$ defined by setting $(fr)(m) = f(rm) \in A$, gives $\mathrm{Hom}_{\mathbb{Z}}(M, A)$ the structure of a right R-module. If A is also a left R-module is the subgroup $\mathrm{Hom}_R(M, A)$ an R-submodule of $\mathrm{Hom}_{\mathbb{Z}}(M, A)$?

2. Let R, S be rings. Recall from p. 234 that an (R, S)-bimodule is an abelian group M having the structure of a left R-module and the structure of a right S-module such that these scalar multiplications commute, i.e., $r(ms) = (rm)s$ for all $m \in M$, $r \in R$ and $s \in S$.[4] Now assume that M is an (R, S)-bimodule and that N is a right S-module. Show that one may give $\mathrm{Hom}_S(M, N)$ the structure of a right R-module by setting $(fr)(m) = f(rm)$, $r \in R$, $m \in M$, $f \in \mathrm{Hom}_S(M, N)$.

3. Let M be a right R-module. Interpret and prove:

$$\mathrm{Hom}_R(R, M) \cong_R M.$$

4. Show that the "functorially induced mapping" of Eq. (8.24) is a homomorphism as left R-modules. [Hint: the only real point is to verify the equation

$$\beta \circ (r\phi) \circ \alpha = r(\beta \circ \phi \circ \alpha)$$

using the left R-action defined for these modules on p. 261 preceding the equation.]

5. Let R be a ring, let M be a right R-module, and let A be an abelian group. Interpret and prove:

$$\mathrm{Hom}_R(M, \mathrm{Hom}_{\mathbb{Z}}(R, A)) \cong_{\mathbb{Z}} \mathrm{Hom}_{\mathbb{Z}}(M, A).$$

6. Let R be an integral domain in which every ideal is a *free* R-module. Prove that R is a *principal ideal domain*.

7. (a) Let M, N_1 and N_2 be right R-modules. Show that

$$\mathrm{Hom}(M, N_1 \oplus N_2) \cong_{\mathbb{Z}} \mathrm{Hom}(M, N_1) \oplus \mathrm{Hom}(M, N_2).$$

More generally, show that if $\{N_\alpha, \ \alpha \in \mathcal{A}\}$ is a family of right R-modules, then

$$\mathrm{Hom}\left(M, \bigoplus_{\alpha \in \mathcal{A}} N_\alpha\right) \cong_{\mathbb{Z}} \bigoplus_{\alpha \in \mathcal{A}} \mathrm{Hom}(M, N_\alpha).$$

---

[4]Perhaps the most important example of bimodules occur as follows: if R is a ring and if $S \subseteq R$ is a subring, then R is an (R, S)-bimodule.

(b) Let N, M_1 and M_2 be right R-modules and show that

$$\text{Hom}(M_1 \times M_2, N) \cong_{\mathbb{Z}} \text{Hom}(M_1, N) \times \text{Hom}(M_2, N).$$

More generally, show that if $\{M_\alpha,\ \alpha \in \mathcal{A}\}$ is a family of R-modules, then

$$\text{Hom}\left(\prod_{\alpha \in \mathcal{A}} M_\alpha, N\right) \cong_{\mathbb{Z}} \prod_{\alpha \in \mathcal{A}} \text{Hom}(M_\alpha, N).$$

(c) Show that if

$$0 \to N' \overset{\mu}{\to} N \overset{\epsilon}{\to} N'' \to 0$$

is a split short exact sequence of R-modules, and if M is a fixed right R-module, then the sequences

$$0 \to \text{Hom}(M, N') \overset{\text{Hom}(1_M, \mu)}{\longrightarrow} \text{Hom}(M, N) \overset{\text{Hom}(1_M, \epsilon)}{\longrightarrow} \text{Hom}(M, N'') \to 0,$$

and

$$0 \to \text{Hom}(N'', M) \overset{\text{Hom}(\mu, 1_M)}{\longrightarrow} \text{Hom}(N, M) \overset{\text{Hom}(\epsilon, 1_M)}{\longrightarrow} \text{Hom}(N', M) \to 0,$$

are both exact (and also split).

8. Show that the sequences which appear in Theorem 8.4.3 exhibiting the left or right exactness of the "hom' functors" are morphisms as left R-modules, not just abelian groups.

9. Let $0 \to M' \to M \to M'' \to 0$ be a short exact sequence of right R-modules. Prove that M is Noetherian (resp. Artinian) if and only if both M', M'' are. [Of course, this is just a restatement of Exercise (3) in Sect. 8.5.2]

10. Prove that if A_α, $\alpha \in \mathcal{A}$, is a family of abelian groups, then

$$\text{Hom}_{\mathbb{Z}}\left(R, \prod_{\alpha \in \mathcal{A}} A_\alpha\right) \cong_R \prod_{\alpha \in \mathcal{A}} \text{Hom}_{\mathbb{Z}}(R, A_\alpha).$$

11. Let P_α, $\alpha \in \mathcal{A}$ be a family of right R-modules. Show that $\bigoplus_{\alpha \in \mathcal{A}} P_\alpha$ is projective if and only if each P_α is projective.

12. Let I_α, $\alpha \in \mathcal{A}$ be a family of right R-modules. Show that $\prod_{\alpha \in \mathcal{A}} I_\alpha$ is injective if and only if each I_α is injective.

13. Let P be an R-module. Prove that P is projective if and only if given any exact sequence $0 \to M' \overset{\mu}{\to} M \overset{\epsilon}{\to} M'' \to 0$, the induced sequence

$$0 \to \mathrm{Hom}_R(P, M') \overset{\mathrm{Hom}(1_P, \mu)}{\longrightarrow} \mathrm{Hom}_R(P, M) \overset{\mathrm{Hom}(1_P, \epsilon)}{\longrightarrow} \mathrm{Hom}_R(P, M'') \to 0$$

is exact.

14. Suppose we have a sequence $0 \to M' \overset{\mu}{\to} M \overset{\epsilon}{\to} M'' \to 0$ of R-modules. Prove that this sequence is exact if and only if the sequence

$$0 \to \mathrm{Hom}_R(P, M') \overset{\mathrm{Hom}(1_P, \mu)}{\longrightarrow} \mathrm{Hom}_R(P, M) \overset{\mathrm{Hom}(1_P, \epsilon)}{\longrightarrow} \mathrm{Hom}_R(P, M'') \to 0$$

is exact for every projective R-module P.

15. Let I be an R-module. Prove that I is injective if and only if given any exact sequence $0 \to M' \overset{\mu}{\longrightarrow} M \overset{\epsilon}{\longrightarrow} M'' \to 0$, the induced sequence

$$0 \longrightarrow \mathrm{Hom}_R(M'', I) \overset{\mathrm{Hom}(\epsilon, 1_I)}{\longrightarrow} \mathrm{Hom}_R(M, I) \overset{\mathrm{Hom}(\mu, 1_I)}{\longrightarrow} \mathrm{Hom}_R(M', I) \to 0$$

is exact.

16. Suppose we have a sequence $0 \to M' \overset{\mu}{\to} M \overset{\epsilon}{\to} M'' \to 0$ of R-modules. Prove that this sequence is exact if and only if the sequence

$$0 \to \mathrm{Hom}_R(M'', I) \overset{\mathrm{Hom}(\epsilon, 1_I)}{\longrightarrow} \mathrm{Hom}_R(M, I) \overset{\mathrm{Hom}(\mu, 1_I)}{\longrightarrow} \mathrm{Hom}_R(M', I) \to 0$$

is exact for every injective R-module I.

17. Give an example of a non-split short exact sequence of the form

$$0 \to P \to M \to I \to 0$$

where P is a projective R-module and where I is an injective R-module.

18. Let F be a field and let $R =$ be the ring of $n \times n$ lower-triangular matrices over F. For each $m \le n$ the F-vector space

$$L_m = \{[\alpha_1 \, \alpha_2 \, \cdots \, \alpha_m \, 0 \, \cdots \, 0] \mid \alpha_1 \, \alpha_2, \ldots, \alpha_m \in F\}$$

is a right R-module. Prove that each L_m, $1 \le m \le n$ is a projective R-module, but that none of the quotients L_k/L_j, $1 \le j < k \le n$ is projective.

19. A ring for which every ideal is projective is called a *hereditary* ring.
    Prove that if F is a field, then the ring $M_n(F)$ of $n \times n$ matrices over F is hereditary. The same is true for the ring of lower triangular $n \times n$ matrices over F.

20. Let A be an abelian group and let $B \le A$ be such that A/B is infinite cyclic. Prove that $A \cong A/B \times B$.

21. Let A be an abelian group and assume that $A = H \times C_1 = K \times C_2$ where C_1 and C_2 are infinite cyclic groups. Prove that $H \cong K$. [Hint: First $H/(H \cap K) \cong HK/K \le A/K \cong C_2$ so $H/(H \cap K)$ is either trivial or infinite cyclic. Similarly

for $K/(H \cap K)$. Next $A/(H \cap K) \cong H/(H \cap K) \times C_1$ and $A/(H \cap K) \cong K/(H \cap K) \times C_2$ so $H/(H \cap K)$ and $K/(H \cap K)$ are either both trivial (in which case $H = K$) or both infinite cyclic. Thus, from the preceding Exercise (20) in Sect. 8.5.3 obtain $H \cong H/(H \cap K) \times H \cap K \cong K/(H \cap K) \times H \cap K \cong K$, done.]

22. Prove *Baer's Criterion:* Let I be a left R-module and assume that for any left ideal $J \subseteq R$ and any R-module homomorphism $\alpha_J : J \to I$, α extends to an R-module homomorphism $\alpha : R \to I$. Show that I is an injective module. [Hint: Let $M' \subseteq M$ be R-modules and assume that there is an R-module homomorphism $\alpha : M' \to I$. Consider the poset of pairs (N, α_N), where $M' \subseteq N \subseteq M$ and where α_N extends α. Apply Zorn's Lemma to obtain a maximal element (N_0, α_0). If $N_0 \neq M$, let $m \in M - N_0$ and let $J = \{r \in R \mid rm \in N_0\}$; note that J is a left ideal of R. Now what?]

Reference

1. Rotman J (1979) An introduction to homological algebra. Academic Press, Boston

Chapter 9
The Arithmetic of Integral Domains

Abstract Integral domains are commutative rings whose non-zero elements are closed under multiplication. If each nonzero element is a unit, the domain is called a field and is shipped off to Chap. 11. For the domains D which remain, divisibility is a central question. A prime ideal has the property that elements outside the ideal are closed under multiplication. A non-zero element $a \in D$ is said to be *prime* if the principle ideal Da which it generates is a prime ideal. D is a *unique factorization domain* (or UFD) if any expression of an element as a product of prime elements is unique up to the order of the factors and the replacement of any prime factor by a unit multiple. If D is a UFD, so is the polynomial ring $D[X]$ where X is a finite set of commuting indeterminates. In some cases, the unique factorization property can be determined by the localizations of a domain. Euclidean domains (like the integers, Gaussian and Eisenstein numbers) are UFD's, but many domains are not. One enormous class of domains (which includes the algebraic integers) is obtained the following way: Suppose K a field which is finite-dimensional over a subfield F which, in turn, is the field of fractions of an integral domain D. One can then define the ring $\mathcal{O}_D(K)$ of elements of K which are integral with respect to D. Under modest conditions, the integral domain $\mathcal{O}_D(K)$, will become a Noetherian domain in which every prime ideal is maximal—a so-called Dedekind domain. Although not UFD's, Dedekind domains offer a door prize: every ideal can be uniquely expressed as a product of prime ideals (up to the order of the factors, of course).

9.1 Introduction

Let us recall that an *integral domain* is a species of commutative ring with the following simple property:

(ID) If D is an integral domain then there are no zero divisors—that is, there is no pair of non-zero elements whose product is the zero element of D.

Of course this is equivalent (in the realm of commutative rings) to the proposition

(ID') The set D^* of non-zero elements of D, is closed under multiplication, and thus $(D^*, \cdot)$ is a commutative monoid under ring multiplication.

© Springer International Publishing Switzerland 2015

E. Shult and D. Surowski, *Algebra*, DOI 10.1007/978-3-319-19734-0_9

As an immediate consequence one has the following:

Lemma 9.1.1 *The collection of all non-zero ideals of an integral domain are closed under intersection, and so form a lattice under either the containment relation or its dual. Under the containment relation, the 'meets' are intersections and the 'joins' are sums of two ideals.*

Proof The submodules of D_D form a lattice under the containment relation with 'sum' and 'intersection' playing the roles of and "join" and "meet". Since such submodules are precisely the ideals of D, it remains only to show that two nonzero ideals cannot intersect at the zero ideal in an integral domain. But if A and B are ideals carrying non-zero elements a and b, respectively, then ab is a non-zero element of $A \cap B$. $\square$

We had earlier introduced integral domains as a class of examples in Chap. 6 on basic properties of rings. One of the unique and identifying properties of integral domains presented there was:

(The Cancellation Law) If $(a, b, c) \in D^* \times D \times D$, then $ab = ac$ implies $b = c$.

It is this law alone which defines the most interesting aspects of integral domains— *their arithmetic*—which concerns who divides who among the elements of D. (It does *not* seem to be an interesting question in general rings with zero divisors—such as full matrix algebras.)

9.2 Divisibility and Factorization

9.2.1 Divisibility

One may recall from Sect. 7.1.3 that the units of a ring R form a multiplicative group, denoted $U(R)$. Two elements a and b of a ring are called *associates* if $a = bu$ for some unit $u \in U(R)$. Since $U(R)$ is a group, the relation of being an associate, is an equivalence relation. This can be seen in the following way: the group of units acts by right multiplication on the elements of the ring R. Two elements are associates if and only if they belong to the same $U(R)$-orbit under this action. Since the $U(R)$-orbits partition the elements of R, the property of belonging to a common orbit is clearly an equivalence relation on the set of ordered pairs $R \times R$. We call these equivalence classes *association classes*.

Equivalence relations are fine, but how can we connect these up with "divisibility"? To make the notion precise, we shall say that element a *divides* element b if and only if $b = ca$ for some element c in R. Notice that this concept is very sensitive to the ambient ring R which contains elements a and b, for that is the "well" from which a potential element c is to be drawn. By this definition,

- Every element divides itself.
- Zero divides only zero, while every element divides zero.
- If element a divides b and element b divides c, then a divides c.

Thus the divisibility relationship is transitive and so we automatically inherit a *pre-order* which is trying to tell us as much as possible about the question of who divides whom.

As one may recall from the very first exercise for Chap. 2 (Sect. 2.7.1), for every pre-order, there is an equivalence relation (that of being less-than-or-equal in both directions) whose equivalence classes become the elements of a partially ordered set. Under the divisibility pre-order for an integral domain, D, equivalent elements a and b should satisfy

$$a = sb, \quad \text{and} \quad b = ta.$$

for certain elements s and t in D. If either one of a or b is zero, then so is the other, so zero can only be equivalent to zero. Otherwise, a and b are both non-zero and so, by the Cancellation Law,

$$a = s(ta) = (st)a \quad \text{and} \quad st = 1.$$

In this case, both s and t are units. That means a and b are associates.

Conversely, if a and b are associates, they divide one another.

Thus in an integral domain D, the equivalence classes defined by the divisibility preorder, are precisely the association classes of D—that is, the multiplicative cosets $xU(D)$ of the group of units in the multiplicative monoid $D^* := (D - \{0\}, \cdot)$.

The resulting poset of association classes of the integral domain D is called the *divisibility poset* and is denoted $\mathrm{Div}(D)$. Moreover:

Lemma 9.2.1 (Properties of the divisibility poset)

1. *The additive identity element 0_D, is an association class that is a global maximum of the poset $\mathrm{Div}(D)$.*
2. *The group of units $U(D)$ is an associate-class comprising the global minimum of the poset $\mathrm{Div}(D)$.*

One can also render this poset in another way. We have met above the lattice of all ideals of D under inclusion—which we will denote here by the symbol $L(D)$. Its dual lattice (ideals under reverse containment) is denoted $L(D)^*$. We are interested in a certain induced sub-poset of $L(D)^*$. Ideals of the form xD which are generated by a single element x are called *principal ideals*. Let $P(D)$ be the subposet of $L(D)$ induced on the collection of all principal ideals of D, including the zero ideal, $0 = 0D$. We call this the *principal ideal poset*. To obtain a comparison with the poset $\mathrm{Div}(D)$, we must pass to the dual. Thus $P(D)^*$ is defined to be the subposet of L^* induced on all principal ideals—that is, the poset of principal ideals ordered by reverse inclusion.

We know the following:

Lemma 9.2.2 (Divisibility and principal ideal posets) *In an integral domain D, the following holds:*

1. *Element a divides element b if and only if $bD \subseteq aD$.*
2. *Elements a and b are associates if and only if $aD = bD$.*
3. *So there is a bijection between the poset $P(D)^*$ of all principal ideals under reverse containment, and the poset $Div(D)$ of all multiplicative cosets of the group of units (the association classes of D) under the divisibility relation, given by the map $f : xD \to xU(D)$.*
4. *Moreover, the mapping*
$$f : P(D)^* \to Div(D),$$

is an isomorphism of posets.

One should be aware that the class of principal ideals of a domain D need not be closed under taking intersections and taking sums. This has a lot to do—as we shall soon see—with the questions of the existence of "greatest" common divisors, "least" common multiples and ultimately the success or failure of unique factorization. Notice that here, the "meet" of two cosets $xU(D)$ and $yU(D)$ would be a coset $dU(D)$ such that $dU(D)$ divides both $xU(D)$ and $yU(D)$ and is the unique coset maximal in $Div(D)$ having this property. We pull this definition back to the realm of elements of D in the following way:

The element d is a *greatest common divisor* of two non-zero elements a and b of an integral domain D if and only

1. d divides both a and b.
2. If e divides both a and b, then e divides d.

Clearly, then, if a greatest common divisor of two elements of a domain exists, it is unique up to taking associates—i.e. up to a specified coset of $U(D)$—and this association class is unchanged upon replacing any of the two original elements by any of *their* associates.

In fact it is easy to see that if d is the greatest common divisor of a and b, then dD is the smallest *principal ideal* containing both aD and bD. Note that this might not be the ideal $aD + bD$, the join in the lattice of all ideals of D.

Similarly, lifting back the meaning of a "join" in the poset $Div(D)$, we say that the element m is a *least common multiple* of two elements a and b in an integral domain D if and only if:

1. Both a and b divide m (i.e. m is a multiple of both a and b).
2. If n is also a common multiple of a and b, then n is a multiple of m

Again, we note that if m is the least common multiple of a and b, then mD is the largest principal ideal contained in both aD and bD. Again, this does not mean that the ideal mD is $aD \cap bD$, the meet of aD and bD in the full lattice L of all ideals of D.

The above discussion calls attention to the special case in which the principal ideal poset $P(D)$ actually coincides with the full lattice of ideals $L(D)$. An integral domain D for which $P(D) = L(D)$ is called a *principle ideal domain*, or PID for short. From our discussion in the preceding paragraphs we have:

Lemma 9.2.3 *For any two elements a and b of a principal ideal domain D, a greatest common divisor $d = gcd(a, b)$ and a least common multiple $m = lcm(a, b)$ exist. For the least common multiple m, one has $Dm = Da \cap Db$, the meet of Da and Db in the lattice $L(D)$. Similarly the greatest common divisor d generates the join of the ideals spanned by a and b. That is,*

$$Da + Db = Dd.$$

Thus there exist elements s and t in D, such that $d = sa + tb$.

9.2.2 Factorization and Irreducible Elements

The next two lemmas display further instances in which a question on divisibility in an integral domain, refers back to properties of the poset of principal ideals.

A *factorization* of an element of the integral domain D is simply a way of representing it as a product of other elements of D. Of course one can write an element as a product of a unit and another associate in as many ways as there are elements in $U(D)$. The more interesting factorizations are those in which none of the factors are units. These are called *proper factorizations*.

Suppose we begin with an element x of D. If it is a unit, it has no proper factorization. Suppose element x is not a unit, and has a proper factorization $x = x_1 y_1$ into two non-zero non-units. We attempt, then, to factor each of these factors into two further factors. If such a proper factorization is not possible for one factor, one proceeds with the other proper factor. In this way one may imagine an infinite schedule of such factorizations. This schedule would correspond to a downwardly growing binary tree in the graph of the divisibility poset $\text{Div}(D)$, An end-node to this tree (that is, a vertex of degree one) results if and only if one obtains a factor y which possesses no proper factorization. We call such an element "irreducible".

Precisely, an element y of D^* is *irreducible* if and only if it is a non-zero non-unit with the property that if $y = ab$ is a factorization, one of the factors a or b, is a unit. Applying the poset isomorphism $f : \text{Div}(D) \to P(D)^*$ given in Lemma 9.2.2 above, these comments force the following:

Lemma 9.2.4 *If D is an integral domain, an element y of D is an irreducible element if and only if the principal ideal yD is non-zero and does not properly lie in any other principal ideal except D itelf (that is, it is maximal in the induced poset $P(D) - \{D, 0\}$ of proper principal ideals of D under containment).*

Here is another important central result of this type which holds for a large class of integral domains.

Lemma 9.2.5 *Let D be an integral domain possessing the ascending chain condition on its poset $P(D)$ of principal ideals. Then every non-zero non-unit is the product of a unit and a finite number of irreducible elements.*

Proof Suppose there were a non-zero element $x \in D - U(D)$ which was not the product of a finite number of irreducible elements. Then the collection $\mathcal{T}$ of principal ideals xD where x is not a product of finitely many irreducible elements is non-empty. We can thus choose an ideal yD which is maximal in $\mathcal{T}$ because $\mathcal{T}$ inherits the ascending chain condition from the poset of principal ideals. Since y itself is not irreducible, there is a factorization $y = ab$ where neither a nor b is a unit. Then aD properly contains yD, since otherwise $aD = yD$ and y is an associate ua of a where u is a unit. Instantly, the cancellation law yields $b = u$ contrary to hypothesis. Thus certainly the principal ideal aD does not belong to $\mathcal{T}$. Then a *is* a product of finitely many irreducible elements of D. Similarly, b is a product of finitely many irreducible elements of D. Hence their product y must also be so, contrary to the choice of y. The proof is complete. $\square$

9.2.3 Prime Elements

So far, we have been discussing irreducible elements. A similar, but distinct notion is that of a *prime* element. A non-zero non-unit r in D is said to be a *prime*, if and only if, whenever r divides a product ab, then either r divides a or r divides b.

One of the very first observations to be made from this definition is that the notion of being prime is stronger than the notion of being irreducible, Thus:

Lemma 9.2.6 *In any integral domain, any prime element is irreducible.*

Proof Let r be a prime element of the integral domain D. Suppose, by way of contradiction that r is not irreducible, so that $r = ab$, where a and b are non-units. Then r divides ab, and so, as r is prime, r divides one of the two factors a or b—say a. Then $r = rvb$ for some $v \in D$. But then, by the Cancellation laws, $1 = vb$, whence b is a unit, contrary to our assumption.

Thus r is irreducible. $\square$

In general, among integral domains, an irreducible element may not be prime.[1] However it is true for principle ideal domains.

Lemma 9.2.7 *In a principle ideal domain, every irreducible element is a prime element.*

Proof Let x be an irreducible element in the principle ideal domain D. By definition x is a non-unit. Assume, by way of contradiction that x is not a prime. Then there exists elements a and b in D, such that x divides ab, but x does not divide either

---

[1] An example is given under the title "a special case" on p. 290.

a of b. Notice that if a were a unit, then x would divide b, against our hypothesis. Thus it follows that neither a nor b are units. Thus Da is a proper ideal that does not lie in Dx. Since x is irreducible, Dx is a maximal ideal in D (Lemma 9.2.4), and is properly contained in the ideal $Dx + Da$. Thus $Dx + Da = D$, and so $1 = d_1a + d_2x$ for some elements d_1, d_2 of D. But since x divides ab, we may write $ab = d_3x$ for $d_3 \in D$. Now

$$b = b \cdot 1 = b(d_1a + d_2x) = d_1ab + d_2xb$$
$$= d_1(d_3x) + d_2xb = (d_1d_3 + d_2b)x.$$

Thus b is a multiple of x, which is impossible, since x does not divide b by hypothesis. This contradiction tells us that x is indeed a prime element. $\square$

Thus, in a principal ideal domain, the set of irreducible elements and the set of prime elements coincide.

The reader will observe, the following:

Corollary 9.2.8 *In a principle ideal domain D, the element a is prime, if and only if the factor ring D/Da is an integral domain.*

9.3 Euclidean Domains

An integral domain D is said to be a *Euclidean Domain* if and only if there is a function $g : D^* \to \mathbb{N}$ into the natural numbers (non-negative integers)[2] satisfying the following:

(ED1) For $a, b \in D^* := D - \{0\}$, $g(ab) \geq g(a)$.
(ED2) If $(a, b) \in D \times D^*$, then there exist elements q and r such that $a = bq + r$, where either $r = 0$ or else $g(r) < g(b)$.

The notation in (ED2) is intentionally suggestive: q stands for "quotient" and r stands for "remainder". The function g is sometimes called the "grading function" of the domain D.

Recall that in a ring R, an ideal I of the form xR is called a *principal ideal*. An integral domain in which every left ideal is a principal ideal is called a *principal ideal domain* or *PID*. We have

Theorem 9.3.1 *Every Euclidean domain is a principal ideal domain.*

Proof Let D be a Euclidean domain with grading function $g : D^* \to \mathbb{N}$. Let J be any ideal in D. Among the non-zero elements of J we can find an element x with $g(x)$ minimal. Let a be any other element of J. Then $a = qx + r$ as in (ED2) where $r = 0$ or $g(r) < g(x)$. Since $r = a - qx \in J$, minimality of $g(x)$ shows that the

---

[2]As remarked several times, in this book, the term "natural numbers" includes zero.

second alternative cannot occur, so $r = 0$ and $a = qx$. Thus $J \subseteq xD$. But $x \in J$, an ideal, already implies $xD \subseteq J$ so $J = xD$ is a principal ideal. Since J was an arbitrary ideal, D is a PID. $\square$

There are, however, principle ideal domains which are not Euclidean domains.

Clearly any Euclidean domain D possesses the *algebraic* property that it is a PID, so that greatest common divisors $d = gcd(a, b)$ always exist along with elements s and t such that $d = sa + bt$ (Lemmas 9.2.3 and 9.3.1). What is really new here is the *compuional-logical* property that the greatest common divisor $d = gcd(a, b)$ as well as the elements s and t can actually be computed! Behold!

EUCLIDEAN ALGORITHM. Given a and b in D and D^*, respectively, by (ED2) we have

$$a = bq_1 + r_1,$$

and if $r_1 \neq 0$,

$$b = r_1 q_2 + r_2,$$

and if $r_2 \neq 0$,

$$r_1 = r_2 q_3 + r_3,$$

etc., until we finally obtain a remainder $r_k = 0$ in

$$r_{k-2} = r_{k-1} q_k + r_k.$$

Such a termination of this process is inevitable for $g(r_1), g(r_2), \ldots$ is a strictly decreasing sequence of non-negative integers which can only terminate at $g(r_{k-1})$ when $r_k = 0$.

We claim the number r_{k-1} (the last non-zero remainder) is a greatest common divisor of a and b. First $r_{k-2} = r_{k-1} q_k$ and also $r_{r-3} = r_{k-2} q_{k-1} + r_{k-1}$ are multiples of r_{k-1}. Inductively, if r_{k-2} divides both r_j and r_{j+1}, it divides r_{j-1}. Thus all of the right hand sides of the equations above are multiples of r_{k-1}; in particular, a and b are multiples of r_{k-1}. On the other hand if d' is a common divisor of a and b, d' divides $r_1 = a - bq_1$ and in general divides $r_j = r_{j-2} - r_{j-1} q_j$ and so d' divides r_{k-1}, eventually. Thus $d := r_{k-1}$ is a greatest common divisor of a and b.

Also an explicit expression of $d = r_{k-1}$ as a D-linear combination $sa + tb$ can be obtained from the same sequence of equations. For

$$d = r_{k-1} = r_{k-3} - r_{k_2} q_{k-1},$$

and each r_j similarly is a D-linear combination of r_{j-2} and r_{j-1}, $j = 1, 2, \ldots$ Successive substitutions for the r_j ultimately produces an expression $d = sa + tb$. All quite computable.

9.3.1 Examples of Euclidean Domains

Example 45 The *Gaussian integers and the Eisenstein integers.*

These are respectively the domains $D_1 := \mathbb{Z} \oplus \mathbb{Z}i$ (where $i^2 = -1 \neq -i$) and $D_2 = \mathbb{Z} \oplus \mathbb{Z}\omega$ (where $\omega = e^{2i\pi/3}$). The complex number ω is a zero of the irreducible polynomial $x^2 + x + 1$, and so is a cube root of unity distinct from 1.

Both of these domains are subrings of $\mathbb{C}$, the complex number field, and each is invariant under complex conjugation. Thus both of them admit a multiplicative norm function $N : \mathbb{C} \to \mathbb{R}$ where $N(z) := z \cdot \bar{z}$ records the square of the Euclidean distance of z from zero in the complex plane. (As usual, $\bar{z}$ denotes the complex conjugate of the complex number z.)

We shall demonstrate that these rings are Euclidean domains with the norm function in the role of g. That its values are taken in the non-negative integers follows from the formulae

$$N(a + bi) = a^2 + b^2$$
$$N(a + b\omega) = a^2 - ab + b^2.$$

As already remarked, the norm function is multiplicative so (ED1) holds. To demonstrate (ED2) we choose elements a and b of $D = D_1$ or D_2 with $b \neq 0$. Now the elements of D_1 form a square tessellation of the complex plane with $\{0, 1, i, i + 1\}$ as a fundamental square. Similarly, the elements of D_2 form the equilateral triangle tessellation of the complex plane with $\{0, 1, -\omega^2\}$ or $\{0, 1, -\omega\}$ at the corners of the fundamental triangles. We call the points where three or more tiles of the tessellation meet *lattice points*.

When we superimpose the ideal bD on either one of these tessellations of the plane, we are depicting a tessellation of the same type on a subset of the lattice points with a possibly larger fundamental tile. Thus for the Gaussian integers, bD is a tessellation whose fundamental square is defined by the four "corner" points $\{0, b, ib, (1 + i)b\}$. The resulting lattice points are a subset of the Gaussian integers closed under addition (vector addition in the geometric picture). Similarly for the Eisenstein numbers, a fundamental triangle of the tessellation defined by the ideal bD is the one whose "corners" are in $\{0, b, -b\omega^2\}$ or $\{(0, b, -b\omega)\}$. In either case the tiles of the tessellation bD cover the plane, and so the element a must fall in some tile T—either a square or an equilateral triangle—of the superimposed tessellation bD. Now let qb be a corner point of T which is nearest a. Since T is a square or an equilateral quadrangle, we have

(Nearest Corner Principle) *The distance from a point of T to its nearest corner is less than the length of a side of T*

Thus the *distance* $|a - qb| = \sqrt{N(a - qb)}$ is less than $|b| = \sqrt{N(b)}$. Using the fact that the square function is monotone increasing on non-negative real numbers, we have

$$g(a - bq) = N(a - bq) < N(b) = g(b).$$

Thus setting $r = a - qb$, we have $r = 0$ or $g(r) < g(b)$. Thus (with respect to the function g), r serves to realize the condition (ED2).[3]

Before passing to the next example, this may be a good place to illustrate how embedding one domain into another sometimes may provide new results for the original domain: in this case we are embedding the integers $\mathbb{Z}$ into the Gaussian integers $\mathbb{Z}[i]$. When we select a prime integer (called a *rational prime*), the question is raised whether it is still a prime element in the ring of Gaussian integers (a *Gaussian prime*). The Theorem 9.3.3 below gives a complete answer.

Lemma 9.3.2 *Let p be any rational prime. If p is not a Gaussian prime then p is the sum of two integer squares.*

Proof Suppose that rational prime p is the product of two non-units in $\mathbb{Z}[i]$, say $p = \zeta\eta$. Taking norms, we have $p^2 = N(\zeta)N(\eta)$. Since a Gaussian integer is a unit if and only if its norm is ± 1, it follows that the last two norms of the previous equation are both positive integers dividing p^2. Since p is a rational prime, each of these norms is equal to p. Thus, writing $\zeta = a + bi$ one has $p = N(\zeta) = a^2 + b^2$. $\square$

Theorem 9.3.3 *Let p be a rational prime. Then p is a Gaussian prime if and only if it leaves a remainder of 3, when divided by 4.*

Proof First, the prime integer 2 is not a Gaussian prime since $2 = (1 + i)(1 - i)$, so we may assume that the rational prime p is odd. Since the square of every odd integer leaves a remainder of 1, when divided by 4, the sum of two integer squares can only be congruent to 0, 1, or 2 modulo 4. Thus if $p \equiv 3 \mod 4$, it is not the sum of two integer squares and so must be a Gaussian prime, by the previous Lemma 9.3.2.

That leaves the case that $p \equiv 1 \mod 4$. In this case $\mathbb{Z}/(p)$ is a field whose multiplicative group of non-zero elements contains a unique cyclic subgroup of order 4 whose generator has a square equal to -1, that is, there exists an integer b such that $b^2 \equiv -1 \mod p$. Let P be the principle ideal generated by p in the domain of Gaussian integers. Since $\mathbb{Z}[i]$ is Euclidean, it is a principle ideal domain, and so, by Corollary 9.2.8, the factor ring $\mathbb{Z}[i]/P$ is an integral domain if and only if p is a prime in $\mathbb{Z}[i]$. But since $b \not\equiv 0 \mod p$, the numbers $1 \pm bi$ cannot be multiples of p in $\mathbb{Z}[i]$. Thus the equation

$$(1 + bi + P)(1 - bi + P) = (1 + bi)(1 - bi) + P = (1 + b^2) + P = P,$$

reveals that the factor ring $\mathbb{Z}[i]/P$ possesses zero divisors, and so cannot be an integral domain. Accordingly p is not a prime element of $\mathbb{Z}[i]$. $\square$

---

[3]Usually boxing matches are held (oxymoronically) in square "boxing rings". Even if (as well) they were held in equilateral triangles, it is a fact that when the referee commands a boxer to "go to a neutral corner", the fighter really does not have far to go (even subject to the requirement of "neutrality" of the corner). But woe the fighter who achieves a knock-down near the center, of a more ring-like N-gon for large N. After the mandatory hike to a neutral corner, the fighter can only return to the boxing match totally exhausted from the long trip. No Euclidean algorithm for these boxers.

Corollary 9.3.4 *If p is a rational prime of the form $4n + 1$, then p is a sum of two integer squares.*

Proof By Theorem 9.3.3 p is not a Gaussian prime. Now apply Lemma 9.3.2. $\square$

This result does not seem easy to prove within the realm of integers alone.

Example 46 Polynomial domains over a field. Let F be a field and let $F[x]$ be the ring of polynomials in indeterminate x and with coefficients from F. We let deg be the degree function, $F[x] \to \mathbb{N}$.[4] Then if $f(x)$ and $g(x)$ are non-zero polynomials,

$$\deg(fg) = \deg f + \deg g,$$

so (ED1) holds. The condition (ED2) is the familiar long division algorithm of College Algebra and grade school.

9.4 Unique Factorization

9.4.1 Factorization into Prime Elements

Let D be any integral domain. Recall that a non-zero non-unit of D is said to be *irreducible* if it cannot be written as the product of two other non-units. This means that if $r = ab$ is irreducible, then either a or b is a unit.

Recall also that a non-zero non-unit r in D is said to be a *prime element*, if and only if, whenever r divides a product ab, then either r divides a or r divides b.

In the case of the familiar ring of integers (indeed for all PID's), the two notions coincide, and indeed they are forced to coincide in an even larger collection of domains.

An integral domain D is said to be a *unique factorization domain* (or UFD) if and only if

(UFD1) *Every non-zero non-unit is a product of a unit and a finite number of irreducible elements.*

(UFD2) *Every irreducible element is prime.*

We shall eventually show that a large class of domains—the principal ideal domains—are UFD's. (See Theorem 9.4.3.) One of the very first observations to be made from these definitions is that the notion of being prime is stronger than the notion of being irreducible. Thus:

---

[4]The student will recall the familiar degree function, that records the highest exponent of x that appears when a polynomial is expressed as a linear combination of powers of x with non-zero coefficients.

Lemma 9.4.1 *The following are equivalent:*

(i) D is a UFD.
(ii) In D every non-zero non-unit is a product of a unit and a finite number of primes.

Proof (i) $\Rightarrow$ (ii) is obvious since (i) implies each irreducible element is a prime.

(ii) $\Rightarrow$ (i). First we show that every irreducible element is prime. Let r be irreducible. Then by (ii), r is a product of a unit and a finite number of primes. But since r is irreducible, $r = up$ where u is a unit and p is a prime.

Now every non-zero non-unit is a product of a unit and finitely many irreducible elements, since each prime is irreducible by Lemma 9.2.6. Thus the two defining properties of a UFD follow from (ii) and the proof is complete. $\square$

Theorem 9.4.2 (The Basic Unique Factorization Theorem) *Let D be an integral domain which is a UFD. Then every non-zero non-unit can be written as a product of finitely many primes. Such an expression is unique up to the order in which of the prime factors appear, and the replacement of any prime by an associate—that is, a multiple of that prime by a unit.*

Proof Let r be a non-zero non-unit of the UFD D. Then by Lemma 9.4.1, r can be written as a unit times a product of finitely many primes. Suppose this could be done in more than one way, say,

$$r = up_1 p_2 \cdots p_s = vq_1 q_2 \cdots q_t$$

where u and v are units and the p_i and q_i are primes. If $s = 0$ or $t = 0$, then r is a unit, contrary to the choice of r. So without loss of generality we may assume $0 < s \leq t$. Now as p_1 is a prime, it divides one of the right hand factors, and this factor cannot be v. Rearranging the indexing if necessary, we may assume p_1 divides q_1. Since q_1 is irreducible, p_1 and q_1 are associates, so $q_1 = u_1 p_1$. Then

$$r = up_1 \cdots p_s = (vu_1)p_1 q_2 \cdots q_t$$

so, by the cancellation law

$$up_2 \cdots p_s = (vu_1)q_2 \cdots q_t,$$

with $t - 1$ factors on the right side. By induction on t, we have $s = t$ and $q_i = u_i p_{\pi(i)}$ for some unit u_i, $i = 2, \ldots t$ and permutation π of these indices. Thus the two factorizations involve the same number of primes with the primes in one of the factorizations being associates of the primes in the other, written in some possibly different order. $\square$

A Special Case

Of course we are accustomed to the domain of the integers, which is a UFD. So it might be instructive to look at a more pathological case.

Consider the ring $D = \{a + b\sqrt{-5} | a, b \in \mathbb{Z}\}$. Being a subring of the field of complex numbers $\mathbb{C}$, it is an integral domain.

The mapping $z \to \bar{z}$, where $\bar{z}$ is the complex conjugate of z, is an automorphism of $\mathbb{C}$ which leaves D invariant (as a set) and so induces an automorphism of D. We define the *norm* of a complex number ζ to be $N(\zeta) := \zeta\bar{\zeta}$. Then $N : \mathbb{C} \to \mathbb{R}^+$, the positive real numbers, and $N(\zeta\psi) = N(\zeta)N(\psi)$ for all $\zeta, \psi \in \mathbb{C}$. Clearly, $N(a + b\sqrt{-5}) = a^2 + 5b^2$ so the restriction of N to D has non-negative integer values. (Moreover, we obtain for free the fact that the integers of the form $a^2 + 5b^2$ are closed under multiplication.)

Let us determine the units of D. Clearly if $\nu \in D$ is a unit, then there is a μ in D so that $\nu\mu = 1 = 1 + 0\sqrt{-5}$, the identity of D. Then

$$N(\nu)N(\mu) = N(\nu\mu) = N(1) = 1^2 + 5 \cdot 0^2 = 1.$$

But how can the integer 1 be expressed as the product of two non-negative integers? Obviously, only if $N(\nu) = 1 = N(\mu)$. But if $a^2 + 5b^2 = 1$, it is clear that $b = 0$ and $a = \pm 1$. Thus

$$U(D) = \{\pm 1\} = \{d \in D | N(d) = 1\}.$$

We can also use norms to locate irreducible elements of D. For example, if ζ is an element of D of norm 9, then ζ is irreducible. For otherwise, one would have $\zeta = \psi\eta$ where ψ and η are non-units. But that means $N(\psi)$ and $N(\eta)$ are both integers larger than one. Yet $9 = N(\psi)N(\eta)$ so $N(\psi) = N(\eta) = 3$, which is impossible since 3 is not of the form $a^2 + 5b^2$.

But now note that

$$9 = 3 \cdot 3 = (2 + \sqrt{-5})(2 - \sqrt{-5})$$

are two factorizations of 9 into irreducible factors (irreducible, because they have norm 9) and, as $U(D) = \{\pm 1\}$, 3 is *not* an associate of either factor on the right hand side.

Thus unique factorization fails in the domain $D = \mathbb{Z} \oplus \mathbb{Z}\sqrt{-5}$. The reason is that 3 is an irreducible element which is *not* a prime. It is not a prime because 3 divides $(2 + \sqrt{-5})(2 - \sqrt{-5})$, but does not divide either factor.

9.4.2 Principal Ideal Domains Are Unique Factorization Domains

The following theorem utilizes the fact that if an integral domain possesses the ascending chain condition on principal ideals, then every element is a product of finitely many irreducible elements (Lemma 9.2.5).

Theorem 9.4.3 *Any PID is a UFD.*

Proof Assume D is a PID. Then by Theorem 8.2.11 of Chap. 8, D has the ascending chain condition on all ideals, and so by Lemma 9.2.5, the condition (UFD1) holds.

It remains to show (UFD2), that every irreducible element is prime. Let x be irreducible. Then Dx is a maximal ideal (Lemma 9.2.4).

Now if x were not a prime, there would exist elements a and b such that x divides ab but x does not divide either a or b. Thus $a \notin xD$, $b \notin xD$, yet $ab \in xD$. Thus xD is not a prime ideal, contrary to the conclusion of the previous paragraph that xD is a maximal ideal, and hence a prime ideal. $\square$

Corollary 9.4.4 *If F is a field, the ring of polynomials $F[x]$ is a UFD.*

Proof We have observed in Example 46 that $F[x]$ is a Euclidean ring with respect to the degree function. Thus by Theorem 9.3.1, $F[x]$ is a PID, and so is a UFD by Theorem 9.4.3 above. $\square$

9.5 If D Is a UFD, Then so Is $D[x]$

We begin with three elementary results for arbitrary integral domains D, regarded as subrings of the polynomial rings $D[x]$.

Lemma 9.5.1 $U(D[x]) = U(D)$.

Proof Since D is an integral domain, degrees add in taking the products of non-zero polynomials. Thus the units of $D[x]$ must have degree zero and so must lie in D. That a unit of D is a unit of $D[x]$ depends only upon the fact that D is a subring of $D[x]$. $\square$

Lemma 9.5.2 *Let p be an element of D, regarded as a subring of $D[x]$. If p divides the polynomial $q(x)$ in $D[x]$, then every coefficient of $q(x)$ is divisible by p.*

Proof If p divides $q(x)$ in $D[x]$, then $q(x) = pf(x)$ for some $f(x) \in D[x]$. Then the coefficients of $q(x)$ are those of $f(x)$ multiplied by p. $\square$

Lemma 9.5.3 (Prime elements of D are prime elements of $D[x]$.) *Let p be a prime element of D. If p divides a product $p(x)q(x)$ of two elements of $D[x]$, then either p divides $p(x)$ or p divides $q(x)$.*

Proof Suppose the irreducible element p divides $p(x)q(x)$ as hypothesized. Write

$$p(x) = \sum_{i=0}^{n} a_i x^i, \text{ and } q(x) = \sum_{j=0}^{m} b_j x^j,$$

where $a_i, b_j \in D$. Suppose by way of contradiction that p does not divide *either* $p(x)$ or $q(x)$. Then by the previous Lemma 9.5.2, p divides each coefficient of $p(x)q(x)$ while there is a *first* coefficient a_r of $p(x)$ not divisible by p, and a first coefficient b_s of $q(x)$ not divisible by p. Then the coefficient of x^{r+s} in $p(x)q(x)$ is

$$c_{r+s} = \sum_{i+j=r+s} a_i b_j, \text{ subject to } 0 \le i \le n \text{ and } 0 \le j \le m \qquad (9.1)$$

Now if $(i, j) \neq (r, s)$, and $i + j = r + s$, then either $i < r$ or $j < s$, and in either case $a_i b_j$ is divisible by p. Thus all summands $a_i b_j$ in the right side of Eq. (9.1) except possibly $a_r b_s$ are divisible by p. But p is a prime element in D that does not divide either a_r or b_s. It follows that p does not divide $a_r b_s$ which would mean that p does not divide c_{r+s}, against Lemma 9.5.2 and the fact that p divides $p(x)q(x)$.

Thus p must divide one of $p(x)$ or $q(x)$, completing the proof. $\square$

We are all familiar with the way that the rational number field $\mathbf{Q}$ is obtained as a system of fractions of integers. In an identical manner, one can form a *field of fractions* F of any integral domain D. Its elements are "fractions"—that is, equivalence classes of pairs in $D \times D^*$ with the equivalence class $[n, d]$ containing the pair (n, d) defined to be the set of pairs (bn, bd) as b ranges over all nonzero elements $b \in D^*$. Addition and multiplication of classes are as they are for rational numbers:

$$[a, b] + [c, d] = [ad + bc, bd] \text{ and } [a, b] \cdot [c, d] = [ac, bd].$$

(This an example of a *localization* $F = D_S$, of the sort studied in the next section, with $S = D^* = D - \{0\}$.) Since D is a subdomain of the field F, $D[x]$, the domain of polynomials with coefficients from D, can be regarded as a subring of $F[x]$ by the device of regarding coefficients d of D as fractions $d/1$. We wish to compare the factorization of elements in $D[x]$ with those in $F[x]$.

We say that a polynomial $p(x) \in D[x]$ is *primitive* if and only if, whenever $d \in D$ divides $p(x)$, then d is a unit.

From this point onward, we assume that D is a UFD, so all irreducible elements are prime and every non-zero element is a unit or a product of prime elements.

We now approach two lemmas that involve $F[x]$ where F is the field of fractions of D.

Lemma 9.5.4 *If $p(x) = \alpha q(x)$, where $p(x)$ and $q(x)$ are primitive polynomials in $D[x]$, and $\alpha \in F$, then α is a unit in D.*

Proof Since D is a UFD, greatest common divisors and least common multiples exist, and so there is a "lowest terms" representation of α as a fraction t/s where any greatest common divisor of s and t is a unit. If s is itself a unit, then t divides each coefficient of $p(x)$, and since $p(x)$ is primitive, t must also be a unit. In that case $\alpha = s/t - st^{-1}$ is a unit of D, as claimed. Otherwise, there is a prime divisor p of s. Then p does not divide t, and so from $sp(x) = tq(x)$, we see that every coefficient of $q(x)$ is divisible by p, against $q(x)$ being primitive. $\square$

Lemma 9.5.5 *If $f(x) \in F[x]$, then $f(x)$ has a factorization*

$$f(x) = rp(x)$$

where $r \in F$ and $p(x)$ is a primitive polynomial in $D[x]$. This factorization is unique up to replacement of each factor by an associate.

Proof We prove this in two steps. First we establish

Step 1. *There exists a scalar γ such that $f(x) = \gamma p(x)$, where $p(x)$ is primitive in $D[x]$.*

Each non-zero coefficient of x^i in $f(x)$ has the form a_i/b_i with $b_i \in D - \{0\}$. Multiplying through by a least common multiple m of the b_i (recall that lcm's exist in UFD's), we obtain a factorization

$$f(x) = (\frac{1}{m}) \sum a_i (m/b_i) x^i,$$

whose second factor is clearly primitive in $D[x]$.

Step 2. *The factorization in Step 1 is unique up to associates.*

Suppose $f(x) = \gamma_1 p_1(x) = \gamma_2 p_2(x)$ with $\gamma_i \in F$, and $p_i(x)$ primitive in $D[x]$. Then $p_1(x)$ and $p_2(x)$ are associates in $F[x]$ so

$$p_1(x) = \gamma p_2(x), \text{ for } \gamma \in F.$$

Then by Lemma 9.5.4, γ is a unit in D. But as $\gamma = \gamma_2 \gamma_1^{-1}$, the result follows. $\square$

A non-zero element of an integral domain D' is said to be *reducible* simply if it is *not* irreducible—i.e. it has a factorization into two non-units of D'.

Lemma 9.5.6 (Gauss' Lemma) *Suppose $p(x) \in D[x]$ is reducible in the ring $F[x]$. Then $p(x)$ is reducible in $D[x]$.*

Proof By hypothesis $p(x) = f(x)g(x)$, where $f(x)$ and $g(x)$ are polynomials of positive degree in $F[x]$. Then by Step 1 of the proof of Lemma 9.5.5 above, we may write

$$f(x) = \mu f_1(x)$$
$$g(x) = \gamma g_1(x).$$

where $\mu, \gamma \in F$ and $f_1(x)$ and $g_1(x)$ are primitive polynomials in $D[x]$. Now by Lemma 9.5.3, $f_1(x)g_1(x)$ is a primitive polynomial in $D[x]$ so

$$p(x) = (\mu\gamma)(f_1(x)g_1(x)),$$

and Lemma 9.5.4 shows that $\mu\gamma$ is a unit in D. Then

$$p(x) = ((\mu\gamma)f_1(x)) \cdot g_1(x)$$

is a factorization of $p(x)$ in $D[x]$, with factors of positive degree. The conclusion thus follows. $\square$

Theorem 9.5.7 *If D is a UFD, then so is $D[x]$.*

Proof Let $p(x)$ be any element of $D[x]$. We must show that $p(x)$ has a factorization into irreducible elements which is unique up to the replacement of factors by associates. Since D is a unique factorization domain, a consideration of degrees shows that it is sufficient to do this for the case that $p(x)$ is primitive of positive degree.

Let $S = D - \{0\}$ and form the field of fractions $F = D_S$, regarding $D[x]$ as a subring of $F[x]$ in the usual way. Now by Corollary 9.4.4, $F[x]$ is a unique factorization domain. Thus we have a factorization

$$p(x) = p_1(x)p_2(x) \cdots p_n(x)$$

where each $p_i(x)$ is irreducible in $F[x]$. Then by Lemma 9.5.5, there exist scalars $\gamma_i, i = 1, \ldots, n$, such that $p_i(x) = \gamma_i q_i(x)$, where $q_i(x)$ is a primitive polynomial in $D[x]$. Then

$$p(x) = (\gamma_1\gamma_2 \cdots \gamma_n)q_1(x) \cdots q_n(x).$$

Since the $q_i(x)$ are primitive, so is their product (Lemma 9.5.3). Then by Lemma 9.5.4 the product of the γ_i is a unit u of D. Thus

$$p(x) = uq_1(x) \cdots q_n(x)$$

is a factorization in $D[x]$ into irreducible elements.

If

$$p(x) = vr_1(x) \cdots r_m(x)$$

were another such factorization, the fact that $F[x]$ is a PID and hence a UFD shows that $m = n$ and the indexing can be chosen so that $r_i(x)$ is an associate of $q_i(x)$ in $F[x]$—i.e. there exist scalars ρ_i in F such that $r_i(x) = \rho_i q_i(x)$, for all i. Then as $r_i(x)$ and $q_i(x)$ are irreducible in $D[x]$, they are primitive, and so by Lemma 9.5.4, each γ_i is a unit in D. Thus the two factorizations of $p(x)$ are alike up to replacement of the factors by associates in $D[x]$. The proof is complete. $\square$

Corollary 9.5.8 *If D is a UFD, then so is the ring*

$$D[x_1, \ldots, x_n].$$

Proof Repeated application of Theorem 9.5.7 to

$$D[x_1, \ldots, x_{j+1}] \simeq (D[x_1, \ldots, x_j])[x_{j+1}]. \qquad \square$$

9.6 Localization in Commutative Rings

9.6.1 Localization by a Multiplicative Set

Section 9.2 revealed that the divisibility structure of D, as displayed by the divisibility poset $\text{Div}(D)$, is in part controlled by the group of units. It is then interesting to know that the group of units can be enlarged by a process described in this section.[5]

We say that a subset S of a ring is *multiplicatively closed* if and only if $SS \subseteq S$ and S does not contain the zero element of the ring. If S is a non-empty multiplicatively closed subset of the ring R, we can define an equivalence relation "$\sim$" on $R \times S$ by the rule that $(a, s) = (b, t)$ if and only if, for some element u in S, $u(at - bs) = 0$.

Let us first show that the relation "$\sim$" is truly an equivalence relation. Obviously the relation "$\sim$" is symmetric and reflexive. Suppose now

$$(a, s) \sim (b, t) \sim (c, r) \text{ for } \{r, s, t\} \subseteq S.$$

Then there exists elements u and v in S such that

$$u(at - bs) = 0 \text{ and } v(br - tc) = 0.$$

Multiplying the first equation by vr we get $vr(uat) = vr(ubs)$. But $vr(ubs) = us(vbr) = us(vtc)$, by the second equation. Hence

$$vut(ar) = vut(sc),$$

so $(a, s) \sim (c, r)$, since $vut \in S$. Thus $\sim$ is a transitive relation.

For any element $s \in S$, we now let the symbol a/s (sometimes written $\frac{a}{s}$) denote the $\sim$—equivalence class containing the ordered pair (a, s). We call these equivalence classes *fractions*.

Next we show that these equivalence classes enjoy a ring structure. First observe that if there is an element u in S such that $u(as' - sa') = 0$ (i.e. $(a, s) \sim (a', s')$) then

---

[5]When this happens, the student may wish to verify that the new divisibility poset so obtained is a homomorphic image of the former poset.

$u(abs't - sta'b) = 0$, so $(ab, st) \sim (a'b, s't)$. Thus we can unambiguously define the product of two equivalence classes—or 'fractions'—by setting $(a/s) \cdot (b/t) = ab/st$.

Similarly, if $(a, s) \sim (a', s')$, so that for some $u \in S$, $u(as' - sa') = 0$, we see that

$$u(at + sb)s't - u(a't + s'b)st = 0,$$

so

$$(at + bs)/s = (a't + s'b)/s't.$$

Thus 'addition', defined by setting

$$\frac{a}{s} + \frac{b}{t} := \frac{at + bs}{st},$$

is well defined, since this is also $(a't + s'b)/s't$.

The set of all $\sim$—classes on $R \times S$ is denoted R_S. Now that multiplication and addition are defined on R_S, we need only verify that

1. $(R_S, +)$ is an abelian group with identity element $0/s$—that is, the $\sim$—class containing $(0, s)$ for any $s \in S$.
2. $(R_S, \cdot)$ is a commutative monoid with multiplicative identity s/s, for any $s \in S$.
3. Multiplication is distributive with respect to addition in R_S.

All of these are left as Exercises in Sects. 9.13.1–9.13.2. The conclusion, then, is that

$$(R_s, +, \cdot) \text{ is a ring.}$$

This ring R_s is called the *localization of R by S*.

9.6.2 Special Features of Localization in Domains

For each element s in the multiplicatively closed set S, there is a homomorphism of additive groups

$$\psi_s : (R, +) \to (R_S, +)$$

given by $\psi_s(r) = r/s$. This need not be an injective morphism. If $a/s = b/s$, it means that there is an element $u \in S$ such that $us(a - b) = 0$. Now if the right annihilator of us is nontrivial, such an $a - b$ exists with $a \neq b$, and ψ_s is not one-to-one.

Conversely, we can say, however,

Lemma 9.6.1 *If s is an element of S such that no element of sS is a "zero divisor"—an element with a non-trivial annihilator—then $\psi_s : R \to R_S$ is an embedding of additive groups. The converse is also true. In particular, we see that in an integral domain, where every non-zero element has only a trivial annihilator, the map ψ_s is injective for each $s \in S$.*

Lemma 9.6.2 *Suppose S is a multiplicatively closed subset of non-zero elements of the commutative ring R and suppose no element of S is a zero divisor in R.*

(i) *Suppose $ac \neq 0$ and b, d are elements of S. Then in the localized ring R_S*

$$(\frac{a}{b})(\frac{c}{d}) \text{ is non-zero in } R_S.$$

(ii) *If a is not a zero divisor in R, then for any $b \in S$, a/b is not a zero divisor in R_S.*

(iii) *If R is an integral domain, then so is R_S.*

(iv) *If R is an integral domain, the mapping $\psi_1 : R \to R_S$ is an injective homomorphism of rings (that is an embedding of rings).*

Proof Part (i). Suppose $ac \neq 0$, but that for some $\{b, d, s\} \subseteq S$,

$$(\frac{a}{b})(\frac{c}{d}) = \frac{0}{s}. \tag{9.2}$$

Then

$$(ac, bD) \sim (0, s).$$

Then by the definition of the relation "$\sim$", there is an element $u \in S$ such that $u(acs - ab \cdot 0) = 0$ which implies $uacs = 0$. But since u and s are elements of S, they are not zero divisors, and so we obtain $ac = 0$, a contradiction. Thus the assumption $ac \neq 0$ forces Eq. (9.2) to be false. This proves Part (i).

Parts (ii) and (iii) follow immediately from Part (i).

Part (iv). Assume R is an integral domain. From the second statement of Lemma 9.6.1, the mapping $\psi_s : R \to R_S$ which takes element r to element r/s is an injective homomorphism of additive groups for each $s \in S$. Now put $s = 1$. We see that $\psi_1(ab) = \psi_1(a)\psi_1(b)$ for all $a, b \in R$. Thus ψ_1 is an injective ring homomorphism. $\square$

9.6.3 A Local Characterization of UFD's

Theorem 9.6.3 (Local Characterization of UFD's) *Let D be an integral domain. Let S be the collection of all elements of D which can be written as a product of a unit and a finite number of prime elements. Then S is multiplicatively closed and does not contain zero; so the localization D_S can be formed.*

The domain D is a UFD if and only if D_S is a field.

Proof ($\Rightarrow$) If D is a UFD, then, by Lemma 9.4.1, S comprises all non-zero non-units of D. One also notes that $D_S = D_{S'}$ where S' is S with the set $U(D)$ of all units of D adjoined. (By Exercise (2) in Sect. 9.13.1 of this chapter, this holds for all domains.) Thus $S' = S - \{0\}$. Then $D_{S'}$ is the field of fractions of D.

($\Leftarrow$) Let $D^* := D - \{0\}$, the non-zero elements of D. It suffices to prove that S is all non-zero non-units of D (Lemma 9.4.1). In fact, as $S \subseteq D - U(D)$, it suffices to show that $D^* - U(D) \subseteq S$. Let r be any non-zero non-unit—i.e. an element of $D^* - U(D)$, and suppose by way of contradiction that $r \notin S$.

Since r is a non-unit, Dr is a proper ideal of D. Then

$$(Dr)_S; = \{a/s | a \in Dr, s \in S\}$$

is clearly a non-zero ideal of D_S. Since D_S is a field, it must be all of D_S. But this means that $1/1$ is an element of $(Dr)_S$—i.e. one can find $(b, s) \in D \times S$, such that

$$br/s = 1/1.$$

This means $br = s$ so

$$Dr \cap S \neq \emptyset.$$

Now from the definition of S, we see that there must be multiples of r which can be expressed as a unit times a non-zero number s of primes. Let us choose the multiple so that the number of primes that can appear in such a factorization attains a minimum m. Thus there exists an element r' of D such that

$$rr' = up_1 \cdots p_m$$

where u is a unit and the p_i are primes. We may suppose the indexing of the primes to be such that $p_1, \ldots, p_d$ do not divide r', while $p_{d+1}, \ldots,$ do divide r'. Then if $d < m$, p_{d+1} divides r' so $r' = bp_{d+1}$. Then we have

$$rbp_{d+1} = up_1 \cdots p_d p_{d+1} \cdots p_m$$

so

$$rb = up_1 \cdots p_d p_{d+2} \cdots p_m \ (m - 1 \text{ prime factors})$$

against the minimal choice of m. Thus $d = m$, and each p_i does not divide r'. Then each p_i divides r. Thus $r = a_1 p_1$ so $a_1 r' = up_2 \cdots p_d$, upon canceling p_1. But again, as p_2 does not divide r', p_2 divides a_1. Thus $a_1 = a_2 p_2$ so $a_2 r' = up_3 \cdots p_d$. As each p_i does not divide r', this argument can be repeated, until finally one obtains $a_d r' = u$ when the primes run out. But then r' is a unit. Then

$$r = ((r')^{-1} u) p_1 \cdots p_m \in S.$$

Thus as r was arbitrarily chosen in $D^* - U(D) \subseteq S$, we are done. $\square$

9.6.4 Localization at a Prime Ideal

Now let P be a *prime ideal* of the commutative ring R—that means P has this property: If, for two ideals A and B of R, one has $AB \subseteq P$, then at least one of A and B lie in P. For commutative rings, this is equivalent to asserting either of the following (see the Exercise (2) in Sect. 7.5.2 at the end of Chap. 7):

(1) For elements a and b in R, $ab \in P$ implies $a \in P$ or $b \in P$.
(2) The set $R - P$ is a multiplicatively closed set.

It should be clear that a prime ideal contains the annihilator of every element outside it.

Now let P be a prime ideal of the integral domain D, and set $S := D - P$, which, as noted, is multiplicatively closed. Then the localization of D by S is called the *localization of D at the prime ideal P*. In the literature the prepositions are important: The localization *by* $S = D - P$ is the localization of D *at* the prime ideal P.

Now we may form the ideal $M := \{p/s | (p, s) \in P \times S\}$ in D_S for which each element of $D_S - M$, being of the form s'/s, $(s, s') \in S \times S$, is a unit of D_S. This forces M to be a maximal ideal of D_S, and in fact, every proper ideal B of D_S must lie in it. Thus we see

$$D_S \text{ has a unique maximal ideal } M. \qquad (9.3)$$

Any ring having a unique maximal ideal is called a *local ring*.

This discussion together with Part (iv) of Lemma 9.6.2 yields

Theorem 9.6.4 *Suppose P is a prime ideal of the integral domain D. Then the localization at P (that is, the ring D_S where $S = D - P$) is a local ring that is also an integral domain.*

Example 47 The zero ideal $\{0\}$ of any integral domain is a prime ideal. Forming the localization at the zero ideal of an integral domain D thus produces an integral domain with the zero ideal as the unique maximal ideal and every non-zero element a unit. This localized domain is called the *field of fractions of the domain D*.[6] (This standard construction was used in Sect. 9.5 in studying $D[x]$ as a subring of $D_S[x]$.)

9.7 Integral Elements in a Domain

9.7.1 Introduction

In Sect. 8.2.5 the notion of an algebraic integer was introduced as a special instance of sets of field elements that are integral over a subdomain. Using elementary properties

---

[6]One finds "field of quotients" or even "quotient field" used in place of "field of fractions" here and there. Such usage is discouraged because the literature also abounds with instances in which the term "quotient ring" is used to mean "factor ring".

of finitely-generated modules and Noetherian rings, it was a fairly simple matter to demonstrate that sums and products of algebraic integers were again algebraic integers. At the beginning of the Appendix to this chapter (which the student should now be able to read without further preparation) these discussions are carried a bit further, wherein the algebraic integers contained in any quadratic extension of the rational field are determined. Furthermore, the factorization properties of these quadratic domains are addressed, with the result that these rings enjoyed varying degrees of good factorization properties, ranging from being Euclidean (as with the Gaussian integers $\mathbb{Z}[i]$) through not even satisfying unique factorization (as with $\mathbb{Z}[\sqrt{-5}]$).

A little historical perspective is in order. The subject of Arithmetic is probably one of the first subjects that a student of Mathematics encounters. Its fundamentals can be taught to anyone. But its mysteries are soon apparent to even the youngest student. Why is any positive prime number that is one more than a multiple of four, the sum of the squares of two integers? Why is every positive integer expressible as the sum of four squares of integers? These problems have been solved. But there are many more unsolved problems. In fact, there is no other field of Mathematics which presents so many unsolved problems that could be easily stated to the man on the street. For example, the famous Goldbach Conjecture that asserts that

Every even integer greater than two is the sum of two prime numbers.

One of these questions concerns an assertion known as "Fermat's Last Theorem". According to tradition, Fermat had jotted in the margin of a book that he had a proof of the following theorem[7]

Suppose x, y, z is a triplet of (not-necessarily distinct) integers. If n is a positive integer greater than 2 , then there exists no such triplet such that

$$x^n + y^n = z^n. \tag{9.4}$$

That is, in the realm of integers, no nth power can be the sum of two other nth powers for $n > 2$. Perhaps the problem became more intriguing because there *are* solutions when $n = 2$. At any rate this problem attracted the attention of many great mathematicians of the 19th century. Here is a case where the *solution* was far less important than the *theory* that was developed to solve the problem. Our earliest mathematical ancestors might not have approached the problem with a full-fledged Galois Theory of field extensions at hand, for they knew that they were dealing with

---

[7]The book in which he wrote this marginal note has never been found. All that we actually have is a third party who initiated the anecdote. So there are two schools of thought: (1) Fermat thought that he had a proof but must have made a mistake. Skeptics believe his proof could not only not fit in a margin, but that his proof (in order to be less than a thousand pages) must have been in error. Then there is the other view: (2) He really did prove the theorem in a relatively simple way—we just haven't discovered how he did it. The authors are personally acquainted with at least one great living research mathematician who does not rule out the second view. That man is not willing to dismiss a mind (like Fermat's) of such proven brilliance. That respect says something.

some "subring" of "integers" in these fields. (Complex numbers had been explicitly developed by Gauss and the idea of Gaussian integers may be seen as an anticipation of the general idea guiding others to this point.) So this was the foil that produced the concept of Algebraic Integers.

It came as quite a revelation to early nineteenth century mathematicians that even in the algebraic integer domains, unique factorization of elements into prime elements could fail. Perhaps our intellectual history (as opposed to the political one) is simply the gradual divestment of unanalyzed assumptions.

The main objects of study for the rest of this chapter will be the rings $\mathcal{O}_E$ consisting of the elements in the field E which are integral with respect to a subdomain D where E is a finite extension of $F := \mathcal{F}(D)$, the field of fractions of D. (The phrase "finite extension" means that the field E contains F and is finite dimensional as a vector space over its subfield F.) In the special case that $D = \mathbb{Z}$, the ring of integers (so that $F = \mathbb{Q}$, the field of the rational numbers), the rings $\mathcal{O}_E$ are the *algebraic integer domains* of the previous paragraph.

Such algebraic integer domains $\mathcal{O}_E$ include the above-mentioned quadratic domains (such as $\mathbb{Z} + \mathbb{Z}\sqrt{-5}$). Thus, while *microscopically* (i.e., element-wise), these rings may not enjoy unique factorization, they all satisfy a *macroscopic* version of unique factorization inasmuch as their ideals will always factor uniquely as a product of prime ideals. This can be thought of as a partial remedy to the fact that the rings $\mathcal{O}_K$ tend not to be UFDs.

9.8 Rings of Integral Elements

Suppose K is a field, and D is a subring of F. Then, of course, D is an integral domain. Recall from Sect. 8.2.5. that an element α of K is *integral over D* if an only if α is a zero of a polynomial in $D[x]$ whose lead coefficient is 1 (the multiplicative identity of D and K). Say that an integral domain D is *integrally closed* if, whenever $\alpha \in \mathcal{F}(D)$ and α is integral over D, then $\alpha \in D$. Here $\mathcal{F}(D)$ is the field of fractions of the integral domain D.

The following is a sufficient, but not a necessary condition, for an integral domain to be integrally closed.

Lemma 9.8.1 *If D is a UFD, then D is integrally closed.*

Proof Graduate students who have taught basic-level courses such as "college algebra" will recognize the following argument. Indeed, given that D is a UFD, with field of fractions F, then elements of F may be expressed in the form $\alpha = a/b$, where a and b have no common prime factors in D. If such an element α were integral over D, then we would have a monic polynomial $f(x) = \sum_{i=0}^{r} a_i x^i \in D[x]$ with $f(\alpha) = 0$: thus

$$\left(\frac{a}{b}\right)^r + a_{r-1}\left(\frac{a}{b}\right)^{r-1} + \cdots + a_1\left(\frac{a}{b}\right) + a_0 = 0.$$

Upon multiplying both sides of the above equation by b^r one obtains

$$a^r + a_{r-1}ba^{r-1} + \cdots + a_1 b^{r-1}a + a_0 b^r = 0.$$

Therefore, any prime divisor p of b will divide $\sum_{i=0}^{r-1} a_i b^{r-i}a^i$ and so would divide a^r.

Since D is a UFD, that would imply that this prime divisor p divides a, as well.[8] This is a contradiction since a and b possess no common prime factors. We are left with the case that b is not divisible by any prime whatsoever. In that case b is a unit, in which case $\alpha = a/b \in D$. $\square$

Remark The above lemma provides us with a very large class of integral domains that are not UFDs. Indeed, Appendix will show that the quadratic domains $\mathcal{O} = A(d)$ consisting of algebraic integers in the quadratic extension $\mathbb{Q}(\sqrt{d}) \supseteq \mathbb{Q}$, where d is a square-free integer, $d \equiv 1 \bmod 4$, have the description

$$A(d) = \left\{ \frac{a + b\sqrt{d}}{2} \mid a, b \in \mathbb{Z}, a, b \text{ are both even or are both odd} \right\}.$$

This implies immediately that the proper subring $\mathbb{Z}[\sqrt{d}] \subsetneq A(d)$ is not integrally closed and therefore cannot enjoy unique factorization. A more direct way of seeing this is that in the domain $D = \mathbb{Z}[\sqrt{d}]$, where d is square-free and congruent to 1 modulo 4, the element 2 is irreducible in D (easy to show directly) but not prime, as $2|(1 + \sqrt{d})(1 - \sqrt{d})$ but 2 doesn't divide either $1 + \sqrt{d}$ or $1 - \sqrt{d}$.

The following lemmas essentially recapitulate much of the discussion of Sect. 8.2.5.

Lemma 9.8.2 *Let $R \subseteq S$ be integral domains.*

1. *If R is Noetherian and S is a finitely-generated R-module, then every element of S is integral over R.*
2. *If $R \subseteq S \subseteq T$ are integral domains with T a finitely-generated S-module and S a finitely-generated R-module, then T is a finitely-generated R-module.*

Proof The first statement is immediate from Theorem 8.2.12 and Lemma 8.2.14.

From the hypothesis of the second statement, $T = \sum t_i S$ and $S = \sum s_j R$, where the summing parameters i and j have finite domains. Thus $T = \sum_{i,j} t_i s_j R$ is generated as an R module by the finite set $\{t_i s_j\}$. $\square$

Note that R need not be Noetherian in part 2 of the above Lemma.

---

[8] A word of caution is in order here. If D is not a UFD, it's quite possible for an element—even an irreducible element—to divide a power of an element without dividing the element itself. See Exercise (5) in Sect. 9.12, below.

Lemma 9.8.3 *Suppose K is a field containing the Noetherian integral domain D. Then we have the following:*

1. *An element $\alpha \in K$ is integral over D if and only if $D[\alpha]$ is a finitely generated D-module* (Theorem 8.2.14).
2. *If $\mathcal{O}$ is the collection of all elements of K which are integral over D, then $\mathcal{O}$ is a subring of K* (Theorem 8.2.15).

Let K be a field containing the integral domain D and let $\mathcal{O}$ be the ring of integral elements (with respect to D) as in the above Lemma.

Theorem 9.8.4 *Let $\mathcal{O}$ be the ring of integral elements with respect to the Noetherian subdomain D of field K. Then $\mathcal{O}$, contains all elements of K that are integral with respect to $\mathcal{O}$. Since the field of fractions $\mathcal{F}(\mathcal{O})$ lies in K, we see that $\mathcal{O}$ is integrally closed.*

Proof Assume that $\alpha \in K$ is integral over $\mathcal{O}$. Thus there exist coefficients $a_0, a_1, \ldots, a_{n-1} \in \mathcal{O}$ with $\alpha^n + a_{n-1}\alpha^{n-1} + \cdots + a_0 = 0$. Therefore, we see that α is integral over the domain $D[a_0, a_1, \ldots, a_{n-1}]$. In turn each a_i is integral over D, so by repeated application of Lemma 9.8.2 (using both parts), we conclude that $D[a_0, a_1, \ldots, a_{n-1}, \alpha]$ is a finitely generated D-module. Since D is Noetherian, the submodule $D[\alpha]$ is also finitely generated. This means α is integral over D, i.e., $\alpha \in \mathcal{O}$. That $\mathcal{O}$ is integrally closed follows immediately. $\square$

The field K is said to be *algebraic over a subfield L* if and only if every element of K is a zero of a polynomial in $L[x]$—equivalently, the set of all powers of any single element of K are L-linearly dependent. Clearly if K has finite dimension as a vector space over L, then K is algebraic over L.

Corollary 9.8.5 *Suppose K is algebraic over the field of fractions $F = \mathcal{F}(D)$ of its subdomain D and let $\mathcal{O}$ denote the ring of elements of K that are integral with respect to D. Then for each element $\alpha \in K$, there exists an element $d_\alpha \in D$, such that $d_\alpha \alpha \in \mathcal{O}$. It follows that $K = F\mathcal{O} = \mathcal{F}(\mathcal{O})$.*

Proof Suppose $\alpha \in K$. Then the powers of K are linearly dependent over F. Thus, for some positive integer m, there are fractions $f_0, f_1, \ldots f_{m-1}$ in $\mathcal{F}(D) = F$, such that

$$\alpha^m = f_{m-1}\alpha^{m-1} + \cdots + f_1\alpha + f_0. \tag{9.5}$$

Let d be the product of all the denominators of the f_i. Then d and each df_i lies in $\mathcal{O}$. Multiplying both sides of Eq. (9.5) by d^m we obtain

$$(d\alpha)^m = df_{m-1}(d\alpha)^{m-1} + \cdots + d^{m-1}f_1(d\alpha) + d^m f_0.$$

Since all coefficients $d^i f_{m-i}$ lie in $\mathcal{O}$, we see from the statement of Theorem 9.8.4 that $d\alpha$ must lie in $\mathcal{O}$. Since $d \in D \subseteq \mathcal{O}$, and $d \neq 0$, by its definition, we see that $\alpha \in F\mathcal{O} \subseteq \mathcal{F}(\mathcal{O})$. Since α was arbitrarily chosen in K, the equations $K = F\mathcal{O} = \mathcal{F}(\mathcal{O})$ follow. $\square$

There is an interesting property that $\mathcal{O}$ inherits from D.

Theorem 9.8.6 *If every prime ideal in D is maximal, then the same is true of $\mathcal{O}$.*

Remark Recall that when we speak of an ideal being maximal, we are referring to the poset of all *proper ideals* of a ring.

Proof of Theorem 9.8.6 Suppose P is a prime ideal of the ring $\mathcal{O}$. By way of contradiction, we assume that P is not a maximal ideal in $\mathcal{O}$. Then P properly lies in a larger ideal J of D, producing the proper containment $P \subset J \subset \mathcal{O}$. Since both P and J are properly contained in $\mathcal{O}$, neither contains the multiplicative identity element 1, which lies in D. Thus P and J intersect D at proper ideals of D. But $P_0 := P \cap D$ is clearly a prime ideal of D and so by hypothesis is a maximal ideal of D. Since $J \cap D$ is a proper ideal of D containing P_0, we must conclude that $P_0 = J \cap P$.

Now choose $\beta \in J \backslash P$. Since β is an algebraic integer, it is the zero of some monic polynomial of $D[x]$. Therefore, there is a non-empty collection of monic polynomials $p(x)$ in $D[x]$, such that $p(\beta)$ lies in P. Among these, we choose $p(x)$ of minimal degree. Thus we have

$$p(\beta) = \beta^m + b_{m-1}\beta^{m-1} + \cdots + b_1\beta + b_0 \in P \qquad (9.6)$$

with all $b_i \in D$ and the degree m minimal. Since β lies in the ideal J, so does

$$\beta(\beta^{m-1} + b_{m-1}\beta^{m-2} + \cdots + b_2\beta + b_1) = p(\beta) - b_0. \qquad (9.7)$$

Since $p(\beta) \in P \subset J$, the above Eq. (9.7) shows that b_0, being the difference of two elements of J, must also lie in J. But then $b_0 \in D \cap J = P_0 \subseteq P$. So it now follows the left side of Eq. (9.7) is a product in $\mathcal{O}$ that lies in the prime ideal P. Since β *does not* lie in P by our choice of β, the second factor

$$\beta^{m-1} + b_{m-1}\beta^{m-2} + \cdots + b_2\beta + b_1$$

is an element of P. But that defies the minimality of m, and so we have been forced into a contradiction.

It follows that P is a maximal ideal of $\mathcal{O}$. $\square$

9.9 Factorization Theory in Dedekind Domains and the Fundamental Theorem of Algebraic Number Theory

Let D be an integral domain. We say that D is a *Dedekind domain* if and only if

(a) D is Noetherian,
(b) Every non-zero prime ideal of D is maximal, and
(c) D is integrally closed.

The reader will notice that if the integral domain D is a field, then D possesses only one proper ideal $\{0\}$, which is both prime and maximal. Such a D is its own field of fractions, and any element $\alpha \in D$ is a zero of the monic polynomial $x - \alpha$ and so is integral over D. Thus, in a trivial way, fields are Dedekind domains. However, the interesting features of a Dedekind domain involve its nonzero ideals. For the rest of this section, we begin our examination of the ideal structure of an arbitrary *Dedekind* domain, D.

The reader is reminded of the following notation that is customary in ring theory: If A and B are ideals in a commutative ring R, one writes AB for the set $\{\sum_1^n a_i b_i | a_i \in A, b_i \in B, n \in \mathbb{N}\}$, the ideal generated by all products of an element of A with an element of B. The ideal AB is called a *product of ideals* A and B.

Lemma 9.9.1 *Assume that* $P_1, P_2, \ldots, P_r$ *are maximal ideals of the integral domain* D, *and that* P *is a prime ideal satisfying*

$$P_1 P_2 \cdots P_r \subseteq P.$$

Then $P = P_i$ *for some* i.

Proof If $P \neq P_i$, $i = 1, 2, \ldots, r$, then, since each P_i is a maximal ideal distinct from P we may find elements $a_i \in P_i \backslash P$, $i = 1, 2, \ldots, r$. Thus, $a_1 a_2 \cdots a_r \in P_1 P_2 \cdots P_r \subseteq P$. Since P is a prime ideal, we must have $a_i \in P$ for some i, which is a contradiction. $\square$

Lemma 9.9.2 *Any non-zero ideal of* D *contains a finite product of non-zero prime ideals.*

Proof We let $\mathcal{A}$ be the family of all non-zero ideals of D for which the desired conclusion is false. Assume $\mathcal{A}$ is non-empty. As D is Noetherian, $\mathcal{A}$ must contain a maximal member I_0. Clearly, I_0 cannot be a prime ideal, since it is a member of $\mathcal{A}$. Accordingly, there must exist a pair of elements α, $\beta \in D \backslash I_0$ with $\alpha\beta \in I_0$. Next, form the ideals $I_0' = I_0 + \alpha D$ and $I_0'' = I_0 + \beta D$. By maximality of I_0, neither of the ideals I_0' and, I_0'' can lie in $\mathcal{A}$. Therefore, both I_0' and I_0'' respectively contain products $P_1 \cdots P_m$ and $P_{m+1} \cdots P_n$ of non-zero prime ideals. Since $I_0' I_0'' \subseteq I_0$, and $I_0' I_0''$ contains $P_1 \cdots P_n$, we have arrived at a contradiction. Thus $\mathcal{A} = \emptyset$ and the conclusion follows. $\square$

At this point it is convenient to introduce some notation. Let D be a Dedekind domain with field of fractions E. If $I \subseteq D$ is an ideal, set

$$I^{-1} = \{\alpha \in E | \alpha \cdot I \subseteq D\}.$$

Note that $D^{-1} = D$, for if $\alpha \cdot D \subseteq D$, then $\alpha = \alpha \cdot 1 \in D$. Next note that $I \subseteq J$ implies that $I^{-1} \supseteq J^{-1}$.

Lemma 9.9.3 *If I is a proper ideal of D, then I^{-1} properly contains D.*

Proof Clearly $D \subseteq I^{-1}$. Let $0 \neq \alpha \in I$; by Lemma 9.9.2, we have prime ideals $P_1, P_2, \ldots, P_r$ with $P_1 P_2 \cdots P_r \subseteq \alpha D \subseteq I$. We may assume, furthermore, that the index r above is minimal. Since D is Noetherian, there is a maximal ideal M containing I; thus we have $P_1 P_2 \cdots P_r \subseteq M$. Applying Lemma 9.9.1 we conclude that $P_1 = M$ for some index i. Re-index if necessary so that $P_1 = M$. This says that

(i) $M P_2 P_3 \cdots P_r \subseteq \alpha D \subseteq M$, and
(ii) $P_2 P_3 \cdots P_r \not\subseteq \alpha D$,

by the minimality of r. Let $\beta \in P_2 P_3 \cdots P_r \backslash \alpha D$ and set $\lambda = \beta/\alpha$. Then $\lambda \in E \backslash D$; yet, by (i) and (ii),

$$\lambda I = \beta \alpha^{-1} I \subseteq \beta \alpha^{-1} M \subseteq \alpha^{-1} M P_2 \cdots P_r \subseteq \alpha^{-1}(\alpha D) \subseteq D,$$

which puts $\lambda \in I^{-1} \backslash D$. $\square$

Lemma 9.9.4 *If $I \subseteq D$ is an ideal then I^{-1} is a finitely generated D-module.*

Proof If $0 \neq \alpha \in I$, then $I^{-1} \subseteq (\alpha D)^{-1} = D[\alpha^{-1}]$, which is a finitely-generated D-module. Since D is Noetherian, $D[\alpha^{-1}]$ is Noetherian (Theorem 8.2.12), and so I^{-1} is finitely generated (Theorem 8.2.11). $\square$

Theorem 9.9.5 *If $I \subseteq D$ is an ideal, then $I^{-1}I = D$.*

Proof Set $B = I^{-1}I \subseteq D$, so B is an ideal of D. Thus, $I^{-1}IB^{-1} = BB^{-1} \subseteq D$; which says that $I^{-1}B^{-1} \subseteq I^{-1}$. But then for any $\beta \in B^{-1}$, $I^{-1}\beta \subseteq I^{-1}$, forcing $I^{-1}[\beta] \subseteq I^{-1}$. Since, by Lemma 9.9.4, I^{-1} is a Noetherian D-module, so is its D-submodule $I^{-1}D[\beta]I^{-1}[\beta]$, From $D[\beta] \subseteq I^{-1}[\beta] \subseteq I^{-1}$ we infer that $D[\beta]$ is a finitely-generated D module and so $\beta \in E$ is integral over D. As D is integrally closed, it follows that $\beta \in D$. Therefore, $B^{-1} \subseteq D \subseteq B^{-1}$ and so $D = B^{-1}$. An application of Lemma 9.9.3 completes the proof. $\square$

Corollary 9.9.6 *If $I, J \subseteq D$ are ideals, then $(IJ)^{-1} = I^{-1}J^{-1}$.*

Proof $D = D^2 = I^{-1}I \cdot J^{-1}J = I^{-1}J^{-1}IJ$. Therefore, $(IJ)^{-1} = D(IJ)^{-1} = I^{-1}J^{-1}(IJ)(IJ)^{-1} = I^{-1}J^{-1}D = I^{-1}J^{-1}$. $\square$

The following theorem gives us basic factorization theory in a Dedekind domain.

Theorem 9.9.7 *Let D be a Dedekind domain and let $I \subseteq D$ be an ideal. Then there exist prime ideals $P_1, P_2, \cdots, P_r \subseteq D$ such that*

$$I = P_1 P_2 \cdots P_r.$$

The above factorization is unique in that if also

$$I = Q_1 Q_2 \cdots Q_s,$$

where the Q_i's are prime ideals, then $r = s$ and $Q_i = P_{\pi(i)}$, for some permutation π of $1, 2, \cdots, r$.

Proof By Lemma 9.9.2 we know that the ideal $I \subseteq D$ contains a product of prime ideals: $P_1 P_2 \cdots P_r \subseteq I$. We shall argue by induction that if $P_1 P_2 \cdots P_r \subseteq I$, and if r is minimal in this respect, then, in fact, $P_1 P_2 \cdots P_r = I$. Since prime ideals are maximal, the result is certainly true when $r = 1$. Next, as D is Noetherian, we may select a maximal ideal M containing I; thus we have

$$P_1 P_2 \cdots P_r \subseteq I \subseteq M.$$

Applying Lemma 9.9.1 we conclude that (say) $M = P_1$. But then

$$M^{-1} M P_2 \cdots P_r \subseteq M^{-1} I \subseteq M^{-1} M = D.$$

That is to say, $M^{-1} I$ is an ideal of D and that

$$P_2 P_3 \cdots P_r = M^{-1} M_1 P_2 \cdots P_r \subseteq M^{-1} I.$$

We apply induction to infer that $P_2 P_3 \cdots P_r = M^{-1} I$. If one multiplies both sides by M, then, noting again that $MM^{-1} = D$, one obtains $P_1 P_2 \cdots P_r = I$. This proves the existence of a factorization of I into a product of prime ideals.

Next we prove uniqueness. Thus, assume that there exist prime ideals $P_1, P_2, \ldots, P_r, Q_1, Q_2, \ldots, Q_s$ with

$$P_1 P_2 \cdots P_r = Q_1 Q_2 \cdots Q_s. \tag{9.8}$$

We argue by induction on the minimum of r and s. If, say, $r = 1$, then we set $P = P_1$ and we have a factorization of the form $P = Q_1 Q_2 \cdots Q_s$. By Lemma 9.9.1 we may assume that $P = Q_1$ and so $P = P Q_2 \cdots Q_s$. Multiply both sides by P^{-1} and infer that $D = Q_2 \cdots Q_s$. If $s > 2$ this is an easy contradiction as then $D = Q_2 \cdots Q_s \subseteq Q_2$.

Thus we may assume that both s and r are at least 2. Since D is Noetherian, we may find a maximal ideal M containing the common ideal in (9.8) above, so

$$P_1 P_2 \cdots P_r = Q_1 Q_2 \cdots Q_s \subseteq M.$$

An application of Lemma 9.9.1 allows us to infer that (again possibly after reindexing) that $M = P_1 = Q_1$. Upon multiplying both sides by M^{-1} one obtains $P_2 \cdots P_r = Q_2 \cdots Q_s$. Induction takes care of the rest. $\square$

We close by mentioning the following result; a proof is outlined in Exercise (8) in Sect. 9.9. It can be viewed as saying that Dedekind domains are "almost" principal ideal domains.

Theorem 9.9.8 *Let $E \supseteq \mathbb{Q}$ be a finite field extension and let $D = \mathcal{O}_E$. Then any ideal $I \subseteq D$ can be expressed as $I = D\alpha + D\beta$ for suitable elements $\alpha, \beta \in I$.*

9.10 The Ideal Class Group of a Dedekind Domain

We continue to assume that D is a Dedekind domain, with fraction field E. A D-submodule $B \subseteq E$ is called a *fractional ideal* if it is a finitely generated D-module.

Lemma 9.10.1 *Let B be a fractional ideal. Then there exist prime ideals $P_1, P_2, \ldots, P_r, Q_1, Q_2, \ldots, Q_s$ such that $B = P_1 P_2 \cdots P_r Q_1^{-1} Q_2^{-1} \cdots Q_s^{-1}$. (It is possible that either $r = 0$ or $s = 0$.)*

Proof Since B is finitely generated, there exist elements $\alpha_1, \alpha_2, \ldots, \alpha_k \in E$ with $B = D[\alpha_1, \ldots, \alpha_k]$. Since E is the fraction field of D, we may choose an element $\beta \in D$ with $\beta\alpha_i \in D$, $i = 1, 2, \ldots, k$. Therefore, it follows immediately that $\beta B \subseteq D$, i.e., βB is an ideal of D. Therefore, apply Theorem 9.9.7 to obtain the factorization $\beta B = P_1 P_2 \cdots P_r$ into prime ideals of D. Next, factor βD as $\beta D = Q_1 Q_2 \cdots Q_s$, prime ideals, and so $(\beta D)^{-1} = Q_1^{-1} Q_2^{-1} \cdots Q_s^{-1}$. Thus, $B = \beta^{-1} P_1 P_2 \cdots P_r = D[\beta^{-1}] P_1 P_2 \cdots P_r = (\beta D)^{-1} P_1 P_2 \cdots P_r = P_1 P_2 \cdots P_r Q_1^{-1} Q_2^{-1} \cdots Q_s^{-1}$. $\square$

Corollary 9.10.2 *The set of fractional ideals in E forms an abelian group under multiplication.*

Proof It suffices to prove that for any collection $P_1, \cdots, P_r, Q_1, \ldots, Q_s$ of prime ideals of D, the D-module $P_1 \cdots P_r Q_1^{-1} \cdots Q_s^{-1}$ is finitely generated over D. Let $\alpha \in Q_1 Q_2 \cdots Q_s$, and so $\alpha D \subseteq Q_1 Q_2 \ldots Q_s$. In turn, it follows that $Q_1^{-1} \cdots Q_s^{-1} \subseteq (\alpha D)^{-1} = \alpha^{-1} D = D[\alpha^{-1}]$. But then

$$P_1 \cdots P_r Q_1^{-1} \cdots Q_s^{-1} \subseteq Q_1^{-1} \cdots Q_s^{-1} \subseteq D[\alpha^{-1}].$$

That is to say, $P_1 \cdots P_r Q_1^{-1} \cdots Q_s^{-1}$ is contained in a finitely-generated module over the Noetherian domain D and hence must be finitely generated. $\square$

A fractional ideal $B \subseteq E$ is called a *principal fractional ideal* if it is of the form αD, for some $\alpha \in E$. Note that in this case, $B^{-1} = \alpha^{-1} D$. It is easy to show that if D is a principal ideal domain, then every fractional ideal is principal (Exercise (1) in Sect. 9.10).

If $\mathcal{F}$ is the set of fractional ideals in E we have seen that $\mathcal{F}$ is an abelian group under multiplication, with identity D. If we denote by $\mathcal{P}$ the set of principal fractional ideals, then it is easy to see that $\mathcal{P}$ is a subgroup of $\mathcal{F}$; the quotient group $\mathcal{C} = \mathcal{F}/\mathcal{P}$ is called the *ideal class group* of D; it is trivial precisely when D is a principal ideal domain. If $D = \mathcal{O}_E$ for a finite extension $E \supseteq \mathbb{Q}$, then it is known that $\mathcal{C}$ is a finite group. The order $h = |\mathcal{C}|$ is called the *class number* of D (or of E) and is a fundamental (though somewhat subtle) invariant of E.

9.11 A Characterization of Dedekind Domains

In this section we'll prove the converse of Theorem 9.9.7, thereby giving a characterization of Dedekind domains.

To begin with, let D be an arbitrary integral domain, with fraction field E. In analogy with the preceding section, if $I \subseteq D$ is an ideal, we set

$$I^{-1} = \{\alpha \in E \mid \alpha I \subseteq D\},$$

and say that I is *invertible* if $I^{-1}I = D$.

Lemma 9.11.1 *Assume that an ideal I of D admits factorizations into invertible prime ideals:*

$$P_1 P_2 \cdots P_r = I = Q_1 Q_2 \cdots Q_s.$$

Then $r = s$, and (possibly after re-indexing) $P_i = Q_i$, $i = 1, 2, \ldots, r$.

Proof We shall apply induction on the total number of ideals $r + s$. Among the finitely many ideals in $\{P_i\} \cup \{Q_j\}$ chose one that is minimal with respect to the subset relationship. By reindexing the ideals and transposing the symbols "P" and "Q", if necessary, we may assume this minimal prime ideal is P_1. Since P_1 is prime and contains $\prod Q_j$, we must have $Q_j \subseteq P_1$ for some index j. After a further reindexing, we may assume $j = 1$. Now, since P_1 was chosen minimal in the finite poset $(\{P_i\} \cup \{Q_j\}, \subseteq)$, one has $P_1 = Q_1$. Then by invertibility of the ideals, we see that

$$P_1^{-1}I = Q_1^{-1}I = P_2 \cdots P_r = Q_2 \cdots Q_s. \tag{9.9}$$

Applying induction to Eq. (9.9) forces $r = s$ and (with further reindexing) $P_i = Q_i$, $i = 2, \ldots, s$. This completes the proof. $\square$

Lemma 9.11.2 *Let D be an integral domain.*

(i) *Any non-zero principal ideal is invertible.*
(ii) *If $0 \neq x \in D$, and if the principal ideal xD factors into prime ideals as $xD = P_1 P_2 \cdots P_r$, then each P_i is invertible.*

Proof Clearly if $x \in D$, then $(xD)^{-1} = x^{-1}D$ and $(xD)(x^{-1}D) = D$, so xD is invertible, proving (i). For (ii), simply note that for any $i = 1, 2, \ldots, r$, that

$$D = P_1 P_2 \cdots P_r \cdot (x^{-1}D) = P_i(P_1 \cdots P_{i-1}P_{i+1} \cdots P_r)(x^{-1}D),$$

forcing P_i to be invertible. $\square$

Now assume that D is an integral domain satisfying the following condition:

(*) If $I \subseteq D$ is an ideal of D, then there exist prime ideals $P_1, P_2, \ldots, P_r \subseteq D$ such that

$$I = P_1 P_2 \cdots P_r.$$

Note that no assumption is made regarding the uniqueness of the above factorization. We shall show not only that uniqueness automatically follows (See Corollary 9.11.6, below), but that D is actually a Dedekind domain, giving us the desired characterization.

Theorem 9.11.3 *Any invertible prime ideal of D is maximal.*

Proof Let P be an invertible prime ideal and let $a \in D \backslash P$. Define the ideals $I = P + aD$, $J = P + a^2 D$, and factor into prime ideals:

$$I = P_1 P_2 \cdots P_r, \ J = Q_1 Q_2 \cdots Q_s.$$

Note that each P_i, $Q_j \supseteq P$. We now pass to the quotient ring $\overline{D} = D/P$ and set $\overline{P}_i = P_i/P$, $i = 1, 2, \ldots, r$, $\overline{Q}_j = Q_j/P$, $j = 1, 2, \ldots, s$. Clearly the ideals $\overline{P}_i, \overline{Q}_j, i = 1, 2, \ldots, r$, $j = 1, 2, \ldots, s$ are prime ideals of $\overline{D}$. Note that where $\overline{a} = a + P$, we have $\overline{I} = \overline{a}\overline{D}$, $\overline{J} = \overline{a}^2 \overline{D}$, principal ideals of $\overline{D}$.

Note that

$$\overline{a}\overline{D} = \overline{I} = \overline{P}_1 \cdots \overline{P}_r, \ \overline{a}^2 \overline{D} = \overline{J} = \overline{Q}_1 \cdots \overline{Q}_s,$$

which, by Lemma 9.11.2 part (ii), imply that the prime ideals $\overline{P}_1, \cdots, \overline{P}_r$ and $\overline{Q}_1, \cdots \overline{Q}_s$, are invertible ideals of $\overline{D}$. However, as $\overline{J} = \overline{I}^2$, then

$$\overline{Q}_1 \cdots \overline{Q}_s = \overline{P}_1^2 \cdots \overline{P}_r^2,$$

by Lemma 9.11.1 we conclude that $s = 2r$ and (possibly after reindexing) $\overline{P}_i = \overline{Q}_{2j-1} = \overline{Q}_{2j}$, $j = 1, 2, \ldots, r$. This implies that $P_i = Q_{2j-1} = Q_{2j}$, $j = 1, 2, \ldots, r$, and so $J = I^2$. Therefore, $P \subseteq J = I^2 = (P + aD)^2 \subseteq P^2 + aD$. If $x \in P$ we can write $x = y + az$, where $y \in P^2$, $z \in D$. Thus, $az = x - y \in P$. As $a \notin P$, and P is prime, we infer that $z \in P$. Therefore, in fact, we have $P \subseteq P^2 + aP \subseteq P$, and so it follows that $P = P^2 + aP$. As P is invertible by hypothesis, we may multiply through by P^{-1} and get $D = P + aD = I$, and so P is maximal. $\square$

Theorem 9.11.4 *Any prime ideal is invertible, hence maximal.*

Proof Let P be a prime ideal of D and let $x \in P$. We may factor the principal ideal xD as $xD = P_1 P_2 \cdots P_r$. By Lemma 9.11.2 (ii) the prime ideals P_i, $i = 1, 2, \ldots, r$ are invertible, and hence, by Theorem 9.11.3, they are maximal. Now apply Lemma 9.9.1 to infer that $P = P_i$, for some i and hence is invertible. Theorem 9.11.3 now forces P to be maximal. $\square$

The following two corollaries are now immediate:

Corollary 9.11.5 *Any ideal of D is invertible.*

Corollary 9.11.6 *Any ideal of D factors uniquely into prime ideals.*

Theorem 9.11.7 *D is Noetherian.*

Proof Let $I \subseteq D$ be an ideal. Being invertible by Corollary 9.11.5, there exist elements $a_1, a_2, \ldots, a_r \in I$, $b_1, b_2, \ldots, b_r \in I^{-1}$ such that $\sum a_i b_i = 1$. If $x \in I$, then $b_i x \in D$, $i = 1, 2, \ldots, r$ and $x = \sum (x b_i) a_i$, i.e., $I = D[a_1, a_2, \ldots, a_r]$, proving that I is finitely generated, and hence D is Noetherian. □

Our task of showing that D is a Dedekind domain will be complete as soon as we can show that D is integrally closed. To do this it is convenient to introduce certain "overrings" of D, described below.

Let D be an arbitrary integral domain and let $E = \mathcal{F}(D)$, the field of fractions of D. If $P \subseteq D$ is a prime ideal of D, we set

$$D_P = \{\alpha/\beta \in E \mid \alpha, \beta \in D, \ \beta \notin P\}.$$

It should be clear (using the fact that P is a prime ideal) that D_P is a subring of E containing D. (The reader will recall from Sect. 9.6.4 that D_P is the *localization of D at the prime ideal P.*) It should also be clear that the same field of fractions emerges: $\mathcal{F}(D_P) = E$.

Lemma 9.11.8 *Let I be an ideal of D, and let P be a prime ideal of D.*

 (i) *If $I \nsubseteq P$ then $D_P I = D_P$.*
(ii) *$D_P P^{-1}$ properly contains D_P.*

Proof Note that $D_P I$ is an ideal of D_P. Since $I \nsubseteq P$, any element $\alpha \in I \backslash P$ is a unit in D_P. It follows that $D_P I = D_P$, proving (i). If $D_P P^{-1} = D_P$, then multiplying through by P, and using the fact that $P^{-1} P = D$, we obtain $D_P = D_P P$. Therefore, there is an equation of the form

$$1 = \sum_{i=1}^{k} \frac{r_i}{s_i} x_i, \quad \text{where each } r_i \in D, \ s_i \in D \backslash P, \ x_i \in P.$$

Multiply the above through by $s_1 s_2 \cdots s_k$, and set $s_i' = s_1 \cdots s_{i-1} s_{i+1} \cdots s_k$, $i = 1, 2, \ldots, k$. Then

$$s_1 \cdots s_k = \sum_{i=1}^{k} r_i s_i' x_i \in P,$$

which is an obvious contradiction as each $s_i \notin P$ and P is prime. □

Lemma 9.11.9 *If $\alpha \in E$ then either $\alpha \in D_P$ or $\alpha^{-1} \in D_P$.*

Proof Write $\alpha = ab^{-1}$, $a, b \in D$, and factor the principal ideals aD and bD as $aD = P^e I$, $bD = P^f J$, where $I, J \nsubseteq P$. Thus, $D_P a = D_P P^e$, $D_P b = D_P P^f$. Assuming that $e \geq f$, we have $ab^{-1} \in D_P P^e P^{-f} = D_P P^{e-f} \in D_P$. Similarly, if $e \leq f$ one obtains that $ba^{-1} \in D_P$. □

Lemma 9.11.10 D_P *is integrally closed.*

Proof Let $\alpha \in E \backslash D_P$ be integral over D_P. Then there is an equation of the form

$$\alpha^m + a_{m-1}\alpha^{m-1} + \cdots + a_1\alpha + a_0 = 0,$$

where $a_0, a_1, \ldots, a_{m-1} \in D_P$. Since $\alpha \notin D_P$, we have, by Lemma 9.11.9 that $\alpha^{-1} \in D_P$; therefore,

$$\alpha = \frac{1}{\alpha^{m-1}}\alpha^m = -\frac{1}{\alpha^{m-1}}\left(a_{m-1}\alpha^{m-1} + \cdots + a_0\right)$$

$$= a_{m-1} + a_{m-2}\frac{1}{\alpha} \cdots a_0\frac{1}{\alpha^{m-1}} \in D_P,$$

a contradiction. $\square$

Theorem 9.11.11 $D = \cap D_P$, *the intersection taken over all prime ideals* $P \subseteq D$.

Proof Let $ab^{-1} \in \cap D_P$, where $a, b \in D$. Factor the principal ideals aD and bD as $aD = P_1^{e_1}P_2^{e_2}\cdots P_r^{e_r}$, $bD = P_1^{f_1}P_2^{f_2}\cdots P_r^{f_r}$; here all exponents are ≥ 0. It suffices to show that $e_i \geq f_i$, $i = 1, 2, \ldots, r$. Fix an index i and set $P = P_i$, $e = e_i$, $f = f_i$. Therefore, $aD_P = D_P P^e$, $bD_P = D_P P^f$, which gives $ab^{-1}D_P = D_P P^{e-f}$. Since $ab^{-1} \in D_P$, and since $D_P P^{-1}$ properly contains D_P by Lemma 9.11.8, we have $ab^{-1}D_P \subseteq D_P \subsetneq P^{-c}$ for all integers $c > 0$. Thus, it follows that $e - f \geq 0$. $\square$

As an immediate result, we get

Corollary 9.11.12 D *is integrally closed.*

Proof Indeed, if $\alpha \in E$ and is integral over D, then it is, a fortiori, integral over D_P. Since D_P is integrally closed by Lemma 9.11.10 we have that $\alpha \in D_P$. Now apply Theorem 9.11.11. $\square$

Combining all of the above we get the desired characterization of Dedekind domains:

Corollary 9.11.13 D *is a Dedekind domain if and only if every ideal of D can be factored into prime ideals.*

We conclude this section with a final remark.

Corollary 9.11.14 *Any Dedekind domain (in particular, any algebraic integer domain) is a PID if and only if it is a UFD.*

Proof By Theorem 9.4.3 any principle ideal domain is a unique factorization domain. So it only remains to show that any Dedekind domain that is a UFD, is also a PID. Assume that the Dedekind domain D is a UFD. We begin by showing that all prime ideals of D are principal. Thus, let $0 \neq P \subset D$ be a prime ideal and let $0 \neq x \in P$. As D is a UFD, we may factor x into primes: $x = p_1 p_2 \cdots p_k \in P$;

this easily implies that $p_i \in P$ for some index i. Set $p = p_i$. The principal ideal pD generated by p is a prime ideal and hence is maximal (as D is Dedekind), forcing $P = pD$.

Next let $I \subseteq D$ be an arbitrary ideal of D. Since D is a UFD, each nonzero non-unit of D can be factored into a unique number of primes in D; we shall denote this number by $l(x)$ (the "length" of x). Now select an element $x \in I$ such that $l(x)$ is a minimum. If also $y \in I$, and x does not divide y, we take z to be the greatest common divisor of x and y; clearly $l(z) < l(x)$. We set $x = za$, $y = zb$, where a, b are relatively prime. If the ideal $D(a, b)$ of D generated by both a and b is not all of D, then since D is Noetherian, the non-empty collection {proper ideals $J \mid J \supseteq D(a, b)$} must contain a maximal ideal M. Since maximal ideals are prime, the assumption guarantees that there is a prime p with $M = pD \supseteq D(a, b)$. But this says that p divides both a and b, which is impossible. Therefore, the ideal $D(a, b)$ must be all of D and so $1 \in D(a, b)$. Therefore, there exist elements $x_0, y_0 \in D$ with $1 = x_0 a + y_0 b$, which leads to $z = x_0 a z + y_0 b z = x_0 x + y_0 y \in I$. Since $l(z) < l(x)$, this contradiction implies that the ideal I was principal in the first place. $\square$

9.12 When Are Rings of Integers Dedekind?

The *tour de force* of the theory of Dedekind Domains given in the preceding section would lead one to suspect that this theory should apply as well to the rings of integral elements of Sect. 9.8—that is, rings $\mathcal{O}_K$ where K is a field and integrality is with respect to an integrally closed sub-domain D in which every prime ideal is maximal.[9]

In order to make such a dream a reality, the definition of Dedekind domain would also demand that $\mathcal{O}_K$ be Noetherian, and that seems to ask that the dimension of K over the subfield $\mathcal{F}(D)$ be finite. Even given that, is it really Noetherian? So, let us say that a ring $\mathcal{O}$ is a *classical ring of integral elements* if and only $\mathcal{O} = \mathcal{O}_K$ is the ring of elements of K that are integral with respect to a subdomain D, having the following properties:

(CRI1) The domain D is integrally closed in its field of fractions $F = \mathcal{F}(D)$. (F is regarded as a subfield of K.)

(CRI2) Every prime ideal of the domain D is a maximal ideal of D.

(CRI3) K has finite dimension as an F-vector space.

Lemma 9.12.1 *If $\mathcal{O}_K$ is a classical ring of integral elements, then*

1. *It is integrally closed.*
2. *Every prime ideal of $\mathcal{O}_K$ is maximal.*

Proof Conclusion 1 is immediate from Theorem 9.8.4. Theorem 9.8.6 implies conclusion 2. $\square$

---

[9]Somehow this historical notation $\mathcal{O}_K$ seems to glorify K at the expense of the subdomain D, which is hardly mentioned.

Remark Notice that the proof did not utilize (CRI3).

The goal of this section is to present a condition sufficient to force a classical ring of integers to be Dedekind. By Lemma 9.12.1 all that is needed is a proof that $\mathcal{O}_K$ is Noetherian.

Since the proof involves bilinear forms and dual bases, it might be useful to review these concepts from linear algebra in the next two paragraphs enclosed in brackets.

[Let V be a finite-dimensional vector space over a field F. In general, a mapping $B : V \times V \to F$ is a *symmetric bilinear form* if and only if:

(i) $B(x, ay + bz) = aB(x, y) + bB(x, z)$
(ii) $B(ax + by, z) = aB(x, z) + bB(y, z)$
(iii) $B(x, y) = B(y, x)$

for each $(x, y, z, a, b) \in V \times V \times K \times F \times F$. Thus for each vector $x \in V$, statement (i) asserts that the mapping $\lambda_x : K \to F$ defined by $y \mapsto B(x, y)$ is a *functional*, that is, a vector of the dual space $V^* := Hom_F(V, F)$. The bilinear form B is said to be *non-degenerate* if and only if the linear transformation $\lambda : V \to V^*$ defined by

$$x \mapsto \lambda_x, \text{ for all } x \in K$$

is an injection, and so, as V is finite-dimensional over F, a bijection. Thus the bilinear form B is *non-degenerate* if and only if ker $\lambda = 0$—that is, the only element x, such that $B(x, y) = 0$ for *every* element $y \in K$, is $x = 0$.

For any finite F-basis $\{x_i\}$ of K, there exists a functional f_i, such that $f_i(x_i) = 1$, while $f_i(x_j) = 0$ for $j \neq i$. If the bilinear form B is non-degenerate, the previous paragraph tells us that the associated mapping $\lambda : V \to V^* := Hom_F(K, F)$ is surjective. Thus for each functional f_i described just above, there exists a vector y_i such that $\lambda(y_i) = f_i$. Thus, for each index i, one has

$$B(y_i, x_j) = \delta_{ij}, \text{ where } \delta_{ij} = 1 \text{ if } i = j \text{ and is zero otherwise.}$$

We call $\{y_i\}$ a *dual basis of* $\{x_i\}$ *with respect to the form* B. Note that a dual basis exists only if the vector space V is finite dimensional, and the form B is non-degenerate.]

Now, as above, consider D, $F = \mathcal{F}(D)$, K a field containing F as a subfield so that $\dim(K_F)$ is finite, and consider $\mathcal{O}_K$, the subring of elements of K that are integral with respect to D. Let $\mathcal{T}$ be the collection of F-linear transformations $K \to F$. This collection contains the "zero" mapping $K \to \{0\}$, and of course many others. We say that a transformation $T \in \mathcal{T}$ is *tracelike* if and only if

$$T(\mathcal{O}_K) \subseteq D.$$

We now offer the following:

Theorem 9.12.2 *Let $\mathcal{O}_K$ be a classical ring of integral elements of a field K, with respect to the subdomain D so that conditions (CID1) (CID2) and (CID3) hold for*

D. Suppose a non-zero tracelike transformation $T : K \to F$ *exists. Then* $\mathcal{O}_K$ *is a Noetherian domain, and so* (by Lemma 9.12.1) *is a Dedekind domain.*

Proof Let $T : K \to F$ be the tracelike transformation whose existence was assumed. From T we define a symmetric bilinear form $B_T : K \times K \to F$, by the following recipe:

$$B_T(x, y) := T(xy), \quad \text{for all } x, y \in K.$$

Next we show that the form B_T is non-degenerate. If this form were degenerate there would exist a non-zero element x in the field K such that $T(xy) = 0$ for all elements $y \in K$. Since xy wanders over all of K as y wanders over K, we see that this means $T(K) = 0$. But that is impossible as T was assumed to be non-zero.

Now by (CID3) K has finite dimension over F and so we may choose an F-basis of K, say $X := \{x_1, \ldots, x_n\}$. By Corollary 9.8.5, we may assume that each basis element x_i lies in $\mathcal{O}_K$. Since K is finite-dimensional, and B_T is a non-degenerate form, there exists a so-called *dual basis* $\{y_i\}$ of K, such that $B_T(y_i, x_j)$ is 0 if $i \neq j$, and is 1, if $i = j$.

Now consider an arbitrary integral element $\beta \in \mathcal{O}_K$. We may write $\beta = \sum a_i y_i$, since $\{y_i\}$ is a basis for K. Then for any fixed index j, we see that $T(\beta \cdot x_j) = a_j$, where $a_j \in F$. Since both x_j and β belong to $\mathcal{O}_K$, so does the product βx_j. Since T is tracelike, we must conclude that $T(\beta \cdot x_j) = a_j$ belongs to D. Thus β is a D-linear combination of the y_i. Since β was an arbitrary element of $\mathcal{O}_K$, we must conclude that

$$\mathcal{O}_K \subseteq Dy_1 \oplus Dy_2 \oplus \cdots \oplus Dy_n := M$$

a finitely generated D-module. Now D is Noetherian, and so, by Noether's Theorem (Theorem 8.2.12), M is also Noetherian. By Lemma 8.2.2, part (i), each D-submodule of M is also Noetherian. Thus $\mathcal{O}_K$ is also Noetherian, and the proof is complete. $\square$

Where does this leave us? Chap. 11 provides a detailed study of fields and field extensions. Among field extensions, $F \subseteq K$, there is a class called *separable extensions*. It will emerge that for any separable extension $F \subseteq K$ with $\dim_F K$ finite, there *does* exist a non-zero tracelike linear transformation $K \to F$ (see Corollary 11.7.6 in Sect. 11.7.4). Moreover, the extension $F \subseteq K$ is always separable if the field F (and hence K) has characteristic zero (Lemma 11.5.4).

Of course, historically, the prize integral domain motivating all of this, is the ring of *algebraic integers*. That would be the ring of integral elements $\mathcal{O}_K$ where $D = \mathbb{Z}$, the ring of integers, and the extension field K has finite-dimension as a vector space over the field of rational numbers $\mathbb{Q}$. The result of the previous paragraph (anticipating results from Chap. 11, to be sure) together with Theorem 9.12.2 yields the following.

Corollary 9.12.3 *Any ring of algebraic integers is a Dedekind domain.*

9.13 Exercises

9.13.1 Exercises for Sects. 9.2, 9.3, 9.4 and 9.5

1. Earlier, in Theorem 7.4.7, we learned that every finite integral domain is a field. Prove the following theorem which replaces finiteness by finite dimension.

Theorem 9.13.1 *Suppose an integral domain D is finite-dimensional over a subfield F. Then D is a field.*

[Hint: If $F = D$ their is nothing to prove, so we may assume $\dim_F D = n > 1$ and proceed by induction on n. Choose an element $\alpha \in D - F$, consider a minimal submodule M for the subring $F[\alpha]$ (Why does it exist?) Prove that it has the form $mF[\alpha]$ (i.e. it is cyclic,) and that there is a right $F[\alpha]$-module isomorphism $\phi : M \rightarrow F[x]/I$, where $\phi(\alpha) = x + I$, and I is the ideal in $F[x]$ consisting all polynomials $r(x) \in F[x]$ such that $mr(\alpha) = 0$. Cite the theorems that force I to be a maximal principal ideal $F[x]p(x)$, and explain why $p(x)$ is an irreducible polynomial. Noting that D is a domain, conclude that $p(\alpha)$ is the zero element of D and so $K := F[\alpha] \simeq F[x]/F[x]p(x)$ as rings, and so K is a field. Finish the induction proof.]

2. Let D be a fixed unique-factorization-domain (UFD), and let p be a prime element in D. Then the principal ideal $(p) := pD$ is a maximal ideal and $D/(p)$ is a field. The next few exercises concern the following surjection between integral domains:

$$m_p : D[x] \rightarrow (D/(p))[x],$$

which reduces mod p the coefficients of any polynomial of $D[x]$.
Show that m_p is a ring homomorphism. Conclude that the principal ideal in $D[x]$ generated by p is a prime ideal and from this, that p is a prime element in the ring $D[x]$.

3. This exercise has two parts. In each of them D is a UFD and is embedded in the polynomial ring $D[x]$ as the polynomials of degree zero together with the zero polynomial. For every prime element p in D we let $F_p := D/(p)$, a field, and let $F := \mathcal{F}(D)$ be the field of fractions of D. We fix a maximal ideal M in $D[x]$ and set $K = D[x]/M$, another field. Our over-riding assumption in this exercise is that $M \cap D = (0)$—that is, M contains no non-zero polynomials of degree zero. Prove the following statements:

   (a) For any prime $p \in D$, show that $m_p(M) = F_p[x]$.
   [Hint: If $m_p(M) \neq F_p[x]$, then $\ker(m_p) + M$ is a proper ideal of $D[x]$. Since $M \cap D = (0)$, we see that $p \in \ker(m_p) - M$, contradicting the maximality of M.]
   (b) Let n be the minimal degree of a non-zero polynomial in M. (Since $M \cap D = (0)$, n is positive.) Then there exists a prime element $b(x)$ in $D[x]$ of degree n such that $M = D[x]b(x)$, a principle ideal of $D[x]$.

[Hint: Choose a polynomial $b'(x) \in M$ such that $\deg b'(x) = n$. Since $D[x]$ is a UFD, $b'(x)$ is a product of prime elements of $D[x]$, and so one of these prime factors $b(x)$, must belong to the prime ideal M. Since $M \cap D = (0)$, $\deg b(x) = n$.

Now consider any non-zero polynomial $a(x) \in M$. Using the fact that $F[x]$ is a Euclidean domain as well as the minimality of n, show that there exists an element $d \in D$ such that $da(x)$ is divisible by $b(x)$. Since $b(x)$ is a prime element of $D[x]$ which does not divide a (because of its degree), $b(x)$ divides $a(x)$, because it is a prime element.]

4. Prove the following result:

Theorem 9.13.2 *Assume D is a UFD with infinitely many association classes of primes. Then, if M is a maximal ideal of $D[x]$, $M \cap D \neq 0$—that is, M contains a non-zero polynomial of degree zero.*

[Hint: By way of contradiction assume $M \cap D = (0)$. Let p be any prime element in D. By Exercise 3, part (a), there exists a polynomial $a(x) \in M$, such that $m_p((a(x)) = \bar{1}$, the multiplicative identity element of $F_p[x]$. By part (ii) of Exercise 3, $a(x) = b(x) \cdot e(x)$, for some $e(x) \in D[x]$. Apply the morphism m_p to this factorization to conclude that $m_p(b(x)$ has degree zero in the polynomial ring $F_p[x]$. Thus, writing $b(x) = \sum_i b_i x^i$, $b_i \in D$, we see that if $i > 0$, then b_i is divisible by p. But this is true for any prime, and since there are infinitely many association classes of primes, this can be true only if $b_i = 0$, for $i > 0$. But then $b(x) \in D \cap M$, a contradiction. Why?]

5. Let D be a UFD with infinitely many association classes of primes. Show that any maximal ideal of $D[x]$ is generated by a prime $p \in D$ and a polynomial $p(x)$ such that $m_p(p(x))$ is irreducible in $F_p[x]$.
6. Let D be a UFD with finitely many association classes of primes with representatives $\{p_1, \ldots, p_n\}$. Set $\pi = \prod p_i$ and form the principle ideal $M = D[x](1+\pi x)$. Then $M \cap D = (0)$. Show that M is a maximal ideal. [Hint: Show that each non-zero element of $D[x]/M$ is a unit of that factor ring, using the unique factorization in D and the identity $x\pi = -1 \mod M$.]
7. (Another proof of Gauss' Lemma.) Let D be a UFD. A polynomial $h(x)$ in $D[x]$ is said to be *primitive* if and only there is no prime element p of D dividing each of the D-coefficients of $h(x)$. Using the morphism m_p of the first exercise, show that if f and g are primitive polynomials in $D[x]$, then the product fg is also primitive. [Hint: by assumption, there is no prime in D dividing f or g in $D[x]$. If a prime p of D divides fg in $D[x]$ then $m_p(fg) = 0 = m_p(f) \cdot m_p(g)$. Then use the fact that $(D/(p))[x]$ is an integral domain.]
8. (Eisenstein's Criterion) Let D be a UFD. Suppose p is a prime in D and

$$f(x) = a_n x^n + \cdots a_1 x + a_0$$

is a polynomial in $D[x]$ with $n > 1$ and $a_n \neq 0$. Further, assume that

(E1) p does not divide the lead coefficient a_n;

(E2) p divides each remaining coefficient a_i for $i = 0, \ldots, n - 1$; and

(E3) p^2 does not divide a_0.

Show that f cannot factor in $D[x]$ into polynomial factors of positive degree. In particular f is irreducible in $F[x]$, where F is the field of fractions of D. [Hint: Suppose by way of contradiction that there was a factorization $f = gh$ in $D[x]$ with f and g of degree at least one. Then apply the ring morphism m_p to get

$$m_p(f) = m_p(g) \cdot m_p(h) = m_p(a_n)x^n \neq 0.$$

Now by hypothesis (E1) the polynomials f and $m_p(f)$ have the same degree, while the degrees of g and h are at least as large as the respective degrees of their m_p images. Since $f = gh$, and $\deg m_p(f) = \deg m_p(g) + \deg m_p(h)$, we must have the last two summands of positive degree. Since $(D/(p))[x]$ is a UFD, x must divide both $m_p(g)$ and $m_p(h)$. This means the constant coefficients of g and h are both divisible p. This contradicts hypothesis (E3).]

9 Show that under the hypotheses of the previous Exercise 8, one cannot conclude that f is an irreducible element of $D[x]$. [Hint: In $\mathbb{Z}[x]$, the polynomial $2x + 6$ satisfies the Eisenstein hypotheses with respect to the prime 3. Yet it has a proper factorization in $\mathbb{Z}[x]$.]

10 Let p be a prime element of D. Using the morphism m_p show that the subset

$$B := \{p(x) = a_0 + a_1 x + \cdots + a_n x^n | a_i \equiv 0 \bmod p, \text{ for } i > 0\}$$

of $D[x]$ is a subdomain.

11 Let E be an arbitrary integral domain. For each non-zero polynomial $p(x) \in E[x]$ let $\ell(p)$ be its *leading coefficient*—that is the coefficient of the highest power of x possessing a non-zero coefficient. Thus $p(x)$ is a monic polynomial if and only if $\ell(p(x)) = 1$, and for non-zero polynomials $p(x)$ and $q(x)$, one has $\ell(p(x)q(x)) = \ell(p(x)) \cdot \ell(q(x))$. Prove the following theorem:

Theorem 9.13.3 *Let D_1 is an integral domain and suppose D is a subring of D_1. Then, of course, D is also an integral domain and we have a containment of polynomial rings: $D[x] \subseteq D_1[x]$. Suppose $a(x), b(x),$ and $c(x)$ are polynomials in $D_1[x]$ such that $c(x) = a(x)b(x)$. Assume the following:*

(i) $a(x)$ and $c(x)$ lie in $D[x]$.

(ii) $a(x)$ is monic.

(iii) $\ell(b(x)) \in D$.

Then $b(x) \in D[x]$.

In particular, if $a(x), b(x),$ and $c(x)$ are monic polynomials in $D_1[x]$ such that $c(x) = a(x)b(x)$, then two of the polynomials lie in $D[x]$ if and only if the third does.

[Hint: By hypothesis $\ell(a(x))\ell(b(x)) = \ell(c(x)) \in D$. Write out the three polynomials $a(x), b(x)$, and $c(x)$ as linear combinations of powers of x with coefficients
a_i, b_i and c_i, with $\ell(a(x))$, $\ell(b(x))$, and $\ell(c(x))$ represented as a_s, b_t and c_n,
respectively (of course $n = s + t$, $a_s = 1$ and $b_t = c_n \in D$). By induction on k,
show that $b_{t-k} \in D$ for all k for which $0 \leq k \leq t$.]

12 Let F be a subfield of the field K. Consider the set

$$L_F(K[x]) = \{p(x) \in K[x] | \ell(p(x)) \in F\} \cup \{0\}$$

consisting of the zero polynomial and all non-zero polynomials whose lead
coefficient lies in the subfield F.

(a) Show that $L_F(K[x])$ is a subring of $K[x]$.
(b) Show that $L_F(K[x])$ is a Euclidean Domain. [Hint: Use the Euclidean algorithm for $F[x]$ and the properties of the lead-coefficient function described
in the preamble of the preceding exercise.]

9.13.2 Exercises on Localization

1. Using the definitions of multiplication and addition for fractions, fill in all the
steps in the proof of the following assertions:

(a) $(D_S, +)$ is an abelian group,
(b) Multiplication of fractions D_S is a monoid.
(c) In D_S multiplication is distributive with respect to addition.

2. Let D be an integral domain with group of units $U(D)$. Show that if S is a
multiplicatively closed subset and S' is either $U(D)S$ or $U(D) \cup U(D)S$, then
the localizations D_S and $D_{S'}$ are isomorphic.

3. Below are given examples of a domain D and a multiplicatively closed subset of
non-zero elements of D. Describe as best you can the local ring D_S.

(a) $D = \mathbf{Z}$; S is all powers of 3. Is D_S a local ring?
(b) $D = \mathbf{Z}$; S consists of all numbers of the form 3^a5^b where a and b are natural
numbers and $a + b > 0$. Is the condition $a + b > 0$ necessary?
(c) $D = \mathbf{Z}$; S is all positive integers which are the sum of two squares.
(d) $D = \mathbf{Z}$; S is all integers of the form $a^2 + 5b^2$.
(e) $D = \mathbf{Z}$; S is all integers which are congruent to one modulo five.
(f) $D = \mathbf{Z}[x]$; S is all primitive polynomials of D.

4. A *valuation ring* is an integral domain D such that if I and J are ideals of D,
then either $I \subseteq J$ or $J \subseteq I$. Prove that for an integral domain D, the following
three conditions are equivalent:

(i) D is a valuation ring.

(ii) if $a, b \in R$, then either $Da \subseteq Db$ or $Db \subseteq Da$.

(iii) If $\alpha \in E := \mathcal{F}(D)$, then either $\alpha \in D$ or $\alpha^{-1} \in D$.

Thus, we see that the localizations D_P defined at the prime ideal P (which were defined on p. 312) are valuation rings.

5. Let D be a *Noetherian* valuation ring.

   (i) Prove that D is a PID.
   (ii) Prove that D contains a unique maximal ideal. (This is true even if D isn't Noetherian.)
   (iii) Conclude that, up to associates, D contains a unique prime element.

   (A ring satisfying the above is often called a *discrete valuation ring*.)

6. Let D be a discrete valuation ring, as in Exercise 5, above, and let π be the prime, unique up to associates. Define $\nu(a) = r$, where $a = \pi^r b$, $\pi \nmid b$.
   Prove that ν is an algorithm for D, giving D the structure of a Euclidean domain.

7. Let D be a Noetherian domain and let P be a prime ideal. Show that the localization D_P is Noetherian.

9.13.3 Exercises for Sect. 9.9

1. Let D be a Dedekind domain and let $I \subseteq D$ be an ideal. Show that $I \subseteq P$ for the prime ideal P if and only if P is a factor in the prime factorization of I. More generally, show that $I \subseteq P^e$ if and only if P^e is a factor in the prime factorization of I. Conclude that $I = P_1^{e_1} P_2^{e_2} \cdots P_r^{e_r}$ is the prime factorization of the ideal I if and only if for each $i = 1, 2, \ldots, r$, $I \subseteq P_i^{e_i}$, but that $I \not\subseteq P_i^{e_i+1}$.

2. Let P and Q be distinct prime ideals of the Dedekind domain D. Show that $PQ = P \cap Q$. (Note that $PQ \subseteq P \cap Q$ in any ring.)

3. Assume that D is a Dedekind domain and that $I = P_1^{e_1} P_2^{e_2} \cdots P_r^{e_r}$, $J = P_1^{f_1} P_2^{f_2} \cdots P_r^{f_r}$. Show that

$$I + J = P_1^{\min\{e_1, f_1\}} \cdots P_r^{\min\{e_r, f_r\}}, \quad I \cap J = P_1^{\max\{e_1, f_1\}} \cdots P_r^{\max\{e_r, f_r\}}.$$

   Conclude that $AB = (A+B)(A \cap B)$. (Use Exercise 1. Note that this generalizes Exercise 2.)

4. (Chinese Remainder Theorem) Let R be an arbitrary ring and let $I, J \subseteq R$ be ideals. Say that I and J are *coprime* if $I + J = R$. Prove that if I, J are coprime ideals in R, then $R/(I \cap J) \cong R/I \times R/J$. [Hint: map $R/(I \cap J) \to R/I \times R/J$ by $r + (I \cap J) \mapsto (r + I, r + J)$. Map $R/I \times R/J \to R(I \cap J)$, as follows. Fix $x \in I$, $y \in J$ with $x + y = 1$, and let $(a + I, b + J) \mapsto (xb + ya) + (I \cap J)$.]

5. Let R be a commutative ring and let $I \subseteq R$ be an ideal. Assume that for some prime ideals $P_1, P_2, \ldots, P_r$ one has $I \subseteq P_1 \cup P_2 \cup \cdots \cup P_r$. Show that $I \subseteq P_i$ for some i

6. Let D be a Dedekind domain and let $I, J \subseteq D$ be ideals. Show that I and J are coprime (see Exercise 4) if and only if they do not have a common prime ideal in their factorizations into prime ideals.

7. Let D be a Dedekind domain and let $I \subseteq D$ be an ideal. If $I = P_1^{e_1} P_2^{e_2} \cdots P_r^{e_r}$ is the factorization of I into a product of prime ideals (where $P_i \neq P_j$ if $i \neq j$) in the Dedekind domain D, show that

$$D/I \cong D/P_1^{e_1} \times D/P_2^{e_2} \times \cdots \times D/P_r^{e_r}.$$

8. Let D be a Dedekind domain with ideal $I \subseteq D$. Factor I as $I = P_1^{e_1} P_2^{e_2} \cdots P_r^{e_r}$ into distinct prime-power factors.

   (i) For each $i = 1, 2, \ldots, r$ select $\alpha_i \in P_i^{e_i} \setminus P_{i+1}^{e_i+1}$, $i = 1, 2 \ldots, r$. Use Exercise 7 to infer that there exists an element $\alpha \in D$ with $\alpha \equiv \alpha_i \bmod P_{i+1}^{e_i+1}$, $i = 1, 2, \ldots, r$.

   (ii) Show that $\alpha \in I$, and so the principal ideal $\alpha D \subseteq I$.

   (iii) Assuming $r > 1$ if necessary, show that the principal ideal αD factors as $\alpha D = IJ$, where I and J are coprime.

   (iv) Factor $J = Q_1^{f_1} Q_2^{f_2} \cdots Q_s^{f_s}$, a product of distinct prime powers. Show that there exists an element $\beta \in I \setminus (Q_1 \cup Q_2 \cup \cdots \cup Q_s)$. (See Exercise 5.)

   (v) Show that the ideal $(\alpha, \beta)D = I$.

   The above shows that every ideal in a Dedekind domain is "almost principal" inasmuch as no ideal requires more than two elements in a generating set, proving Theorem 9.9.8.

9. In the Dedekind domain $D = \mathbb{Z}[\sqrt{-5}]$ show that $(3) = (3, 4 + \sqrt{-5})(3, 4 - \sqrt{-5})$ is the factorization of the principal ideal (3) into a product of prime ideals.

10. Let E be a finite extension of the rational field $\mathbb{Q}$, and set $R = \mathcal{O}_E$. Let P be a prime ideal of R. Then $P \cap \mathbb{Z}$ is a prime ideal of $\mathbb{Z}$ so that $P \cap \mathbb{Z} = p\mathbb{Z}$, for some prime number p. Show that we may regard $\mathbb{Z}/p\mathbb{Z}$ as a subfield of R/P, and that dimensions satisfy the inequality

$$\dim_{(\mathbb{Z}/p\mathbb{Z})}(R/P) \leq \dim_{\mathbb{Q}} E,$$

with equality if and only if p remains prime in $\mathcal{O}_E$. [Hint: Don't forget to first consider why R/P is a field.]

9.13.4 Exercises for Sect. 9.10

1. If D is a PID prove that every fractional ideal of E is principal.

2. Let D be a Dedekind domain with fraction field E. Prove that E itself is not a fractional ideal (except in the trivial case in which D is a field to be begin with).

[Hint: If E properly contains D, one must prove that E is not a finitely generated D-module.]

3. Let D be a Dedekind domain with ideal class group C. Let $P \subseteq D$ be a prime ideal and assume that the order of the element $[P] \in C$ is $k > 1$. If $P^k = (\pi) := D\pi$, for some $\pi \in D$, show that π is irreducible but not prime.

4. An integral domain D such that every non-unit $a \in D$ can be factored into finitely many irreducible non-units is called an *atomic domain*. Prove that every Noetherian domain is atomic. (Note that uniqueness of factorization will typically fail.)

5. Let D be a Dedekind domain with ideal class group of order at most 2. Prove that the number of irreducible factors in a factorization of an element $a \in D$ depends only on a.[10] [Hint: Note first that by Exercise 4, any non-unit of D can be factored into finitely many irreducible elements. By induction on the minimal length of a factorization of $a \in D$ into irreducible elements, we may assume that a has no prime factors. Next assume that $\pi \in D$ is a non-prime irreducible element. If we factor the principal ideal into prime ideals: $(\pi) = Q_1 Q_2 \cdots Q_r$ then the assumption guarantees that $Q_1 Q_2 = (\alpha)$, for some $\alpha \in D$. If $r > 2$, then (π) is properly contained in $Q_1 Q_2 = (\alpha)$ and so α is a proper divisor of π, a contradiction. Therefore, it follows that a principal ideal generated by a non-prime irreducible element factors into the product of two prime ideals. Now what?]2593

6. Let D be as above, i.e., a Dedekind domain with ideal class group of order at most 2. Let $\pi_1, \pi_2 \in D$ be irreducible elements. As seen in Exercise 5 above, any factorization of $\pi_1 \pi_2$ will involve exactly two irreducible elements. Show that, up to associates, there can be at most three distinct factorizations of $\pi_1 \pi_2$ into irreducible elements. (As a simple illustration, it turns out that the Dedekind domain $\mathbb{Z}[\sqrt{-5}]$ has class group of order 2; correspondingly we have distinct factorizations: $21 = 3 \cdot 7 = (1 + 2\sqrt{-5})(1 - 2\sqrt{-5}) = (4 + \sqrt{-5})(4 - \sqrt{-5})$.)

9.13.5 Exercises for Sect. 9.11

1. Let D be a ring in which every ideal $I \subseteq D$ is invertible. Prove that D is a Dedekind domain. [Hint: First, as in the Proof of Theorem 9.11.7, show that D is Noetherian. Now let C be the set of all ideals that are not products of prime ideals. Since D is Noetherian, $C \neq \emptyset$ implies that C has a maximal member J. Let $J \subseteq P$, where P is a maximal ideal. Clearly $J \neq P$. Then $JP^{-1} \subseteq PP^{-1} = D$ and so JP^{-1} is an ideal of D; clearly $J \subseteq JP^{-1}$. If $J = JP^{-1}$, then $JP^{-1} = P_1 P_2 \cdots P_r$ so $J = PP_1 P_2 \cdots P_r$. Thus $J = JP^{-1}$ so $JP = J$. This is a contradiction. Why?]

2. Here is an example of a non-invertible ideal in an integral domain D. Let

---

[10]See L. Carlitz, A characterization of algebraic number fields with class number two, *Proc. Amer. Math. Soc.* **11** (1960), 391–392. In case R is the ring of integers in a finite extension of the rational field, Carlitz also proves the converse.

$$D = \{a + 3b\sqrt{-5}|\ a, b \in \mathbb{Z}\},$$

and let $I = (3, 3\sqrt{-5})$, i.e., I is the ideal generated by 3 and $3\sqrt{-5}$. Show that I is not invertible. (An easy way to do this is to let $J = (3)$, the principal ideal generated by 3, and observe that despite the fact that $I \neq J$, we have $I^2 = IJ$.)

3. This exercise gives a very important class of projective R-modules; see Sect. 8.4.4. Let R be an integral domain with field of fractions F and let I be an ideal of R. Prove that if I is invertible, then I is a projective R-module. Conversely, prove that if the ideal I is finitely generated and projective as an R-module, then I is an invertible ideal. [Hint: Assuming that I is invertible, there must exist elements $\alpha_1, \alpha_2, \ldots, \alpha_n \in I$, $\beta_1, \beta_2, \ldots, \beta_n \in I^{-1}$ with $\sum \alpha_i \beta_i = 1$. Let F be the free R-module with basis $\{f_1, f_2, \ldots, f_n\}$ and define $\sigma : I \to F$ by $\sigma(\alpha) = \sum \alpha \beta_i f_i \in F$. Show that $\sigma(I) \cong_R I$ and that $\sigma(I)$ is a direct summand of F. Apply Theorem 8.4.5 to infer that I is projective. For the converse, assume that $I = R[\alpha_1, \alpha_2, \ldots, \alpha_n]$ and let F be the free R-module with basis $\{f_1, f_2, \ldots, f_n\}$. Let $\epsilon : F \to I$ be the homomorphism given by $\epsilon(f_i) = \alpha_i$, $i = 1, 2, \ldots, n$. Show that since I is projective, there is a homomorphism $\sigma : I \to F$ such that $\epsilon \circ \sigma = 1_I$. For each $\alpha \in I$, write $\sigma(\alpha) = \sum a_i(\alpha) f_i$ and show that for each $i = 1, 2, \ldots, n$ the elements $\beta_i := a_i(\alpha)/\alpha \in F$ are independent of $0 \neq \alpha \in I$. Next, show that each $\beta_i \in I^{-1}$ and finally that $\sum \alpha \beta_i = 1$, proving that I is invertible. Note, incidentally, that this exercise shows that Dedekind rings are hereditary in the sense of Exercise (19) in Sect. 8.5.3. Also, by Exercise 1 of this subsection we see that a Noetherian domain is a Dedekind domain if and only if every ideal is projective.]

4. Let R be a Dedekind domain with field of fractions E. If $I, J \subseteq E$ are fractional ideals, and if $0 \neq \phi \in \text{Hom}_R(I, J)$, prove that ϕ is injective. [Hint: Argue that if $J_0 = \text{im}\ \phi$, then J_0 is a projective R-module. Therefore one obtains $I = \ker \phi \oplus J'$, where $J \cong_R J_0$. Why is such a decomposition a contradiction?]

5. Let R be a Dedekind domain with field of fractions E, and let $I, J \subseteq E$ be fractional ideals representing classes $[I], [J] \in \mathcal{C}_R$, the ideal class group of R. If $[I] = [J]$, prove that $I \cong_R J$. (The converse is also true; see Exercise (9) in Sect. 13.13.4 of Chap. 13.)

9.13.6 Exercises for Sect. 9.12

1. Let F be a field and let x be an indeterminate. Prove that the ring $R = F[x^2, x^3]$ is not integrally closed, hence is not a UFD.

2. Prove the assertion that if a is an algebraic integer, so is $\sqrt{a}$.

3. Let E be a finite field extension of the rational numbers. Suppose τ is an automorphism of E. Show that τ leaves the algebraic integer ring $\mathcal{O}_E$ invariant. [Hint: Since the multiplicative identity element 1 is unique, show that any automorphism of E fixes $\mathbb{Z}$ and $\mathbb{Q}$ element-wise. The rest of the argument examines the effect of such an automorphism on the definition of an algebraic integer.]

4. Show that $A(-6) := \mathbb{Z}[\sqrt{-6}]$ is not a UFD.

5. As we have seen, the ring $A(-5) = \mathbb{Z}[\sqrt{-5}]$ is not a UFD. Many odd things happen in this ring. For instance, find an example of an irreducible element $\pi \in \mathbb{Z}[\sqrt{-5}]$ and an element $a \in \mathbb{Z}[\sqrt{-5}]$ such that π doesn't divide a, but π divides a^2. [Hint: look at factorizations of $9 = 3^2$.]

6. The following result is well-known to virtually every college student. Let $f(x) \in \mathbb{Z}[x]$, and let $\frac{a}{b}$ be a rational root of $f(x)$. If the fraction $\frac{a}{b}$ is in lowest terms, then a divides the constant term of $f(x)$ and b divides the leading coefficient of $f(x)$. If we ask the same question in the context of the ring $\mathbb{Z}[\sqrt{-5}]$, then the answer is negative. Indeed, if we consider the polynomial $f(x) = 3x^2 - 2\sqrt{-5}\,x - 3 \in \mathbb{Z}[\sqrt{-5}][x]$, then the zeroes are $\frac{2+\sqrt{-5}}{3}$ and $\frac{-2+\sqrt{-5}}{3}$. Since both 3 and $\pm 2 + \sqrt{-5}$ are non-associated irreducible elements, then the fractions can be considered to be in lowest terms. Yet neither of the numerators divide the constant term of $f(x)$.

7. We continue on the theme set in Exercise 6, above. Let D be an integral domain with field of fractions $\mathcal{F}(D)$. Assume the following condition on the domain D:

   For every polynomial $f(x) = a_n x^n + \cdots + a_0 \in D[x]$, with $a_0, a_n \neq 0$, and a zero $\frac{a}{b} \in \mathcal{F}(D)$ which is a fraction in lowest terms—i.e., a and b have no common non-unit factors and $f(a/b) = 0$— then a divides a_0 and b divides a_n.

   Now prove that for such a ring every irreducible element is actually prime. [Hint: Let $\pi \in D$ be an irreducible element and assume that $\pi | uv$, but that π doesn't divide either u or v. Let $uv = r\pi$, $r \in D$, and consider the polynomial $ux^2 - (\pi + r)x + v \in R[x]$.]

8. Let K be a field such that K is the field of fractions of both subrings $R_1, R_2 \subseteq K$. Must it be true that K is the field of fractions of $R_1 \cap R_2$? [Hint: A counter-example can be found in the rational function field $K = F(x)$.]

9. Let $\mathbb{Q} \subseteq K$ be a finite-degree extension of fields. Prove that if K is the field of fractions of subrings $R_1, R_2 \subseteq K$, then K is also the field of fractions of $R_1 \cap R_2$.

10. Again, let $\mathbb{Q} \subseteq K$ be a finite-degree extension of fields. This time, let $\{R_\alpha \,|\, \alpha \in \mathcal{A}\}$ consist of *all* the subrings of K having K as field of fractions. Show that K is *not* the field of fractions of $\cap_{\alpha \in \mathcal{A}} R_\alpha$. (In fact, $\cap_{\alpha \in \mathcal{A}} R_\alpha = \mathbb{Z}$.)

Appendix: The Arithmetic of Quadratic Domains

Introduction

We understand the term *quadratic domain* to indicate the ring of algebraic integers of a field K of degree two over the rational number field $\mathbb{Q}$. The arithmetic properties of interest are those which tend to isolate Euclidean domains: Are these principal ideal domains (or PID's)? Are they unique factorization domains (or UFD's, also called "factorial domains" in the literature)?

Quadratic Fields

Throughout $\mathbb{Q}$ will denote the field of rational numbers. Suppose K is a subfield of the complex numbers. Then it must contain $1, 2 = 1 + 1$, all integers and all fractions of such integers. Thus K contains $\mathbb{Q}$ as a subfield and so is a vector space over $\mathbb{Q}$. Then K is a *quadratic field* if and only of $\dim_{\mathbb{Q}}(K) = 2$.

Suppose now that K is a quadratic field, so as a $\mathbb{Q}$-space,

$$K = \mathbb{Q} \oplus \omega\mathbb{Q}, \text{ for any chosen } \omega \in K - \mathbb{Q}.$$

Then $\omega^2 = -b\omega - c$ for rational numbers b and c, so ω is a root of the monic polynomial

$$p(x) = x^2 + bx + c$$

in $\mathbb{Q}[x]$. Evidently, $p(x)$ is irreducible, for otherwise, it would factor into two linear factors, one of which would have ω for a rational root, against our choice of ω. Thus $p(x)$ has two roots ω and $\bar{\omega}$ given by

$$(-1 \pm \sqrt{b^2 - 4c})/2.$$

We see that K is a field generated by $\mathbb{Q}$ and either ω or $\bar{\omega}$—a fact which can be expressed by writing

$$K = \mathbb{Q}(\omega) = \mathbb{Q}(\bar{\omega}).$$

Evidently

$$K = \mathbb{Q}(\sqrt{d}), \text{ where } d = b^2 - 4c.$$

Observe that substitution of ω for the indeterminate x in each polynomial of $\mathbb{Q}[x]$ produces an onto ring homomorphism,

$$v_\omega : \mathbb{Q}[x] \to K,$$

whose kernel is the principal ideal $p(x)\mathbb{Q}[x]$.[11] But as $v_{\bar{\omega}} : \mathbb{Q}[x] \to K$ has the same kernel, we see that the $\mathbb{Q}$-linear transformation

$$\sigma : \mathbb{Q} \oplus \omega\mathbb{Q} \to \mathbb{Q} \oplus \bar{\omega}\mathbb{Q},$$

which takes each vector $a + b\omega$ to $a + b\bar{\omega}$, $a, b \in \mathbb{Q}$, is an automorphism of the field K. Thus $\sigma(\omega) = \bar{\omega}$, and $\sigma^2 = 1_K$ (why?).

Associated with the group $\langle \sigma \rangle = \{1_K, \sigma\}$ of automorphisms of K are two further mappings.

---

[11] This was the "evaluation homomorphism" of Theorem 7.3.3, p. 207.

First is the *trace mapping*,
$$Tr : K \to \mathbb{Q},$$

which takes each element ζ of K to $\zeta + \zeta^\sigma$. Clearly this mapping is $\mathbb{Q}$-linear.

The second mapping is the *norm mapping*,

$$N : K \to \mathbb{Q},$$

which takes each element ζ of K to $\zeta\zeta^\sigma$. This mapping is not $\mathbb{Q}$-linear, but it follows from its definition (and the fact that multiplication in K is commutative) that it is *multiplicative*—that is
$$N(\zeta\psi) = N(\zeta)N(\psi),$$

for all ζ, ψ in K.

Just to make sure everything is in place, consider the roots ω and $\bar{\omega}$ of the irreducible polynomial $p(x) = x^2 + bx + c$. Since $p(x)$ factors as $(x - \omega)(x - \bar{\omega})$ in $K[x]$, we see that the trace is

$$T(\omega) = T(\bar{\omega}) = -b \in \mathbb{Q},$$

and that the norm is

$$N(\omega) = N(\bar{\omega}) = c, \text{ also in } \mathbb{Q}.$$

Quadratic Domains

We wish to identify the ring A of algebraic integers of K. These are the elements $\omega \in K$ which are roots of at least one monic irreducible polynomial in $Z[x]$. At this stage the reader should be able to prove:

Lemma 9.A.1 *An element $\zeta \in K - \mathbb{Q}$ is an algebraic integer if and only if its trace $T(\zeta)$ and its norm $N(\zeta)$ are both rational integers.*

Now putting $\zeta = \sqrt{d}$, where d is square-free and $K = \mathbb{Q}(\sqrt{d})$, the lemma implies:

Lemma 9.A.2 *If d is a square-free integer, then the ring of algebraic integers of $K = \mathbb{Q}(\sqrt{d})$ is the $\mathbb{Z}$-module spanned by*

$$\{1, \sqrt{d}\} \quad \text{if } d \equiv 2, 3 \bmod 4,$$
$$\{1, (1 + \sqrt{d})/2\} \quad \text{if } d \equiv 1 \bmod 4.$$

Thus the ring of integers for $\mathbb{Q}(\sqrt{7})$ is the set $\{a + b\sqrt{7} | a, b \in \mathbb{Z}\}$. But if $d = -3$, then $(1 + \sqrt{-3})/2$ is a cube root of unity, and the ring of integers for $\mathbb{Q}(\sqrt{-3})$ is the domain of Eisenstein numbers.

Which Quadratic Domains Are Euclidean, Principal Ideal or Factorial Domains?

Fix a square free integer d and let $A(d)$ denote the full ring of algebraic integers of $K = \mathbb{Q}(\sqrt{d})$. A great deal of effort has been devoted to the following questions:

1. When is $A(d)$ a Euclidean domain?
2. More generally, when is $A(d)$ a principal ideal domain?

In this chapter we showed that the Gaussian integers, $A(-1)$ and the Eisenstein numbers, $A(-3)$ were Euclidean by showing that the norm function could act as the function $g : A(d) \to \mathbb{Z}$ with respect to which the Euclidean algorithm is defined. In the case that the norm can act in this role, we say that the *norm is algorithmic*.

The two questions have been completely answered when d is negative.

Proposition 9.A.1 *The norm is algorithmic for $A(d)$ when $d = -1, -2, -3, -7$ and -11. For all remaining negative values of d, the ring $A(d)$ is not even Euclidean. However, $A(d)$ is a principal ideal domain for*

$$d = -1, -2, -3, -7, -11, -19, -43, -67, -163,$$

and for no further negative values of d.

There are ways of showing that $A(d)$ is a principal ideal domain, even when the norm function is not algorithmic. For example:

Lemma 9.A.3 *Suppose d is a negative square-free integer and the (positively valued) norm function of $A(d)$ has this property:*

- *If $N(x) \geq N(y)$ then either y divides x or else there exist elements u and v (depending on x and y) such that*

$$0 < N(ux - yv) < N(y).$$

Then $A(d)$ is a principal ideal domain.

This lemma is used, for example, to show that $A(-19)$ is a PID.[12]
For positive d (where $\mathbb{Q}(\sqrt{d})$ is a real field) the status of these questions is not so clear. We can only report the following:

Proposition 9.A.2 *Suppose d is a positive square-free integer. Then the norm function is algorithmic precisely for*

$$d = 2, 3, 5, 6, 7, 11, 13, 17, 19, 21, 29, 33, 37, 41, 57, 73.$$

---

[12] See an account of this in Pollard's *The Theory of Algebraic Numbers*, Carus Monograph no. 9, Math. Assoc. America, p. 100, and in Wilson, J.C., A principal ideal domain that is not Euclidean, *Mathematics Magazine*, vol. 46 (1973), pp. 34–48.

The domain $A(d)$ is known to be a principal ideal domain for many further values of d. But we dont even know if this can occur for infinitely many values of d! Moreover, among these PID's, how many are actually Euclidean by some function g other than the norm function? Are there infinitely many?

Non-Factorial Subdomains of Euclidean Domains

Here it is shown that standard Euclidean domains such as the Gaussian integers contain subdomains for which unique factorization fails. These nice examples are due to P. J. Arpaia ("A note on quadratic Euclidean Domains", Amer. Math. Monthly, vol. **75**(1968), pp 864–865).

Let $A_0(d)$ be the subdomain of $A(d)$ consisting of all integral linear combinations of 1 and $\sqrt{d}$, where d is a square-free integer. Thus $A_0(d) = A(d)$ if $d \equiv 2, 3 \bmod 4$, and is a proper subdomain of $A(d)$ only if $d \equiv 1 \bmod 4$.

Let p be a rational prime. Let $A_0/(p)$ be the ring whose elements belong to the vector space

$$(\mathbb{Z}/(p))1 \oplus (\mathbb{Z}/(p))\sqrt{d},$$

where multiplication is defined by setting $(\sqrt{d})^2$ to be the residue class modulo p containing d. Precisely, $A_0/(p)$ is the ring $\mathbb{Z}[x]/I$ where I is the ideal generated by the prime p and the polynomial $x^2 - d$. It should be clear that $A_0/(p)$ is a field if d is not zero or is not a quadratic residue $\bmod p$, and is the direct (ring) sum of two fields if d *is* a quadratic residue $\bmod p$. If p divides d, it has a 1-dimensional nilpotent ideal. But in all cases, it contains $\mathbb{Z}/(p)$ as a subfield containing the multiplicative identity element.

There is a straightforward ring homomorphism

$$n_p : A_0 \to A_0/(p),$$

which takes $a + b\sqrt{d}$ to $\bar{a} + \bar{b}\sqrt{d}$ where "bar" denotes the taking of residue classes of integers $\bmod p$. Now set

$$B(d, p) := n_p^{-1}(\mathbb{Z}/(p)).$$

Then $B(d, p)$ is a ring, since it is the preimage under n_p of a subfield of the image $A_0/(p)$. In fact, as a set,

$$B(d, p) := \{a + b\sqrt{d} \in A_0 | b \equiv 0 \bmod p\}.$$

Clearly it contains the ring of integers and so is a subdomain of $A(p)$. Note that the domains A_0 and $B(d, p)$ are both σ-invariant.

In the subdomain $B(d, p)$, we will show that under a mild condition, there are elements which can be factored both as a product of two irreducible elements and as a product of three irreducible elements. Since d and p are fixed, we write B for $B(d, p)$ henceforward.

Lemma 9.A.4 *An element of B of norm one is a unit in B.*

Proof Since $B = B^\sigma$, $N(\zeta) = \zeta\zeta^\sigma = 1$ for $\zeta \in B$ implies ζ^σ is an inverse in B for ζ.

Lemma 9.A.5 *Suppose ζ is an element A_0 whose norm $N(\zeta)$ is the rational prime p. The ζ does not belong to B.*

Proof By hypothesis, $N(\zeta) = a^2 - b^2 d = p$. If p divided integer b, it would divide integer a, so p^2 would in fact divide $N(\zeta)$, an absurdity. Thus p does not divide b and so ζ is not an element of B.

Corollary 9.A.1 *If an element of B has norm p^2, then it is irreducible in B.*

Proof By Lemma 9.A.4, elements of norm one in B are units in B. So if an element of norm p^2 in B were to have a proper factorization, it must factor into two elements of norm p. But the preceding lemma shows that B contains no such factors.

Theorem 9.A.1 *Suppose there exist integers a and b such that $a^2 - db^2 = p$ (that is, A_0 contains an element of norm p). Then the number p^3 admits these two factorizations into irreducible elements of B:*

$$p^3 = p \cdot p \cdot p$$
$$p^3 = (pa + pb\sqrt{d}) \cdot (pa - pb\sqrt{d}).$$

Proof The factors are clearly irreducible by Corollary 9.A.1.

We reproduce Arpaia's table displaying $p = a^2 - db^2$ for all d for which the norm function on $A(d)$ is algorithmic.

d	$p = a^2 - db^2$	d	$p = a^2 - db^2$	d	$p = a^2 - db^2$
-11	$47 = 6^2 - (-11)1^2$	5	$11 = 4^2 - 5 \cdot 1^2$	21	$-17 = 2^2 - 21 \cdot 1^2$
-7	$11 = 2^2 - (-3)1^2$	6	$-5 = 1^2 - (6)1^2$	29	$-13 = 4^2 - (26)1^2$
-3	$7 = 2^2 - (-3)1^2$	7	$-3 = 2^2 - 3 \cdot 1^2$	33	$-29 = 2^2 - 33 \cdot 1^2$
-2	$3 = 1^2 - (-2)1^2$	11	$-7 = 2^2 - 11 \cdot 1^2$	37	$107 = 12^2 - 37 \cdot 1^2$
-1	$2 = 1 - (-1)1^2$	13	$3 = 4^2 - 13 \cdot 1^2$	41	$-37 = 2^2 - 41 \cdot 1^2$
2	$-7 = 1^2 - 2 \cdot 2^2$	17	$-13 = 2^2 - 17 \cdot 1^2$	57	$-53 = 2^2 - 57 \cdot 1^2$
3	$-2 = 1^2 - 3 \cdot 1^2$	19	$-3 = 4^2 - 19 \cdot 1^2$	73	$-37 = 6^2 - 73 \cdot 1^2$

A Student Project

In the preceding subsection, a non-factorial subdomain B was produced for each case that the domain $A(d) = $ alg. int.$(\mathbb{Q}(\sqrt{d}))$ was Euclidean. But there are only finitely many of these domains $A(d)$. This raises the question whether one can find such a subdomain B in more standard generic Euclidean domains? We know that no such subdomain can exist in the domain of integers $\mathbb{Z}$ due to the paucity of it subdomains. But are there non-factorial subdomains B of the classical Euclidean domain of polynomials over a field?

This might be done by duplicating the development of the previous section replacing the ring of integers $\mathbb{Z}$, by an arbitrary unique factorization domain D which is distinct from its field of fractions k. Suppose further that K is a field of degree two over k generated by a root ω of an irreducible quadratic polynomial $x^2 - bx - c$ in $D[x]$ which has distinct roots $\{\omega, \bar{\omega}\}$. (One must assume that the characteristic of D is not 2.) Then the student can show that, as before, $K = k \oplus \omega k$ as a vector space over k and that the mapping σ defined by

$$\sigma(u + \omega v) = u + \bar{\omega} v, \text{ for all } u, v \in k,$$

is a field automorphism of K of order two so that the trace and norm mappings T and N can be defined as before.

Now let A be the set of elements in K whose norm and trace are in D. The student should be able to prove that if D is Noetherian, A is a ring. In any event, $A_0 := \{a + \omega b | a, b \in D\}$ is a subring.

Now since D is a UFD not equal to its quotient field, there is certainly a prime p in D. It is easy to see that

$$B_p = \{a + \omega b | a, b \in D, b \in pD\} = D \oplus \omega p D$$

is a σ-invariant integral domain. Then the proofs of Lemmas 9.A.4 and 9.A.5 and Corollary 9.A.1 go through as before (note that the fact that D is a UFD is used here). The analogue of the Theorem will follow:

If A_0 contains an element of norm p, B is not factorial.

Can we do this so that between A and B lies the Euclidean ring $F[x]$? The answer is "yes", if the characteristic of the field F is not 2.

In fact we can pass directly from $F[x]$ to the subdomain B and fill in the roles of D, k, K, A, and A_0 later.

Let $p(x)$ be an irreducible polynomial in $F[x]$ where the characteristic of F is not 2. Then (using an obvious ring isomorphism) $p(x^2)$ is an irreducible element of the subdomain $F[x^2]$. Now the student may show that the set

$$B_p : F[x^2] \oplus xp(x^2)F[x^2]$$

is a subdomain of $F[x]$.

Now $D = F[x^2]$ is a UFD with quotient field $k = F(x^2)$—the so called "field of rational functions" over F. The quotient field $K = F(x)$ of $F[x]$, is degree two over k, with the indeterminate x playing the role of ω. Thus

$$F(x) = F(x^2) \oplus xF(x^2)$$

since $\omega = x$ is a zero of the polynomial $y^2 - x^2$ which is irreducible in $F[x^2][y] = D[y]$. Since the characteristic is not 2, the roots x and $-x$ are distinct. Thus we obtain our field automorphism σ which takes any element $\zeta = r(x^2) + xs(x^2)$ of $F(x)$, to $r(x^2) - xs(x^2)$ (r and s are quotients of polynomials in $F[x^2]$). (Note that this is just the substitution of $-x$ for x in $F(x)$.) Now the student can write out the norm and trace of such a generic ζ. For example

$$N(\zeta) = r^2 - x^2 s^2. \tag{9.10}$$

At this point the ring B_p given above is not factorial if there exists a polynomial $\zeta = a(x^2) + xb(x^2)$ in $F[x]$ of norm $p(x^2)$. That is,

$$a(x^2)^2 - x^2 b(x^2)^2 = p(x^2)$$

is irreducible in $F[x^2]$.

Here is an example: Take $a(x^2) = 1 = b(x^2)$, so $\zeta = x + 1$. Then $N(1 + x) = x^2 - 1$ is irreducible in $F[x^2]$. Then in $B_p = F[x^2] \oplus x(x^2 - 1)F[x^2]$, we have the factorizations

$$(1 - x^2)^3 = (1 + x - x^2 - x^3)(1 - x - x^2 + x^3)$$

into irreducible elements.

Chapter 10
Principal Ideal Domains and Their Modules

Abstract An integral domain in which every ideal is generated by a single element is called a *principle ideal domain* or PID. Finitely generated modules over a PID are completely classified in this chapter. They are uniquely determined by a collection of ring elements called the *elementary divisors*. This theory is applied to two of the most prominent PIDs in mathematics: the ring of integers, $\mathbb{Z}$, and the polynomial rings $F[x]$, where F is a field. In the case of the integers, the theory yields a complete classification of finitely generated abelian groups. In the case of the polynomial ring one obtains a complete analysis of a linear transformation of a finite-dimensional vector space. The rational canonical form, and, by enlarging the field, the Jordan form, emerge from these invariants.

10.1 Introduction

Throughout this chapter, the integral domain D (and sometimes R) will be a principal ideal domain (PID). Thus it is a commutative ring for which a product of two elements is zero only if one of the factors is already zero—that is, the set of non-zero elements is closed under multiplication. Moreover every ideal is generated by a single element— and so has the form Dg for some element g in the ideal.

Of course, as rings go, principal ideal domains are hardly typical. But two of their examples, the ring of integers $\mathbb{Z}$, and the ring $F[x]$, of all polynomials in a single indeterminate x over a field F, are so pervasive throughout mathematics that their atypical properties deserve special attention.

In the previous chapter we observed that principal ideal domains were in fact unique factorization domains. In this chapter, the focus is no longer on the internal *arithmetic* of such rings, but rather on the structure of all finitely-generated submodules over these domains. As one will soon see, the structure is rather precise.

© Springer International Publishing Switzerland 2015

E. Shult and D. Surowski, *Algebra*, DOI 10.1007/978-3-319-19734-0_10

10.2 Quick Review of PID's

Principal Ideal Domains (PID's) are integral domains D for which each ideal I has the form $I = aD$ for some element a of D. One may then employ Theorem 8.2.11 to infer from the fact that every ideal is finitely generated, that the Ascending Chain Condition (ACC) holds for the poset of all ideals.

Recall that a non-zero element a *divides* element b if and only if

$$aD \supseteq bD.$$

It follows—as we have seen in the arguments that a principal ideal domain (PID) is a unique factorization domain (UFD)—that a is irreducible if and only if aD is a maximal ideal, and that this happens if and only if a is a prime element (see Lemma 9.2.4).

We also recall that for ideals,
$$xD = yD$$

if and only if x and y are associates. For this means $x = yr$ and $y = xs$ for some r and s in D, so $x = xrs$, so $sr = 1$ by the cancellation law. Thus r and s, being elements of D, are both units.

The *least common multiple* of two elements $a, b \in D$ is a generator m of the ideal

$$aD \cap bD = mD$$

where m is unique up to associates.

The *greatest common divisor* of two non-zero elements a and b is a generator d of

$$aD + bD = dD,$$

again up to the replacement of d by an associate. Two elements a and b are *relatively prime* if and only if

$$aD + bD = D.$$

Our aim in this chapter will be to find the structure of finitely-generated modules over a PID. Basically, our main theorem will say that if M is a finitely-generated module over a principal ideal domain D, then

$$M \simeq (D/De_1) \oplus (D/De_2) \oplus \cdots \oplus (D/De_k)$$

as right D-modules, where k is a natural number and for $1 \leq i \leq k - 1$, e_i divides e_{i+1}.

This is a very useful theorem. Its applications are (1) the classification of finite abelian groups, and (2) the complete analysis of a single linear transformation T : $V \to V$ of a finite-dimensional vector space into itself. It will turn out that the set of elements e_i which are not units is uniquely determined up to associates. But all of that still lies ahead of us.

10.3 Free Modules over a PID

We say that an R-module M has a basis if and only if there is a subset B of M (called an R-*basis* or simply a *basis*, if R is understood) such that every element m of M has a unique expression

$$\sum b_i r_i, \quad b_i \in B, r_i \in R,$$

as a finite R-linear combination of elements of B. The uniqueness is the key item here.

Recall that in Chap. 8, p. 243, we have defined a free (right) module over the ring R to be a (right) R module M satisfying any of the following three equivalent conditions:

- M possesses an R-basis, as defined just above.
- M is generated by an R-linearly independent set B. (In this case, B is indeed a basis).
- M is a direct sum of submodules M_σ each of which is isomorphic to R_R as right R-modules.

The equivalence of these conditions was the content of Lemma 8.1.8.

In general, there is no result asserting that two bases of a free R-module have the same cardinality. However, the result is true when R is commutative (see Theorem 8.1.11), and so it is true for an integral domain D. We call the unique cardinality of a basis for a free D-module F, the D-*rank* of F and denote it $\mathrm{rk}_D(F)$ or just $\mathrm{rk}(F)$ if D is understood.

10.4 A Technical Result

Theorem 10.4.1 *Let n and m be natural numbers and let A be an $n \times m$ matrix with coefficients from a principal ideal domain D. Then there exists an $n \times n$ matrix P invertible in $(D)^{n \times n}$, and an $m \times m$ matrix Q, invertible in $(D)^{m \times m}$, such that*

$$PAQ = \begin{pmatrix} d_1 & 0 & \dots & 0 \\ 0 & d_2 & \dots & 0 \\ \cdot & \cdot & \dots & \cdot \\ \cdot & \cdot & \dots & \cdot \end{pmatrix} \text{ where } d_i \text{ divides } d_{i+1}.$$

The elements $\{d_1, d_2, \ldots, d_k\}$, where $k = \min(n, m)$, are uniquely determined up to associates by the matrix A.

Proof For each non-zero element d of D let $\rho(d)$ be the number of prime divisors occurring in a factorization of d into primes. (Recalling that a PID is always a UFD, the function ρ is seen to be well-defined.) We call $\rho(d)$, the *factorization length* of the element d.

Our goal is to convert A into a diagonal matrix with successively dividing diagonal elements by a series of right and left multiplications by invertible matrices. These will allow us to perform several so-called *elementary row and column operations* namely:

(I) *Add b times the ith row (column) to the jth row (column), $i \neq j$.* This is left (right) multiplication by a suitably sized matrix having b in the (i, j)th (or (j, i)th) position while having all diagonal elements equal to 1_D, and all further entries equal to zero.

(II) *Multiply a row (or column) throughout by a unit u of D.* This is left (right) multiplication by a diagonal matrix $\mathrm{diag}(1, \ldots, 1u, 1, \ldots, 1)$ of suitable size.

(III) *Permute rows (columns).* This is left (right) multiplication by a suitable permutation matrix.

First observe that there is nothing to do if A is the $n \times m$ matrix with all entries zero. In this case each $d_i = 0_D$, and they are unique.

Thus for $A \neq 0$, we consider the entire collection of matrices

$$S = \{PAQ \mid P, Q \text{ units in } D^{(n \times n)} \text{ and } D^{(m \times m)}, \text{respectively}\}.$$

Among the non-zero entries of the matrices in S, there can be found a matrix element b_{ij} (in some $B \in S$) with factorization length $\rho(b_{ij})$ minimal. Then by rearranging the rows and columns of B if necessary, we may assume that this b_{ij} is in the $(1, 1)$-position of B—that is, $(i, j) = (1, 1)$.

We now claim

$$b_{11} \text{ divides } b_{1k} \text{ for } k = 2, \ldots, m. \tag{10.1}$$

To prove the claim, assume b_{11} does not divide b_{1k}. Permuting a few columns if necessary, we may assume $k = 2$. Since D is a principal ideal domain, we may write

$$b_{11}D + b_{12}D = Dd.$$

Then as d divides b_{11} and b_{12}, we see that

$$b_{12} = ds \text{ and } -b_{11} = dt, \text{ with } s, t \in D.$$

Then, as $d = b_{11}x + b_{12}y$, for some $x, y \in D$, we see that

$$d = (-dt)x + (ds)y \text{ so } 1 = sy - tx,$$

by the cancellation law. Thus

$$\begin{pmatrix} -t & s \\ y & -x \end{pmatrix} \begin{pmatrix} x & s \\ y & t \end{pmatrix} = \begin{pmatrix} 1 & 0 \\ 0 & 1 \end{pmatrix}.$$

So, after multiplying B on the right by the invertible

$$Q_1 = \begin{pmatrix} x & s & 0 \dots & 0 \\ y & t & 0 \dots & 0 \\ 0 & 0 & 1 & 0 \dots \\ & \cdot & \cdot & \cdots & \cdot \\ 0 & 0 & \dots & 0 & 1 \end{pmatrix}$$

the first row of BQ_1 is $(d, 0, b_{13}, \dots, b_{1m})$. Thus, as d now occurs as a non-zero element of D in a matrix in S, we see that $\rho(b_{11}) \le \rho(d) \le \rho(b_{11})$ so that b_{11} is an associate of d. Thus b_{11} divides b_{12} as claimed.

Similarly, we see that b_{11} divides each entry b_{k1} in the first column of B. Thus, subtracting suitable multiples of the first row from the remaining rows of B, we obtain a matrix in S with first column, $(b_{11}, 0, \dots, 0)^T$. Then subtracting suitable multiples of the first column from the remaining columns of this matrix, we arrive at a matrix

$$B_0 := \begin{pmatrix} b_{11} & 0 & 0 \dots 0 \\ 0 & & & \\ \cdot & & B_1 & \\ \cdot & & & \\ 0 & & & \end{pmatrix},$$

where $B_0 \in S$. We set $B_1 = (c_{ij})$, where $1 \le i \le n - 1$ and $1 \le j \le m - 1$.

We now expand on this: we even claim that

$$b_{11} \text{ divides each entry } c_{ij} \text{ of matrix } B_1.$$

This is simply because addition of the jth row to the first row of the matrix B_0 just listed above, yields a matrix B' in S whose first row is $(b_{11}, c_{j2}, c_{j3}, \dots c_{jm})$. But we have just seen that in any matrix B_0 in S containing an element of minimal possible factorization length in the $(1, 1)$-position, that this element divides every element of the first row and column. Since B' is such a matrix, b_{11} divides c_{ij}.

Now by induction on $n + m$, there exist invertible matrices P_1 and Q_1 such that $P_1 B_1 Q_1 = \text{diag}(d_2, d_3, \dots,)$ with d_2 dividing d_3 dividing $\dots d_k$.[1] Augmenting P_1 and Q_1 by adding a row of zeros above, and then a column to the left whose

---

[1] We apologize for the notation here: $\text{diag}(d_2, d_3, \dots,)$ should not be presumed to be a square matrix. Of course it is still an $(n - 1) \times (m - 1)$-matrix, with all entries zero except possibly those that bear a d_i at the (i, i)-position. If $n < m$ the main diagonal hits the "floor" before it hits the "east wall", and the other way round if $n > m$.

top entry is 1, the remaining entries 0, one obtains matrices P_1' and Q_1' such that $P_1' B_0 Q_1' = \mathrm{diag}(d_1, d_2, \ldots)$ where we have written d_1 for b_{11}.

Let us address the uniqueness. Now the element $d_1 = b_{11}$ with minimal prime length among the entries in B was completely determined by B. Moreover, the pair (b_{11}, B) determined B_0 and as well as its submatrix B_1. Finally, by induction, on $n + m$, the numbers $d_2, \ldots d_k$, are uniquely determined up to associates by the matrix B_1. Thus all of the d_i are uniquely determined by the initial matrix B. $\square$

10.5 Finitely Generated Modules over a PID

Theorem 10.5.1 (The Invariant Factor Theorem) *Let D be a principal ideal domain and let M be a finitely generated D-module. Then as right D-modules, we have*

$$M \simeq (D/Dd_1) \oplus (D/Dd_2) \oplus \cdots \oplus (D/Dd_n),$$

where d_i divides d_{i+1}, $i = 1, \ldots, n - 1$.

Proof Let M be a finitely generated D-module. Then by Theorem 8.1.9, there is a finitely generated free module F and a D-epimorphism

$$f : F \to M.$$

Then F has a D-basis $\{x_1, \ldots, x_n\}$, and, as D is Noetherian, its submodule $\ker f$ is finitely generated. Thus we have

$$M = x_1 D \oplus x_2 D \oplus \cdots \oplus x_n D \tag{10.2}$$
$$\ker f = y_1 D + y_2 D + \cdots + y_m D. \tag{10.3}$$

We then have m unique expressions

$$y_i = x_1 a_{i1} + x_2 a_{i2} + \cdots + x_n a_{in}, \ i = 1, \ldots, m,$$

and an $m \times n$ matrix $A = (a_{ij})$ with entries in D. By the technical result of the previous subsection, there exist matrices $P = (p_{ij})$ and $Q = (q_{ij})$ invertible in $D^{m \times m}$ and $D^{n \times n}$, respectively, such that

$$PAQ = \begin{pmatrix} d_1 & 0 & \ldots & 0 \\ 0 & d_2 & \ldots & 0 \\ & & \ddots & \end{pmatrix} \text{ (where } d_i \mid d_{i+1}),$$

an $m \times n$ matrix whose only non-zero entries are on the principal diagonal and proceed until they cant go any further (that is, they are defined up to subscript $\min(m, n)$).

If we set

$$Q^{-1} = (q_{ij}^*), \text{ and}$$
$$y_i' = \sum y_j p_{ij} \text{ and}$$
$$x_i' = \sum x_j q_{ij}^*$$

then PAQ expresses the y_i' in terms of the x_i', which now comprise a new D-basis for F. Thus we have

$$y_1' = x_1' d_1, \ y_2' = x_2' d_2, \ \dots \text{ etc.,}$$

(where, by convention the expression d_k is taken to be zero for $\min(m, n) < k \le m$). Also, since P and Q^{-1} are invertible, the y_i' also span $\ker f$. So, bearing in mind that some of the $y_i' = x_i' d_i$ may be zero, and extending the definition of the d_k by setting $d_k := 0$ if $m < k \le n$, we may write

$$F = x_1' D \oplus x_2' D \oplus \cdots \oplus x_n' D$$
$$\ker f = x_1' d_1 D \oplus x_2' d_2 D \oplus \cdots \oplus x_n' d_n D.$$

Then

$$M = f(F) \simeq F/\ker f \simeq (D/d_1 D) \oplus (D/d_2 D) \oplus \cdots \oplus (D/d_m D)$$

and the proof of the theorem is complete. $\square$

Remark The essential uniqueness of the elements $\{d_i\}$ appearing in the above theorem will soon emerge in Sect. 9.6 below. These elements are called the *invariant factors* (or sometimes the *elementary divisors*) of the finitely generated D-module M.

At this stage their uniqueness does not immediately follow from the fact that they are uniquely determined by the matrix A since this matrix itself depended on the choice of a finite spanning set for $\ker f$ and it is not clear that all such finite spanning sets can be moved to one another by the action of an invertible transformation Q in $\hom_D\{\ker f, \ker f\}$.

An element m of a D-module M is called a *torsion element* if and only if there exists a non-zero element $d \in D$ such that $md = 0$. Recall from Exercise (3) in Sect. 8.5.1 that the annihilator $\text{Ann}(m) = \{d \in D | md = 0\}$ is an ideal and that we have just said that m is a torsion element if and only if $\text{Ann}(m) \ne 0$. In an integral domain, the intersection of two non-zero ideals is a non-zero ideal; it follows from this that the sum of two torsion elements of an R-module is a torsion element. In fact the set of all torsion elements of M forms a submodule of M called the *torsion submodule* and denoted $\text{tor}M$.

Corollary 10.5.2 *A finitely generated D-module over a principal ideal domain D, is a direct sum of its torsion submodule $\text{tor}M$ and a free module.*

Proof By the Invariant Factor Theorem (Theorem 10.5.1),

$$M = (D/d_1 D) \oplus \cdots \oplus (D/d_n D)$$

where d_i divides d_{i+1}. Choose index t so that $i \le t$ implies $d_i \ne 0$, and, for $i > t$ we have $d_i = 0$. Then

$$\text{tor} M = (D/d_1 D) \oplus \cdots \oplus (D/d_t D)$$

and so

$$M = \text{tor} M \oplus D_D \oplus D_D \oplus \cdots \oplus D_D,$$

with $n - t$ copies of D_D. $\square$

If $M = \text{tor} M$, we say that M is a *torsion module*.

Let M be a module over a principal ideal domain, D. Let p be a prime element in D. We define the *p-primary part of M* as the set of elements of M annihilated by some power of p, that is

$$M_p = \{m \in M | mp^k = 0 \text{ for some natural number } k = k(m)\}.$$

It is easy to check that M_p is a submodule of M. We now have

Theorem 10.5.3 (Primary Decomposition Theorem) *If M is a finitely generated torsion D-module, where D is a principal ideal domain, then there exists a finite number of primes $p_1, p_2, \ldots, p_N$ such that*

$$M \simeq M_{p_1} \oplus \cdots \oplus M_{p_N}.$$

Proof Since M is finitely generated, we have $M = x_1 D + \cdots x_m D$ for some finite generating set $\{x_i\}$. Since M is a torsion module, each of the x_i is a torsion element, and so all of the ideals $A_i := \text{Ann}(x_i)$ are non-zero; and since D is a principal ideal domain,

$$A_i = \text{Ann}(x_i) = d_i D, \text{ where } d_i \ne 0, i = 1, \ldots, m.$$

Also, as D is a unique factorization domain as well, each d_i is expressible as a product of primes:

$$d_i = p_{i1}^{a_{i1}} p_{i2}^{a_{i2}} \cdots p_{if(i)}^{a_{if(i)}},$$

for some function f of the index i. We let $\{p_1, \ldots, p_N\}$ be a complete re-listing of all the primes p_{ij} which appear in these factorizations. If $f(i) = 1$, then d_i is a prime power and so $x_i D \subseteq M_{p_j}$ for some j. If, on the other hand, $f(i) \ne 1$, then, forming the elements $v_{ij} = d_i/p_{ij}^{a_{ij}}$, we see that the greatest common divisor of the elements $\{v_{i1}, \ldots, v_{if(i)}\}$ is 1, and this means

$$v_{i1}D + v_{i2}D + \cdots + v_{if(i)}D = D.$$

Thus there exist elements $b_1, \ldots, b_{f(i)}$ such that

$$v_{i1}b_1 + v_{i2}b_2 + \cdots + v_{if(i)}b_{f(i)} = 1.$$

Note that

$$b_j v_{ij} x_i \in M_{p_{ij}} \text{ since } p_{ij}^{a_{ij}} v_{ij} b_j x_i = 0,$$

so

$$x_i = x_i \cdot 1 = \sum b_j v_{ij} x_i \in M_{p_{i1}} + \cdots + M_{p_{if(i)}}.$$

Thus in all cases

$$x_i D \subseteq M_{p_1} + \cdots + M_{p_N}$$

for $i = 1, \ldots, m$, and the sum on right side is M. Now it remains only to show that this sum is direct and we do this by the criterion for direct sums of D-modules (see Theorem 8.1.6). Consider an element

$$a \in (M_{p_1} + \cdots + M_{p_k}) \cap M_{p_{k+1}}.$$

Then the ideal $\text{Ann}(a)$ contains an element m which on the one hand is a product of primes in $\{p_1, \ldots, p_k\}$, and on the other hand is a power of the prime p_{k+1}. Since $\gcd(m, p_{k+1}) = 1$, we have

$$\text{Ann}(a) \supseteq mD + p_{k+1}^{\rho}D = D,$$

so $1 \cdot a = a = 0$. Thus we have shown that

$$(M_{p_1} + \cdots + M_{p_k}) \cap M_{p_{k+1}} = (0).$$

This proves that

$$M = M_{p_1} \oplus M_{p_2} \oplus \cdots \oplus M_{p_N},$$

and the proof is complete. $\square$

10.6 The Uniqueness of the Invariant Factors

We have seen that a finitely generated module M over a principal ideal domain has a direct decomposition as a torsion module and a finitely generated free module (corresponding to the number of invariant factors which are zero). This free module is isomorphic to $M/\text{tor}M$ and so its rank is the unique rank of $M/\text{tor}M$ (Theorem 8.1.11), and consequently is an invariant of M. The torsion submodule, in turn, has a unique decomposition into primary parts whose invariant factors determine those of $\text{tor}(M)$. Thus, in order to show that the so-called invariant factors earn their name as genuine invariants of M, we need only show this for a module M which is primary—i.e. $M = M_p$ for some prime p in D.

In this case, applying the invariant factor theorem, we have

$$M \simeq D/p^{s_1}D \oplus \cdots \oplus D/p^{s_t}D \qquad (10.4)$$

where $s_1 \le s_2 \le \cdots \le s_t$, and we must show that the sequence $S = (s_1, \ldots, s_t)$ is uniquely determined by M. We shall accomplish this by producing another sequence $\Omega = \{\omega_i\}$, whose terms are manifestly invariants determined by the isomorphism type of M, and then show that S is determined by Ω.

For any p-primary D-module A, set

$$\Omega_1(A) := \{a \in A | ap = 0\}.$$

Then $\Omega_1(A)$ is a D-submodule of A which can be regarded as a vector space over the field $F := D/pD$. We denote the dimension of this vector space by the symbol $\text{rk}(\Omega_1(A))$. Clearly, $\Omega_1(A)$ and its rank are uniquely determined by the isomorphism type of A.

Now let us apply these notions to the module M with invariants $S = (s_1, \ldots, s_t)$ as in Eq. (10.4). First we set $\omega_1 := \text{rk}(\Omega_1(M))$, and $K_1 := \Omega_1(M)$. In general we set

$$K_{i+1}/K_i := \Omega_1(M/K_i), \text{ and } \omega_{i+1} := \text{rk}(K_{i+1}/K_i).$$

The K_i form an ascending chain of submodules of M which ascends properly until it stabilizes at $i = s_t$, the maximal element of S (Note that $\text{Ann}(M) = Dp^{s_t}$, so that p^{s_t} can be thought of as the "exponent" of M). Moreover, each of the D-modules K_i, is completely determined by the isomorphism type of M/K_{i-1}, which is determined by K_{i-1}, and ultimately by M alone. Thus we obtain a sequence of numbers

$$\Omega = (\omega_1, \omega_2, \ldots, \omega_{s_t}),$$

which are completely determined by the isomorphism type of M. One can easily see

(*) ω_j is the number of elements in the sequence S which are at least as large as j.

Now the sequence S can be recovered from Ω as follows:

(**) For $j = 0, 1, \ldots, t - 1$, s_{t-j} is the number of elements ω_r of Ω of cardinality at least $j + 1$.

Example 48 Suppose p is a prime in the principal ideal domain D and A is a primary D-module with invariant factors:

$$p, \ p, \ p^2, \ p^3, \ p^3, \ p^4, \ p^7, \ p^7, \ p^{10},$$

so $S = (1, 1, 2, 3, 3, 4, 7, 7, 10)$. Then, setting $F = D/pD$, K_1 is an F-module of rank 9, K_2/K_1 is an F-module of rank 7, and in general $\Omega = (9, 7, 6, 4, 3, 3, 3, 1, 1, 1)$. Then $S = (1, 1, 2, 3, 3, 4, 7, 7, 10)$ is recovered by the recipe in (**).

We conclude:

Theorem 10.6.1 *The non-unit invariant factors of a finitely-generated module over a principal ideal domain are completely determined by the isomorphism type of the module—i.e., they are really invariants.*

10.7 Applications of the Theorem on PID Modules

10.7.1 Classification of Finite Abelian Groups

As remarked before, an abelian group is simply a $\mathbb{Z}$-module. Since $\mathbb{Z}$ is a PID, any finitely generated $\mathbb{Z}$-module A has the form

$$A \simeq Z_{d_1} \times Z_{d_2} \times \cdots \times Z_{d_m}$$

—that is, A is uniquely expressible as the direct product of cyclic groups, the order of each dividing the next. The numbers $(d_1, \ldots d_n)$ are thus invariants of A and the decomposition just presented is called the *Jordan decomposition of A*. The number of distinct isomorphism types of abelian groups of order n is thus the number of ordered sets $(d_1, \ldots, d_m)$ such that d_i divides d_{i+1} and $d_1 d_2 \cdots d_m = n$.

Example 49 The number of abelian groups of order 36 is 4. The four possible ordered sets are: (36), $(2, 18)$, $(3, 12)$, $(6, 6)$.

This number is usually much easier to compute if we first perform the primary decomposition. If we assume A is a finite abelian group of order n, and if p is a prime dividing n, then $A_p = \{a \in A | ap^k = 0, k \geq 0\}$ is just the set of elements of A of p-power order. Thus A_p is the unique p-Sylow subgroup of A. Then the primary decomposition

$$A = A_{p_1} \oplus \cdots \oplus A_{p_t}$$

is simply the additive version of writing A as a direct product of its Sylow subgroups. Since the A_{p_i} are unique, the isomorphism type of A is uniquely determined by the isomorphism types of the A_{p_i}. Thus if A_{p_i} is an abelian group of order p^s, we can ask for *its* Jordan decomposition:

$$Z_{p^{s_1}} \times Z_{p^{s_2}} \times \cdots \times Z_{p^{s_m}}$$

where p^{s_i} divides $p^{s_{i+1}}$ and $p^{s_1} p^{s_2} \cdots p^{s_m} = p^s$. But we can write this condition as: $s_i \leq s_{i+1}$ and $s_1 + s_2 + \cdots + s_m = s$. Thus each unordered set renders s as a sum of natural numbers without regard to the order of the summands—called an *unordered partition* of the integer s. The number of these is usually denoted $p(s)$, and $p(s)$ is called the *partition function*. For example $p(2) = 2$, $p(3) = 3$, $p(4) = 5$, $p(5) = 7$, etc. Here, we see that the number of abelian groups of order p^s is just $p(s)$, and that:

Corollary 10.7.1 *The number of isomorphism classes of abelian groups of order* $n = p_1^{t_1} \cdots p_r^{t_r}$ *is*

$$p(t_1) p(t_2) \cdots p(t_r).$$

Again, we can compute the number of abelian groups of order $36 = 2^2 3^2$ as $p(2) p(2) = 2^2 = 4$.

10.7.2 The Rational Canonical Form of a Linear Transformation

Let F be any field and let V be an n-dimensional vector space over F so $V = F \oplus \cdots \oplus F$ (n copies) as an F-module. Suppose $T : V \to V$ is a linear transformation on V, which we regard as a right operator on V. Now in $\hom_F(V, V)$, the ring of linear transformations of V into V, multiplication is composition of transformations. Thus for any such transformation T, the transformations T, $T^2 := T \circ T$ and in fact any polynomial in T are well-defined linear transformations of V.

As is customary, $F[x]$ will denote the ordinary ring of polynomials in an indeterminate x. We wish to convert V into an $F[x]$-module by defining

$$v \cdot p(x) := vp(T) \text{ for any } v \in V \text{ and } p(x) \in F[x].$$

This module is finitely generated since an F-basis of V will certainly serve as a set of generators of V as an $F[x]$-module.

It now follows from our main theorem (Theorem 10.5.1) that there are polynomials $p_1(x), p_2(x), \ldots, p_m(x)$ with $p_i(x)$ dividing $p_{i+1}(x)$ in $F[x]$ such that

$$V_{F[x]} \simeq F[x]/(p_1(x)) \oplus \cdots \oplus F[x]/(p_m(x)),$$

where as usual $(p_i(x))$ denotes the principal ideal $p_i(x)F[x]$. Note that none of the $p_i(x)$ is zero since otherwise $V_{F[x]}$ would have a (free) direct summand $F[x]_{F[x]}$ which is infinite dimensional over F. Also, as usual, those $p_i(x)$ which are units— that is, of degree zero—contribute nothing to the direct decomposition of V. In fact we may assume that the polynomials $p_i(x)$ are monic polynomials of degree at least one. Since they are now uniquely determined polynomials, they have a special name: *the invariant factors of the transformation T*.

Now since each $p_i(x)$ divides $p_m(x)$, we must have that $Vp_m(x) = 0$ and so

$$p_m(x)F[x] = \text{Ann}_{F[x]}(V).$$

For this reason, $p_m(x)$, being a polynomial of smallest degree such that $p_m(T) : V \to 0$, is called the *minimal polynomial* of T and denoted $m_T(x)$. Such a polynomial is determined up to associates, but bearing in mind that we are taking $p_m(x)$ to be monic, the word "the" preceeding "minimal polynomial" is justified.

The product $p_1(x)p_2(x)\cdots p_m(x)$ is called the *characteristic polynomial* of T (denoted $\chi_T(x)$) and we shall soon see that the product of the constant terms of these polynomials is, up to sign, the determinant $\det T$, familiar from linear algebra courses.

How do we compute the polynomials $p_i(x)$? To begin with, we must imagine the transformation $T : V \to V$ given to us in such a way that it can be explicitly determined, that is, by a matrix A describing the action of T on V with respect to a basis $\mathcal{B} = (v_1, \ldots, v_n)$. Specifically, as T is regarded as a right operator,[2]

$$[T]_\mathcal{B} = A = (a_{ij}) \text{ so } T(v_i) = \sum_{j=1}^n v_j a_{ij}.$$

Then in forming the $F[x]$-module $V_{F[x]}$, we see that application of the transformation T to V corresponds to right multiplication of every element of V by the indeterminate x. Thus, for any fixed row index i:

$$v_i x = \sum v_j a_{ij}$$

so

$$0 = v_1 a_{i1} + \cdots + v_{i-1} a_{i,i-1} + v_i(a_{ii} - x) + v_{i+1} a_{i,i+1} \cdots v_n a_{in}.$$

Now as in the proof of the Invariant Factor Theorem, we form the free module

$$Fr = x_1 F[x] \oplus \cdots \oplus x_n F[x]$$

---

[2]Note that the rows of A record the fate of each basis vector under T. Similarly, if T had been a left operator rather than a right one, the columns of A would have been recording the fate of each basis vector. The difference is needed only for the purpose of making matrix multiplication represent composition of transformations. It has no real effect upon the invariant factors.

and the epimorphism $f : Fr \to V$ defined by $f(x_i) = v_i$. Then the elements

$$r_i := x_1 a_{i1} + \cdots + x_{i-1} a_{i,i-1} + x_i(a_{ii} - x) + x_{i+1} a_{i,i+1} \cdots x_n a_{in}.$$

all lie in ker f.

Now for any subscript i, x times the element x_i of Fr is

$$x_i x = r_i + \sum x_j a_{ij}, \ i = 1, \ldots, n.$$

Thus for any polynomial $p(x) \in F[x]$, we can write

$$x_i p(x) = \sum_k r_k h_{ik}(x) + \sum x_j b_j$$

for some polynomials $h_{ik}(x) \in F[x]$ and coefficients $b_j \in F$.

Now if

$$w = x_1 g_1(x) + x_2 g_2(x) + \cdots + x_n g_n(x)$$

were an arbitrary element of ker f, then it would have the form given in the following:

$$w = \sum_{i=1}^n x_i g_i(x) \tag{10.5}$$

$$= \left(\sum_{i=1}^n r_i h_i(x)\right) + \sum_{i=1}^n x_i c_i \tag{10.6}$$

for appropriate $h_i(x) \in F[x]$ and $c_i \in F$. Then since $f(w) = 0 = f(r_i)$ and $f(x_i) = v_i$ we see that

$$f(w) = 0 = f\left(\sum r_i h_i(x)\right) + \sum_{i=1}^n v_i c_i \tag{10.7}$$

$$= 0 + \sum v_i c_i. \tag{10.8}$$

So each $c_i = 0$. This means

$$w \in r_1 F[x] + \cdots + r_n F[x].$$

Thus the elements $r_1, \ldots, r_n$ are a set of generators for ker f.

Thus we have:

Corollary 10.7.2 *To find the invariant factors of the linear transformation represented by the square matrix $A = (a_{ij})$, it suffices to diagonalize the following matrix of $F[x]^{(n \times n)}$,*

$$A - Ix = \begin{pmatrix} a_{11} - x & a_{12} & a_{13} & \cdots & a_{1n} \\ a_{21} & a_{22} - x & a_{23} & \cdots & a_{2n} \\ a_{31} & a_{32} & a_{33} - x & \cdots & a_{3n} \\ \cdot\cdot & \cdot\cdot & \cdot\cdot & \cdot\cdot & \cdot\cdot \\ a_{n1} & \cdots & \cdots & \cdots & a_{nn} - x \end{pmatrix}$$

by the elementary row and column operations to obtain the matrix

$$\mathrm{diag}(p_1(x), p_2(x), \ldots, p_n(x))$$

with each $p_i(x)$ monic, and with $p_i(x)$ dividing $p_{i+1}(x)$. Then the $p_i(x)$ which are not units are the invariant factors of A.

Now, re-listing the non-unit invariant factors of T as $p_1(x), \ldots, p_r(x)$, $r \leq n$, the invariant factor theorem says that T acts on V exactly as right multiplication by x acts on

$$V' := F[x]/(p_1(x)) \oplus \cdots \oplus F[x]/(p_r(x)).$$

The space V' has a very nice basis. Each $p_i(x)$ is a monic polynomial of degree $d_i > 0$, say

$$p_i(x) = b_{0i} + b_{1i}x + b_{2i}x^2 + \cdots + b_{d_i-1,i}x^{d_i-1} + x^{d_i}.$$

Then the summand $F[x]/(p_1(x))$ of V' is non-trivial, and has the F-basis

$$1 + (p_i(x)), \ x + ((p_i(x)), \ x^2 + (p_i(x)), \ldots, x^{d_i-1} + (p_i(x)),$$

and right multiplication by x effects the action:

$$x^j + (p_i(x)) \rightarrow x^{j+1} + (p_i(x)), \ 1 \leq j < d_i - 1 \tag{10.9}$$
$$x^{d_i-1} + (p_i(x)) \rightarrow -b_{0i}(1 + (p_i(x))) - b_{1i}(x + (p_i(x))) - \cdots. \tag{10.10}$$

Thus, with respect to this basis, right multiplication of $F[x]/(p_i(x))$ by x is represented by the matrix

$$C_{p_i(x)} := \begin{pmatrix} 0 & 1 & 0 & \cdots & 0 \\ 0 & 0 & 1 & 0\cdots & 0 \\ \cdot\cdot & \cdot\cdot & \cdot\cdot & \cdots & \cdot\cdot \\ 0 & 0 & \cdots & 0 & 1 \\ -b_{0i} & -b_{1i} & -b_{2i} & \cdots & -b_{(d_i-1)i} \end{pmatrix} \tag{10.11}$$

called the *companion matrix* of $p_i(x)$. If $p_i(x)$ is a unit in $F[x]$—that is, it is just a nonzero scalar—then our convention is to regard the companion matrix as the "empty" (0×0) matrix. That convention is incorporated in the following

Corollary 10.7.3 *If $p_1(x), \ldots, p_n(x)$ are the invariant factors of the square matrix A, then A is conjugate to the matrix*

$$P^{-1}AP = \begin{pmatrix} C_{p_1(x)} & 0 & \cdots & 0 \\ 0 & C_{p_2(x)} & 0 \cdots & 0 \\ \cdot\cdot & & \cdot\cdot & \cdots & \cdot\cdot \\ 0 & & \cdots & 0 & C_{p_n(x)} \end{pmatrix}.$$

This is called the rational canonical form *of A.*

Remark Note that each companion matrix $C_{p_i(x)}$ is square with $d_i := \deg(p_i(x))$ rows. Using decomposition along the first column, one calculates that $C_{p_i(x)}$ in Eq. (10.11) has determinant $(-1)^{d_i-1}(-b_{0i}) = (-1)^{d_i}b_{0i}$. Thus the determinant of A is completely determined by the invariant factors. The determinant is just the product

$$\det A = \prod_{i=1}^{r}(-1)_i^d b_{0i} = (-1)^n \prod_{i=1}^{r} b_{0i},$$

which is the product of the constant terms of the invariant factors times the parity of $n = \sum d_i$, the dimension of the original vector space V.

It may be of interest to know that the calculation of the determinant of matrix A by ordinary means (that is, iterated column expansions) uses on the order of $n!$ steps. Calculating the determinant by finding the invariant factors (diagonalizing $A - xI$) uses on the order of n^3 steps. The "abstract" way is ultimately faster.

10.7.3 The Jordan Form

This section involves a special form that is available when the ground field of the vector space V contains all the roots of the minimal polynomial of transformation T.

Suppose now F is a field containing all roots of the minimal polynomial of a linear transformation $T : V \rightarrow V$. Then the minimal polynomial, as well as all of the invariant factors, completely factors into prime powers $(x - \lambda_i)^{a_i}$. Applying the primary decomposition to the right $F[x]$-module V, we see that a "primary part" $V_{(x-\lambda)}$ of V is a direct sum of modules, each isomorphic to $F[x]$ modulo a principal ideal I generated by a power of $x - \lambda$. This means that any matrix representing the linear transformation T is equivalent to a matrix which is a diagonal matrix of submatrices which are companion matrices of powers of linear factors. So one is reduced to considering $F[x]$-modules of the form $F[x]/I = F[x]/((x - \lambda)^m)$, for some root λ. Now $F[x]/I$ has this F-basis:

$$1 + I, (x - \lambda) + I, (x - \lambda)^2 I, \ldots (x - \lambda)^{m-1} + I.$$

Now right multiplying $F[x]/I$ by $x = \lambda + (x - \lambda)$ takes each basis element to λ times itself plus its successor basis element—except for the last basis element, which is merely multiplied by λ. That basis element, $(x - \lambda)^{m-1} + I$, and its scalar multiples comprise a 1-space of all "eigenvectors" of $F[x]/I$—that is, vectors v such that $vT = \lambda v$. The resulting matrix, with respect to this basis has the form:

$$\begin{pmatrix} \lambda & 1 & 0 & \ldots & 0 & 0 \\ 0 & \lambda & 1 & 0 & \ldots & 0 \\ \ldots & \ldots & \ldots & \ldots & \ldots & \ldots \\ 0 & 0 & 0 & \ldots & \lambda & 1 \\ 0 & 0 & 0 & \ldots & 0 & \lambda \end{pmatrix}.$$

It is called a *Jordan block* and is denoted $J_m(\lambda)$.

The matrix which results from assembling blocks according to the primary decomposition, and rewriting the resulting companion matrices in the above form, is called the *Jordan form of the transformation T*.

Example 50 Suppose the invariant factors of a matrix A were given as follows:

$$x, x(x - 1), x(x - 1)^3(x - 2)^2.$$

Here, F is any field in which the numbers 0, 1 and 2 are distinct. What is the Jordan form?

First we separate the invariant factors into their elementary divisors:

prime x: $\{x, x, x\}$.
prime $x - 2$: $\{(x - 2)^2\}$.
prime $x - 1$: $\{x - 1, (x - 1)^3\}$.

Next we write out each Jordan block as above: the result being

$$\begin{pmatrix} 0 & & & & & & & & \\ & 0 & & & & & & & \\ & & 0 & & & & & & \\ & & & 2 & 1 & & & & \\ & & & & 2 & & & & \\ & & & & & 1 & & & \\ & & & & & & 1 & 1 & \\ & & & & & & & 1 & 1 \\ & & & & & & & & 1 \end{pmatrix}.$$

(All unmarked entries in the above matrix are zero).

10.7.4 Information Carried by the Invariant Factors

What are some things one would like to know about a linear transformation $T : V \to V$, and how can the invariant factors provide this information?

1. We have already seen that the last invariant factor $p_r(x)$ is the minimal polynomial of T, that is—the smallest degree monic polynomial $p(x)$ such that $p(T)$ is the zero transformation $V \to 0_V$. (We recall that the ideal generated by $p(x)$ is $\mathrm{Ann}(V_{F[x]})$ when V is converted to an $F[x]$-module via T.)

2. Sometimes it is of interest to know the *characteristic polynomial* of T, which is defined to be the product $p_1(x) \cdots p_r(x)$ of the invariant factors of T. One must bear in mind that by definition the invariant factors of T are monic polynomials of positive degree. The characteristic polynomial must have degree n, the F-dimension of V. This is because the latter is the sum of the dimensions of the $F[x]/(p_i(x))$, that is, the sum of the numbers $\deg p_i(x) = d_i$.

3. Since V has finite dimension, T is onto if and only if it is one-to-one.[3] The dimension of $\ker T$ is called the *nullity of* T, and is given by the number of invariant factors $p_i(x)$ which are divisible by x.

4. The *rank of* T is the F-dimension of $T(V)$. It is just n minus the nullity of T (see the footnote just above).

5. V is said to be a *cyclic* $F[x]$-module if and only if there exists a module element v such that the set $vF[x]$ spans V—i.e. v generates V as an $F[x]$-module. This is true if and only if the only non-unit invariant factor is $p_r(x)$ the minimal polynomial (equivalently: the minimal polynomial is equal to the characteristic polynomial).

6. Recall that an R-module is said to be *irreducible* if and only if there are no proper submodules. (We met these in the Jordan-Hölder theorem for R-modules.) Here, we say that T *acts irreducibly on* V if and only if the associated $F[x]$-module is irreducible. This means there exists no proper sub-vector space W of V which is T-*invariant* in the sense that $T(W) \subseteq W$. One can then easily see that T acts irreducibly on V if and only the characteristic polynomial is irreducible in $F[x]$.[4]

7. An *eigenvector of* T *associated with the root* λ or λ-*eigenvector* is a vector v of V such that $vT = \lambda v$. A scalar is called an *eigen-root* of T if and only if there exists a non-zero eigenvector associated with this scalar. Now one can see that the eigenvectors associated with λ for right multiplication of $V_{F[x]}$ by x, are precisely the module elements killed by the ideal $(x - \lambda)F[x]$. Clearly they form a T-invariant subspace of V whose dimension is the number of invariant factors of T which are divisible by $(x - \lambda)$.

---

[3]This is a well-known consequence of the so-called "nullity-rank" theorem for transformations but follows easily from the fundamental theorems of homomorphisms for R-modules applied when $R = F$ is a field. Specifically, there is an isomorphism between the poset of subspaces of $T(V)$ and the poset of subspaces of V which contain $\ker T$. Since dimension is the length of an unrefinable chain in such a poset one obtains $\mathrm{codim}_V(\ker T) = \dim(T(V))$.

[4]This condition forces the minimal polynomial to equal the characteristic polynomial (so that the module is cyclic) and both are irreducible polynomials.

8. A matrix is said to be *nilpotent* if and only some power of it is the zero matrix. Similarly a linear transformation $T : V \rightarrow V$ is called a *nilpotent transformation* if and only if some power (number of iterations) of T annihilates V. Quite obviously the following are equivalent:

   - T is a nilpotent transformation.
   - T can be represented by a nilpotent matrix with respect to some basis of V.
   - The characteristic polynomial of T is a power of x.
   - The minimal polynomial of T is a power of x.

10.8 Exercises

10.8.1 Miscellaneous Exercises

Bezout Domains

1. An integral domain is said to be a *Bezout domain* if and only if all of its finitely-generated ideals are principal ideals. Give a proof that an integral domain is a Bezout domain if and only if every ideal generated by two elements is a principal ideal.

2. Of course the class of Bezout domains includes all principal ideal domains, so it would be interesting to have an example of a Bezout domain which was not a principal ideal domain. This and the subsequent four exercises are designed to produce examples of Bezout domains which are not principal ideal domains. Our candidate will be the integral domain D described in the next paragraph.

   Let E be a PID and let F be the field of fractions of E. Let D be the set of all polynomials of $F[x]$ whose constant coefficient belongs to the domain E. The first elementary result is to verify that D is a subdomain of $F[x]$. [The warmup result is elementary. But the reader might in general like to prove it using Exercise (1) in Sect. 7.5.2 on p. 223, as applied to the evaluation homomorphism $e_0 : F[x] \rightarrow F$.]

3. Let $g(x)$ and $h(x)$ be two non-zero polynomials lying in D. The greatest common divisors of these two polynomials are unique up to scalar multiplication.

   (i) Show that if x divides both $g(x)$ and $h(x)$ then

   $$g(x)D + h(x)D = Dd(x)$$

   where $d(x)$ is any greatest common divisor of $g(x)$ and $h(x)$.

   (ii) Suppose at least one of the two polynomials is not divisible by x. Then any greatest common divisor of $g(x)$ and $h(x)$ has a non-zero constant coefficient. We let $d(x)$ be the appropriate scalar multiple of a greatest common divisor such that the constant term $d(0)$ is a greatest common divisor of $g(0)$ and $h(0)$ in the PID E. (The latter exists since not both of $g(0)$ and $h(0)$ are zero). Thus $d(x) \in D$. Moreover, we have $g(x) = d(x)p(x)$ and $h(x) = d(x)q(x)$. Then $p(0) = (g(0)/d(0))$ is an element of E by our choice of $d(0)$ so $p(x) \in D$. Similarly $q(x) \in D$. Moreover these polynomials $p(x)$ and $q(x)$ are relatively prime in $F[x]$. Show that

$$Dg(x) + Dh(x) = Dd(x)$$

if and only if
$$Dp(x) + Dq(x) = Dd(0).$$

4. From now on we assume $p(x)$ and $q(x)$ elements of D which are relatively prime polynomials in $F[x]$. As above, d will denote a greatest common divisor of $p(0)$ and $q(0)$ in the PID E. We set $I = Dp(x) + Dq(x)$. Show that there exists a non-zero element $e \in E$ such that $ed \in I$ [Hint: Since $p(x)$ and $q(x)$ are relatively prime in $F[x]$, there exists polynomials $A(x)$ and $B(x)$ in $F[x]$ such that $d = A(x)p(x) + B(x)q(x)$. Then, since F is the field of fractions of the domain E, there exists a non-zero element $e \in E$ such that $eA(0)$ and $eB(0)$ are both elements of E. Then, observe that $eA(x)$ and $eB(x)$ are elements of D and the result follows].

5. Show that $d \in I$ [Hint: Since d is a greatest common divisor of elements $p(0)$ and $q(0)$ in the domain E, there exist elements A and B of E such that $d = Ap(0) + Bq(0)$. Write $p(x) = p(0) + p'(x)$ where $p'(x)$ is a multiple of x. Noting that $p'(x)/ed$ lies in D and ed lies in I we see that

$$p'(x) = ed \cdot (p'(x)/ed)$$

lies in I, and so $p(0)$ is the difference of two elements of I. Similarly argue that $q(0)$ lies in I so that $d = Ap(0) + Bq(0)$ is also in I].

6. Show that $I \subseteq Dd$ [Hint: We have $p'(x)/d \in D$ (since its constant term is zero) and $p_0/d \in E \subseteq D$ (since d is a divisor of p_0 in E by its definition). It follows that

$$p(x) = d \cdot ((p(0)/d) + p'(x)/d) \in I.$$

Similarly argue that $q(x) \in I$. Thus $I \subseteq Dd$].

7. Assemble the preceding problems into steps that show that D is a Bezout domain.

8. Suppose the domain E is not equal to its field of fractions F. Then it contains a prime element p. Show that the ideal J of D generated by the set

$$X := \{x, x/p, x/p^2, x/p^3, \ldots\}$$

is not a principle ideal [Hint: Let J_k be the ideal of D generated by the first k elements listed in the set X. Show that $J_1 \subset J_2 \subset \cdots$ is an infinite properly ascending chain of ideals of D].

9. Consider the following two integral domains:

   (a) The domain D_1 of all polynomials in $\mathbb{Q}[x]$ whose constant term is an integer.
   (b) Let K be a field and let S be the set of non-zero polynomials in $K[x]$, the latter being regarded as a subring of $K[x, y]$. Let $K[x, y]_S$ be the localization of $K[x, y]$ relative to the multiplicative set S. Observe that $K[x, y]_S = K(x)[y]$ where $K(x)$ is the so-called "field of rational functions" in the indeterminate x. Let D_2 be the subdomain of those polynomials $p(x, y) \in K[x, y]_S$ for which $p(x, 0)$ is a polynomial.

   Show that both domains D_1 and D_2 are Bezout domains which are not principal ideal domains.

10.8.2 Structure Theory of Modules over $\mathbb{Z}$ and Polynomial Rings

1. Suppose A is an additive abelian group generated by three elements a, b, and c. Find the order of A when A is presented by the following relations

$$2a + b + c = 0$$
$$a + b + c = 0$$
$$a + b + 2c = 0$$

2. Answer the same question when

$$2a + 2b + c = 0$$
$$2a + 3b + c = 0$$
$$2a + 4b + c = 0$$

3. Suppose A is presented by the following relations

$$2a + 2b + 2c = 0$$
$$3b + 2c = 0$$
$$2a + c = 0.$$

   Write A as a direct product of cyclic groups.
4. Find the invariant factors of the linear transformation which is represented by the following matrix over the field of rational numbers:

$$\begin{pmatrix} 2 & 1 & 1 & 1 \\ 1 & 1 & 1 & 1 \\ 0 & 0 & 0 & 0 \\ 1 & 0 & 1 & 0 \end{pmatrix}$$

5. Let V be a vector space over the rational field $\mathbb{Q}$ and suppose $T : V \to V$ is a linear transformation. We are presented with three possibilities for the non-trivial invariant factors of T:

   (i) $x, x^4, x^4(x+1)$.
   (ii) $x+1, x^3+1$.
   (iii) $1+x+x^2+x^3 = (1+x)(1+x^2)$,

   (a) What is the dimension of V in each of the three cases (i)–(iii)?
   (b) In which cases is the module V a cyclic module?
   (c) In which cases is the module V irreducible?
   (d) In case (ii) write the rational canonical form of the matrix representing T.
   (e) Write out the full Jordan form of a matrix representing T is case (i).

6. Suppose $t : V \to V$ is a linear transformation of a vector space over the field $\mathbb{Z}/(3)$ represented by the matrix:

$$\begin{pmatrix} 1 & 1 & 0 & 0 \\ 0 & 1 & 1 & 0 \\ 0 & 0 & 1 & 1 \\ 0 & 0 & 0 & 1 \end{pmatrix}$$

Find the order of t as an element of the group $GL(4, 3)$.

Chapter 11
Theory of Fields

Abstract If F is a subfield of a field K, then K is said to be an *extension* of the field F. For $\alpha \in K$, $F(\alpha)$ denotes the subfield generated by $F \cup \{\alpha\}$, and the extension $F \subseteq F(\alpha)$ is called a *simple* extension of F. The element α is *algebraic* over F if $\dim_F F(\alpha)$ is finite. Field theory is largely a study of field extensions. A central theme of this chapter is the exposition of Galois theory, which concerns a correspondence between the poset of intermediate fields of a finite normal separable extension $F \subseteq K$ and the poset of subgroups of $Gal_F(K)$, the group of automorphisms of K which leave the subfield F fixed element-wise. A pinnacle of this theory is the famous Galois criterion for the solvability of a polynomial equation by radicals. Important side issues include the existence of normal and separable closures, the fact that trace maps for separable extensions are non-zero (needed to show that rings of integral elements are Noetherian in Chap. 9), the structure of finite fields, the Chevalley-Warning theorem, as well as Luroth's theorem and transcendence degree. Attached are two appendices that may be of interest. One gives an account of fields with valuations, while the other gives several proofs that finite division rings are fields. There are abundant exercises.

11.1 Introduction

In this chapter we are about to enter one of the most fascinating and historically interesting regions of abstract algebra. There are at least three reasons why this is so:

1. Fields underlie nearly every part of mathematics. Of course we could not have vector spaces without fields (if only as the center of the coefficient division ring). In addition to vector spaces, fields figure in the theory of sesquilinear and multilinear forms, algebraic varieties, projective spaces and the theory of buildings, Lie algebras and Lie incidence geometry, representation theory, number theory, algebraic coding theory as well as many other aspects of combinatorial theory and finite geometry, just to name a few areas.
2. There are presently many open problems involving fields—some motivated by some application of fields to another part of mathematics, and some entirely intrinsic.

© Springer International Publishing Switzerland 2015

E. Shult and D. Surowski, *Algebra*, DOI 10.1007/978-3-319-19734-0_11

3. Third, there is the strange drama that so many field-theoretic questions bear on old historic questions, and indeed, its own history has a dramatic story of its own. These questions for fields involve some of the oldest and longest-standing problems in mathematics—they all involve the existence or non-existence of roots of polynomial equations in one indeterminate.

11.2 Elementary Properties of Field Extensions

Recall that a *field* is an integral domain in which every non-zero element is a unit. Examples are:

1. The field of rational numbers—in fact, the "field of fractions" of an arbitrary integral domain D (that is, the localization D_S where $S = D - \{0\}$). A special case is the field of fractions of the polynomial domain $F[x]$. This field is called the *field of rational functions over F* in the indeterminate x (even though they are not functions at all).
2. The familiar fields $\mathbb{R}$ and $\mathbb{C}$ of real and complex numbers, respectively.
3. The field of integers modulo a prime—more generally, the field D/M where M is a maximal ideal of an integral domain D.

Of course there are many further examples.

Let K be a field. A *subfield* is a subset F of K which contains a non-zero element, and is closed under addition, multiplication, and the taking of multiplicative inverses of the non-zero elements. It follows at once that F is a field with respect to these operations inherited from K, and that F contains the multiplicative identity of K as its own multiplicative identity. Obviously, the intersection of any collection of subfields of K is a subfield, and so one may speak of the *subfield generated by a subset X of K* as the intersection of all subfields of K containing set X. The subfield of K generated by a subfield F and a subset X is denoted $F(X)$.[1]

The subfield generated by the multiplicative identity element, 1, is called the *prime subfield* of K. (By definition, it contains the multiplicative identity as well as zero.) Since a field is a species of integral domain, it has a characteristic. If the multiplicative identity element 1, as an element of the additive group $(K, +)$, has finite order, then its order is a prime number p which we designate as the *characteristic of F*. Otherwise 1 has infinite order as an element of $(K, +)$ and we say that K *has characteristic zero*. We write $\mathrm{char}(F) = p$ or 0 in the respective cases. In the former case, it is clear that the prime subfield of F is just $\mathbb{Z}/p\mathbb{Z}$; in the latter case the prime subfield is isomorphic to the field $\mathbb{Q}$ of rational numbers. One last observation is the following:

Any subfield F of K contains the prime subfield and possesses the same characteristic as K.

---

[1]Of course this notation suffers the defect of leaving K out of the picture. Accordingly we shall typically be using it when the ambient overfield K of F is clearly understood from the context.

Since we shall often be dealing in this chapter with cases in which the subfield is known but the over-field is not, it is handy to reverse the point of view of the previous paragraph. We say that K is an *extension* of F if and only if F is a subfield of K. A chain of fields $F_1 \leq F_2 \leq \cdots \leq F_n$, where each F_{i+1} is a field extension of F_i, is called a *tower of fields*.

If the field K is an extension of the field F, then K is a right module over F, that is, a vector space over F. As such, it possesses a dimension (which might be an infinite cardinal) as an F-vector space. We call this dimension the *degree (or index) of the extension* $F \leq K$, and denote it by the symbol $[K : F]$.

11.2.1 Algebraic and Transcendental Extensions

Let K be a field extension of the field F, and let $p(x)$ be a polynomial in $F[x]$. We say that an element α of K is *a zero of $p(x)$* (or a *root of the equation $p(x) = 0$*) if and only it $p(x) \neq 0$ and yet $p(\alpha) = 0$.[2] If α is a zero in K of a polynomial in $F[x]$, then α is said to be *algebraic over F*. Otherwise α is said to be *transcendental over F*.

If K is an extension of F, and if α is an element of K, then the symbol $F[\alpha]$ denotes the subring of K generated by $F \cup \{\alpha\}$. This would be the set of all elements of K which are F-linear combinations of powers of α. Note that as this subring contains the field F, it is a vector space over F whose dimension is again denoted $[F[\alpha] : F]$.

There is then a ring homomorphism

$$\psi : F[x] \to F[\alpha] \subseteq K$$

which takes the polynomial $p(x)$ to the field element $p(\alpha)$ in K. Clearly $\psi(F[x]) = F[\alpha]$.

Suppose now $\ker \psi = 0$. Then $F[x] \simeq F[\alpha]$ and α is transcendental since any non-trivial polynomial of which it is a zero would be a polynomial of positive degree in $\ker \psi$.

---

[2]Two remarks are appropriate at this point. First note that the definition makes it "un-grammatic" to speak of a zero of the zero polynomial. Had the non-zero-ness of $p(x)$ not been inserted in the definition of root, we should be saying that every element of the field is a zero of the zero polynomial. That is such a bizarre difference from the situation with non-zero polynomials over infinite fields that we should otherwise always be apologizing for that exceptional case in the statements of many theorems. It thus seems better to utilize the definition to weed out this awkwardness in advance.

Second, there are natural objections to speaking of a "root" of a polynomial. Some have asked whether it might not be better to follow the German example and write "zero" (Nullstelle) for "root" (Wurzel), despite usage of the latter term in some English versions of Galois Theory. However we shall follow tradition by speaking of a "root" of an *equation*, $p(x) = 0$, while we speak of a "zero" of a *polynomial*, $p(x)$.

On the other hand, if ker $\psi \neq 0$, then, as $F[x]$ is a principal ideal domain, we have ker $\psi = F[x]p(x) = (p(x))$, for some polynomial $p(x)$. Then

$$F[\alpha] = \psi(F[x]) \simeq F[x]/\text{ker } \psi = F[x]/(p(x)).$$

Now, as $F[\alpha]$ is an integral domain, the principal ideal $(p(x))$ is a prime ideal, and so $p(x)$ is a prime element of $F[x]$. Therefore, since $F[x]$ is a PID, we infer immediately that the ideal $(p(x)) = \text{ker } \psi$ is actually a *maximal* ideal of $F[x]$, and so the subring $F[\alpha]$ is a field, which is to say, $F[\alpha] = F(\alpha)$.

The irreducible polynomial $p(x)$ determined by α is unique up to multiplication by a non-zero scalar. The unique associate of $p(x)$ which is monic is denoted $\text{Irr}_F(\alpha)$. It is the the the unique monic polynomial of smallest possible degree which has α for a zero.

Note that in this case, the F-dimension of $F[\alpha]$ is the F-dimension of $F[x]/(p(x))$, which, by the previous chapter can be recognized to be the degree of the polynomial $p(x)$.

Now conversely, assume that $[F[\alpha] : F]$ is finite. Then there exists a non-trivial finite F-linear combination of elements in the infinite list

$$\{1, \alpha, \alpha^2, \alpha^3, \ldots\}$$

which is zero. It follows that α is algebraic in this case.

Summarizing, we have the following:

Theorem 11.2.1 *Let K be an extension of the field F and let α be an element of K.*

1. *If α is transcendental over F, then the subring $F[\alpha]$ which it generates is isomorphic to the integral domain $F[x]$. Moreover, the subfield $F(\alpha)$ which it generates is isomorphic to the field of fractions $F(x)$ of $F[x]$.*
2. *If α is algebraic over F, then the subring $F[\alpha]$ is a subfield intermediate between K and F. As an extension of F its degree is*

$$[F(\alpha) : F] = [F[\alpha] : F] = \deg \text{Irr}_F(\alpha).$$

3. *Thus $\dim_F F[\alpha]$ is finite or infinite according as α is algebraic or transcendental over F.*

11.2.2 Indices Multiply: Ruler and Compass Problems

Lemma 11.2.2 *If $F \leq K \leq L$ is a tower of fields, then the indices "multiply" in the sense that*

$$[L : F] = [L : K][K : F].$$

Proof Let A be basis of L as a vector space over K, and let B be a basis of K as a vector space over F. Let AB be the set of all products $\{ab|a \in A, b \in B\}$. We shall show that AB is an F-basis of L.

If $\alpha \in L$, then α is a finite K-linear combination of elements of A, say

$$\alpha = a_1\alpha_1 + \cdots + a_m\alpha_m.$$

Now each coefficient α_i, being an element of K, is a finite linear combination

$$\alpha_j = b_{j1}\beta_{j1} + \cdots + b_{jm_j}\beta_{jm_j}$$

of elements b_{ji} of B (with the β_{ji} in F). Substituting these expressions for α_j in the L-linear combination for α, we have expressed α as an F-linear combination of elements of AB. Thus AB is an F-spanning set for L.

It remains to show that AB is an F-linearly independent set. So suppose for some finite subset S of $A \times B$, that

$$\sum_{(a,b)\in S} ab\beta_{a,b} = 0.$$

Then as each $\beta_{a,b}$ is in F, and each b is in K, the left side of the presented formula may be regarded as a K-linear combination of a's equal to zero, and so, by the K-linear independence of A, each coefficient

$$\sum_b b\beta_{a,b}$$

of each a is equal to zero. Hence each $\beta_{a,b} = 0$, since B is F-linearly independent. Thus AB is F-linearly independent and hence is an F-basis for L. Since this entails that all the products in AB are pairwise distinct elements, we see that

$$|AB| = |A||B|,$$

which proves the lemma. $\square$

Corollary 11.2.3 *If $F_1 \le F_2 \le \cdots \le F_n$ is a finite tower of fields, then*

$$[F_n : F_1] = [F_n : F_{n-1}] \cdot [F_{n-1} : F_{n-2}] \cdots [F_2 : F_1].$$

We can use this observation to sketch a proof of the impossibility of certain ruler and compass constructions. Given a unit length 1, using only a ruler and compass we can replicate it m times, or divide it into n equal parts, and so can form all lengths which are rational numbers. We can also form right-angled triangles inscribed in a circle and so with ruler and compass, we can extract the square root of the differences of squares, and hence any square root because of the formula

$$c = \left(\frac{c+1}{2}\right)^2 - \left(\frac{c-1}{2}\right)^2.$$

If α is such a square root, where c is a rational length, then by ruler and compass we can form all the lengths in the ring $\mathbb{Q}[\alpha]$, which (since α is a zero of $x^2 - c$) is a field extension of $\mathbb{Q}$ of degree 1 or 2. Iterating these constructions a finite number of times, the possible lengths that we could encounter all lie in the uppermost field of a tower of fields

$$\mathbb{Q} = F_1 < F_2 < \cdots < F_n = K$$

with each F_{i+1} an extension of degree two over F_i. By the Corollary above, the index $[K : \mathbb{Q}]$ is a power of 2.

Now we come to the arm-waving part of the proof: One needs to know that once a field L of constructible distances has been achieved, the *only* new number not in K constructed entirely from old numbers already in K is in fact obtained by *producing a missing side of a right triangle, two of whose sides have lengths in K*. (A really severe proof of that fact would require a formalization of exactly what "ruler and compass" constructions are—a logical problem beyond the field-theoretic applications whose interests are being advanced here.) In this sketch, we assume that that has been worked out.

This means, for example, that we *cannot* find by ruler and compass alone, a length α which is a zero of an irreducible polynomial of degree n where n contains an odd prime factor. For if so, $\mathbb{Q}[\alpha]$ would be a subfield of a field K of degree 2^m over $\mathbb{Q}$. On the other hand, the index $n = [F[\alpha] : F]$ must divide $[K : \mathbb{Q}] = 2^m$, by the Theorem 11.2.1. But the odd prime in n cannot divide 2^m. That's it! That's the whole amazingly simple argument!

Thus one cannot solve the problem posed by the oracle of Delos, to "duplicate" (in volume) a cubic altar—i.e., find a length α such that $\alpha^3 - 2 = 0$—at least not with ruler and compass.

Similarly, given angle α, trisect it with ruler and compass. If one could, then one could construct the length $\cos \beta$ where $3\beta = \alpha$. But $\cos(3\beta) = 4\cos^3 \beta - 3\cos \beta = \lambda = \cos \alpha$. This means we could always find a zero of $4x^3 - 3x - \lambda$ and when $\alpha = 60^o$, so $\lambda = 1/2$, setting $y = 2x$ yields a constructed zero of $y^3 - 3y - 1$, which is irreducible over $\mathbb{Q}$. As observed above, this is impossible.

11.2.3 The Number of Zeros of a Polynomial

Lemma 11.2.4 *Let K be an extension of the field F and suppose the element α of K is a zero of the polynomial $p(x)$ in $F[x]$. Then in the polynomial ring $F[\alpha][x]$, $x - \alpha$ divides $p(x)$.*

Proof Since $p(\alpha) = 0$ we may write

$$
\begin{aligned}
p(x) &= a_0 + a_1 x + \cdots + a_n x^n \\
&= p(x) - p(\alpha) \\
&= a_1(x - \alpha) + a_2(x^2 - \alpha^2) + \cdots + a_n(x^n - \alpha^n) \\
&= (x - \alpha)[a_1 + a_2(x + \alpha) + a_3(x^2 + \alpha x + \alpha^2) + \cdots \\
&\qquad \cdots + a_n(x^{n-1} + \cdots \alpha^{n-1})].
\end{aligned}
$$

$\square$

Theorem 11.2.5 *Let K be an extension of the field F and let $p(x)$ be a polynomial in $F[x]$. Then $p(x)$ possesses at most $\deg p(x)$ zeros in K.*

Proof We may assume $p(x)$ is a monic polynomial. If $p(x)$ has no zeroes in K, we are done. So assume α is a zero of $p(x)$. Then by Lemma 11.2.4 we obtain a factorization $p(x) = (x - \alpha)p_1(x)$ in $K[x]$, where $\deg p_1(x)$ is one less than the degree of $p(x)$. Suppose β is any zero of $p(x)$ in K which is distinct from α. Then

$$0 = p(\beta) = (\beta - \alpha)p_1(\alpha).$$

Since the first factor on the right is not zero, and F is a domain, β is forced to be a zero of the polynomial $p_1(x)$. By induction on the degree of $p(x)$, we may conclude that the number of distinct possibilities for β does not exceed the degree of $p_1(x)$. Thus, as any zero of $p(x)$ is either α or one of the zeroes β of $p_1(x)$, we see that the total number of zeroes of $p(x)$ cannot exceed $1 + \deg p_1(x)$, that is, the degree of $p(x)$. $\square$

Of course, when we write $p(x) = (x - \alpha)p_1(x)$, it may happen that α is also a zero of $p_1(x)$, as well. In that case, we can again write $p_1(x) = (x - \alpha)p_2(x)$, and ask whether α is a zero of $p_2(x)$. Pushing this procedure as far as we can, we eventually obtain a factorization $p(x) = (x - \alpha)^k p_k(x)$ where $p_k(x)$ does not have α for a zero. Repeating this for the finitely many zeroes α_i of $p(x)$, one obtains a factorization

$$p(x) = (x - \alpha_1)^{n_1}(x - \alpha_2)^{n_2} \cdots (x - \alpha_m)^{n_m} r(x) \qquad (11.1)$$

in $K[x]$ where $r(x)$ possesses no "linear factors"—that is, factors of degree one.

The number n_i is called the *multiplicity* of the zero α_i, and clearly from Eq. (11.1), $\sum n_i \leq \deg p(x)$. Thus the following slight improvement of Theorem 11.2.5 emerges:

Corollary 11.2.6 *If K is an extension of the field F and $p(x)$ is a non-zero polynomial in $F[x]$, then the number of zeroes—counting multiplicities—is at most the degree of $p(x)$.*[3]

---

[3]The student should notice that when we count zeroes with their multiplicities we are not doing anything mysterious. We are simply forming a multiset of zeroes. The Corollary just says that the degree of the polynomial bounds the weight of this multiset.

There are a number of other elementary but important corollaries of Theorem 11.2.5.

Corollary 11.2.7 (Polynomial Identities) *Suppose $a(x)$, $b(x)$, are polynomials in $F[x]$. Suppose $a(\alpha) = b(\alpha)$ for all $\alpha \in F$. If F contains infinitely many elements, then*

$$a(x) = b(x)$$

is a polynomial identity—that is, both sides are the same element of $F[x]$.

Proof We need only show that the polynomial $J(x) := a(x) - b(x)$ has degree zero. If it has positive degree, then by Theorem 11.2.5 it possesses only finitely many zeros. But then by hypothesis, the infinitely many elements of F would all be zeros of $J(x)$, a contradiction. $\square$

Corollary 11.2.8 (Invertibility of the Vandermonde matrix) *Let n be a positive integer. Suppose $z_1, z_2, \ldots, z_n$ are pairwise distinct elements of a field F. Then the $n \times n$ matrix*

$$M = \begin{bmatrix} 1 & z_1 & z_1^2 & \cdots & z_1^{n-1} \\ 1 & z_2 & z_2^2 & \cdots & z_2^{n-1} \\ \vdots & \vdots & \vdots & \cdots & \vdots \\ 1 & z_n & z_n^2 & \cdots & z_n^{n-1} \end{bmatrix}$$

is invertible.

Proof The matrix M is invertible if and only if its columns $C_1, \ldots, C_n$ are F-linearly independent. So, if M were not invertible, there would exist coefficients $a_0, \ldots, a_{n-1}$ in F such that

$$\sum_{i=0}^{n-1} a_i C_{i+1} = [0], \text{ the } n \times 1 \text{ zero column vector.} \tag{11.2}$$

But each entry on the left side of Eq. (11.2) is $p(z_i)$ where $p(x)$ is the polynomial $a_0 + a_1 x + \cdots, a_{n-1}x^{n-1} \in F[x]$. This conflicts with Theorem 11.2.5, since the polynomial $p(x)$ has degree at most $n-1$, but yet possesses n distinct zeroes—the z_i.

Thus the columns of M are F-linearly independent and so M is invertible. $\square$

Remark Notice that the proof of Corollary 11.2.8 was determinant-free. Matrices of the form displayed in Corollary 11.2.8 are called *Vandermonde matrices*.

For the next result, we require the following definition: a group G is said to be *locally cyclic* if and only if every finitely-generated subgroup of G is cyclic. Clearly such a group is abelian. Recall from Exercise (2) in Sect. 5.6.1 that a group is a *torsion group* if and only each of its elements has finite order. In any abelian group A, the set of elements of finite order are closed under multiplication and the taking of inverses, and so form a subgroup which we call the *torsion subgroup* of A.

Corollary 11.2.9 *Let F be any field and let F^* be its multiplicative group of non-zero elements. Then the torsion subgroup of F^* is locally cyclic.*

Proof Any finitely-generated subgroup of the torsion subgroup of F^* is a finite abelian group A. If A were not cyclic, it would contain a subgroup isomorphic to $Z_p \times Z_p$, for some prime p. In that case, the polynomial $x^p - 1$ would have at least p^2 zeros in F, contrary to Theorem 11.2.5. $\square$

11.3 Splitting Fields and Their Automorphisms

11.3.1 Extending Isomorphisms Between Fields

Suppose E is a field extension of F and that α is an element of E which is algebraic over F. Then the development in the previous section (p. 395) showed that there is a unique irreducible monic polynomial,

$$g(x) = \mathrm{Irr}_F(\alpha) \in F[x],$$

having α as a zero. The subfield $F(\alpha)$ of E generated by $F \cup \{\alpha\}$, is the subring $F[\alpha]$, which we have seen is isomorphic to the factor ring $F[x]/(g(x))$.

We remind the reader of a second principle. Suppose $f : F_1 \to F_2$ is an isomorphism of fields. Then f can be extended to a ring isomorphism

$$f^* : F_1[x] \to F_2[x],$$

which takes the polynomial

$$p := a_0 + a_1 x + \cdots + a_n x^n$$

to

$$f^*(p) := f(a_0) + f(a_1)x + \cdots + f(a_n)x^n.$$

It is obvious that f^* takes irreducible polynomials of $F_1[x]$ to irreducible polynomials of $F_2[x]$.

With these two principles in mind we record the following:

Lemma 11.3.1 (Fundamental Lemma on Extending Isomorphisms) *Let E be a field extension of F, and suppose $f : F \to \bar{E}$ is an embedding of F as a subfield $\bar{F}$ of the field $\bar{E}$. Let f_1 be the induced isomorphism $f_1 : F \to \bar{F}$ obtained by resetting the codomain of f. Next let f_1^* be the extension of this field isomorphism f_1 to a ring isomorphism $F[x] \to \bar{F}[x]$, as described above.*

Finally, let α be an element of E which is algebraic over F, and let g be its monic irreducible polynomial in $F[x]$.

Then the following assertions hold:

(i) *The embedding (injective morphism) $f : F \to \bar{E}$ can be extended to an embedding*
$$\hat{f} : F(\alpha) \to \bar{E},$$

if and only if the polynomial $f_1^*(g) := \bar{g}$ has a zero in $\bar{E}$.

(ii) *Moreover, for each zero $\bar{\alpha}$ of $\bar{g}$ in $\bar{E}$, there is a unique embedding $\hat{f} : F(\alpha) \to \bar{E}$, taking α to $\bar{\alpha}$ and extending f.*

(iii) *The number of extensions $\hat{f} : F(\alpha) \to \bar{E}$ of the embedding f is equal to the number of distinct zeros of $\bar{g}$ to be found in $\bar{E}$.*

Proof (i) If there is an embedding $\hat{f} : F(\alpha) \to \bar{E}$ extending the embedding f, then $\hat{f}(\alpha)$ is a zero of $\bar{g}$ in $\bar{E}$.

Conversely, if $\bar{\alpha}$ in E, then $\bar{F}(\bar{\alpha})$ is a subfield of E isomorphic to $\bar{F}[x]/(\bar{g}(x))$. Similarly, $F(\alpha)$ is isomorphic to $F[x]/(g(x))$. But the ring isomorphism $f_1^* : F[x] \to \bar{F}[x]$ takes the maximal ideal $(g(x))$ of $F[x]$ to the maximal ideal $(\bar{g}(x))$ of $\bar{F}[x]$, and so induces an isomorphism of the corresponding factor rings:

$$f' : F[x]/(g(x)) \to \bar{F}[x]/(\bar{g}(x)).$$

Linking up these three isomorphisms

$$F(\alpha) \to F[x]/(g(x)) \xrightarrow{f'} \bar{F}[x]/(\bar{g}(x)) \to \bar{F}(\bar{\alpha}) \subseteq \bar{E},$$

yields the desired embedding.

(ii) If there were two embeddings $h_1, h_2 : F(\alpha) \to \bar{E}$, taking α to $\bar{\alpha}$ and extending f, then the composition of the inverse of one with the other would fix $\bar{\alpha}$, would fix $\bar{F}$ element-wise, and so would be the identity mapping on $\bar{F}(\bar{\alpha})$. Thus, for any $\beta \in F(\alpha)$,
$$h_1(\beta) = (h_2 \circ h_1^{-1})(h_1(\beta) = h_2(\beta).$$

Thus h_1 and h_2 would be identical mappings.

(iii) This part follows from parts (i) and (ii). $\square$

11.3.2 Splitting Fields

Suppose E is an extension field of F. Then, of course, any polynomial in $F[x]$ can be regarded as a polynomial in $E[x]$, and any factorization of it in $F[x]$ is a factorization in $E[x]$. Put another way, $F[x]$ is a subdomain of $E[x]$.

A non-zero polynomial $p(x)$ in $F[x]$ of positive degree is said to *split over* E if and only if $p(x)$ factors completely into linear factors (that is, factors of degree 1) in $E[x]$. Such a factorization has the form

$$p(x) = a(x - a_1)^{n_1}(x - a_2)^{n_2} \cdots (x - a_r)^{n_r} \qquad (11.3)$$

where a is a nonzero element of F, the exponents n_i are positive integers and the collection $\{a_1, \ldots, a_r\}$ of *zeros of $p(x)$ in E* is contained in E.

Obviously, if E' is a field containing E and the polynomial $p(x)$ splits over E, then it splits over E'. Also, if E_0 were a subfield of E containing F together with all of the zeros $\{a_1, \ldots, a_r\}$, then $p(x)$ would split over over E_0, since the factorization in Eq. (11.3) can take place in $E_0[x]$. In particular, this is true if $E_0 = F(a_1, \ldots, a_r)$, the subfield of E generated by F and the zeros a_i, $i = 1, \ldots, r$.

An extension field E of F is called a *splitting field of the polynomial* $p(x) \in F[x]$ *over F* if and only if

(S1) $F \subseteq E$.
(S2) $f(x)$ factors completely into linear factors in $E[x]$ as

$$f(x) = a(x - a_1)(x - a_2) \cdots (x - a_n).$$

(S3) $E = F(a_1, \ldots, a_n)$, that is, E is generated by F and the zeros a_i.

The following observations follow directly from the definition just given and the proofs are left as an exercise.

Lemma 11.3.2 *The following hold:*

(i) *If E is a splitting field for $p(x) \in F[x]$ over F, and L is a subfield of E containing F, then E is a splitting field for $p(x)$ over L.*
(ii) *Suppose $p(x)$ and $q(x)$ are polynomials in $F[x]$. Suppose E is a splitting field for $p(x)$ over F and K is a splitting field for $q(x)$ over E. Then K is a splitting field for $p(x)q(x)$ over F.*

Our immediate goals are to show that splitting fields for a fixed polynomial exist, to show that they are unique up to isomorphism, and to show that between any two such splitting fields, the number of isomorphisms is bounded by a function of the number of distinct zeros of $f(x)$.

Theorem 11.3.3 (Existence of Splitting Fields) *If $f(x) \in F[x]$, then a splitting field for $f(x)$ over F exists. Its degree over F is at most the factorial number $d!$, where d is the sum of the degrees of the non-linear irreducible polynomial factors of $f(x)$ in $F[x]$.*

Proof Let $f(x) = f_1(x)f_2(x) \cdots f_m(x)$ be a factorization of $f(x)$ into irreducible factors in $F[x]$. (This is possible by Theorem 7.3.2, Part 3.) Set $n = \deg f(x)$, and proceed by induction on $k = n - m$. If $k = 0$, then $n = m$, so each factor $f_i(x)$ is linear and clearly $E = F$ satisfies all the conclusions of the theorem. So assume $k > 0$. Then some factor, say $f_1(x)$, has degree greater than 1. Form the field

$$L = F[x]/(f_1(x)),$$

(this *is* a field since $f_1(x)$ is irreducible and it is an extension of F since it is a vector-space over F). Observe that the coset $a := x + F[x]f_1(x)$ is a zero of $f_1(x)$ (and hence $f(x)$) in the field L. Thus in $L[x]$, we have the factorizations:

$$f_1(x) = (x - a)h_1(x),$$

and

$$f(x) = (x - a)h_1(x)f_2(x) \cdots f_m(x),$$

with $\ell > m$ irreducible factors. Thus $n - \ell < n - m$ and so by induction, there is a splitting field E of $f(x)$ over L. Moreover, this degree is at most $(d - 1)!$ since the sum of the degrees of the non-linear irreducible factors of $f(x)$ over $L[x]$ has been reduced by at least one because of the appearance of the new linear factor $x - a$.

Now we claim that E is a splitting field of $f(x)$ over F. We must verify the three defining properties of a splitting field given at the beginning of this subsection (p. 365).

Property (S1) holds since $L \subseteq E$ implies $F \subseteq E$. We already have (S2) from the definition of E. It remains to see that E is generated by the zeros of f. Since $(x - a)$ is one of the factors of $f(x)$ in the factorization

$$f(x) = (x - a_1)(x - a_2) \cdots (x - a_n)$$

in $E[x]$, we may assume without loss of generality that $a = a_1$. We have $E = L(a_2, \ldots, a_n)$ from the definition of E. But $L = F(a_1)$, so

$$E = L(a_2, \ldots, a_m) = F(a_1)(a_2, \ldots, a_n) = F(a_1, \ldots, a_n),$$

and so (S3) holds. Thus E is indeed a splitting field for $f(x)$ over F.

Finally, since $[L : F] = \deg f_1(x) \le d$, and $[E : L] \le (d - 1)!$ we obtain $[E : F] = [E : L][L : F] \le (d - 1)! \cdot d = d!$ as required. $\square$

Theorem 11.3.4 *Let $\eta : F \to \bar{F}$ be an isomorphism of fields, which we extend to a ring isomorphism $\eta^* : F[x] \to \bar{F}[x]$, and let us write $\bar{f}(x)$ for $\eta^*(f(x))$ for each $f(x) \in F[x]$. Suppose E and $\bar{E}$ are splitting fields of $f(x)$ over F, and $\bar{f}(x)$ over $\bar{F}$, respectively. Then η can be extended to an isomorphism $\hat{\eta}$ of E onto $\bar{E}$, and the number of ways of doing this is at most the index $[E : F]$. If the irreducible factors of $f(x)$ have no multiple zeros in E, then there are exactly $[E : F]$ such isomorphisms.*

Proof We use induction on $[E : F]$. If $[E : F] = 1$, $f(x)$ factors into linear factors, and there is precisely one isomorphism extending η, namely η itself.

Assume $[E : F] > 1$. Then $f(x)$ contains an irreducible factor $g(x)$ of degree greater than 1. Since η^* is an isomorphism of polynomial rings, $\bar{g}(x)$ is an irreducible factor of $\bar{f}(x)$ with the same degree as $g(x)$. Let $\alpha \in E$ be a zero of $g(x)$, and set $K = F(\alpha)$. Since the irreducible polynomial $\bar{g}(x)$ splits completely in $\bar{E}$, we may apply the Fundamental Lemma on Extensions (Lemma 11.3.1), to infer that the isomorphism η has k extensions $\zeta_1, \ldots, \zeta_k : F(\alpha) \to \bar{E}$, where k is the number of distinct zeros of $\bar{g}(x)$ in $\bar{E}$.

Note that if $m = \deg g(x)$, then $[K : F] = m$ and $k \leq m$. If the zeros of $\bar{g}(x)$ are distinct, $k = m$.

Now clearly E is a splitting field of $f(x)$ over K, and $\bar{E}$ is a splitting field of $\bar{f}(x)$ over each $\zeta_i(K)$. Since $[E : K] < [E : F]$, induction implies that each ζ_i can be extended to an isomorphism $E \to \bar{E}$ in at most $[E : K]$ ways, and in exactly $[E : K]$ ways if the irreducible factors of $f(x) \in K[x]$ have distinct zeros in E. However if the irreducible factors of $f(x) \in F[x]$ have distinct zeros in E, the same is obviously true for the irreducible factors of $g(x) \in K[x]$. This yields at most $k[E : K] = [K : F][E : K] = [E : F]$ isomorphisms in general, and exactly $[E : F]$ isomorphisms if the irreducible factors of $f(x)$ have distinct zeros in E.

This proves the theorem. $\square$

Corollary 11.3.5 *If E_1 and E_2 are two splitting fields for $f(x) \in F[x]$ over F, there exists an F-linear field isomorphism:*

$$\sigma : E_1 \to E_2.$$

Proof This follows immediately from Theorem 11.3.4 for the case $\sigma = 1_F$, the identity mapping on $F = F_1 = F_2$. $\square$

Example 51 Suppose $F = \mathbb{Q}$, the rational field, and $p(x) = x^4 - 5$, irreducible by Eisenstein's criterion. One easily has that the complex zeros of $p(x)$ are $\pm\sqrt[4]{5}, \pm i\sqrt[4]{5} \in \mathbb{C}$, where, of course, i is the imaginary complex unit ($i^2 = -1$), and $\sqrt[4]{5}$ is the real fourth root of 5. This easily implies that the splitting field $E \subseteq \mathbb{C}$ can be described by setting $E = \mathbb{Q}(i, \sqrt[4]{5})$. Since $p(x) \in \mathbb{Q}[x]$ is irreducible by Eisenstein, we conclude that $[\mathbb{Q}(\sqrt[4]{5}) : \mathbb{Q}] = 4$. Since E is the complex splitting field over $\mathbb{Q}(\sqrt[4]{5})$ of the polynomial $x^2 + 1 \in \mathbb{Q}(\sqrt[4]{5})[x]$, and since $\mathbb{Q}(\sqrt[4]{5}) \subsetneq E$, we infer that $[E : \mathbb{Q}(\sqrt[4]{5})] = 2$, giving the splitting field extension degree:

$$[E : \mathbb{Q}] = [E : \mathbb{Q}(\sqrt[4]{5})] \cdot [\mathbb{Q}(\sqrt[4]{5}) : \mathbb{Q}] = 2 \cdot 4 = 8.$$

This shows that the bound $[E : F] \leq d!$ need not be obtained. Note, finally, by Theorem 11.3.4, that there are exactly $8 = [E : \mathbb{Q}]$ distinct $\mathbb{Q}$-automorphisms of E.

We close this subsection with a useful observation:

Lemma 11.3.6 *Suppose K is an extension of a field F, and that K contains a subfield E which is a splitting field for a polynomial $p(x) \in F[x]$ over F. Then any automorphism of K fixing F point-wise must stabilize E.*

Remark The Lemma just says that the splitting field E is "characteristic" among fields in K which contain F—that is, E is invariant under all F-linear automorphisms of K.

Proof It suffices to note that if σ is an F-linear automorphism of K, then for any zero α of $p(x)$ in K, $\sigma(\alpha)$ is also a zero of $p(x)$ in K. Since E is generated by F and the zeroes of $f(x)$ in K, we have $E = E^{\sigma}$, for all F-automorphisms σ of K. $\square$

11.3.3 Normal Extensions

So far the notion of splitting field is geared to a particular polynomial. The purpose of this section is to show that the polynomial-splitting property can be seen as a property of a field extension itself, independent of any particular polynomial.

Before proceeding further, let us streamline our language concerning field automorphisms.

Definition Let K and L be fields containing a common subfield F. An isomorphism $K \to L$ of K onto L is called an F-*isomorphism* if and only if it fixes the subfield F element-wise. (Heretofore, we have been calling these F-*linear isomorphisms*.) Of course, if $K = L$, an F-isomorphism, $K \to L$, is called an F-*automorphism*. Finally, an F-isomorphism of K onto a subfield of L is called an F-*embedding*.

We say that a finite extension E of F is *normal over* F if and only if E is the splitting field over F of some polynomial of $F[x]$. Just as a reminder, recall that this means that there is a polynomial $p(x)$ which splits completely into linear factors in $E[x]$ and that E is generated by F and all the zeros of $p(x)$ that lie in E.

Note that if E is a normal extension and K is an intermediate field,—that is, $F \leq K \leq E$—then E is a normal extension of K.

We have a criterion for normality.

Theorem 11.3.7 (A characterization of normal field extensions[4]) *The following are equivalent for the finite extension* $F \subseteq E$:

 (i) *E is normal over F;*
 (ii) *every irreducible polynomial $g(x) \in F[x]$ having a zero in E must split completely into linear factors in $E[x]$.*

Proof Suppose E is normal over F, so that E is the splitting field of a polynomial $f(x) \in F[x]$. Let $g(x)$ be an irreducible polynomial in $F[x]$ with a zero a in E. Let $K \supseteq E$ be a splitting field over E for $g(x)$ and let b be an arbitrary zero of $g(x)$ in K. Since $g(x)$ is irreducible in $F[x]$, there is an isomorphism $\sigma : F(a) \to F(b)$ which is the identity mapping when restricted to F. Furthermore, E is clearly a splitting field for $f(x)$ over $F(a)$; likewise $E(b)$ is a splitting field for $f(x)$ over $F(b)$. Therefore, we may apply Theorem 11.3.4, to obtain an isomorphism $\tau : E \to E(b)$ extending $\sigma : F(a) \to F(b)$. In particular, this implies that $[E : F(a)] = [E(b) : F(b)]$.

---

[4]In many books, the characterizing property (ii) given in this theorem is taken to be the definition of "normal extension". This does not alter the fact that the equivalence of the two distinct notions must be proved.

Since $[F(a) : F] = [F(b) : F]$, it follows that $[E : F] = [E(b) : F]$, and so $b \in E$, forcing E to be a normal extension of F.

Next, suppose that the extension $F \subseteq E$ satisfies condition (ii). Since E is a finite extension of F, it has an F-basis, say $\{u_1, \ldots, u_n\}$. Set $g_i(x) = \mathrm{Irr}_F(u_i)$, and let $p(x)$ be the product of $g_1(x), g_2(x), \ldots, g_n(x)$. Then by hypothesis, every zero of $p(x)$ is in E and $p(x)$ splits into linear factors in $E[x]$. Since the $\{u_i\}$ are among these zeros and generate E over F, we see *a fortiori* that the zeros of $p(x)$ will generate E over F. That is, E is a splitting field for $p(x)$ over F. $\Box$

Corollary 11.3.8 *If E is a (finite) normal extension of F and K is any intermediate field, then any F-embedding*

$$\sigma : K \to E.$$

can be extended to an automorphism of E fixing F element-wise.

Proof E is a splitting field over F for a polynomial $f(x) \in F[x]$. Thus E is a splitting field for $f(x)$ over K as well as over $\sigma(K)$. The result then follows from Theorem 11.3.4. $\Box$

Suppose K/F is a finite extension. Then $K = F(a_1, \ldots, a_n)$ for some finite set of elements $\{a_i\}$ of K (for example, an F-basis of K). As in the proof of Theorem 11.3.7 we let $p(x)$ be the product of the polynomials $g_i(x) := \mathrm{Irr}_F(a_i)$, $i = 1, 2, \ldots, n$. Now any normal extension $E \supseteq F$ capturing K as an intermediate field must contain every zero of $g_i(x)$ and hence every zero of $p(x)$. Thus, between E and K there exists a splitting field L of $p(x)$ over F. The splitting field L is the "smallest" normal extension of F containing K in the sense that any other finite normal extension $E' \supseteq F$ which contains K also contains an isomorphic copy of L—that is, there is an embedding $L \to E'$ whose image contains K. Since this global description of the field L is independent of the polynomial $p(x)$, we have the following:

Corollary 11.3.9 (Normal Closure) *If K is a finite extension of F, then there is a normal extension L of F, unique up to F-isomorphism, containing K and having the property that for every normal extension $E \supseteq F$ containing K, there is an F-isomorphism of L onto a subfield of E which is normal over F and contains K.*

The extension L of F so defined is called the *normal closure of K over F*. The reader should bear in mind that this notion depends critically on F. For example it may happen that K is not normal over F, but is normal over some intermediate field N. Then the normal closure of K over N is just K itself while the normal closure of K over F could be larger than K.

Another consequence of Corollary 11.3.8 is this:

Corollary 11.3.10 (Normality of invariant subfields) *Let K be a normal extension of the field F of finite degree, and let G be the full group of F-automorphisms of K. If L is a subfield of K containing F, then L is normal over F if and only if it is G-invariant.*

Proof If L is normal over F, then L is the splitting field for some polynomial $f(x) \in F[x]$. But the zeroes of $f(x)$ in K are permuted among themselves by G, and so the subfield over F that they generate, namely L, is G-invariant.

On the other hand, assume L is G-invariant. Suppose $p(x)$ is an irreducible polynomial in $F[x]$ with at least one zero, α in L. Suppose β were another zero of $p(x)$ in K. Then there is an F-isomorphism $F[\alpha] \to F[\beta]$ of subfields of K, which, by Corollary 11.3.8, can be extended to an element of G. Since L is G invariant, $\beta \in L$. Thus $p(x)$ has all its zeros in L and so splits completely over L. By Theorem 11.3.7, L is normal over F. $\square$

11.4 Some Applications to Finite Fields

Suppose now that F is a finite field—that is, one which contains finitely many elements. Then, of course, its prime subfield P is the field $\mathbb{Z}/p\mathbb{Z}$, where p is a prime. Furthermore, F is a finite-dimensional vector space over P—say, of dimension n. This forces $|F| = p^n = q$, a prime power.

It now follows from Corollary 11.2.9 that the finite multiplicative group F^* of non-zero elements of F is a cyclic group of order $q - 1$. This means that F contains all the $q - 1$ zeroes of the polynomial $x^{q-1} - 1$ in $P[x]$. Since 0 is also a root of the equation $x^q - x = 0$, the following is immediate:

Lemma 11.4.1 *F is a splitting field of the polynomial $x^q - x \in P[x]$.*

It follows immediately, that F is uniquely determined up to P-isomorphism by the prime-power q alone.

Corollary 11.4.2 *For any given prime-power q, there is, up to isomorphism, exactly one field with q elements.*

One denotes any member of this isomorphism class by the symbol $\mathbf{GF}(q)$.

11.4.1 Automorphisms of Finite Fields

Suppose F is a finite field with exactly q elements, that is, $F \simeq \mathbf{GF}(q)$. We have seen that char $(F) = p$, a prime number, where, for some positive integer n, $q = p^n$. We know from Lemma 7.4.6 of Sect. 7.4.2, p. 221 that the "pth power mapping" $\sigma : F \to F$ is a ring endomorphism. Its kernel is trivial since F contains no non-zero nilpotent elements. Thus σ is injective, and so, by the pigeon-hole principle, is bijective—that is, it is an automorphism of F. Finally, since the multiplicative subgroup P^* of the prime subfield $P \cong \mathbb{Z}/p\mathbb{Z}$ has order $p-1$, it follows immediately that $a^{p-1} = 1$ for all $0 \neq a \in P$. This implies that $a^p = a$ for all $a \in P$, and hence the above p-power automorphism of F is a P-automorphism.[5]

Now suppose pth power mapping, σ, had order k. Then, $\alpha^{p^k} = \alpha$ for all elements α of F. But as there are at most p^k roots of $x^{p^k} - x = 0$, we see that $k \geq n$. On the other hand, we have already seen above that σ^n is the identity on F, forcing $k \leq n$. Thus $k = n$ and σ generates a cyclic subgroup of $\mathrm{Aut}(F)$ of order n.

We proceed now to show that, in fact, $\langle \sigma \rangle = \mathrm{Aut}(F)$. Let τ be an arbitrary automorphism of F, and fix a generator θ of the multiplicative group F^*. Then there exists an integer t, $0 < t < q-1$, such that $\tau(\theta) = \theta^t$. Then, as τ is an automorphism, $\tau(\theta^i) = (\theta^i)^t$. Since also $0^t = 0 = \tau(0)$, it follows that the automorphism τ is a power mapping $\alpha \to \alpha^t$ on F. Next, since τ must fix 1 and preserve addition, one has

$$\alpha^t + 1 = (\alpha + 1)^t = \alpha^t + t\alpha^{t-1} + \binom{t}{2} t^{k-2} + \cdots + 1^t.$$

Thus, if $t > 1$, all of the $q - 1$ elements α in C are roots of the equation

$$\sum_{k=1}^{t-1} \binom{t}{k} x^k = 0. \tag{11.4}$$

If the polynomial on the left were not identically zero, its degree $t - 1$ would be at least as large as the number of its distinct zeros which is at least $q - 1$. That is outside the range of t. Thus each of the binomial coefficients in the equation is zero. Thus either p divides t or $t = 1$.

Now if $t = p^k \cdot s$, where s is not divisible by p, we could apply the argument of the previous paragraph to $\rho := \tau^{p^k}$, to conclude that $s = 1$. Thus t is a prime power and so τ is a power σ.

We have shown the following:

Corollary 11.4.3 *If F is the finite field* $\mathbf{GF}(q)$, $q = p^n$, *then the full automorphism group of F is cyclic of order n and is generated by the pth power mapping.*

Remark The argument above, that $\mathrm{Aut}(F) = \langle \sigma \rangle$, could be achieved in one stroke by the "Dedekind Independence Lemma;" see Sect. 11.6.2. However, the above argument uses only the most elementary properties introduced so far.

We see at this point that $[F : P] = |\mathrm{Aut}(F)|$. In the language of a later section, we would say that F is a "Galois extension" of P. It will have very strong consequences for us. For one thing, it will eventually mean that no irreducible polynomial of positive degree in $\mathbf{GF}(q)[x]$ can have repeated zeros in some extension field.

---

[5]Actually, it's pretty easy to see directly that if E is any field and P is its prime subfield, then any automorphism of E is a P-automorphism.

11.4.2 Polynomial Equations Over Finite Fields: The Chevalley-Warning Theorem

The main result of this subsection is a side-issue in so far as it doesnt really play a role in the development of the Galois theory. However, it represents a property of finite fields which is too important not to mention before leaving a section devoted to these fields. The student wishing to race on to the theory of Galois extensions and solvability of equations by radicals may safely skip this section, hopefully for a later revisitation.

For this subsection, fix a finite field $K = \mathrm{GF}(q)$ of characteristic p and cardinality $q = p^{\ell}$. Now for any natural number k, let

$$S(k, q) := \sum_{a \in K} a^k.$$

the sum of the kth powers of the elements of the finite field K of q elements. Clearly, if $k = 0$, the sum is $S(0, q) = q = 0$, since the field has characteristic p dividing q. If k is a multiple of $q - 1$ distinct from 0, then as K^* is a cyclic group of order $q - 1$,

$$S(k, q) = \sum_{K - \{0\}} 1 = q - 1 = -1.$$

Now suppose k is not a multiple of $q - 1$. Then there exists an element b in K^* with $b^k \neq 1$. Since multiplying on the left by b simply permutes the elements of K, we see that

$$S(k, q) = \sum_{a \in K} a^k = \sum_{a \in K} (ba)^k = b^k S(k, q).$$

So $(b^k - 1)S(k, q) = 0$. Since the first factor is not zero, we have $S(k, q) = 0$ in this case.

We summarize this in the following

Lemma 11.4.4
$$S(k, q) = \begin{cases} 0 & \text{if } k = 0 \\ -1 & \text{if } q - 1 \text{ divides } k \\ 0 & \text{otherwise} \end{cases}$$

Now fix a positive integer n. Let $V = K^{(n)}$, the vector space of n-tuples with entries in K. For any vector $v = (a_1, \dots, a_n) \in V$ and polynomial

$$p = p(x_1, \dots x_n) \in K[x_1, \dots, x_n],$$

the ring of polynomials in the indeterminates $x_1, \dots x_n$, we let the symbol $p(v)$ denote the result of substituting a_i for x_i in the polynomial, that is,

$$p(v) := p(a_1, \dots, a_n).$$

If $p(v) = 0$ for a vector $v \in V$, we say that v is a *zero* of the polynomial p.

Now suppose $p = x_1^{k_1} x_2^{k_2} \cdots x_n^{k_n}$ is a monomial of degree $\sum k_i < n(q \dot- 1)$. Then

$$\sum_{v \in V} p(v) = \prod_{i=1}^{n} S(k_i, q) = 0,$$

since at least one of the exponents k_i is less than $q - 1$ and so by Lemma 11.4.4 introduces a factor of zero in the product. Since every polynomial p in $K[x_1, \ldots, x_n]$ is a sum of such monomials of degree less than $n(q - 1)$, we have

Lemma 11.4.5 *If $p \in K[x_1, \ldots, x_n]$ has degree less than $n(q - 1)$, then*

$$S(p) := \sum_{v \in V} p(v) = 0.$$

Now we can prove the following:

Theorem 11.4.6 (Chevalley-Warning) *Let K be a finite field of characteristic p with exactly q elements. Suppose $p_1, \ldots, p_m$ is a family of polynomials of $K[x_1, \ldots, x_n]$, the sum of whose degrees is less than n, the number of variables. Let X be the collection*

$$\{v \in K^{(n)} | p_i(v) = 0, i = 1, \ldots, m\}$$

of common zeroes of these polynomials. Then

$$|X| \equiv 0 \bmod p.$$

Proof Let $R := \prod_{i=1}^{m}(1 - p_i^{q-1})$. Then R is a polynomial whose degree $(q - 1)$ $\sum \deg p_i$ is less than $n(q - 1)$. Now if $v \in X$, then v is a common zero of the polynomials p_i, so $R(v) = 1$. But if $v \notin X$, then for some i, $p_i(v) \neq 0$, so $p_i(v)^{q-1} = 1$, introducing a zero factor in the definition of $R(v)$. Thus we see that the polynomial R induces the characteristic function of X—that is, it has value 1 on elements of X and value 0 outside of X. It follows that

$$S(R) := \sum_{v \in V} R(v) \equiv |X| \bmod p. \tag{11.5}$$

But since $\deg R < n(q - 1)$, Lemma 11.4.5 forces $S(R) = 0$, which converts (11.5) into the conclusion. $\square$

Corollary 11.4.7 *Suppose $p_1, \ldots p_m$ is a collection of polynomials over $K = GF(q)$ in the indeterminates $x_1, \ldots, x_n$, each with a zero constant term. Then there exists a common non-trivial zero for these polynomials. (Non-trivial means one that is not the zero vector of V.)*

In particular, if p is a homogeneous polynomial over K in more indeterminates than its degree, then it must have a non-trivial zero.

11.5 Separable Elements and Field Extensions

11.5.1 Separability

A polynomial $p(x) \in F[x]$ is said to be *separable* if and only if its irreducible factors in $F[x]$ each have no repeated zeros in any splitting field.

Our discussion on separability of polynomials will be greatly facilitated by the following concept. Let F be a field and define the *formal derivative*

$$\partial : F[x] \to F[x], \quad f(x) \mapsto f'(x)$$

by setting

$$f'(x) = \sum_{k=0}^{n} k a_k x^{k-1}, \quad \text{whenever } f(x) = \sum_{k=0}^{n} a_k x^k.$$

The formal derivative satisfies the familiar "product rule:" $\partial(f(x)g(x)) = (\partial f(x))g(x) + f(x)\partial g(x)$, and hence, its generalization, the "Leibniz rule":

$$\partial(f_1(x)f_2(x) \cdots f_r(x)) = \sum_{i=1}^{r} f_1(x) \cdots f_{i-1}(x)(\partial f_i(x))f_{i+1}(x) \cdots f_r(x).$$

Furthermore, the formal derivative is "independent of its domain" inasmuch as if $F \subseteq E$ is an extension of fields, then the following diagram commutes:

$$
\begin{array}{ccc}
E[x] & \xrightarrow{\ \partial\ } & E[x] \\
\uparrow & & \uparrow \\
F[x] & \xrightarrow{\ \partial\ } & F[x]
\end{array}
$$

where the vertical arrows are obvious inclusions.

The following simple result will be useful in the sequel.

Lemma 11.5.1 *Let $f(x), g(x) \in F[x]$, and let $F \subseteq E$ be a field extension. Then $f(x)$ and $g(x)$ are relatively prime in $F[x]$ if and only if they are relatively prime in $E[x]$.*

Proof If $f(x), g(x)$ are relatively prime in $F[x]$, then there exist polynomials $s(x), t(x) \in F[x]$ with $s(x)f(x) + t(x)g(x) = 1$. Since this equation is obviously valid in $E[x]$ we infer that $f(x), g(x)$ are relatively prime in $E[x]$, as well.

If $f(x)$ and and $g(x)$ are not relatively prime in $F[x]$, their greatest common divisor in $F[x]$ has positive degree, and so this is also true in $E[x]$. Thus not being relatively prime in $F[x]$ implies not being relatively prime in $E[x]$. $\square$

There is a simple way to tell whether a polynomial $f(x)$ in $F[x]$ is separable.

Lemma 11.5.2 *The polynomial $f(x)$ has no repeated zeros in its splitting field if and only if $f(x)$ and $f'(x)$ are relatively prime.*

Proof Let $E \supseteq F$ be a splitting field for $f(x)$ over F, so that $f(x)$ splits into distinct linear factors in $E[x]$:

$$f(x) = (x - \alpha_1)(x - \alpha_2) \cdots (x - \alpha_r) \in E[x],$$

where $\alpha_1, \alpha_2, \ldots, \alpha_r \in E$.

First we assume that $f(x)$ has no repeated zeroes so that $\alpha_i \neq \alpha_j$ whenever $i \neq j$. Using the above-mentioned Leibniz rule, we have

$$f'(x) = \sum_{i=1}^{r} (x - \alpha_1) \cdots (x - \alpha_{i-1}) \partial (x - \alpha_i)(x - \alpha_{i+1}) \cdots (x - \alpha_r)$$

$$= \sum_{i=1}^{r} (x - \alpha_1) \cdots (x - \alpha_{i-1})\widehat{(x - \alpha_i)}(x - \alpha_{i+1}) \cdots (x - \alpha_r),$$

where the notation $\widehat{(x - \alpha_i)}$ simply means that the indicated factor has been removed from the product. From the above, it is obvious that $f'(\alpha_i) \neq 0$, $i = 1, 2, \ldots, r$,— that is to say, $f(x)$ and $f'(x)$ share no common factors in $E[x]$. From Lemma 11.5.1, it follows that $f(x)$ and $f'(x)$ are relatively prime in $F[x]$, as required.

Conversely, assume that $f(x)$ and $f'(x)$ are relatively prime in $F[x]$. Write

$$f(x) = (x - \alpha_1)^{e_1} (x - \alpha_2)^{e_2} \cdots (x - \alpha_r)^{e_r},$$

for positive integral exponents $e_1, e_2, \ldots, e_r$. Again applying the Leibniz rule, we obtain

$$f'(x) = \sum_{i=1}^{r} (x - \alpha_1)^{e_1} \cdots (x - \alpha_{i-1})^{e_{i-1}} e_i (x - \alpha_i)^{e_i - 1} (x - \alpha_{i+1})^{e_{i+1}} \cdots (x - \alpha_r)^{e_r}.$$

If some exponent e_j is greater than 1, then the above shows clearly that $f'(\alpha_j) = 0$, i.e., $f(x)$ and $f'(x)$ share a common zero, and hence cannot be relatively prime in $E[x]$. In view of Lemma 11.5.1, this is a contradiction. $\square$

The preceding lemma has a very nice application when $f(x)$ is irreducible.

Lemma 11.5.3 *If $f(x) \in F[x]$ is irreducible, then $f(x)$ has no repeated zeros in its splitting field if and only if $f'(x) \neq 0$.*

As an immediate corollary to the above, we see that if F is a field of characteristic 0, then $F[x]$ contains no irreducible inseparable polynomials. On the other hand if F has positive characteristic p, then by Lemma 11.5.2, an irreducible polynomial $f(x) \in F[x]$ has a repeated root only when $f'(x) = 0$, forcing $f(x) = g(x^p)$, for some polynomial $g(x) \in F[x]$. In fact, a moment's thought reveals that, in fact, if $f(x) \in F[x]$ is irreducible and inseparable, then we may write $f(x) = g(x^{p^e})$, where e is a positive integral exponent and where $g(x) \in F[x]$ is an irreducible *separable* polynomial.

Let $F \subseteq E$ be an extension and let $\alpha \in E$. We say that α is *separable* over F if it is algebraic over F and if $\mathrm{Irr}_F(\alpha)$ is a separable polynomial. The extension E of F is said to be *separable* if and only if every element of E which is algebraic over F is separable. We may already infer the following:

Lemma 11.5.4 *The algebraic extension $F \subseteq E$ is separable whenever F has characteristic 0 or is a finite field.*

Proof If F has characteristic 0, the result is obvious by the above remarks. If F is a finite field, the p-power mapping $\sigma : F \to F$ is an automorphism, and hence is surjective. Now let $f(x) \in F[x]$ be irreducible and assume that $f(x)$ is inseparable. Write $f(x) = g(x^p)$, where $g(x) = \sum_{i=0}^{m} a_i x^i$. For each $i = 0, 1, \ldots, m$, let $b_i \in F$ satisfy $b_i^p = a_i$; thus,

$$f(x) = g(x^p) = \sum_{i=0}^{m} b_i^p x^{ip} = \left(\sum_{i=0}^{m} b_i x^i \right)^p,$$

contrary to the irreducibility of $f(x)$. Thus, any finite extension of a finite field is also separable. $\square$

We see, therefore, that if $F \subseteq E$ is an inseparable algebraic extension, then F must be an infinite field of positive characteristic p. We shall take up this situation in the section to follow. Before doing this, it shall be helpful to consider two rather typical examples.

Example 52 Let $P = \mathbf{G}F(2)$ be the field of two elements, let x be an indeterminate over P, and set $F = P(x)$, the field of "rational functions" over P. Obviously, F is an infinite field of characteristic 2. Note also that F is the field of fractions of the PID $P[x]$. Now set $f(y) = y^2 + x \in F[y]$; note that since x is prime in $P[x]$, we may apply Eisenstein's criterion (see Exercise (8) in Sect. 9.13.1, p. 318) to infer that $f(y)$ is irreducible in $F[y]$. Thus, if α is a zero of $f(y)$ in a splitting field over F for $f(y)$, then $[F(\alpha) : F] = 2$. Furthermore, since $f'(y) = 2y + 0 = 0 \in F[x]$, we see that α is inseparable over F. Therefore $F(\alpha) \supseteq F$ is an inseparable extension (that is to say, not a separable extension). However, we can argue that every element of $F(\alpha) \backslash F$ is inseparable over F, as follows. Since $\{1, \alpha\}$ is an F-basis of $F(\alpha)$, we see that every element of $F(\alpha) \backslash F$ can be written in the form $\beta = a + b\alpha$, $a, b \in F$, $b \neq 0$. We set

$$g(y) = (y - (a + b\alpha))^2 = y^2 + (a + b\alpha)^2 = y^2 + a^2 + b^2 x \in F[y];$$

since $a+b\alpha$ is a zero of $g(y)$, and since $a+b\alpha \notin F$, we see that $g(y) = \text{Irr}_F(a+b\alpha)$. But as $g'(y) = 0$ we see that β is an inseparable element over F. Thus proves that *every element of* $F(\alpha) \setminus F$ *is* inseparable over F; we shall come to call such extensions *purely inseparable*.

Example 53 As a hybrid example, we take $P = \mathbf{G}F(3)$, the 3-element field, and define $F = P(x)$. Set $f(y) = y^6 + x^2y^3 + x \in F[y]$. Again, an application of Eisenstein reveals that $f(y)$ is irreducible in $F[y]$. Since $f'(y) = 0$ we infer that $f(y)$ is inseparable over F. We take α to be a zero (in a splitting field) of $f(y)$, from which we infer that $[F(\alpha) : F] = 6$. Furthermore, we know that this is not a separable extension since α is not separable over F. We note that $f(y) = g(y^3)$, where $g(y)$ is the irreducible separable polynomial $g(y) = y^2 + x^2y + x \in F[y]$. This says that α^3, being a root of $g(y)$, is separable over F (note that $\alpha^3 \notin F$). Therefore, we see that $F(\alpha) \setminus F$ contains both separable and inseparable elements over F. In fact, we have a tower $F \subseteq F(\alpha^3) \subseteq F(\alpha)$, where $[F(\alpha^3) : F] = 2$ and so $[F(\alpha) : F(\alpha^3)] = 3$. Our work in the next section will show that, in fact, $F(\alpha^3)$ contains all of the separable elements of $F(\alpha)$ over F, and that the extension $F(\alpha^3) \subseteq F(\alpha)$ is purely inseparable.

11.5.2 Separable and Inseparable Extensions

The primary objective of this subsection is to show that a finite-degree extension $F \subseteq E$ can be factored as $F \subseteq E_{\text{sep}} \subseteq E$, where E_{sep} consists precisely of those elements of E separable over F, and where $E_{\text{sep}} \subseteq E$ is a *purely inseparable* extension (i.e., no elements of $E - E_{\text{sep}}$ are separable).

Suppose that F is a field of positive characteristic p, and that $K \supseteq F$ is an extension of finite degree. We have the pth power mapping $K \to K$, $a \mapsto a^p$, $a \in K$. Note that if F is not finite, K is not finite, and we cannot infer that this mapping is an automorphism of K—we can only infer that it gives an embedding of K into itself (as explained in Lemma 7.4.6 of Sect. 7.4.2, p. 221). We denote the image by K^p, and denote by FK^p the subfield of K generated by the subfields F and K^p.

We shall have need of the following technical result:

Theorem 11.5.5 *Let F be a field of positive characteristic p, and let $K \supseteq F$ be an extension of finite degree. If $K = FK^p$, then the pth power mapping on K preserves F-linear independence of subsets.*

Proof Since any F-linearly independent subset of K can be completed to an F-basis for K, it suffices to prove that the pth power mapping preserves the linear independence of any F-basis for K. Suppose, then, that $\{a_1, \ldots, a_n\}$ is a F-basis of K, and that c is an element of K. Then c can be written as an F-linear combination of the basis elements:

$$c = \alpha_1 a_1 + \cdots + \alpha_n a_n, \text{ all } \alpha_i \in F,$$

from which we conclude that

$$c^p = \alpha_1^p a_1^p + \cdots + \alpha_n^p a_n^p.$$

Therefore,

$$K^p = F^p a_1^p + \cdots + F^p a_n^p,$$

and so, by hypothesis,

$$K = FK^p = Fa_1^p + \cdots + Fa_n^p.$$

Thus the pth powers of the basis elements a_i form a F-spanning set of K of size n. Since $n = \dim_F K$, these n spanning elements must be F-linearly independent. $\square$

Theorem 11.5.6 (Separability criterion for fields of prime characteristic) *Let K be any algebraic field extension of the field F, where F (and hence K) have prime characteristic p.*

(i) *If K is a separable extension of F, then $K = FK^p$. (Note that the degree $[K : F]$ need not be finite here.)*

(ii) *If K is a finite extension of F such that $K = FK^p$, then K is separable over F.*

Proof Assume, as in Part (i), that $K \supseteq F$ is a separable extension. Then it is clear that every element of K is also separable over the intermediate subfield $L := FK^p$. If $a \in K$, then $b = a^p \in L$ and so a is a zero of the polynomial $x^p - b \in L[x]$. Thus, if $p(x) = \mathrm{Irr}_L(a) \in L[x]$, we have that $p(x)$ divides $x^p - b$ in $L[x]$. However, in $K[x]$, $x^p - b = x^p - a^p = (x - a)^p$ and so $p(x)$ cannot be separable unless it has degree 1. This forces $p(x) = x - a \in L[x]$, i.e., that $a \in L$, proving that $K = FK^p$.

For Part (ii), assume that $[K : F]$ is finite, and that $K = FK^p$. Suppose, by way of contradiction that a is an element of K which is not separable over F. Then if $f(x) := \mathrm{Irr}_F(a)$, we have that $f(x) = g(x^p)$, where $g(x) \in F[x]$ is irreducible. Write $g(x) = \sum_{j=0}^{m} a_j x^j$ and conclude that

$$0 = f(a) = g(a^p) = a_0 + a_1 a^p + \cdots + a_m a^{pm},$$

which says that $\{1, a^p, a^{2p}, \ldots, a^{mp}\}$ is an F-linearly dependent subset of K On the other hand, if $\{1, a, a^2, \ldots, a^m\}$ were F-linearly dependent, then there would exist a polynomial $p(x)$ in $F[x]$ of degree at most m having element a for a zero, contrary to the fact that $f(x) = \mathrm{Irr}_F(a)$ has degree $pm > m$, which is a contradiction. Therefore, the pth power mapping has taken the F-linearly independent set $\{1, a, a^2, \ldots, a^m\}$ to the F-linearly *dependent* set $\{1, a^p, a^{2p}, \ldots, a^{mp}\}$, a violation of Theorem 11.5.5. Therefore, $a \in K$ must have been separable over F in the first place. $\square$

Corollary 11.5.7 (Separability of simple extensions) *Fix a field F of prime characteristic p, let $F \subseteq K$ be a field extension, and let $a \in K$ be algebraic over F. The following are equivalent:*

(i) $F(a) \supseteq F$ is a separable extension;
(ii) a is separable over F;
(iii) $F(a) = F(a^p)$.

Proof (i)$\Rightarrow$(ii) is, of course, obvious.

Assume (ii). Since a is a zero of the polynomial $x^p - a^p \in F(a^p)[x]$, we see that $\mathrm{Irr}_{F(a^p)}(a)$ divides $x^p - a^p$. But $x^p - a^p = (x - a)^p \in F(a)[x]$. Since a is separable over F, it is separable over $F(a^p)$ and so it follows that $a \in F(a^p)$, forcing $F(a) = F(a^p)$, which proves (ii)$\Rightarrow$(iii).

Finally, if $F(a) = F(a^p)$, then $F \cdot F(a)^p = F \cdot F^p(a^p) = F(a^p) = F(a)$, by hypothesis. Apply Theorem 11.5.6 part (ii) to infer that $F(a)$ is separable over F, which proves that (iii)$\Rightarrow$(i). $\square$

Corollary 11.5.8 (Transitivity of separability among finite extensions) *Suppose K is a finite separable extension of L and that L is a finite separable extension of F. Then K is a finite separable extension of F.*

Proof That $[K : F]$ is finite is known by Lemma 11.2.2, so we only need to prove the separability of K over F. We may assume that F has prime characteristic p, otherwise K is separable by Lemma 11.5.4. By Part (ii) of Theorem 11.5.6, it suffices to prove that $K = FK^p$. But, since K is separable over L, and since L is separable over F, we have

$$K = LK^p = (FL^p)K^p = F(LK)^p = FK^p.$$

That K is separable over F now follows from Theorem 11.5.6, part (ii). $\square$

Corollary 11.5.9 *Let K be an arbitrary extension of F. Then*

$$K_{\mathrm{sep}} = \{a \in K \,|\, a \text{ is separable over } F\}$$

is a subfield of K containing F.

Proof It clearly suffices to show that K_{sep} is a subfield of K. However, if $a, b \in K_{\mathrm{sep}}$, then by Corollary 11.5.7 we have that $F(a)$ is separable over F and that $F(a, b)$ is separable over $F(a)$. Apply Corollary 11.5.8. $\square$

The field K_{sep} is called *the separable closure of F in K*.

Finally, recall that we have called an algebraic extension $F \subseteq K$ *purely inseparable* if every element of $K \setminus F$ is inseparable over F. Therefore, we see already that if $F \subseteq K$ is an algebraic extension, then K_{sep} is a separable extension of F, and, by Corollary 11.5.8, K is a purely inseparable extension of K_{sep}.

We conclude this section with a characterization of purely inseparable extensions.

Theorem 11.5.10 *Let $F \subseteq K$ be an algebraic extension. The following propositions are equivalent:*

(1) K is a purely inseparable extension of F;
(2)   For all $a \in K \backslash F$, $a^{p^e} \in F$, for some positive exponent e of the positive charac-
      teristic p;
(3)   For all $a \in K \backslash F$, $\mathrm{Irr}_F(a) = x^{p^e} - b$, for some $b \in F$ and some positive
      exponent e.

As a corollary to Theorem 11.5.10 we extract the following useful corollary for
simple extensions.

Corollary 11.5.11 *Let F be a field of characteristic p, contained in some field K.
Assume that $a \in K$ satisfies $a^{p^e} \in F$. Then the subfield $F(a) \subseteq K$ is a purely
inseparable extension of F.*

Proof If $[F(a) : F] = r$, then any element $b \in F(a)$ can be expressed as a polynomial in a: $b = \sum_{j=0}^{r-1} a_j a^j$, where the coefficients $a_0, a_1, \ldots, a_{r-1} \in F$. But then

$$b^{p^e} = \left(\sum_{j=0}^{r-1} a_j a^j \right)^{p^e} = \sum_{j=0}^{r-1} a_j^{p^e} a^{jp^e} \in F.$$

Now apply Theorem 11.5.10. $\square$

11.6   Galois Theory

11.6.1   Galois Field Extensions

Let K be any field, and let k be a subfield of K. The k-isomorphisms of K with itself
are called *k-automorphisms* of K. Under composition they form a group which we
denote by $\mathrm{Gal}(K/k)$. The group of all automorphisms of K is denoted $\mathrm{Aut}(K)$, as
usual. It is obvious that $\mathrm{Gal}(K/k)$ is a subgroup of $\mathrm{Aut}(K)$.

Suppose now that G is any group of automorphisms of the field K. The elements
fixed by every automorphism of G form a subfield called the *fixed subfield of G*
(and sometimes the *field of invariants of G*), accordingly denoted $\mathrm{inv}_G(K)$. If $G \leq$
$\mathrm{Gal}(G/k)$ then clearly k is a subfield of $\mathrm{inv}_G(K)$.

Now let Ω_G be the poset of all subgroups of G, and let $\Omega_{K/k}$ be the poset of
all subfields of K which contain k; in both cases we take the partial order to be
containment. To each subfield L with $k \leq L \leq K$, there corresponds a subroup
$\mathrm{Gal}(K/L)$. Similarly, for each subgroup H of $\mathrm{Gal}(K/k)$, there corresponds the
subfield $\mathrm{inv}_H(K)$ containing k. These correspondences are realized as two mappings:

$$\mathrm{Gal}(K/\bullet) : \Omega_{K/k} \longrightarrow \Omega_G, \ \ \mathrm{inv}_\bullet(K) : \Omega_G \longrightarrow \Omega_{K/k},$$

which are obviously inclusion reversing. Furthermore, for all $E \in \Omega_{K/k}$ and for all $H \in \Omega_G$, one has that

$$\mathrm{inv}_\bullet(K) \circ \mathrm{Gal}(K/\bullet)(E) = \mathrm{inv}_{\mathrm{Gal}(K/E)}(K) \supseteq E, \quad \text{and}$$

$$\mathrm{Gal}(K/\bullet) \circ \mathrm{inv}_\bullet(E)(H) = \mathrm{Gal}(E/\mathrm{inv}_H(E)) \geq H.$$

In plain language, the composition of the two poset morphisms in either order, is monotone on its defined poset. Therefore, we see that the quadruple $(\Omega_{K/k}, \Omega_G, \mathrm{Gal}(K/\bullet), \mathrm{inv}_\bullet(K))$ is a Galois connection in the sense of Sect. 2.2.15.

Next, if we assume, as we typically shall, that $[K : k] < \infty$, then every interval in $\Omega_{K/k}$ is algebraic in the sense of Sect. 2.3.2.[6] In fact, the mapping which takes the algebraic interval $[E_1, E_2]$ of $\Omega_{K/k}$ to the index $[E_2 : E_1]$ is a $\mathbb{Z}$-*valued interval measure* in the sense of Sect. 2.5.2. We shall show presently that when $[K : k] \leq \infty$, $|\mathrm{Gal}(K/k)| \leq \infty$, and so similar comments apply to the poset Ω_G, where, if $H_2 \leq H_1$ are subgroups of $G = \mathrm{Gal}(K/k)$, then the mapping which takes the interval $[H_2, H_1]$ to the index $[H_1 : H_2]$ is the appropriate $\mathbb{Z}$-valued interval measure.

11.6.2 The Dedekind Independence Lemma

We begin with the following observation. Let K be a field; let S be a set; and let K^S be the set of mappings $S \to K$. We may give K^S a K-vector space structure by pointwise operations. Thus, if $f_1, f_2 \in K^S$, and if $\alpha \in K$, then we set $\alpha(f_1 + f_2)(s) := \alpha f_1(s) + \alpha f_2(s)$.

In terms of the above, we state the following important result.

Lemma 11.6.1 (Dedekind Independence Lemma)

1. *Let E, K be fields, and let $\sigma_1, \sigma_2, \ldots, \sigma_r$ be distinct monomorphisms $E \to K$. Then $\sigma_1, \sigma_2, \ldots, \sigma_r$ are K-linearly independent in K^E, the K-vector space of all functions from E to K.*
2. *Let E be a field, and let G be a group of automorphisms of E. We may regard each $\alpha \in E$ as an element of E^G via $\sigma \mapsto \sigma(\alpha) \in E$, $\sigma \in G$. Now set $K = \mathrm{inv}_G(E)$ and assume that we are given K-linearly independent elements $\alpha_1, \alpha_2, \ldots, \alpha_r \in E$. Then $\alpha_1, \alpha_2, \ldots, \alpha_r$ are E-linearly independent elements of E^G.*

Proof For Part 1, suppose, by way of contradiction, that there exists a nontrivial linear dependence relation of the form $a_1\sigma_1 + \cdots + a_r\sigma_r = 0 \in K^E$, $a_1, \ldots, a_r \in K$. Among all such relations we may assume that we have chosen one in which the number of summands r is as small as possible.

We have, for all $\alpha \in E$, that

$$a_1\sigma_1(\alpha) + a_2\sigma_2(\alpha) + \cdots + a_r\sigma_r(\alpha) = 0, \tag{11.6}$$

---

[6]Recall that a poset is algebraic if its "zero" and "one" are connected by a finite unrefinable chain.

where each coefficient a_i is non-zero by the minimality of r. Since $\sigma_1 \neq \sigma_2$, we may choose $\alpha' \in E$ such that $\sigma_1(\alpha') \neq \sigma_2(\alpha')$. If we replace the argument α in Eq. (11.6) by $\alpha'\alpha$ and use the fact that each σ_i, $i = 1, 2, \ldots, r$ is a homomorphism, we obtain the following:

$$a_1\sigma_1(\alpha')\sigma_1(\alpha) + a_2\sigma_2(\alpha')\sigma_2(\alpha) + \cdots + a_r\sigma_r(\alpha')\sigma_r(\alpha) = 0. \tag{11.7}$$

Next, multiply both sides of Eq. (11.6) by the scalar $\sigma_1(\alpha')$:

$$a_1\sigma_1(\alpha')\sigma_1(\alpha) + a_2\sigma_1(\alpha')\sigma_2(\alpha) + \cdots + a_r\sigma_1(\alpha')\sigma_r(\alpha) = 0. \tag{11.8}$$

Subtracting Eq. (11.8) from Eq. (11.7) yields

$$a_2(\sigma_2(\alpha') - \sigma_1(\alpha'))\sigma_2(\alpha) + \cdots + a_r(\sigma_r(\alpha') - \sigma_1(\alpha'))\sigma_r(\alpha) = 0.$$

Since $a_2(\sigma_2(\alpha') - \sigma_1(\alpha')) \neq 0$, and α was arbitrary, we have produced a non-trivial dependence relation among the σ_i, $i \neq 1$, against the minimal choice of r. Thus no such dependence relation among the maps $\{\sigma_i\}$ exists and Part 1 is proved.

For part 2, we again argue by considering an E-linear dependence relation among the functions $\alpha_i : G \to E$ with a minimal number of terms r. Thus one obtains a relation

$$a_1\sigma(\alpha_1) + a_2\sigma(\alpha_2) + \cdots + a_r\sigma(\alpha_r) = 0, \tag{11.9}$$

valid for all $\sigma \in G$, and where we may assume that the elements $a_1, a_2, \ldots, a_r \in E$ are all nonzero. We may as well assume that $a_1 = 1$. Then setting $\sigma = 1 \in G$ in Eq. (11.9) yields

$$\alpha_1 + a_2\alpha_2 + \cdots a_r\alpha_r = 0.$$

Since $\alpha_1, \alpha_2, \ldots, \alpha_r$ are linearly independent over K, the preceding equation implies that at least one of the elements $a_2, \ldots, a_r$ is not in K. Re-indexing the a_i if necessary we may assume that $a_2 \notin K$. Therefore, there exists an element $\sigma' \in G$ such that $\sigma'(a_2) \neq a_2$. Equation (11.9) with σ replaced by $\sigma'\sigma$ then reads as:

$$\sigma'\sigma(\alpha_1) + a_2\sigma'\sigma(\alpha_2) + \cdots + a_r\sigma'\sigma(\alpha_r) = 0, \tag{11.10}$$

still valid for all $\sigma \in G$. Applying σ' to both sides of Eq. (11.9) yields

$$\sigma'\sigma(\alpha_1) + \sigma'(a_2)\sigma'\sigma(\alpha_2) + \cdots + \sigma'(a_r)\sigma'\sigma(\alpha_r) = 0, \tag{11.11}$$

and subtracting Eq. (11.11) from Eq. (11.10) yields

$$(a_2 - \sigma'(a_2))\sigma'\sigma(\alpha_2) + \cdots + (a_r - \sigma'(a_r))\sigma'\sigma(a_r) = 0.$$

Since this is true for all $\sigma \in G$, and since $a_2 - \sigma'(a_2) \neq 0$, we have produced a dependence relation on a smaller set of functions $\{\sigma \mapsto \sigma(\alpha_i) | i > 1\}$, which contradicts the minimality of r. Thus no dependence relation as described can exist and Part 2 must hold. $\square$

From the Dedekind Independence Lemma, we extract the following lemma, which summarizes the relationships between field extension degrees and group indices.

Theorem 11.6.2 *Let $k \subseteq K$ be a field extension, with Galois group $G = \mathrm{Gal}(K/k)$.*

1. *Assume that $k \subseteq E_1 \subseteq E_2 \subseteq K$ is a tower of fields, and set $H_i = \mathrm{Gal}(K/E_i)$, $i = 1, 2$. If $[E_2 : E_1] < \infty$, then $[H_1 : H_2] \leq [E_2 : E_1]$.*
2. *Assume that $H_2 \leq H_1 \leq G$ are subgroups, and set $E_i = \mathrm{inv}_{H_i}(K)$, $i = 1, 2$. If $[H_1 : H_2] < \infty$, then $[E_2 : E_1] \leq [H_1 : H_2]$.*

Proof For part 1, we shall assume, by way of contradiction, that $[H_1 : H_2] > [E_2 : E_1]$. Set $r = [E_2 : E_1]$, and let $\{\alpha_1, \ldots, \alpha_r\}$ be an E_1-basis of E_2. Assume that $\{\sigma_1, \sigma_2, \ldots, \sigma_s\}$, $s > r$, is a set of distinct left H_2-coset representatives in H_1. Since $s > r$, we may find elements $a_1, a_2, \ldots, a_s \in K$, not all zero, such that

$$\sum_{i=1}^{s} a_i \sigma_i(\alpha_j) = 0, \ j = 1, 2, \ldots, r.$$

Since any element of E_2 can be written as an E_1-linear combination of $\alpha_1, \ldots, \alpha_r$, we conclude that $\sum_{i=1}^{s} a_i \sigma_i : E_2 \to K$ is the 0-mapping. Since $\sigma_1, \ldots, \sigma_s$ are distinct coset representatives, they are distinct mappings $E_2 \to K$. This contradicts part 1 of the Dedekind Independence Lemma.

For part 2 of the Theorem, we shall assume, by way of contradiction, that $[E_2 : E_1] > [H_1 : H_2]$. Let $\{\sigma_1, \ldots, \sigma_r\}$ be a complete set of H_2-coset representatives in H_1, and assume that $\{\alpha_1, \ldots, \alpha_s\}$ is an E_1-linearly independent subset of E_2, where, by assumption, $s > r$. Again, we may find elements $a_1, a_2, \ldots, a_s \in K$, not all zero, such that

$$\sum_{i=1}^{s} a_i \sigma_j(\alpha_i) = 0, \ j = 1, 2, \ldots, r.$$

If $\sigma \in H_1$, then $\sigma = \sigma_j \tau$, for some index j, $1 \leq j \leq r$ and for some $\tau \in H_2$. Therefore, as E_2 is fixed point-wise by H_2, we have

$$\sum_{i=1}^{s} a_i \sigma(\alpha_i) = \sum_{i=1}^{s} a_i \sigma_j \tau(\alpha_i) = \sum_{i=1}^{s} a_i \sigma_j(\alpha_i) = 0.$$

Since not all a_i are zero in the first term of the equation just presented, the mappings $\sigma \mapsto \sigma(\alpha_i) \in K$ are not K-linearly independent, against part (ii) of the Dedekind Independence Lemma. $\square$

Corollary 11.6.3 *Let $k \subseteq K$ be a finite degree extension, and set $G = \mathrm{Gal}(K/k)$. If $k_0 = \mathrm{inv}_G(K)$, then*

$$|G| = [K : k_0].$$

Proof Note first that $G = \mathrm{Gal}(K/k_0)$. We have

$$[K : k_0] \geq |G| \geq [K : k_0],$$

where the first inequality is Theorem 11.6.2, part (1), and the second inequality is Theorem 11.6.2, part (2). The result follows. $\square$

11.6.3 Galois Extensions and the Fundamental Theorem of Galois Theory

One may recall from Sect. 2.2.15 that with any Galois connection between two posets, there is a closure operator for each poset. That notion of closure holds for the two posets that we have been considering here: the poset of subgroups of $\mathrm{Gal}(K/k)$ and the poset of subfields of K that contain k. Accordingly, we say that a subfield k of K is *Galois closed in K* if and only if

$$k = \mathrm{inv}_{\mathrm{Gal}(K/k)}(K).$$

We define an algebraic extension $k \subseteq K$ to be a *Galois extension* if k is Galois closed in K. Note that from the property that K/k is a Galois extension, we can infer immediately that every subfield $E \in \Omega_{K/k}$ of finite degree over k is also Galois closed in K. Indeed, we can set $G = \mathrm{Gal}(K/k)$, $\overline{E} = \mathrm{inv}_{\mathrm{Gal}(K/E)}(K)$ (the Galois closure of E in K), and use Theorem 11.6.2 to infer that

$$[E : k] \geq [G : \mathrm{Gal}(K/E)] \geq [\overline{E} : k].$$

Since we already have $E \subseteq \overline{E}$, the result that $E = \overline{E}$ follows immediately.

We characterize the Galois extensions as follows.

Theorem 11.6.4 *Let $k \subseteq K$ be a finite extension. The following are equivalent:*

(i) $k \subseteq K$ is a Galois extension;
(ii) $k \subseteq K$ is a separable, normal extension.

Proof Assume that $k \subseteq K$ is a Galois extension. Assume that $f(x) \in k[x]$ is an irreducible polynomial having a zero $\alpha \in K$. Let $\{\alpha_1 = \alpha, \alpha_2, \ldots, \alpha_r\}$ be the G-orbit of α in K, where $G = \mathrm{Gal}(K/k)$. Set

$$g(x) = \prod_{i=1}^{r}(x - \alpha_r) \in K[x].$$

Note that since each $\sigma \in G$ simply permutes the elements $\alpha_1, \ldots, \alpha_r$, we infer immediately that for each $\sigma \in G, \sigma^* g(x) = g(x)$. (Here σ^* is the ring automorphism of $K[x]$ that applies σ to the coefficients of the polynomials. See Sect. 11.3.1, p. 363.) Therefore the coefficients of $g(x)$ are all in $\text{inv}_G(K) = k$, as k is closed in K. Therefore $g(x) \in k[x]$; since $g(\alpha) = 0$, $f(x)$ must divide $g(x)$, which implies that $f(x)$ splits completely in $K[x]$. Since $f(x)$ was arbitrarily chosen in $k[x]$, we conclude that K is a normal extension of k.

Note that the above also proves that the arbitrarily-chosen irreducible polynomial $f(x) \in k[x]$ is separable. Applying this to $f(x) = \text{Irr}_k(\alpha)$, $\alpha \in K$, we see that K is a separable extension of k, as well.

Conversely, assume that the finite extension $k \subseteq K$ is a separable normal extension. Let $\alpha \in K \backslash k$ and set $f(x) = \text{Irr}_k(x)$. Then $f(x)$ is of degree at least two and splits into distinct linear factors in $K[x]$. Thus, if $\beta \in K$ is another zero of $f(x)$, then by Lemma 11.3.1 there exists a k-isomorphism $k(\alpha) \to k(\beta)$ taking α to β. Next, by Theorem 11.3.4 this isomorphism can be extended to one defined on all of K. Therefore, we have shown that for all $\alpha \in K \backslash k$, there is an element of $\sigma \in G = \text{Gal}(K/k)$ such that $\sigma(\alpha) \neq \alpha$. It follows that $k = \text{inv}_G(K)$, proving the result. $\square$

Theorem 11.6.5 (The Fundamental Theorem of Galois Theory) *Suppose K is a finite separable normal extension of the the field k. Let $G := \text{Gal}(K/k)$, let $S(G)$ be the dual of the poset of subgroups of G, and let $\Omega_{K/k}$ be the poset of subfields of K which contain k. Then the following hold:*

(i) (the Galois Connection) *The mappings of the Galois correspondence*

$$S(G) \quad \overset{\text{Gal}(K/\bullet)}{\underset{\text{inv}_\bullet(K)}{\leftrightarrows}} \quad \Omega_{K/k}$$

are inverse to each other, and hence are bijections.

(ii) (The connection between upper field indices and group orders) *For each intermediate field $L \in \Omega_{K/k}$*

$$[K : L] = |\text{Gal}(K/L)|.$$

(iii) (The connection between group indices and lower field indices) *If $H \leq G$, then $[\text{inv}_H(K) : k] = [G : H]$.*

(iv) (The correspondence of normal fields and normal subgroups).

  1. $L \in \Omega_{K/k}$ *is normal over k if and only if $\text{Gal}(K/L)$ is a normal subgroup of $G = \text{Gal}(K/k)$.*

  2. $N \unlhd G$ *if and only if $\text{inv}_N(K)$ is a normal extension of k.*

(v) (Induced groups and normal factors) *Suppose L is a normal extension of k contained in K. Then $\text{Gal}(L/k)$ is isomorphic to the factor group $\text{Gal}(K/k)/ \text{Gal}(K/L)$.*

Proof Thanks to our preparation, the statement of this Theorem is almost longer than its proof. Part (i) follows by the discussion at the beginning of this subsection. By part (i) together with Theorem 11.6.2, one has, for any subfield $L \in \Omega_{K/k}$, that

$$[K : L] \geq [\mathrm{Gal}(K/L) : 1] = |\mathrm{Gal}(K/L)| \geq [K : \mathrm{inv}_{\mathrm{Gal}(K/L)}(K)] = [K : L].$$

Likewise, for any subgroup $H \leq G$, one has

$$[G : H] \geq [\mathrm{inv}_H(K) : k] \geq [G : \mathrm{Gal}(K/\mathrm{inv}_H(K))] = [G : H],$$

proving both parts (ii) and (iii).

We prove part (iv) part 1: If $L \in \Omega_{K/k}$ is normal over k, then by Corollary 11.3.8 K is G-invariant. This gives a homomorphism $G \to \mathrm{Gal}(L/k)$, $\sigma \mapsto \sigma|_L$; and the kernel of this homomorphism is obviously $\mathrm{Gal}(K/L) \trianglelefteq G$.

For (iv), part 2, suppose $N \trianglelefteq G$, and choose $\alpha \in \mathrm{inv}_N(K)$. Then for any $\sigma \in G$ and $n \in N$ we have $n\sigma(\alpha) = \sigma(\sigma^{-1}n\sigma)(\alpha) = \sigma(\alpha)$, where we have used the fact that $\sigma^{-1}n\sigma \in N$ and N fixes point-wise the elements of $\mathrm{inv}_N(K)$. This proves that $\sigma(\alpha) \in \mathrm{inv}_N(K)$. Now apply Corollary 11.3.8 to conclude that $\mathrm{inv}_N(K)$ is a normal extension of k.

Finally, we prove part (v): Now we have observed above that when $L \in \Omega_{K/k}$ is a normal extension of k, there a homomorphism $G \to \mathrm{Gal}(L/k)$ having kernel $\mathrm{Gal}(K/L)$.

This homomorphism can be seen to be surjective by two distinct arguments. (1) First a direct application of Corollary 11.3.8 (with (K, L, k) playing the role of (E, K, F) of that Corollary) shows that any automorphism in $\mathrm{Gal}(L/k)$ lifts to an automorphism of $\mathrm{Gal}(K/k)$. (2) A second argument achieves the surjectivity by showing that the order of the image of the homomorphism is at least as large as the codomain $\mathrm{Gal}(L/k)$. First, by the fundamental theorem of group homomorphisms, the order of the image is $[G : \mathrm{Gal}(K/L)]$. By (iii),

$$[G : \mathrm{Gal}(K/L)] = [\mathrm{inv}_{\mathrm{Gal}(K/L)}(K) : k].$$

But by definition, $\mathrm{inv}_{\mathrm{Gal}(K/L)}(K)$ contains L, and so the field index on the right side is at least $[L : k]$. But since L/k is also a Galois extension, one has $[L : k] = |\mathrm{Gal}(L/k)|$, by (ii) applied to L/k. Putting these equations and inequalities together one obtains

$$|G/\mathrm{Gal}(K/L)| \geq |\mathrm{Gal}(L/k)|$$

and so the homomorphism $G \to \mathrm{Gal}(L/k)$ is again onto.

Now, since the homomorphism is onto, the fundamental theorem of group homomorphisms shows that the factor group $G/\mathrm{Gal}(K/L)$ is isomorphic to the homomorphic image $\mathrm{Gal}(L/k)$. $\square$

11.7 Traces and Norms and Their Applications

11.7.1 Introduction

Throughout this section, F is a separable extension of a field k, of finite degree $\dim_k(F) = [F : k]$.

Let E be the normal closure of F over k. Then E is a finite normal separable extension of k—that is, it is a Galois extension. Accordingly, if $G = \mathrm{Gal}(E/k)$ and H is the subgroup of G fixing the subfield F element-wise, we have $[G : H] = [F : k]$. Let $\sigma_1 = 1, \sigma_2, \ldots, \sigma_n$ be any right transversal of H in G.[7] We regard each σ_i as an isomorphism $F \to E$.

The *trace* and *norm* are two functions $T_{F/k}$ and $N_{F/k}$ from F into k, which are defined as follows: for each $\alpha \in F$,

$$T_{F/k}(\alpha) = \sum_{i=1}^{n} \alpha^{\sigma_i}, \tag{11.12}$$

$$N_{F/k}(\alpha) = \prod_{i=1}^{n} \alpha^{\sigma_i}. \tag{11.13}$$

The elements $\{\alpha^{\sigma_i}\}$ list the full orbit α^G of α under the (right) action of G, and so does not depend on the particular choice of coset representatives $\{\sigma_i\}$. Since the orbit sum $T_{F/k}(\alpha)$ and orbit product $N_{F/k}(\alpha)$ are both fixed by G, and E/K is a Galois extension, these elements must lie in $k = \mathrm{inv}_G(E)$.

When the extension $k \subseteq F$ is understood, one often writes $T(\alpha)$ for $T_{F/k}(\alpha)$ and $N(\alpha)$ for $N_{F/k}(\alpha)$.

The formulae (11.12) and (11.13) imply the following:

$$T(\alpha + \beta) = T(\alpha) + T(\beta), \alpha, \beta \in F \tag{11.14}$$
$$T(\alpha c) = T(\alpha)c, \alpha \in F, c \in k, \tag{11.15}$$

so that T is a k-linear transformation $F \to k$.

Similarly, for the norm (11.12) and (11.13) yield

$$N(\alpha\beta) = N(\alpha)N(\beta), \alpha, \beta \in F \tag{11.16}$$
$$N(\alpha c) = N(\alpha)c^n, \alpha \in F, c \in k. \tag{11.17}$$

---

[7]Recall from Chap. 3 that a right transversal of a subgroup is simply a complete system of right coset representatives of the subgroup in its parent group. In this case $\{H\sigma_i\}$ lists all right cosets of H in G.

11.7.2 Elementary Properties of the Trace and Norm

Theorem 11.7.1 (The Transitivity Formulae) *Suppose $k \subseteq F \subseteq L$ is a tower of fields with both extension L/F and F/k finite separable extensions. Then for any element $\alpha \in L$,*

$$T_{L/k}(\alpha) = T_{F/k}(T_{L/F}(\alpha)) \tag{11.18}$$
$$N_{L/k}(\alpha) = N_{F/k}(N_{L/F}(\alpha)). \tag{11.19}$$

Proof Let E be the normal closure of L so that E/k is a Galois extension. Set $G = \mathrm{Gal}(E/k)$, $H = \mathrm{Gal}(E/F)$ and $U = \mathrm{Gal}(E/L)$. Then $U \leq H \leq G$, and

$$n = [G : H] = [K : k]$$
$$m = [H : U] = [L : F] \text{ and}$$
$$nm = [G : U] = [L : k].$$

If X is a right transversal of H in G, and Y is right transversal of U in H, then

$$T_{L/k}(\alpha) = \sum_{\sigma \in XY} \alpha^\sigma = \sum_{\sigma \in X} \left(\sum_{\tau \in Y} \alpha^\tau \right)^\sigma \tag{11.20}$$

$$= \sum_{\sigma \in X} (T_{L/F}(\alpha))^\sigma = T_{F/k}(T_{L/F}(\alpha)). \tag{11.21}$$

The anagolous formula for norms is obtained upon replacing sums by products in the preceding Eq. (11.21). $\square$

Corollary 11.7.2 *If $k \subset F$ is a separable extension of degree $n = [F : k]$, and if $\alpha \in F$ has monic irreducible polynomial*

$$\mathrm{irr}(\alpha) = x^d + a_{d-1}x^{d-1} + \cdots + a_1 x + a_0 \in k[x],$$

then

$$T_{F/k}(\alpha) = -(n/d)a_{d-1}. \tag{11.22}$$

Similarly,

$$N_{F/k}(\alpha) = ((-1)^d a_0)^{n/d}. \tag{11.23}$$

Proof Let E be the normal closure of F, so that we have the factorization

$$\mathrm{irr}(\alpha) = (x - \theta_1) \cdots (x - \theta_d)$$

in $E[x]$. Then $-a_{d-1} = \sum \theta_i$ and $\prod \theta_i = (-1)^d a_0$. But $\sum \theta_i = T_{k(\alpha)/k}(\alpha)$ and $\prod \theta_i = N_{k(\alpha)/k}(\alpha)$. Now applying the transitivity formulae (11.18) and (11.19) for the extension tower $k \subseteq k(\alpha) \subseteq F$ one obtains

$$T_{E/k(\alpha)}(\alpha) = T_{k(\alpha)/k}(T_{F/k(\alpha)}(\alpha) = T_{k(\alpha)/k}(\alpha) \cdot [F : k(\alpha)]$$

since $\alpha \in k(\alpha)$. Similarly

$$N_{F/k}(\alpha) = N_{k(\alpha/k}(\alpha))^{[F:k(\alpha)]}.$$

$\square$

11.7.3 The Trace Map for a Separable Extension Is Not Zero

Theorem 11.7.3 *Suppose $k \subseteq F$ is a finite separable extension. Then the trace function $T_{F/k} : F \to k$ is not the zero function.*

Proof Let E be the normal closure of F; it then follows that $k \subseteq E$ is a Galois extension. Let $G = \mathrm{Gal}(E/k)$, and let $\{\sigma_1, \ldots, \sigma_n\}$ be a complete listing of its elements. By the Dedekind Independence Lemma (Lemma 11.6.1), part 1, the functions $\sigma_i : E \to E$ are k-linearly independent. In, particular, the mapping $\sum \sigma_i$ defined by $\alpha \mapsto \sum_{i=1}^{n} \alpha^{\sigma_i}$ cannot be the zero function. So there is an element $\beta \in E$ such that

$$0 \neq \sum_i \beta^{\sigma_i} = T_{E/k}(\beta). \tag{11.24}$$

Set $\beta' = T_{E/F}(\beta)$. Then β' lies in F since $F \subseteq E$ is also a Galois extension. Now if $T_{F/k}(\beta') = 0$, then

$$T_{E/k}(\beta) = T_{F/k}(T_{E/F}(\beta) = T_{F/k}(\beta') = 0,$$

against Eq. (11.24). Thus we have found $\beta' \in F$ such that $T_{F/k}(\beta') \neq 0$ showing that the trace mapping $T_{F/k}$ is not the zero mapping. $\square$

Associated with the trace function $T_{F/k} : F \to k$ is a symmetric bilinear form $B_T : F \times F \to k$ called the *trace form* defined by

$$B_T(\alpha, \beta) = T_{F/k}(\alpha\beta), \alpha, \beta \in F.$$

This form is said to be *non-degenerate* if $B_T(\alpha, \beta) = 0$ for all $\beta \in F$ implies $\alpha = 0$.[8]

---

[8]The reader is referred to Sect. 9.12 where these concepts were applied to all k-linear maps $T : F \to k$.

Corollary 11.7.4 *Let F be a finite separable extension of the field k. Then the trace form is non-degenerate.*

Proof If $B_T(\alpha, \beta) = 0$ for all $\beta \in F$, then $T_{F/k}(\alpha\beta) = 0$ for all $\beta \in F$. If $\alpha \neq 0$, this would imply $T_{F/k}(F) = 0$, contrary to the conclusion of the preceding Theorem 11.7.3. $\square$

11.7.4 An Application to Rings of Integral Elements

Suppose D is an integral domain, which is integrally closed in its field of fractions k. (Recall that this means that any fraction formed from elements of D that is also a zero of a monic polynomial in $D[x]$, is already an element of D.) Suppose F is a finite separable extension of k, and let $\mathcal{O}_F$ be the ring of integral elements of F with respect to D—that is, the elements of D which are a zero of at least one monic polynomial in $D[x]$.

In Sect. 9.12 of Chap. 9, a k-linear transformation $t : F \to k$ was said to be *tracelike* if and only $t(\mathcal{O}_F) \subseteq D$. Theorem 9.12.2 then asserted the following:

> If there exists a non-zero tracelike transformation $t : F \to k$, then the ring $\mathcal{O}_F$ is a Noetherian D-module.

But now we have

Lemma 11.7.5 *Let F be a finite separable extension of k, the field of fractions of the integrally closed domain D, and let $\mathcal{O}_F$ be the ring of integral elements as in the introductory paragraph of this subsection. Then the trace function $T_{F/k} : F \to k$ is a tracelike k-linear transformation.*

Proof It is sufficient to show that if $\alpha \in \mathcal{O}_F$, then $T_{F/k}(\alpha) \in \mathcal{O}_F$, for in that case $T_{F/k}(\alpha) \in \mathcal{O}_F \cap k = D$, since D is integrally closed. Let E be the normal closure of F and set $G := \mathrm{Gal}(E/k)$. For each $\sigma \in G$, and $\alpha \in \mathcal{O}_F$, α^σ is also a zero of the same monic polynomial in $D[x]$ that α is; so it follows that $\sigma(\mathcal{O}_k) \subseteq \mathcal{O}_E$. Now $T_{F/k}(\alpha)$ is the sum of the elements in the orbit α^G and so, being a finite sum of elements of $\mathcal{O}_E$, must lie in $\mathcal{O}_E$ as well as k. Thus $T_{F/k}(\alpha) \in \mathcal{O}_E \cap k \subseteq \mathcal{O}_E \cap F = \mathcal{O}_F$. $\square$

Corollary 11.7.6 *Let D be an integral domain that is integrally closed in its field of fractions k. Let $k \subseteq F$ be a finite separable extension, and let $\mathcal{O}_F$ be the ring of integral elements (with respect to D) in the field F. Then $\mathcal{O}_F$ is a Noetherian D-module.*

Proof By Theorem 9.12.2 it is sufficient to observe that the trace function $T_{F/k}$ is a non-zero tracelike transformation $F \to k$. But as F is a separable extension of k, these two features of the trace function are established in Theorem 11.7.3 and Lemma 11.7.5. $\square$

11.8 The Galois Group of a Polynomial

11.8.1 The Cyclotomic Polynomials

Let n be a positive integer, and let ζ be the complex number $\zeta = e^{2\pi i/n}$. Set

$$\Phi_n(x) = \prod_d (x - \zeta^d),$$

where $1 \leq d \leq n$, and $\gcd(d, n) = 1$. We call $\Phi_n(x)$ the nth *cyclotomic polynomial*. Thus, we see that the zeros of $\Phi_n(x)$ are precisely the generators of the unique cyclic subgroup of order n in the multiplicative group $\mathbb{C}^*$ of the complex numbers, also called the *primitive nth roots of unity*. It follows that the degree of $\Phi_n(x)$ is the number $\phi(n)$ of residue classes mod n which are relatively prime to n.[9] Note in particular that

$$x^n - 1 = \prod_{d|n} \Phi_d(x). \tag{11.25}$$

Since each nth root of unity is a power of the primitive root ζ we see that the field $K = \mathbb{Q}(\zeta)$ is the splitting field over $\mathbb{Q}$ for $x^n - 1$. If we set $G = \mathrm{Gal}(K/\mathbb{Q})$, then G clearly acts on the zeros of $\Phi_n(x)$ (though we don't know yet that this action is transitive!), and so the coefficients of $\Phi_n(s)$ are in $\mathrm{inv}_G(K) = \mathbb{Q}$. In fact, however,

Lemma 11.8.1 *For each positive integer n, $\Phi_n(x) \in \mathbb{Z}[x]$.*

Proof We shall argue by induction on n. First one has $\Phi_1(x) = x - 1 \in \mathbb{Z}[x]$, so the assertion holds for $n = 1$. Assume $n > 1$. In Eq. (11.25), $c(x) := x^n - 1$ is written as a product of cyclotomic polynomials, $\Phi_d(x)$, all of which are monic by definition. By induction

$$a(x) := \prod_{\substack{d|m \\ 1 \leq d < m}} \Phi_d(x) \in \mathbb{Z}[x],$$

(where d ranges over proper divisors of n) is a product of monic polynomials in $\mathbb{Z}[x]$, and so itself is such a polynomial. Setting $b(x) := \Phi_n(x)$, we see that $a(x)$, $b(x)$ and $c(x)$ are monic polynomials with $c(x) = a(x)b(x)$, with $a(x)$ and $c(x)$ in $\mathbb{Z}[x]$ and with $b(x)$ monic in $\mathbb{Q}[x]$. Now we apply Theorem 9.13.3 (with $\mathbb{Z}$ and $\mathbb{Q}$ in the roles of the domains D and D_1, respectively—see Exercise (11) in Sect. 9.13.1, Chap. 9), to conclude that $b(x) = \Phi_n(x)$ is also in $\mathbb{Z}[x]$. The induction proof is complete. $\square$

Theorem 11.8.2 *For each positive integer n, the nth cyclotomic polynomial $\Phi_n(x)$ is irreducible in $\mathbb{Q}[x]$.*

---

[9]The function: $\phi : \mathbb{Z} \to \mathbb{Z}$ is called *Euler's totient function* or simply the *Euler phi-function*.

Proof Since $\Phi_n(x) \in \mathbb{Z}[x]$, we may invoke Gauss's lemma and be satisfied with proving that $\Phi_n[x]$ is irreducible in $\mathbb{Z}[x]$. Thus, assume that $\Phi_n(x) = h(x)k(x)$, where $h(x), k(x) \in \mathbb{Z}[x]$ and $h(x)$ is monic and irreducible. Let p be a prime not dividing n, and let ζ be a zero of $h(x)$ in a splitting field F for $\Phi_n(x)$ over $\mathbb{Q}$. We shall show that ζ^p is also a zero of $h(x)$. Note that since p and n are relatively prime, ζ^p is another zero of $\Phi_n(x)$. Assuming that ζ^p is not a zero of $h(x)$, it must be a zero of $k(x)$, forcing ζ to be a zero of the polynomial of $k(x^p)$. This implies that $h(x)$ divides $k(x^p)$, and so we may now write

$$k(x^p) = h(x)l(x) \tag{11.26}$$

for some monic polynomial $l(x) \in \mathbb{Z}[x]$.

At this point, we may invoke the ring homomorphism

$$m_p : \mathbb{Z}[x] \rightarrow (\mathbb{Z}/(p))[x],$$

which preserves degrees but reads the integral coefficients of all polynomials modulo p. For each polynomial $p(x) \in \mathbb{Z}[x]$, we write $\bar{p}(x)$ for $m_p(p(x))$.

Since n is relatively prime to p, there exists an integer b such that $bn \equiv 1$ mod p. Since $(b\bar{x})\partial(\bar{x}^n - 1) - (\bar{x}^n - 1) = 1$ in $\mathbb{Z}/(p))[x]$, by Lemma 11.5.3, the polynomial $\bar{x}^n - 1$ must have distinct zeroes in its splitting field K over $\mathbb{Z}/p$. So this also must be true of any factor of the polynomial $\bar{x}^n - 1$. We now have $\bar{x}^n - 1 = \bar{\Phi}_n(x)\bar{f}(x) = \bar{h}(x)\bar{k}(x)\bar{f}(x)$, so $\bar{\Phi}_n(x)$ is such a factor.

Now, applying m_p to each side of Eq. (11.26), one obtains

$$\bar{h}(x)\bar{l}(x) = \bar{k}(x^p) = \bar{k}(x)^p \in (\mathbb{Z}/p)[x].$$

Thus the zeroes of $\bar{h}$ in K can be found among those of $\bar{k}(x)$. Since $\bar{\Phi}_n(x) = \bar{h}(x)\bar{k}(x)$, we see that $\bar{\Phi}_n(x)$ has repeated zeroes in K, contrary to the observation in the previous paragraph.

What the above has shown is that if ζ is a zero of $h(x)$ in the splitting field F, then so is ζ^p, for every prime p not dividing n. Finally, let η be any primitive n-root of unity (i.e., a zero of $\Phi_n(x)$ in F). Therefore, $\eta = \zeta^r$ for some integer r relatively prime to n. We factor r as $r = p_1^{e_1} p_2^{e_2} \cdots p_s^{e_s}$; then as each p_i is relatively prime to n, and since the p_ith power of a zero of $h(x)$ is another zero of $h(x)$, we conclude that $\eta = \zeta^r$ is also a zero of $h(x)$. It follows that all zeros of $\Phi_n(x)$ are zeroes of its irreducible monic factor $h(x)$, whence $h(x) = \Phi(x)$, completing the proof. $\square$

11.8.2 The Galois Group as a Permutation Group

Let F be a field and let $f(x) \in F[x]$ be a polynomial. If $K \subseteq F$ is a splitting field over F for $f(x)$, and if $G = \mathrm{Gal}(K/F)$, we call G the *Galois group of the polynomial* $f(x)$. If $\alpha_1, \alpha_2, \ldots, \alpha_k$ are the distinct zeros of $f(x)$ in K, then, since

$K = F(\alpha_1, \alpha_2, \ldots, \alpha_k)$, we see that the automorphisms in G are determined by their effects on the elements $\alpha_1, \ldots, \alpha_k$. Furthermore, as the elements of G clearly permute these zeros, we have an injective homomorphism $G \to S_k$ (where S_k is identified with the symmetric group on the k zeros of $f(x)$) thereby embedding G as a subgroup of S_k. Note finally that if $f(x)$ is *irreducible*, then the above embedding represents the Galois group G as a *transitive* subgroup of S_k.

For example, in the previous subsection we saw that the nth cyclotomic polynomial $\Phi_n(x) \in \mathbb{Z}[x]$ is irreducible and of degree $\phi(n)$, where ϕ is Euler's "totient" function. Setting $k = \phi(n)$, $\zeta = e^{2\pi i/n}$, and $G = \mathrm{Gal}(\mathbb{Q}(\zeta)/\mathbb{Q})$, we have an embedding of G into S_k. However, as G must act as a group of automorphisms of $K = \mathbf{Q}(\mathbf{1})$, we see in particular that it must restrict to a group of automorphisms of the cyclic group $\langle \zeta \rangle$ of order n. Therefore, we have a (faithful) homomorphism $G \to \mathrm{Aut}(\langle \zeta \rangle)$, the latter being abelian of order $\phi(n)$ (see p. 392). Since $|G| = \phi(n) = \deg \Phi_n(x)$, and since the splitting field of $\Phi_n(x)$ is the same as that of $x^n - 1$, we conclude that:

Theorem 11.8.3 *The Galois group of the polynomial $x^n - 1$ is isomorphic with the automorphism group of a cyclic group of order n and is therefore abelian of order $\phi(n)$, where ϕ is Euler's totient function.*

The following is a useful summary of what has been obtained thus far.

Theorem 11.8.4 *Let $f(x) \in F[x]$ and let G be the corresponding Galois group. Assume that $f(x)$ factors into irreducibles as*

$$f(x) = \prod f_i(x)^{e_i} \in F[x].$$

Let E be a splitting field over E of $f(x)$, and let Λ_i be the set of zeros of $f_i(x)$ in E. Then G acts transitively on each Λ_i.

Proof If α, α' are distinct zeros of $f_i(x)$ in E the by Lemma 11.3.1 there is an isomorphism $F(\alpha) \to F(\alpha')$; by Theorem 11.3.4, this can be extended to an automorphism of E. $\square$

From the above, we see that if Λ is the set of zeros of a polynomial $f(x)$ in a splitting field E, we have an embedding $G \to \mathrm{Sym}(\Lambda)$ of the Galois group into the symmetric group on Λ. Identifying G with its image in $\mathrm{Sym}(\Lambda)$, an interesting question that naturally occurs is whether $G \leq \mathrm{Alt}(\Lambda)$, the alternating group on Λ. To answer this, we introduce the *discriminant* of the separable polynomial $f(x)$. Thus let $f(x) \in F[x]$, where *char* $F \neq 2$, and let E be a splitting field over F for $f(x)$. Let $\{\alpha_1, \alpha_1, \ldots, \alpha_k\}$ be the set of distinct roots of $f(x)$ in E. Set

$$\delta = \prod_{1 \leq j < i \leq k} (\alpha_i - \alpha_j) = \det \begin{bmatrix} 1 & \alpha_1 & \alpha_1^2 & \cdots & \alpha_1^{k-1} \\ 1 & \alpha_2 & \alpha_2^2 & \cdots & \alpha_2^{k-1} \\ \vdots & \vdots & \vdots & & \vdots \\ \vdots & \vdots & \vdots & & \vdots \\ \vdots & \vdots & \vdots & & \vdots \\ 1 & \alpha_k & \alpha_k^2 & \cdots & \alpha_k^{k-1} \end{bmatrix},$$

and let $D_f = \delta^2$. We call D_f the *discriminant* of the polynomial $f(x)$, sometimes denoted disc $f(x)$. Note that $D_f \in \mathrm{inv}_G(E)$; when $f(x)$ is separable, this implies that $D_f \in F$.

Example 54 (*Discriminant of a quadratic*) Suppose that $f(x) \in F[x]$ is a monic separable quadratic; thus $f(x) = x^2 + bx + c = (x - \alpha_1)(x - \alpha_2)$ in a splitting field over F for $f(x)$. Therefore, we have

(i) $\alpha_1 + \alpha_2 = -b$, and
(ii) $\alpha_1 \alpha_2 = c$.

It follows that

$$
\begin{aligned}
\mathrm{disc}\, f(x) &= (\alpha_2 - \alpha_1)^2 \\
&= \alpha_1^2 + \alpha_2^2 - 2\alpha_1\alpha_2 \\
&= (\alpha_1 + \alpha_2)^2 - 4\alpha_1\alpha_2 \\
&= b^2 - 4c \in F[x],
\end{aligned}
$$

a formula for the discriminant familiar to every high school student.

Before tackling the $n = 3$ case, we make a few observations of general interest. Relative to $\alpha_1, \alpha_2, \ldots, \alpha_k$, we define the ith *power sum*, $i = 0, 1, 2, \ldots$:

$$
s_i = \alpha_1^i + \alpha_2^i \cdots + \alpha_k^i.
$$

Since the determinant of a matrix is the same as the determinant of its transpose, we may express the discriminant of the polynomial $f(x)$ thus:

$$
\mathrm{disc}\, f(x) = \det
\begin{bmatrix}
1 & 1 & 1 & \cdots & 1 \\
\alpha_1 & \alpha_2 & \alpha_3 & \cdots & \alpha_k \\
\alpha_1^2 & \alpha_2^2 & \alpha_3^2 & \cdots & \alpha_k^2 \\
\vdots & \vdots & \vdots & \cdots & \vdots \\
\alpha_1^{k-1} & \alpha_2^{k-1} & \alpha_3^{k-1} & \cdots & \alpha_k^{k-1}
\end{bmatrix}
\begin{bmatrix}
1 & \alpha_1 & \alpha_1^2 & \cdots & \alpha_1^{k-1} \\
1 & \alpha_2 & \alpha_2^2 & \cdots & \alpha_2^{k-1} \\
1 & \alpha_3 & \alpha_3^2 & \cdots & \alpha_3^{k-1} \\
\vdots & \vdots & \vdots & \cdots & \vdots \\
1 & \alpha_k & \alpha_k^2 & \cdots & \alpha_k^{k-1}
\end{bmatrix},
$$

from which one concludes that

$$
\mathrm{disc}\, f(x) = \det
\begin{bmatrix}
1 & s_1 & s_2 & \cdots & s_{k-1} \\
s_1 & s_2 & s_3 & \cdots & s_k \\
s_2 & s_3 & s_4 & \cdots & s_{k+1} \\
\vdots & \vdots & \vdots & \cdots & \vdots \\
s_{k-1} & s_k & s_{k+1} & \cdots & s_{2k-2}
\end{bmatrix}.
\tag{11.27}
$$

Example 55 (*Discriminant of a cubic*) Suppose that $f(x) = x^3 + a_2x^2 + a_1x + a_0 \in F[x]$ is a monic separable cubic, with zeros $\alpha_1, \alpha_2, \alpha_3$ in a splitting field. In this case, using Eq. 11.27 one obtains

$$\text{disc } f(x) = \det \begin{bmatrix} 1 & s_1 & s_2 \\ s_1 & s_2 & s_3 \\ s_2 & s_3 & s_4 \end{bmatrix}.$$

Next, note that the coefficients of $f(x)$ are given by

$$-\sigma_1 := -(\alpha_1 + \alpha_2 + \alpha_3) = a_2,$$
$$\sigma_2 := \alpha_1\alpha_2 + \alpha_1\alpha_3 + \alpha_2\alpha_3 = a_1,$$
$$-\sigma_3 := -\alpha_1\alpha_2\alpha_3 = a_0.$$

Furthermore, one has $s_1 = \sigma_1 = -a_2$; furthermore,

$$\begin{aligned} s_2 &= \alpha_1^2 + \alpha_2^2 + \alpha_3^2 \\ &= (\alpha_1 + \alpha_2 + \alpha_3)^2 - 2(\alpha_1\alpha_2 + \alpha_1\alpha_3 + \alpha_2\alpha_3) \\ &= \sigma_1^2 - 2\sigma_2 \\ &= a_2^2 - 2a_1. \end{aligned}$$

$$\begin{aligned} s_3 &= \alpha_1^3 + \alpha_2^3 + \alpha_3^3 \\ &= (\alpha_1 + \alpha_2 + \alpha_3)(\alpha_1^2 + \alpha_2^2 + \alpha_3^2) \\ &\quad -(\alpha_1 + \alpha_2 + \alpha_3)(\alpha_1\alpha_2 + \alpha_1\alpha_3 + \alpha_2\alpha_3) + 3\alpha_1\alpha_2\alpha_3 \\ &= \sigma_1(\sigma_1^2 - 2\sigma_2) - \sigma_1\sigma_2 - 3\sigma_3 \\ &= \sigma_1^3 - 3\sigma_1\sigma_2 + 3\sigma_3 \\ &= -a_2^3 + 3a_1a_2 - 3a_0. \end{aligned}$$

$$\begin{aligned} s_4 = &= \alpha_1^4 + \alpha_2^4 + \alpha_3^4 \\ &= (\alpha_1 + \alpha_2 + \alpha_3)(\alpha_1^3 + \alpha_2^3 + \alpha_3^3) \\ &\quad -(\alpha_1\alpha_2 + \alpha_1\alpha_3 + \alpha_2\alpha_3)(\alpha_1^2 + \alpha_2^2 + \alpha_3^2) \\ &\quad +\alpha_1\alpha_2\alpha_3(\alpha_1 + \alpha_2 + \alpha_3) \\ &= -\sigma_1(\sigma_1^3 - 3\sigma_1\sigma_2 + 3\sigma_3) \\ &\quad -\sigma_2(\sigma_1^2 - 2\sigma_2) + \sigma_1\sigma_3 \\ &= a_2^4 - 4a_1a_2^2 + 4a_0a_2 + 2a_1^2. \end{aligned}$$

From all of the above, one obtains, therefore, that

$$\text{disc } f(x) = \det \begin{bmatrix} 3 & -a_2 & a_2^2 - 2a_1 \\ -a_2 & a_2^2 - 2a_1 & -a_2^3 + 3a_1a_2 - 3a_0 \\ a_2^2 - 2a_1 & -a_2^3 + 3a_1a_2 - 3a_0 & a_2^4 - 4a_1a_2^2 + 4a_0a_2 + 2a_1^2 \end{bmatrix}$$

$$= a_1^2a_2^2 - 4a_1^3 - 4a_0a_2^3 - 27a_0^3 + 18a_0a_1a_2,$$

after admittedly odious calculations!

The above examples reveal a general trend, viz., that the discriminant of a polynomial can always be expressed in terms of the coefficients of the polynomial. To see why this is the case, we assume, for the moment that $x_1, x_2, \ldots, x_n$ are commuting indeterminates, and define the *elementary symmetric polynomials* $\sigma_1, \sigma_2, \ldots, \sigma_n$, and the *power sum polynomials* $s_1, s_2, \ldots, s_n$ by setting

$$\sigma_k = \sum_{i_1 < i_2 < \cdots < i_k} x_{i_1} x_{i_2} \cdots x_{i_k}, \quad s_l = \sum_{i=1}^{n} x^l, \; k = 1, 2, \ldots, n, \; l = 1, 2, \ldots$$

Therefore,

$$\sigma_1 = s_1 = x_1 + x_2 + \ldots x_n,$$
$$\sigma_2 = \sum_{i<j} x_i x_j = x_1 x_2 + x_1 x_3 + \cdots x_{n-1} x_n,$$
$$s_2 = x_1^2 + x_2^2 + \ldots + x_n^2,$$

and so on.

Next, note that

$$\prod_{i=1}^{n} (x - x_i) = x^n - \sigma_1 x^{n-1} + \sigma_2 x^{n-2} + \cdots + (-1)^n \sigma_n.$$

From this, it follows immediately that if the power sum polynomials $s_1, s_2, \ldots$ can be written as polynomials in the elementary symmetric polynomials $\sigma_1, \sigma_2, \ldots, \sigma_n$, then from Eq. (11.27) we infer immediately that the discriminant of a polynomial can be expressed as a polynomial in its coefficients. Our detailed calculations of the discriminant of quadratics and cubics hint that this might be possible. We turn now to a demonstration that this can always be done.

First, suppose that $g(\mathbf{x}) = g(x_1, x_2, \ldots, x_n) \in F[x_1, x_2, \ldots, x_n]$. We say that $g(\mathbf{x})$ is a *symmetric* polynomial if it remains unchanged upon any permutation of the indices $1, 2, \ldots, n$. Therefore, we see immediately that the elementary symmetric polynomials as well as the power sum polynomials are symmetric polynomials.

Theorem 11.8.5 (Fundamental Theorem on Symmetric Polynomials) *Any symmetric polynomial $g(x) \in F[x_1, x_2, \ldots, x_n]$ can be expressed as a polynomial in the elementary symmetric polynomials.*

Proof We clearly may assume that $g = g(\mathbf{x})$ is homogeneous, i.e., that it is composed of monomials each of the same degree, say k. Next, we introduce into the set of monomials of the form $x_1^{i_1} x_2^{i_2} \cdots x_n^{i_n}$, where $i_1 + i_2 + \cdots + i_n = k$, the so-called *lexicographic ordering*. Thus we say that $x_1^{i_1} \cdots x_n^{i_n} < x_1^{j_1} \cdots x_n^{j_n}$ if $i_1 = j_1, i_2 = j_2, \ldots, i_l = j_l, i_{l+1} < j_{l+1}$. Thus, if $n = 3$ we have $x_1^2 x_2 x_3^3 < x_1^2 x_2^2 x_3^2 < x_1^3 x_2 x_3^2$. Next, let $x_1^{m_1} x_2^{m_2} \cdots x_n^{m_n}$ be the highest monomial occurring in g with nonzero coefficient. Since g is symmetric, it must also contain all monomials obtained from $x_1^{m_1} x_2^{m_2} \cdots x_n^{m_n}$ by permuting the indices. It follows, therefore, that we must have $m_1 \geq m_2 \geq \cdots \geq m_n$.

Next, let $d_1, d_2, \ldots, d_n$ be exponents; we wish to identify the highest monomial in the symmetric polynomial $\sigma_1^{d_1} \sigma_2^{d_2} \cdots \sigma_n^{d_n}$. Here it is useful to observe that if M_1, M_2 are monomials of the same degree with $M_1 < M_2$, and if M is any monomial, then $M_1 M < M_2 M$. Having observed this, we note next that the highest monomial in σ_i is clearly $x_1 x_2 \cdots x_i$. Therefore, the highest monomial in $\sigma_i^{d_i}$ is $x_1^{d_i} x_2^{d_i} \cdots x_n^{d_i}$. In turn, the highest monomial in $\sigma_1^{d_1} \sigma_2^{d_2} \cdots \sigma_n^{d_n}$ is therefore

$$x_1^{d_1+d_2+\cdots d_n} x_2^{d_2+d_3+\cdots+d_n} \cdots x_n^{d_n}.$$

From this, we see that the highest degree monomial in both g and and the polynomial $\sigma_1^{m_1-m_2} \sigma_2^{m_2-m_3} \cdots \sigma_n^{m_n}$ is $x_1^{m_1} x_2^{m_2} \cdots x_n^{m_n}$. This implies that if

$$g = a x_1^{m_1} x_2^{m_2} \cdots x_n^{m_n} + \text{lower monomials} ,$$

then the symmetric polynomial $g - a\sigma_1^{m_1-m_2} \sigma_2^{m_2-m_3} \cdots \sigma_n^{m_n}$ will only involve monomials lower than $x_1^{m_1} x_2^{m_2} \cdots x_n^{m_n}$. A simple induction finishes the proof. $\square$

Corollary 11.8.6 *The discriminant of the separable polynomial $f(x) \in F[x]$ can be expressed as a polynomial (with coefficients in F) in the coefficients of $f(x)$.*

The permutation-group theoretic importance of the discriminant is the following.

Theorem 11.8.7 *Let $f(x) \in F[x]$, where char $F \neq 2$ be a polynomial with discriminant $D_f = \delta^2 = \text{disc } f(x)$ defined as above. Let G be the Galois group of $f(x)$, regarded as a subgroup of $Sym(n)$, acting on the zeros $\{\alpha_1, \cdots, \alpha_n\}$ in a splitting field E over F for $f(x)$. If $A := G \cap Alt(n)$, then $inv_A(E) = F(\delta)$.*

Proof Note that any odd permutation of the zeros $\alpha_1, \ldots, \alpha_k$ will transform δ into $-\delta$ and even permutations will fix δ. $\square$

The following is immediate.

Corollary 11.8.8 *Let G be the Galois group of $f(x) \in F[x]$, where char $F \neq 2$. If the discriminant D_f is the square of an element in F, then $G \leq Alt(n)$.*

The following is occasionally useful in establishing that the Galois group of a polynomial is the full symmetric group.

Theorem 11.8.9 *Let $f(x) \in \mathbb{Q}[x]$ be irreducible, of prime degree p, and assume that $f(x)$ has exactly 2 non-real roots. Then $G_f = Sym(p)$.*

We close by mentioning that for a general "trinomial," there is a formula for the discriminant, due to R.G. Swan,[10] given as follows.

Theorem 11.8.10 *Let $f(x) = x^n + ax^k + b$, and let $d = g.c.d.(n, k)$, $N = \frac{n}{d}$, $K = \frac{k}{d}$. Then*

$$D_f = (-1)^{\frac{1}{2}n(n-1)} b^{k-1} [n^N b^{N-k} - (-1)^N (n-k)^{N-K} k^K a^N]^d.$$

11.9 Solvability of Equations by Radicals

11.9.1 Introduction

Certainly one remembers that point in one's education that one first encountered the quadratic equation. If we are given the polynomial equation

$$ax^2 + bx + c = 0,$$

where $a \neq 0$ (to ensure that the polynomial is indeed "quadratic"), then the roots of this equation are given by the formula

$$(-b \pm \sqrt{b^2 - 4ac})/2a.$$

Later, perhaps, the formula is justified by the procedure known as "completing the square". One adds some constant to both sides of the equation so that the left side is the square of a linear polynomial, and then one takes square roots. It is fascinating to realize that this idea of completing the square goes back at least two thousand years to the Near East and India. It means that at this early stage, there is the suggestion that there could be a realm where square roots could always be taken, and the subtlety that there are cases in which square roots can never be taken because the radican is negative.

A second method of solving the equation involves renaming the variable x in a suitable way. One first divides the equation through by a so that it has the form $x^2 + b'x + c' = 0$. Then setting $y := x - b'/2$, the equation becomes

$$y^2 + c' - (b')^2/4 = 0,$$

---

[10]R.G. Swan, Factorization of polynomials over finite fields, *Pacific Journal*, vol 12, pp. 1099–1106, **MR** 26 #2432. (1962); see also Gary Greenfield and Daniel Drucker, On the discriminant of a trinomial, *Linear Algebra Appl.* **62** (1984), 105-112.

from which extraction of the square root (if that can be done) yields $y = \sqrt{(4c' - (b')^2)/2}$. The original zero named by x, is obtained by $x = y + b'/2$, and the equations $b' = b/a$ and $c' = c/a$.

Certainly the polynomials in these equations were seen as generic expressions representing relations between some sort of possible numbers. But these numbers were known not to be rational numbers in many cases. One could say that solutions don't exist in those cases, or that there is a larger realm of numbers (for example fields like $\mathbb{Q}(\sqrt{2})$ or the reals), and the latter view only began to emerge in the last three centuries.

Of course the quadratic formula is not valid over fields of characteristic 2 since dividing by 2 was used in deriving and expressing the formula. The same occurs for the solutions of the cubic and quartic equations: fields of characteristic 2 and 3 must be excluded.

Dividing through by the coefficient of u^3, the general cubic equation has the form $u^3 + bu^2 + cu + d = 0$, where u is the generic unkown root. Replacing u by $x - b/3$ yields the simpler equation,

$$x^3 + qx + r = 0.$$

Next, one puts $x = y + z$ to obtain

$$y^3 + z^3 + (3yz + q)x + r = 0. \tag{11.28}$$

There is still a degree of freedom allowing one to demand that $yz = -q/3$, thus eliminating the coefficient of x in (11.28). Then z^3 and y^3 are connected by what amounts to the norm and trace of a quadratic equation:

$$z^3 + y^3 = -r,$$

and

$$z^3 y^3 = -q^3/27.$$

One can then solve for z^3 and y^3 by the quadratic formula. Taking cube roots (assuming that is possible) one gets a value for z and y. The possible zeroes of $x^3 + qx + r$ are:

$$y + z, \quad \omega y + \omega^2 z, \quad \omega^2 y + \omega z,$$

where ω is a complex primitive cube root of unity. The formula seems to have been discovered in 1515 by Scipio del Ferro and independently by Tartaglia.

Just thirty years later, when a manuscript of Cardan published the cubic formula, Ferrari discovered the formula for solving the general quartic equation. Here one seeks the roots of a monic fourth-degree polynomial, whose cubic term can be eliminated by a linear substitution to yield an equation

$$x^4 + qbx^2 + rx + s = 0.$$

We would like to re-express the left side as

$$(x^2 - kx + u)(x^2 + kx + v),$$

for suitable numbers k, u and v. From the three equations arising from equating coefficients, the first two allow u and v to be expressed in terms of rk, and a substitution of these expressions for u and v into the third equation produces a cubic in k^2, which can be solved by the cubic formula. Then u and v are determined and so the roots can be obtained from two applications of the quadratic formula.

It is hardly surprising, with all of this success concentrated in the early sixteenth century, that the race would be on to solve the general quintic equation. But this effort met with complete frustration for the next 270 years.

Looking back we can make a little better sense of this. First of all, where are we finding these square roots and cube roots that appear in these formulae? That can be answered easily from the concepts of this chapter. If we wish to find the nth root of a number w in a field F, we are seeking a root of the polynomial $x^n - w \in F[x]$. And of course we need only form the field $F[\alpha] \simeq F[x]/p(x)F[x]$ where $p(x)$ is any irreducible factor of $x^n - w$. But in that explanation, why should we start with a general field F? Why not the rational numbers? The answer is that in the formula for the general quartic equation, one has to take square roots of numbers which themselves are roots of a cubic equation. So in general, if one wishes to consider the possibility of having a formula for an nth degree equation, he or she must be prepared to extract pth roots of numbers which are already in an uncontrollably large class of field extensions of the rational numbers, or whatever field we wish to begin with.

Then there is this question of actually having an explicit formula that is good for every nth degree equation. To be sure, the sort of formula one has in mind would involve only the operations of root extraction, multiplication and addition and taking multiplicative inverses. But is it not possible that although there is not one formula good for all equations, it might still be possible that roots of a polynomial $f(x) \in F[x]$ can be found in a splitting field K which is obtained by a series of root extractions—put more precisely, K lies in some field F_r which is the terminal member of a tower of fields

$$F = F_0 < F_1 < \cdots < F_r,$$

where $F_{i+1} = F_i(\alpha_i)$ and $\alpha_i^{n_i} \in F_i$, for positive integers n_i, $i = 0, \ldots, r-1$? In this case we call F_r a *radical extension of F*. Certainly, if there is a formula for the roots of the general nth degree equation, there must be a radical extension F_r/F which contains a copy of the the splitting field $K_{f(x)}$ for *every* polynomial $f(x)$ of degree n. Yet something more general might happen: Conceivably, for each polynomial $f(x)$, there is a radical extension F_f/F, containing a copy of the splitting field of $f(x)$, which is special for that polynomial, without there being one universal radical extension which does the job for everybody. Thus we say that $f(x)$ *is solvable by radicals* if and only if its splitting field lies in some radical extension. Then, if there is a polynomial not solvable by radicals, there can be no universal formula. Indeed that turned out to be the case for polynomials of degree at least five.

This field-theoretic view-point was basically the invention of a French teenager, Evariste Galois. His account of it appears in a final letter frantically written the night before he died in a duel, just before his twenty first birthday, we are told.[11] The discovery of his theory might have been delayed for many decades had this letter not come to light some years after his death.[12] For further reading on Galois, the authors recommend the following references:

Livio, Mario, *The Equation That Couldn't Be Solved: How Mathematical Genius Discovered the Language of Symmetry*, Simon and Schuster, New York, 2005

Rothman T., "The short life of Evariste Galois", *Scientific American*, **246**, no.4, (1982), p. 136.

Rothman, Tony, "Genius and biographers: the fictionalization of Evariste Galois," *Amercan Math. Monthly*, **84** (1982), p. 89

The next sections will attempt to describe this theory.

11.9.2 Roots of Unity

Any zero of $x^n - 1 \in F[x]$ lying in some extension field of F is called an nth *root of unity*. If F has characteristic p, a prime, we may write $n = m \cdot p^a$ where p does not divide m and observe that

$$x^n - 1 = (x^m - 1)^{p^a}. \tag{11.29}$$

Thus all nth roots of unity are in fact mth roots of unity in this case.

Of course if α and β are nth roots of unity lying in some extension field of K of F, then $(\alpha\beta)^n = \alpha^n \beta^n = 1$ so $\alpha\beta$ is also an nth root of unity. Thus the nth roots of unity lying in K always form a finite multiplicative subgroup of K^*, which, by Corollary 11.2.9, is necessarily cyclic (of order m, if $n = mp^e$ as above). Any generator of this cyclic group is called a *primitive nth root of unity*.

Suppose now that n is relatively prime to the characteristic p of F, or that F has characteristic zero. Let K be the splitting field of $x^n - 1$ over F. Then $x^n - 1$ is relative prime to its formal derivative, as $nx^{n-1} \neq 0$. Therefore, in this case $x^n - 1$ splits completely in $K[x]$. But even if F had positive characteristic p and if $n = mp^e$ where p does not divide m, the nth roots of unity are, by Eq. (11.29), just the mth roots of unity, and so the splitting field of $x^n - 1$ is just the splitting field of the

---

[11] There is some doubt about this, but his birthday nonetheless seems to be near this date.

[12] What appears to be an early hand-written copy of this letter (in French) now exists as a photocopy in some American University libraries, for example, the Morris Library at Southern Illinois University.

separable polynomial $x^m - 1$. Thus in either case, the splitting field K of $x^n - 1$ over the field F is separable (and, of course normal), and hence is Galois over F.

We denote the splitting field over F of $x^n - 1$ by K, and set $G = \mathrm{Gal}(K/F)$. Since K is generated over F by the nth roots of unity, and since G clearly acts on these roots of unity, this action is necessarily faithful. Finally, it is clear that G acts as a group of automorphisms of the (cyclic) group of nth roots of unity; since the full automorphism group of a cyclic group is abelian, G is abelian, as well.

We summarize these observations in the following way:

Lemma 11.9.1 *Let n be any positive integer and let F be any field. Let K be the splitting field of $x^n - 1$ over F. Then the following hold:*

 (i) *K is separable over F and is a simple extension $K = F[\zeta]$, where ζ is a primitive nth root of unity.*
(ii) *$\mathrm{Gal}(K/F)$ is an abelian group.*

11.9.3 Radical Extensions

Suppose n is a positive integer. We say that a field F *contains all nth roots of unity* if and only if the polynomial $x^n - 1$ splits completely into linear factors in $F[x]$.

Lemma 11.9.2 *Let F be a field, let n be a positive integer, and suppose F contains all nth roots of unity. Suppose α is an element of some extension field of F, where $\alpha^n \in F$. Then*

 (i) *the simple extension $F[\alpha]$ is normal over F, and*
(ii) *the Galois group $\mathrm{Gal}(F[\alpha]/F)$ is cyclic of order equal to a divisor of n.*

Proof Note that if K is a splitting field over F for $f(x) = x^n - \alpha^n \in F[x]$, then for any zero $\alpha' \in K$ of $f(x)$, we have $(\alpha'/\alpha)^n = 1$, and so, by hypothesis, $\eta = \alpha'/\alpha \in F$, and so $\alpha' = \eta\alpha \in F(\alpha)$, proving that $K = F(\alpha)$, and so $F(\alpha)$ is a splitting field over F for $x^n - \alpha^n$, hence is normal.

Next, the multiplicative group of the nth roots of unity in F is a cyclic group; let ζ be a generator of this group. Where $G = \mathrm{Gal}(F(\alpha)/F)$ is the Galois group of $F(\alpha)$ over F, we define a mapping $\phi : G \to \langle \zeta \rangle$ as follows. We know that if $\sigma \in G$, then $\sigma(\alpha) = \zeta^{i_\sigma}\alpha$, for some index i_σ; we set $\phi(\sigma) = \zeta^{i_\sigma}$. Next, if $\sigma, \sigma' \in G$, then $\sigma\sigma'(\alpha) = \sigma(\zeta^{i_{\sigma'}}\alpha) = \zeta^{i_{\sigma'}}\sigma(\alpha) = \zeta^{i_{\sigma'}}\zeta^{i_\sigma}\alpha$, and so it follows that $\phi(\sigma\sigma') = \zeta^{i_{\sigma'}}\zeta^{i_\sigma} = \zeta^{i_\sigma}\zeta^{i_{\sigma'}} = \phi(\sigma)\phi(\sigma')$, i.e., ϕ is a homomorphism of groups. If $\sigma \in \ker\phi$, then $\sigma(\alpha) = \alpha$. But then, $\sigma(\zeta^j\alpha) = \zeta^j\sigma(\alpha) = \zeta^j\alpha$, forcing $\sigma = 1$. Therefore, ϕ embeds G as a subgroup of the cyclic group $\langle \zeta \rangle$, proving the result. $\square$

Recall from p. 401 that a polynomial $p(x) \in F[x]$ is said to be *solvable by radicals* if and only its splitting field lies in a radical extension of F.

Theorem 11.9.3 *Suppose $p(x)$ is a polynomial in $F[x]$ which is solvable by radicals. Then if E is a splitting field for $p(x)$ over F, $\mathrm{Gal}(E/F)$ is a solvable group.*

Proof By hypothesis, there is a radical extension $F = F_0 \subseteq F_1 \subseteq \cdots \subseteq F_r$ such that $E \subseteq F_r$. Assume that $F_s = F_{s-1}(\alpha_s)$, $s = 1, 2, \ldots, r$, where each $\alpha_s^{j_s} \in F_{s-1}$. Let m be the least common multiple of $j_1, j_2, \ldots, j_r$, and let K be a splitting field over F_r of $x^m - 1$. It suffices to show that $\mathrm{Gal}(K/F)$ is solvable, for then $\mathrm{Gal}(E/F)$ would be a quotient of $\mathrm{Gal}(K/F)$, and hence would also be solvable.

We argue by induction that $\mathrm{Gal}(K/F)$ is solvable. Let η be a generator for the cyclic group of mth roots of unity in K. We have the tower

$$F \subseteq F(\eta) \subseteq F_1(\eta) \subseteq \cdots \subseteq F_{r-1}(\eta) \subseteq F_r(\eta) = K \supseteq E.$$

By Lemma 11.9.2 part (i) each of the intermediate subfields $F_l(\eta)$ is normal over F. We get a short exact sequence

$$1 \to \mathrm{Gal}(F_r(\eta)/F_{r-1}(\eta)) \to \mathrm{Gal}(K/F) \to \mathrm{Gal}(F_{r-1}(\eta)/F) \to 1.$$

By Lemma 11.9.2 part (ii) $\mathrm{Gal}(F_r(\eta)/F_{r-1}(\eta))$ is solvable, and by induction $\mathrm{Gal}(F_{r-1}(\eta)/F)$ is solvable. It follows that $\mathrm{Gal}(K/F)$, is solvable. The proof is complete. $\square$

11.9.4 Galois' Criterion for Solvability by Radicals

In this subsection we will prove a converse to Theorem 11.9.3 for polynomials over a ground field in characteristic zero. (In characteristic p the converse statement is not even true.) In order to do this, we first require a partial converse to Lemma 11.9.2.

Lemma 11.9.4 *Let $F \subseteq K$ be a Galois extension of prime degree q, and assume that the characteristic of F is distinct from q. Assume further that $x^q - 1$ splits completely in $F[x]$. Then K is a simple radical extension of F—that is $K = F[\zeta]$ where $\zeta^q \in F$.*

Proof Our assumption about the characteristic guarantees that the polynomial $x^q - 1$ splits completely into q distinct linear factors in $F[x]$. Let $z_1, z_2, \ldots, z_q$ be the distinct qth roots of unity in F. We know that the Galois group $G = \mathrm{Gal}(K/F)$ has order dividing the prime q, and so it is cyclic. Fix a generator $\sigma \in G$, so that $G = \langle \sigma \rangle$ where $\sigma^q = 1 \in G$. (Note that σ may be the identity automorphism.) Now, for each element $\theta \in K \backslash F$ we set $\theta_i = \sigma^{i-1}(\theta)$, $i = 1, 2, \ldots, q$ and define the corresponding *Lagrange resolvents*:

$$(z_i, \theta) = \theta_1 + \theta_2 z_i + \theta_3 z_i^2 + \cdots + \theta_q z_i^{q-1} \tag{11.30}$$

$$= \sum_{j=1}^{q} \sigma^{j-1}(\theta) z_i^{j-1} \tag{11.31}$$

$$= \sum_{j=0}^{q-1} \sigma^j(\theta) z_i^j. \tag{11.32}$$

Note that $\sigma(z_i, \theta) = \theta_2 + \theta_3 z_i + \cdots + \theta_1 z_i^{q-1} = z_i^{-1}(z_i, \theta)$, and so it follows that $\sigma(z_i, \theta)^q = (z_i, \theta)^q$, which says, of course, that $(z_i, \theta)^q \in F$. Thus, we'll be finished as soon as we can show that one of the resolvents $(z_i, \theta) \notin F$. We display Eq. (11.30) in matrix form:

$$\begin{bmatrix} 1 & z_1 & z_1^2 & \cdots & z_1^{q-1} \\ 1 & z_2 & z_2^2 & \cdots & z_2^{q-1} \\ \vdots & \vdots & \vdots & \cdots & \vdots \\ 1 & z_q & z_q^2 & \cdots & z_q^{q-1} \end{bmatrix} \begin{bmatrix} \theta_1 \\ \theta_2 \\ \vdots \\ \theta_q \end{bmatrix} = \begin{bmatrix} (z_1, \theta) \\ (z_2, \theta) \\ \vdots \\ (z_q, \theta) \end{bmatrix}$$

As we have already observed, the qth roots of unity are all distinct (by our assumption that the characteristic of F does not divide q), and so by Corollary 11.2.8, this Vandermonde matrix on the left can be inverted to solve for each of the coefficients $\theta_1, \theta_2, \ldots, \theta_q$ as F-linear combinations of the resolvents (z_1, θ), $i = 1, 2, \ldots, q$. Since $\theta_1 = \theta \notin F$, we conclude immediately that at least one of the resolvents $(z_i, \theta) \notin F$, and we are finished. $\square$

We are now ready for the main result of this section.

Theorem 11.9.5 *Suppose F is a field of characteristic zero and that $f(x)$ is a polynomial in $F[x]$. Let K be the splitting field of $f(x)$ over F. If $G = \mathrm{Gal}(K/F)$ is a solvable group then the equation $f(x) = 0$ is solvable in radicals.*

Proof Let E be a splitting field over K for the polynomial $x^n - 1$, where $n = |G|$, and let ζ be a primitive nth root of unity so that $E = K(\zeta)$. Now both K and and $F(\zeta)$ are splitting fields and so E is a normal extension of each of the subfields. These extensions are Galois extensions since they are also separable for the reason that all fields in sight have characteristic zero.

Since K is invariant under $\mathrm{Gal}(E/F)$, there is a restriction homomorphism $\phi : \mathrm{Gal}(E/F) \to G$, $\sigma \mapsto \sigma|_K$, with kernel $\mathrm{Gal}(E/K)$, mapping each F-automorphism of E to the automorphism it induces on K. Let $H = \mathrm{Gal}(E/F(\zeta))$, a normal subgroup of $\mathrm{Gal}(E/F)$. The restriction of the homomorphism ϕ to H is injective since any element $\tau \in H$ fixing K point-wise fixes ζ and hence fixes $E = K(\zeta)$ point-wise. Thus ϕ embeds H into the solvable group $G = \mathrm{Gal}(K/F)$, and so H itself is solvable.

Next, form the subnormal series

$$H = H_0 \trianglerighteq H_1 \trianglerighteq \cdots \trianglerighteq H_m = 1,$$

where each quotient H_i/H_{i+1}, $i = 0, 1, \ldots, m - 1$, is cyclic of prime order. Next set $E_i = \mathrm{inv}_{H_i}(E)$; then E_{i+1} is Galois over E_i and $\mathrm{Gal}(E_{i+1}/E_i) \cong H_i/H_{i+1}$, which is cyclic of prime order. Apply Lemma 11.9.4 to infer that E_{i+1} is a radical extension of E_i, $i = 0, 1, \ldots, m - 1$. Since the field at the bottom of this chain, $E_0 = \text{inv}_H(E) = F(\zeta)$ is patently a radical extension of F, we see that E is a radical extension of F. The result now follows. $\square$

Corollary 11.9.6 (Galois' solvability criterion) *Assume $f(x) \in F[x]$ where F has characteristic zero. Then the polynomial $f(x)$ is solvable by radicals if and only if the Galois group $\text{Gal}(K/F)$ of the splitting field K for $f(x)$ is a solvable group.*

Proof This is just a combination of Theorems 11.9.3 and 11.9.5. $\square$

Example 56 Let $E = \mathbf{GF}(2)$ and let $F = E(x)$ be the field of rational functions in the indeterminate x. Let $f(y) = y^2 + xy + x \in F[y]$, irreducible by Eisenstein (see Exercise (8) in Sect. 9.13.1, p. 318) and separable as $\partial f(y) = x \neq 0$. Thus, the splitting field K over F for $f(y)$ is a Galois extension of degree 2, and therefore has Galois group $G = \text{Gal}(K/F)$ cyclic of order 2. We shall argue, however, that $f(y)$ is not solvable by radicals. Thus, assume that $F = F_0 \subseteq F_1 \subseteq \cdots \subseteq F_r \supseteq K$ is a root tower over F, say $F_i = F_{i-1}(a_i)$, where $a_i^{n_i} \in F_{i-1}$. Next let m be the least common multiples of the exponents $n_1, n_2, \ldots, n_r$, and let $L \supseteq F_r$ be the splitting field over F_r for the polynomial $x^n - 1$. Let ζ be a generator of the cyclic group of zeros of $x^n - 1$ in L.

Note that $F(\zeta) = E(\zeta)(x)$ and—again using Eisenstein—$f(y)$ continues to be separable and irreducible over $F(\zeta)$. Therefore $\text{Gal}(F(\zeta)(\alpha)/F(\zeta))$ is cyclic of order 2. We set $E' = E(\zeta)$, a finite field of characteristic 2 over which the polynomial $x^n - 1$ splits completely. At the same time, we set $F_i' = F_i(\zeta)$, $i = 0, 1, \ldots, r$, so that $F_0' = F(\zeta) = E'(\zeta)$.

Next, note that if $n_r = m2^e$, where m is odd, then set $b_r = a_r^{2^e}$ and obtain the subextension:

$$F_{r-1}' \subseteq F_{r-1}'(b_r) \subseteq F_{r-1}'(a_r) = F_r'.$$

By Corollaries 11.5.7 and 11.5.11 we may conclude that $F_{r-1}'(b_r)$ is precisely the subfield $(F_r')_{\text{sep}}$ of F_r'. If $\alpha \in K$ is a zero of $f(y)$, then α is separable over F', and hence is separable over every subfield of F_r'. This puts $\alpha \in (F_r')_{\text{sep}} = F_{r-1}'(b_r)$. Next, as F_{r-1}' contains all nth roots of unity, we may apply Lemma 11.9.2 to infer that $\text{Gal}(F_{r-1}'(b_r)/F_{r-1}')$ is not only cyclic but is of order a divisor of m, which is odd. But, since α is separable over F', it is separable over F_{r-1}'. As α satisfies a polynomial of degree 2, and since

$$F_{r-1}' \subseteq F_{r-1}'(\alpha) \subseteq F_{r-1}'(b_r),$$

we conclude immediately that $\alpha \in F_{r-1}'$. Continuing this argument, we eventually reach the conclusion that $\alpha \in F_0' = F'$, which is false. Therefore the polynomial $f(y) = y^2 + xy + x \in F[x]$ is not solvable by radicals despite having a Galois group which is cyclic of order 2.

11.10 The Primitive Element Theorem

It seems to be a tradition that a course in higher algebra must include the "primitive element theorem" (Theorem 11.10.2 below). Of course it is of interest, but its usefulness seems vastly over-rated. It is not used to prove any further theorem in this book and the reader who has better things to do is invited to skip this subsection.

In many of our concrete discussions of splitting fields of polynomials, we obtained such fields over the base field by adjoining several field elements. For example, an oft-quoted example is that of the splitting field K over the rational field $\mathbb{Q}$ of the irreducible polynomial $x^4 - 2 \in \mathbb{Q}[x]$. Since the (complex) zeros of this polynomial are $\pm \omega, \pm i\omega$, where $\omega = \sqrt[4]{2}$, the definition of splitting field might compel us to describe K by listing all of the zeros, adjoined to $\mathbb{Q}$: $K = \mathbb{Q}(\omega, -\omega, i\omega, -i\omega)$. However, it's obvious that the second and fourth roots listed above are superfluous, and so we can write K more simply as $K = \mathbb{Q}(\omega, i\omega)$. Of course, there are variants of this representation, as a moment's thought reveals that also $K = \mathbb{Q}(i, \omega)$. At this stage, one might inquire as to whether one can write K as a *simple* extension of $\mathbb{Q}$, that is, as $K = \mathbb{Q}(\alpha)$, for some judiciously chosen element $\alpha \in K$. In this case, we shall take a somewhat random stab at this question and ask whether $K = \mathbb{Q}(i + \omega)$. In this situation, we can argue in the affirmative, as follows. If $G = \text{Gal}(K/\mathbb{Q})$, and if $\sigma \in G$, then clearly $\sigma(i + \omega) = \pm i \pm \omega$, or $\sigma(i + \omega) = \pm i \pm i\omega$. However, one concludes very easily that the elements i, ω and $i\omega$ are $\mathbb{Q}$-linearly independent. Therefore, $\sigma(i + \omega) = i + \omega$ if and only if $\sigma = 1$. That is to say, $\text{Gal}(K/\mathbb{Q}(i + \omega)) = 1$, forcing $\mathbb{Q}(i + \omega) = K$, and we have succeeded in representing K as a simple extension of $\mathbb{Q}$.

Henceforth, if $F \subseteq K$ is a simple field extension, then any element $\alpha \in K$ with $K = F(\alpha)$ is called a *primitive element* of K over F.

The theorem below gives a somewhat surprising litmus test for a finite extension to be a simple field extension.

Theorem 11.10.1 *Let $F \subseteq K$ be a field extension of finite degree. Then K has a primitive element over F if and only if there are only a finite number of subfields between F and K.*

Proof Assume first that $K = F(\alpha)$ for some $\alpha \in K$. Set $f(x) = \text{Irr}_F(\alpha)$, and assume that E is an intermediate subfield: $F \subseteq E \subseteq K$. Let $g(x) = \text{Irr}_E(\alpha)$; clearly $g(x)|f(x)$. Furthermore, if $E' \subseteq K$ is the subfield generated over F by the coefficients of $g(x)$, then it is clear that $E' \subseteq E$. Furthermore, we also have that $g(x) = \text{Irr}_{E'}(\alpha)$, and so

$$[K : E] = \deg g(x) = [K : E'],$$

forcing $E = E'$. Therefore, we see that subfields of K containing F are generated by coefficients of factors of $f(x)$ in $K[x]$. There are clearly finitely many such factors, so we are finished with this case.

Conversely, assume that there are only finitely many subfields of K containing F. As $[K : F] < \infty$, we clearly have $K = F(\alpha_1, \alpha_2, \ldots, \alpha_r)$ for suitable elements

$\alpha_1, \alpha_2, \ldots, \alpha_r \in K$. By induction, we shall be finished provided that we can show that for any pair of elements $\alpha, \beta \in K$, $F(\alpha, \beta)$ has a primitive element over F. Clearly, we may assume that the field F is infinite. Our hypothesis therefore implies that there must exist elements $a \neq b \in F$ with $F(\alpha + a\beta) = F(\alpha + b\beta)$. But then,

$$\beta = (a - b)^{-1}(\alpha + a\beta - \alpha - b\beta) \in F(\alpha + a\beta).$$

Thus $\alpha = \alpha + a\beta - a\beta \in F(\alpha + a\beta)$. Thus $F(\alpha, \beta) = F(\alpha + b\beta)$, proving the desired result. $\square$

Remark The assumption above that $F \subseteq K$ be a finite-degree extension is crucial. Indeed, if F is any field, if x is indeterminate over F, then $K = F(x)$ has infinitely many subfields, including $F(x^n)$, $n = 1, 2, \ldots$.

From Theorem 11.10.1 we extract the following main result.

Theorem 11.10.2 (Primitive Element Theorem) *Let $F \subseteq K$ be a finite-degree, separable field extension. Then K contains a primitive element over F.*

Proof Given that the extension $F \subseteq K$ has finite degree and is separable, we may write $K = F(\alpha_1, \alpha_2, \ldots, \alpha_r)$ where $f_i(x) = \mathrm{Irr}_F(\alpha_i)$, $i = 1, 2, \ldots, r$ are separable polynomials. Therefore, if E is a splitting field over F for the product $f_1(x)f_2(x) \cdots f_r(x)$, then E is Galois over F, is of finite degree over F, and hence its Galois group $G = \mathrm{Gal}(E/F)$ is a finite group. Since the subfields of E containing F are in bijective correspondence with the subgroups of G, we see, a fortiori, that there can only be finitely many subfields between F and K. Now apply Theorem 11.10.1. $\square$

11.11 Transcendental Extensions

11.11.1 Simple Transcendental Extensions

Suppose K is some extension field of F. Recall that an element α of K is said to be algebraic over F if and only if there exists a polynomial $p(x) \in F[x]$ such that $p(\alpha) = 0$. Otherwise, if there is no such polynomial, we say that α is *transcendental over F* and that $F(\alpha)$ is a *simple transcendental extension of F*.

Recall further that a *rational function* over F is any element in the field of quotients $F(x)$ of the integral domain $F[x]$.[13] Thus every rational function $r(x)$ is a quotient $f(x)/g(x)$ of two polynomials in $F[x]$, where $g(x)$ is not the zero polynomial, and $f(x)$ and $g(x)$ are relatively prime. For each element $\alpha \in K$, we understand $r(\alpha)$ to be the element

---

[13]Note that round parentheses distinguish the field $F(x)$ from the polynomial ring $F[x]$.

$$r(\alpha) := f(\alpha) \cdot (g(\alpha))^{-1},$$

provided α is not a zero of $g(x)$. If α is transcendental over F, then $r(\alpha)$ is always defined.

Lemma 11.11.1 *If $F(\alpha)$ is a transcendental extension of F, then there is an isomorphism $F(\alpha) \to F(x)$ taking α to x.*

Proof Define the mapping $\epsilon : F(x) \to F(\alpha)$ by setting $\epsilon(r(x)) := r(\alpha)$, for every $r(x) \in F(x)$. Clearly ϵ is a ring homomorphism. Suppose

$$r_1(x) = \frac{a(x)}{b(x)}, \quad r_2(x) = \frac{c(x)}{d(x)},$$

are rational functions with $r_1(\alpha) = r_2(\alpha)$. We may assume $a(x), b(x), c(x)$ and $d(x)$ are all polynomials with $b(x)d(x) \neq 0$. Then

$$a(\alpha)d(\alpha) - b(\alpha)c(\alpha) = 0.$$

Since α is transcendental, $a(x)d(x) = b(x)c(x)$ so $r_1(x) = r_2(x)$. Thus ϵ is injective. But its image is a subfield of $F(\alpha)$ containing F and the element $\epsilon(x) = \alpha$ and so it is also surjective. Thus ϵ is bijective, and so ϵ^{-1} is the desired isomorphism. $\square$

Now suppose $F(\alpha)$ is a simple transcendental extension of F, and let β be any element of $F(\alpha)$. By Lemma 11.11.1, there exist a coprime pair of polynomials $(f(x), g(x))$, with $g(x) \neq 0 \in F[x]$, such that

$$\beta = f(\alpha)(g(\alpha))^{-1}.$$

Suppose

$$f(x) = a_0 + a_1 x + \cdots + a_n x^n$$
$$g(x) = b_o + b_1 x + \cdots + b_n x^n$$

where n is the maximum of the degrees of $f(x)$ and $g(x)$, so at least one of a_n or b_n is non-zero. Then $f(\alpha) - \beta g(\alpha) = 0$, so

$$(a_0 - \beta b_0) + (a_1 - \beta b_1)\alpha + \cdots + (a_n - \beta b_n)\alpha^n = 0.$$

Thus α is a zero of the polynomial

$$h_\beta(x) := (a_0 - \beta b_0) + (a_1 - \beta b_1)x + \cdots + (a_n - \beta b_n)x^n$$

in $F(\beta)[x]$.

Claim *The polynomial $h_\beta(x)$ is the zero polynomial if and only if $\beta \in F$.*

First, if $\beta \in F$, then $h_\beta(x) = f(x) - \beta g(x)$ is in $F[x]$. If $h_\beta(x)$ were a non-zero polynomial, then its zero, α, would be algebraic over F, which it is not. Thus $h_\beta(x)$ is the zero polynomial.

On the other hand, if $h_\beta(x)$ is the zero polynomial, each coefficient $a_i - \beta b_i$ is zero. Now, since $g(x)$ is not the zero polynomial, for some index j, $b_j \neq 0$. Then $\beta = a_j \cdot b_j^{-1}$, an element of F.

The claim is proved.

Now suppose β is not in F. Then the coefficient $(a_i - \beta b_i)$ is zero if and only if $a_i = 0 = b_i$, and this cannot happen for $i = n$ as n was chosen. Thus $h_\beta(x)$ has degree n. So certainly α is algebraic over the intermediate field $F(\beta)$. Since $[F(\alpha) : F]$ is infinite, and $[F(\alpha) : F(\beta)]$ is finite, $[F(\beta) : F]$ is infinite, and so β is transcendental over F.

We shall now show that $h_\beta(x)$ is irreducible in $F(\beta)[x]$.

Let $D := F[y]$ be the domain of polynomials in the indeterminate y, with coefficients from F. (This statement is here only to explain the appearance of the new symbol y.) Then $F(y)$ is the quotient field of D. Since β is transcendental over F, Lemma 11.11.1 yields an F-isomorphism

$$\epsilon : F(\beta) \to F(y)$$

of fields which can be extended to a ring isomorphism

$$\epsilon^* : F(\beta)[x] \to F(y)[x].$$

Then $h(x) := \epsilon^*(h_\beta(x)) = f(x) - yg(x)$ is a polynomial in $D[x] = F[x, y]$. Moreover, $h_\beta(x)$ is irreducible in $F(\beta)[x]$ if and only if $h(x)$ is irreducible in $F(y)[x]$. Now since $D = F[y]$ is a UFD, by Gauss' Lemma, $h(x) \in D[x]$ is irreducible in $F(y)[x]$ if and only if it is irreducible in $D[x]$. Now, $D[x] = F[y, x]$ is a UFD and $h(x)$ is degree one in y. So, if $h(x)$ had a non-trivial factorization in $D[x]$, one of the factors $k(x)$ would be of degree zero in y and hence a polynomial of positive degree in $F[x]$. The factorization is then

$$h(x) = f(x) - yg(x) = k(x)(a(x) + yb(x)),$$

where $k(x)$, $a(x)$ and $b(x)$ all lie in $F[x]$. Then $f(x)$ and $g(x)$ possess the common factor $k(x)$ of positive degree, against the fact that $f(x)$ and $g(x)$ were coprime in $F[x]$. We conclude that $h(x)$ is irreducible in $F(y)[x]$, and so $h_\beta(x)$ is irreducible of degree n in $F(\beta)[x]$ as promised.

We now have the following

Theorem 11.11.2 (Lüroth's Theorem) *Suppose $F(\alpha)$ is a transcendental extension of F. Let β be an element of $F(\alpha) - F$. Then we can write $\beta = f(\alpha)/g(\alpha)$ where $f(x)$ and $g(x)$ are a pair of coprime polynomials in $F[x]$, unique up to multiplying* *the pair through by a scalar from F. Let n be the maximum of the degrees of $f(x)$ and $g(x)$. Then $[F(\alpha) : F(\beta)] = n$. In particular, α is algebraic over $F(\beta)$.*

Remark A general consequence is that if E is an intermediate field distinct from F— that is, $F < E \leq F(\alpha)$—then $[E : F]$ is infinite. In particular, $F(\alpha)$—F contains no elements which are algebraic over F (a fact that has an obvious direct proof from first principles).

Lüroth's theorem gives us a handle on the Galois group $\mathrm{Gal}(F(\alpha)/F)$ of a transcendental extension. Suppose now that σ is an F-automorphism $F(\alpha) \rightarrow F(\alpha)$. Then σ is determined by what it does to the generator α. Thus, setting $\beta = \sigma(\alpha)$, for any rational function $t(x)$ of F, we have $\sigma(t(\alpha)) = t(\beta)$. Now in particular $\beta = r(\alpha)$ for a particular rational function $r(x) = f(x)/g(x)$ where $(f(x), g(x))$ is a coprime pair of polynomials in $F[x]$. Since σ is onto, $F(\alpha) = F(\beta)$. On the other hand, by Lüroth's Theorem, $[F(\alpha) : F(\beta)]$ is the maximum of the degrees of $f(x)$ and $g(x)$. Thus we can write

$$\beta = \frac{a\alpha + b}{c\alpha + d}, \tag{11.33}$$

where a, b, c and d are in F, a and c are not both zero, and $ad - bd \neq 0$ to keep the polynomials $ax + b$ and $cx + d$ coprime. Actually the latter condition implies the former, that a and c are not both zero.

Now let $G = \mathrm{Gal}(F(\alpha)/F)$ and set $O := \alpha^G$, the orbit of α under G. Our observations so far are that there exists a bijection between any two of the following three sets:

1. the elements of the group $\mathrm{Gal}(F(\alpha)/F)$,
2. the orbit O, and
3. the set $LF(F)$ of so-called "linear fractional" transformations,

$$\omega \rightarrow \frac{a\omega + b}{c\omega + d}, \ a, b, c, d \in F, \ ad - cb \ \text{non-zero},$$

viewed as a set of permutations of the elements of O.

A perusal of how these linear factional transformations compose shows that $LF(F)$ is a group, and that in fact, there is a surjective group homomorphism

$$GL(2, F) \rightarrow LF(2, F),$$

defined by

$$\begin{pmatrix} a & b \\ c & d \end{pmatrix} \rightarrow \left[\text{ the transformation } z \rightarrow \frac{az + b}{cz + d} \text{ on set } O \right],$$

where the matrix has non-zero determinant—i.e. is invertible. The kernel of this homomorphism is the subgroup of two-by-two scalar matrices γI, comprising the center of $GL(2, F)$. As the reader will recall, the factor group, $GL(2, F)/Z(GL(2, F))$ is called the "one-dimensional projective group over F" and is denoted $PGL(2, F)$. Our observation, then, is that

$$\text{Gal}(F(\alpha)/F) \simeq LF(F) \simeq PGL(2, F).$$

Notice how the fundamental theorem of Galois Theory falls to pieces in the case of transcendental extensions. Suppose F is the field $\mathbb{Z}/p\mathbb{Z}$, the ring of integers mod p, a prime number. If α is transcendental over F, then $[F(\alpha) : F]$ is infinite, while $\text{Gal}(F(\alpha)/F)$ is $PGL(2, p)$, a group of order $p(p^2 - 1)$ which, when $p \geq 5$, contains at index two, a simple subgroup.

Let K be an extension of the field F. We define a notion of *algebraic dependence over F* on the elements of K as follows: We say that element α *algebraically depends* (over F) on the finite subset $\{\beta_1, \ldots, \beta_n\}$ if and only if

- α *is algebraic over the subfield* $F(\beta_1, \ldots, \beta_n)$.

Theorem 11.11.3 *Algebraic dependence over F is an abstract dependence relation on K in the sense of Chap. 2, Sect. 2.6.*

Proof We must show that the relation of algebraic dependence over F satisfies the three required properties of a dependence relation: reflexivity, transitivity and the exchange condition.

Clearly if $\beta \in \{\beta_1, \ldots, \beta_n\}$, then β lies in $F(\beta_1, \ldots, \beta_n)$ and so is algebraic over it. So the relation is reflexive.

Now suppose γ is algebraic over $F(\alpha_1, \ldots, \alpha_m)$, and each α_i is algebraic over $F(\beta_1, \ldots, \beta_n)$. Then certainly, α_i is algebraic over the field

$$F(\beta_1, \ldots, \beta_n, \alpha_1, \ldots, \alpha_{i-1}).$$

So, setting $B := \{\beta_1, \ldots, \beta_n\}$, we obtain a tower of fields,

$$F(B) \leq F(B, \alpha_1) \leq F(B, \alpha_1, \alpha_2) \leq \cdots$$
$$\leq F(B, \alpha_1, \ldots, \alpha_n) \leq F(\beta_1, \ldots, \beta_m, \alpha_1, \ldots, \alpha_n, \gamma)$$

with each field in the tower of finite degree over its predecessor. Thus the top field in the tower is a finite extension of the bottom field, and so γ is algebraic over $F(B)$.

We now address the exchange condition: we are to show that if γ is algebraic over $F(\beta_1, \ldots, \beta_n\}$ but is not algebraic over $F(\beta_1, \ldots, \beta_{n-1})$, then β_n is algebraic over the field $F(\beta_1, \ldots, \beta_{n-1}, \gamma\}$.[14]

---

[14]It is interesting to observe that in this context the Exchange Condition is essentially an "implicit function theorem". The hypothesis of the condition, transferred to the context of algebraic geometry, says that the "function" γ is expressible locally in terms of $\{\beta_1, \ldots, \beta_n\}$ but not in terms of $\{\beta_1, \ldots, \beta_{n-1}\}$. Intuitively, this means $\frac{\partial \gamma}{\partial \beta_n} \neq 0$, from which we anticipate the desired conclusion.

By hypothesis, there is a polynomial relation

$$0 = a_k \gamma^k + a_{k-1} \gamma^{k-1} + \cdots + a_1 \gamma + a_0 \qquad (11.34)$$

where a_k is non-zero, and each a_j is in $F(\beta_1, \ldots, \beta_n)$, and so, at the very least, can be expressed as a rational function, n_j/d_j, where the numerators n_j and denominators d_j are polynomials in $F[\beta_1, \ldots, \beta_n]$. We can then multiply Eq. (11.34) through by the product of the d_i, from which we see that without loss of generality, we may assume the coefficients a_i are polynomials—explicitly:

$$a_i = \sum_j b_{ij} \beta_n^j, \quad b_{ij} \in F[\beta_1, \ldots, \beta_{n-1}], \quad a_k \neq 0. \qquad (11.35)$$

Then substituting the formulae of Eq. (11.35) into Eq. (11.34), and collecting the powers of β_n, we obtain the polynomial relation

$$0 = p_m \beta_n^m + p_{m-1} \beta_n^{m-1} + \cdots p_1 \beta_n + p_0, \qquad (11.36)$$

with each coefficient

$$p_t = \sum_{s=0}^k b_{st} \gamma^s, \qquad (11.37)$$

a polynomial in the ring $F[\beta_1, \ldots, \beta_{n-1}, \gamma]$.

Now we get the required algebraic dependence of β_n on $\{\beta_1, \ldots, \beta_{n-1}, \gamma\}$ from Eq. (11.36), provided not all of the coefficients p_t are zero. Suppose, by way of contradiction, that they were all zero. Then as γ does not depend on $\{\beta_1, \ldots, \beta_{n-1}\}$, having each of the expressions in Eq. (11.37) equal to zero forces each coefficient b_{st} to be zero, so by Eq. (11.35), each a_i is zero. But that is impossible since certainly a_k is non-zero.

Thus the Exchange Condition holds and the proof is complete. □

So, as we have just seen, if K is an extension of the field F, then the relation of being algebraically dependent over F is an abstract dependence relation, and we call any subset X of K which is independent with respect to this relation *algebraically independent (over F)*. By the basic development of Chap. 1, Sect. 1.3, maximal algebraically independent sets exist, they "span" K, and any two such sets have the same cardinality. We call such subsets of K *transcendence bases* of K over F, and their common cardinality is called the *transcendence degree* of K over F, and is denoted

$$\mathrm{tr.deg}(K/F).$$

If X is a transcendence basis for K over F, then the "spanning" property means that every element of K is algebraic over the subfield $F(X)$ generated by X. One calls $K(X)$ a "purely transcendental" extension of F, although this notion conveys little since it is completely relative to X. For example, if x is a real number that is transcendental over $\mathbb{Q}$, the field of rational numbers, then both $\mathbb{Q}(x)$ and $\mathbb{Q}(x^2)$ are purely transcendental extensions of $\mathbb{Q}$, while the former is algebraic over the latter.[15]

Next we observe that transcendence degrees add (in the sense of cardinal numbers) under iterated field extension just as ordinary degrees multiply. Specifically:

Corollary 11.11.4 *Suppose we have a tower of fields*

$$F \leq K \leq L.$$

Then

$$tr.deg[K : F] + tr.deg[L : K] = tr.deq[L : F].$$

Proof Let X be a transcendence basis of K over F and let Y similarly be a transcendence basis of L over K. Clearly then, $X \cap Y = \emptyset$, every element of K is algebraic over $F(X)$, and every element of L is algebraic over $K(Y)$.

The corollary will be proved if we can show that $X \cup Y$ is a transcendence basis of L over F.

First, we claim that $X \cup Y$ spans L—that is, every element of L is algebraic over $F(X \cup Y)$. Let λ be an arbitrary element of L. Then there is an algebraic relation

$$\lambda^m + a_m \lambda^{m-1} + \cdots + a_1 \lambda + a_0 = 0,$$

where $m \geq 1$, and the coefficients a_i are elements of $K(Y)$. This means that each of them can be expressed as a quotient of two polynomials in $K[Y]$. Let B be the set of all coefficients of the monomials in Y appearing in the numerators and denominators of these quotients representing a_i, as i ranges over $0, 1, \ldots, m$. Then B is a finite subset of K, and λ is now algebraic over $F(X \cup Y \cup B)$. Since each element of B is algebraic over $F(X)$, $F(X \cup B)$ is a finite extension of $F(X)$, and so $F(X \cup Y \cup B)$ is a finite extension of $F(X \cup Y)$. It then follows that $F(X \cup Y \cup B, \lambda)$ is a finite extension of $F(X \cup Y)$, and so λ is algebraic over $F(X \cup Y)$ as desired.

It remains to show that $X \cup Y$ is algebraically independent over Y. But if not, there is a polynomial $p \in F[z_1, \ldots, z_{n+m}]$ and finite subsets $\{x_1, \ldots, x_m\}$ and $\{y_1, \ldots, y_n\}$ of X and Y, respectively, such that upon substitution of the m x_i for the first m z_i's, and the y_i for the remaining n z_i's, one obtains

$$0 = p(x_1, \ldots, x_m, y_1, \ldots y_m).$$

Now this can be rewritten as a polynomial whose monomial terms are monomials in the y_i, with coefficients which are polynomials c_k in $F[z_1, \ldots, z_n]$ evaluated at $x_1, \ldots, x_n$. Since the y_i are algebraically independent over K, the coefficient polynomials c_k are each equal to zero when evaluated at $x_1, \ldots, x_n$. But again,

---

[15]In this connection, there is a notion that is very useful in looking at the subject of Riemann Surfaces from an algebraic point of view: A field extension K of F of finite transcendence degree over F is called an *algebraic function field* if and only if, for any transcendence basis X, K is a finite extension of $F(X)$.

since the x_j are algebraically independent over F, we see that each polynomial c_k is identically 0. This means that the original polynomial p was the zero polynomial of $F[z_1, \ldots, z_{m+n}]$, establishing the algebraic independence of $X \cup Y$. The proof is complete. $\square$.

Corollary 11.11.5 *An algebraic extension of an algebraic extension is algebraic. Precisely, if $F \leq K \leq L$ is a tower of fields with K an algebraic extension of F and L an algebraic extension of K, then L is an algebraic extension of F.*

Proof This follows from the previous theorem and the fact that zero plus zero is zero.

Remark Before, we knew this corollary anyway, but with the word "finite degree" replacing the word "algebraic" in the statement.

11.12 Exercises

11.12.1 Exercises for Sect. 11.2

1. Compute the minimal polynomials in $\mathbb{Q}[x]$ of each of the following complex numbers. (Recall that if a is a positive real number and n is a positive integer, the symbol $\sqrt[n]{a}$ refers to the positive nth root of a.)

   (a) $\sqrt{2} + \sqrt{3}$.
   (b) $\sqrt{2} + \zeta$, where $\zeta = e^{2\pi i/3}$.
   (c) $\sqrt{2 + \sqrt{2}}$
   (d) $\zeta + \zeta^{-1}$, where $\zeta = e^{2\pi i/16}$
   (e) $\zeta + \zeta^{-1}$, where $\zeta = e^{2\pi i/7}$

2. Let $F \subseteq K$ be a field extension with $[K : F]$ odd. If $\alpha \in K$, prove that $F(\alpha^2) = F(\alpha)$.

3. Assume that $\alpha = a + bi \in \mathbb{C}$ is algebraic over $\mathbb{Q}$, where a is rational and b is real. Prove that $\mathrm{Irr}_{\mathbb{Q}(\alpha)}$ has even degree.

4. Let $K = \mathbb{Q}(\sqrt[3]{2}, \sqrt{2}) \subseteq \mathbb{C}$. Compute $[K : \mathbb{Q}]$.

5. Let $f(x) = x^5 - 9x^3 + 3x + 3 \in \mathbb{Q}[x]$.

   (i) Show that $f(x)$ is irreducible over $\mathbb{Q}$.
   (ii) Show that $f(x)$ is irreducible over $\mathbb{Q}(i)$.

6. Let $K = \mathbb{Q}(\sqrt[4]{2}, i) \subseteq \mathbb{C}$. Show that

   (a) K contains all roots of $x^4 - 2 \in \mathbb{Q}[x]$.
   (b) Compute $[K : \mathbb{Q}]$.

7. Compute

$$\left[\mathbb{Q}\left(\sqrt{2+\sqrt{2+\sqrt{2}}}\right):\mathbb{Q}\right]$$

8. Let $F \subseteq E \subseteq K$ be fields, let $\alpha \in K$, and let $f(x) = \mathrm{Irr}_F(\alpha)$. Assume that $[E:F]$ and $\deg f(x)$ are relatively prime. Prove that $f(x) = \mathrm{Irr}_E(\alpha)$.

9. Let F be any field, and prove that there are infinitely many irreducible polynomials in $F[x]$. [Hint: Euclid's proof of the corresponding result for the ring $\mathbb{Z}$ works here, too.]

10. Let $F = \mathbb{C}(x)$, where $\mathbb{C}$ is the complex number field and x is an indeterminate. Assume that $F \subseteq K$ and that K contains an element y such that $y^2 = x(x-1)$. Prove that there exists an element $z \in F(y)$ such that $F(y) = \mathbb{C}(z)$, i.e., $F(y)$ is a "simple transcendental extension" of $\mathbb{C}$.

11. Let $F \subseteq K$ be a field extension. If the subfields of K containing F are totally ordered by inclusion, prove that K is a simple extension of F. (Is the converse true?)

12. Let $\mathbb{Q} \subseteq K$ be a field extension. Assume that K is closed under taking square roots, i.e., if $\alpha \in K$, then $\sqrt{\alpha} \in K$. Prove that $[K:\mathbb{Q}] = \infty$.

13. Suppose the field F is a subring of the integral domain R. If every element of R is algebraic over F, show that R is actually a field. Give an example of a non-integral domain R containing a field F such that every element of R is algebraic over F. Obviously, R cannot be a field.

14. Let $F \subseteq K$ be fields and let $f(x),\ g(x) \in F[x]$ with $f(x)|g(x)$ in $K[x]$. Prove that $f(x)|g(x)$ in $F[x]$.

15. Let $F \subseteq K$ be fields and let $f(x),\ g(x) \in F[x]$. If $d(x)$ is the greatest common denominator of $f(x)$ and $g(x)$ in $F[x]$, prove that $d(x)$ is the greatest common denominator of $f(x)$ and $g(x)$ in $K[x]$.

16. Let $F \subseteq E_1, E_2 \subseteq E$ be fields. Define $E_1 E_2 \subseteq E$ to be the smallest field containing both E_1 and E_2. $E_1 E_2$ is called the *composite* (or *compositum*) of the fields E_1 and E_2. Prove that if $[E:F] < \infty$, then $[E_1 E_2 : F] \le [E_1 : F] \cdot [E_2 : F]$.

17. Given a complex number α it can be quite difficult to determine whether α is algebraic or transcendental. It was known already in the nineteenth century that π and e are transcendental, but the fact that such numbers as e^π and $2^{\sqrt{2}}$ are transcendental is more recent, and follows from the following deep theorem of Gelfond and Schneider: *Let α and β be algebraic numbers. If*

$$\eta = \frac{\log \alpha}{\log \beta}$$

is irrational, then η is transcendental. (See E. Hille, American Mathematical Monthly, vol. 49 (1942), pp. 654–661.) Using this result, prove that $2^{\sqrt{2}}$ and e^π are both transcendental. [Hint: For $2^{\sqrt{2}}$, set $\alpha = 2^{\sqrt{2}}$, $\beta = 2$.]

11.12.2 Exercises for Sect. 11.3

1. Let $f(x) = x^n - 1 \in \mathbb{Q}[x]$. In each case below, construct a splitting field K over $\mathbb{Q}$ for $f(x)$, and compute $[K : \mathbb{Q}]$.

   (i) $n = p$, a prime.
   (ii) $n = 2p$, where p is prime.
   (iii) $n = 6$.
   (iv) $n = 12$.

2. Let $f(x) = x^n - 2 \in \mathbb{Q}[x]$. Construct a splitting field for $f(x)$ over $\mathbb{Q}$.
3. Let $f(x) = x^3 + x^2 - 2x - 1 \in \mathbb{Q}[x]$.

   (a) Prove that $f(x)$ is irreducible.
   (b) Prove that if $\alpha \in \mathbb{C}$ is a root of $f(x)$, so is $\alpha^2 - 2$.
   (c) Let $K \supseteq \mathbb{Q}$ be a splitting field over $\mathbb{Q}$ for $f(x)$. Using part (b), compute $[K : \mathbb{Q}]$.

4. Let $\zeta = e^{2\pi i/7} \in \mathbb{C}$, and let $\alpha = \zeta + \zeta^{-1}$. Show that $\mathrm{Irr}_{\mathbb{Q}}(\alpha) = x^3 + x^2 - 2x - 1$ (as in Exercise 3 above), and that $\alpha^2 - 2 = \zeta^2 + \zeta^{-2}$.
5. Write the complex splitting field for $x^3 - 2 \in \mathbb{Q}[x]$ in the form $\mathbb{Q}(\alpha)$, for some $\alpha \in \mathbb{C}$.
6. Let $\zeta = e^{2\pi i/n} \in \mathbb{C}$ and set $\omega = \zeta + \zeta^{-1}$. Show that $\mathbb{Q}(\omega)$ is a normal extension of $\mathbb{Q}$.
7. Give an example of a normal extension $\mathbb{Q} \subseteq K$ such that $[K : \mathbb{Q}] = 3$.
8. Let $F = \mathbb{C}(x)$ be the field of rational functions over the complex field $\mathbb{C}$ and let $f(y) = y^3 - x \in F[y]$. Let α be a zero of $f(y)$ in some splitting field over F and show that $F(\alpha)$ is a normal extension of F.
9. Let $E = \mathbb{Q}(\sqrt[5]{2})$, and let K be a normal closure of E over $\mathbb{Q}$. Compute $[K : \mathbb{Q}]$.
10. Which of the following simple extensions of the rational field are normal?

   (a) $\mathbb{Q}(\sqrt{2} + \sqrt{3})$,
   (b) $\mathbb{Q}\left(\sqrt{2 + \sqrt{2}}\right)$,
   (c) $\mathbb{Q}\left(\sqrt{2 + \sqrt{2 + \sqrt{2}}}\right)$,
   (d) $\mathbb{Q}(\zeta\sqrt[4]{2})$, where $\zeta = e^{2\pi i/8}$.

11.12.3 Exercises for Sect. 11.4

1. Let q be a prime power, set $F = GF(q)$ and let $E \supseteq F$ be a field extension of degree n. Let σ be the qth power automorphism of E (sometimes called the *Frobenius map* of E), given by $F(\alpha) = \alpha^q$. Define the *norm map*

$$N = N_{E/F} : E \longrightarrow F$$

by setting

$$N(\alpha) = \alpha \cdot \sigma(\alpha) \cdot \sigma^2(\alpha) \cdots \sigma^{n-1}(\alpha).$$

Note that N restricts to a mapping

$$N : E^* \longrightarrow F^*,$$

where E^* and F^* are the multiplicative groups of no-zero elements of E and F.

(a) Show that $N : E^* \to F^*$ is a group homomorphism.

(b) Show that $|\ker N| = \frac{q^n - 1}{q - 1}$.

2. Let p be a prime and let r be a positive integer. Prove that there exists an irreducible polynomial of degree r over $F = \mathbf{GF}(p)$. [Hint: Isn't this equivalent to the existence of an extension field $K \supseteq F$ of degree r?]

3. Let p be prime, n a positive integer and set $q = p^n$. Let $F = \mathbf{GF}(p)$ and show that if $f(x) \in F[x]$ is irreducible of degree n, then $f(x) | x^q - x$. More generally, show that if $f(x)$ is irreducible of degree m, where $m | n$, then again, $f(x) | x^q - x$.

4. Show that if $F = \mathbf{GF}(q)$, then the polynomial $x^2 + x + 1 \in F[x]$ is irreducible if and only if $3 \nmid q - 1$.

5. For any integer n, let D_n be the number of irreducible polynomials of degree n in $F[x]$, where $F = \mathbf{GF}(q)$. Prove that

$$D_n = \frac{1}{n} \sum_{k|n} \mu\left(\frac{n}{k}\right) q^k.$$

[Hint: Note that, by Exercise 3, $q^n = \sum_{k|n} k \cdot D_k$; now use the Möbuis Inversion Theorem (see Example 43, Theorem 7.4.3, p. 217).]

6. Here's another application of Möbius inversion. Let F be a field and let C be a finite multiplicative subgroup of the multiplicative group F^* of non-zero elements of F. We know C to be cyclic. Assume that $|C| = n$, and $d | n$, and let N_d be the sum in F of the elements of order d in C. Thus N_n is the sum in F of the $\phi(n)$ generators of C. Prove that, in fact,

$$N_n = \mu(n).$$

[Hint: Study $f(n) = \sum_{d|n} N_d$; how does this relate to the polynomial $x^n - 1 \in F[x]$?)]

7. Let $F = \mathbf{GF}(q)$, where $q = p^n$ and p is an odd prime.

(i) Show that for any $\alpha \in F$, and any prime power $r = p^k$, one has

$$(x - \alpha)^{r-1} = x^{r-1} + \alpha x^{r-2} + \cdots + \alpha^{r-2}x + \alpha^{r-1}.$$

(ii) Suppose m is an integer between 1 and $p-1$, inclusive. Prove that if α is a zero of a polynomial $f(x) \in F[x]$ with multiplicity exactly m, then α is a zero of the formal derivative $f'(x)$ of multiplicity exactly $m-1$. [Hint: Write $f(x) = (x-\alpha)^m g(x)$ where $g(\alpha) \neq 0$, then apply the product rule for the formal derivative.]

(iii) Suppose m is as it is in part (ii) of this problem. For fixed non-zero $\alpha \in F$, define the polynomial

$$f_{m,\alpha}(x) := \sum_{i=1}^{p-1} i^m x^i \alpha^{p-i}.$$

(When $\alpha = 1$ and $m = (p-1)/2$ these are known as the *Fekete polynomials*.) Show that α is a root of $f_{m,\alpha}(x)$ with multiplicity exactly m. [Hint: Use (i) to observe that

$$f_{0,\alpha}(x) + \alpha^{p-1} = (x_\alpha)^{p-1}.$$

Then observe that for $m \geq 1$, $f_{m+1,\alpha}(x) = x f'_{m,\alpha}(x)$ and apply part (ii) of this problem.]

11.12.4 Exercises for Sect. 11.5

1. Let $k = \mathbf{GF}(3)$, and set $F = k(x)$, the field of rational functions over k in x. Let $f(y) = y^6 + x^2 y^3 + x \in F[y]$.

   (a) Show that $f(y)$ is irreducible over F.
   (b) Let α be a root of $f(y)$ in some splitting field over F. Is $F(\alpha)$ separable over F?
   (c) If $K = F(\alpha)$ as above, determine $[K : F]$ and $[K_{\text{sep}} : F]$.

2. Let F be a field of characteristic p. Determine the number of roots of the polynomial $x^n - 1 \in F[x]$ in some splitting field. [Hint: write $n = mp^e$, where p does not divide m.]

11.12.5 Exercises for Sect. 11.6

1. Let $F \subseteq K$ be a finite extension of fields, and let E_1, E_2 be two intermediate subfields. Assume that $F \subseteq E_1$ is a Galois extension and that K is generated over F by E_1 and E_2. Prove that

   (a) $E_2 \subseteq K$ is a Galois extension, and that
   (b) $\mathrm{Gal}(K/E_2) \cong \mathrm{Gal}(E_1/E_1 \cap E_2)$.

2. Assume that $F \subseteq K$ is a field extension of finite degree, and let E_1, E_2 be intermediate subfields, both Galois over over F, and generating K. Prove that K is Galois over E_2 and that $\mathrm{Gal}(K/E_2)$ is isomorphic to a subgroup of $\mathrm{Gal}(E_1/F)$. (This is the so-called *Lemma on Accessory Irrationalities*.)

11.12.6 Exercises for Sect. 11.8

1. Using Möbius inversion, prove that

$$\Phi_n(x) = \prod_{d \mid n} (x^d - 1)^{\mu(n/d)}.$$

2. Consider the polynomial $x^4 + 3 \in \mathbb{Q}[x]$, and let $K \subseteq \mathbb{C}$ be its splitting field.

   (a) Show that $K = \mathbb{Q}(i, \omega)$, where $\omega = \zeta \sqrt[4]{3}$, and where $\zeta = e^{2\pi i/8}$.
   (b) Compute $[K : \mathbb{Q}]$.
   (c) If we write the zeros of $f(x)$ as $a_1 = \omega, a_2 = -\omega, a_3 = i\omega, a_4 = -i\omega$, calculate the stabilizer in G of a_1, with its elements written in cycle notation. (So for instance, an element $\sigma = (2\ 3\ 4)$ would fix a_1 and do this:

   $$a_2 \overset{\sigma}{\mapsto} a_3 \overset{\sigma}{\mapsto} a_4 \overset{\sigma}{\mapsto} a_2.$$

   Incidentally, does such an element exist in G? Why or why not?)
   (d) Is G abelian? Why or why not?
   (e) Assuming that you determined that G is, in fact, non-abelian, give a non-normal subfield E, with $\mathbb{Q} \subseteq E \subseteq K$.
   (f) List all the elements of G, using cycle notation.

3. Let $\alpha = \sqrt{2} + \sqrt{3}$, and set $p(x) = \mathrm{Irr}_{\mathbb{Q}}(\alpha)$. Compute the Galois group of $p(x)$.

4. Let $\alpha = \sqrt{2 + \sqrt{2 + \sqrt{2}}}$, let $q(x) = \mathrm{Irr}_{\mathbb{Q}}()$, and compute the Galois group of $q(x)$.

5. Are the Galois groups of the polynomials $x^8 - 2$, $x^8 - 3$, and $x^8 - 5$ in $\mathbb{Q}[x]$ all isomorphic? Investigate.

6. Let $\zeta = e^{2\pi i/32}$, and set $K = \mathbb{Q}(\zeta)$.

   (a) Show that K is a separable normal extension of $\mathbb{Q}$.
   (b) Let α be as in Problem 4, above. Show that $\mathbb{Q}(\alpha) \subseteq K$ and compute $[K : \mathbb{Q}(\alpha)]$. [Hint: Look at $\zeta + \zeta^{-1}$.]
   (c) Compute $\mathrm{Irr}_{\mathbb{Q}}(\zeta) \in \mathbb{Q}[x]$.

7. Let $f(x) = x^6 - 4x^3 + 1 \in \mathbb{Q}[x]$.

   (a) Show that if $\omega = \sqrt[3]{2 + \sqrt{3}}$ is real and $\zeta = e^{2\pi i/3} \in \mathbb{C}$, then the complex splitting field K of $f(x)$ is $\mathbb{Q}(\zeta, \omega)$.

(b) Show that $f(x)$ is irreducible. [Hint: If it's not, it has a linear, quadratic, or cubic factor. Use the factorization of $f(x) \in K[x]$ into linear factors and show that the product of at most three of these factors cannot be in $\mathbb{Q}[x]$.]

(c) Label the zeros of $f(x)$ as $a_1 = \omega$, $a_2 = \zeta\omega$, $a_3 = \zeta^2\omega$, $a_4 = \omega^{-1}$, $a_5\zeta\omega^{-1}$, $a_6 = \zeta^2\omega^{-1}$. Show

   i. With the above correspondence, show that the following elements are in G:

$$\gamma = (2\,3)(5\,6) \text{ (complex conjugation)},$$

$$\sigma = (1\,2\,3)(4\,6\,5), \ \tau = (1\,4)(2\,5)(3\,6).$$

   ii. Show that

$$\mathrm{Gal}(K/\mathbb{Q}(\zeta)) = \langle \tau, \sigma \rangle, \ \mathrm{Gal}(K/\mathbb{Q}(\omega + \omega^{-1})) = \langle \gamma, \tau \rangle,$$

$$G = \langle \gamma, \ \sigma, \ \tau \rangle \cong D_{12}, \ \ \mathbb{Z}(G) = \langle \gamma\tau \rangle,$$

   where D_{12} is the dihedral group of order 12.

   iii. Show that $\mathbb{Q} \subseteq \mathbb{Q}(\sqrt{3}) \subseteq K$ and that $\mathrm{Gal}(K/\mathbb{Q}(\sqrt{3})) = \langle \gamma, \sigma \rangle$.

   iv. Show that $\mathbb{Q} \subseteq \mathbb{Q}(i) \subseteq K$ and that $\mathrm{Gal}(K/\mathbb{Q}(i)) = \langle \sigma\gamma\tau \rangle$.

8. Let $f(x) = x^6 - 2x^3 - 2 \in \mathbb{Q}[x]$.

   (a) Let $\alpha = \sqrt[3]{1 + \sqrt{3}}$, $\beta = \sqrt[3]{1 - \sqrt{3}} \in \mathbb{R}$, and let $\zeta = e^{2\pi i/3} \in \mathbb{C}$. Show that the zeros of $f(x)$ are $\zeta^j\alpha$, $\zeta^j\beta$, $j = 0, 1, 2$.

   (b) Show that the complex splitting field K of $f(x)$ over $\mathbb{Q}$ contains the field $L = \mathbb{Q}(i, \sqrt[3]{2}, \sqrt{3})$.

   (c) Show that $[L : \mathbb{Q}] = 12$.

   (d) Show that $[K : \mathbb{Q}] = 12$ or 36.

   (e) Let $\omega = \sqrt[3]{2 + \sqrt{3}}$, exactly as in Exercise 7. Show that $\omega + \omega^{-1} \in K$. (Take the quotient of the two real zeros of $f(x)$.)

   (f) Show that $L \supseteq \mathbb{Q}$ is a Galois extension and that $\mathrm{Gal}(L/\mathbb{Q}) \cong D_{12}$, the dihedral group of order 12. (Arguments similar to those given for Exercise 7 will do.) Note that D_{12} has three 2-Sylow subgroups.

   (g) Show that if $K = L$, then there are exactly three intermediate subfields of extension degree 3 over $\mathbb{Q}$, viz., $\mathbb{Q}(\sqrt[3]{2})$, $\mathbb{Q}(\zeta\sqrt[3]{2})$, and $\mathbb{Q}(\zeta^2\sqrt[3]{2})$.

   (h) Show that $\mathbb{Q}(\omega + \omega^{-1}) \neq \mathbb{Q}(\sqrt[3]{2}), \mathbb{Q}(\zeta\sqrt[3]{2}), \mathbb{Q}(\zeta^2\sqrt[3]{2})$.

   (i) Conclude that $[K : \mathbb{Q}] = 36$ and that $\mathrm{Gal}(K/\mathbb{Q}) \cong S_3 \times S_3$.

9. Let $K = \mathbb{Q}(\sqrt{2}, \sqrt{3}, u)$, where $u = \sqrt{(9 - 5\sqrt{3})(2 - 2\sqrt{2})}$.

   (a) Show that $K \supseteq \mathbb{Q}$ is a Galois extension.

   (b) Find an irreducible polynomial $f(x) \in \mathbb{Q}[x]$ for which K is the splitting field over $\mathbb{Q}$.

   (c) Show that $\mathrm{Gal}(K/\mathbb{Q})$ is nonabelian but is *not* dihedral.

(d) Conclude that $\mathrm{Gal}(K/\mathbb{Q})$ must be isomorphic with Q_8, the quaternion group of order 8.

10. Let F be a field of characteristic $\neq 2$, and let $x_1, x_2, \ldots, x_n$ be commuting indeterminates over F. Let $K = F(x_1, x_2, \ldots, x_n)$, the field of fractions of the polynomial ring $F[x_1, x_2, \ldots, x_n]$ (or the *field of rational functions* in $x_1, x_2, \ldots, x_n$).

(a) Let $\sigma_1, \sigma_2, \ldots, \sigma_n$ be the elementary symmetric polynomials in $x_1, x_2, \ldots, x_n$, and set $E = F(\sigma_1, \sigma_2, \ldots, \sigma_n)$. Show that K is a splitting field over E for the polynomial $x^n - \sigma_1 x^{n-1} + \cdots + (-1)^n \sigma_n \in E[x]$.
(b) Show that $[K : E] \leq n!$.
(c) Noting that $G = S_n$ acts as a group of automorphisms of K in the obvious way, show that $E = \mathrm{inv}_G(K)$ and thereby conclude that $[K : E] = n!$.
(d) Show that K is a Galois extension of E.
(e) Set $\delta = \prod_{i<j}(x_j - x_i) \in K$ and show that $E(\delta) = \mathrm{inv}_{A_n}(K)$.

11. Let $f(x) \in \mathbb{Q}[x]$ be an irreducible polynomial and assume that the discriminant of $f(x)$ is a perfect square in $\mathbb{Q}$.

(a) Show that if $\deg f(x) = 3$, then $f(x)$ must have three real zeros in the complex field $\mathbb{C}$.
(b) Show that if $\deg f(x) = 3$, and if $\alpha \in \mathbb{R}$ is a fixed zero of $f(x)$, show that the other two zeros of $f(x)$ are polynomials in α with rational coefficients.
(c) If $\deg f(x) = 4$, how many real zeros can $f(x)$ have?
(d) If $\deg f(x) = 4$, can the Galois group have an element that acts as a 4-cycle on the four zeros of $f(x)$?

12. Suppose that F is a field and that $f(x) \in F[x]$. Suppose, moreover, that in some splitting field, $f(x)$ factors as

$$f(x) = (x - \alpha_1)^2 (x - \alpha_2) \cdots (x - \alpha_{n-1}),$$

where $\alpha_1, \alpha_2, \ldots, \alpha_{n-1}$ are distinct. Show that $\alpha_1 \in F$.
13. Let $f(x) = x^3 - 9x - 9 \in \mathbb{Q}[x]$ and compute the Galois group of this polynomial.
14. Using Swan's formula, compute the discriminants of $x^7 - 7x + 3$, $x^5 - 14x^2 - 42 \in \mathbb{Q}[x]$.

11.12.7 Exercises for Sects. 11.9 and 11.10

1. (Euler) Show that if F is a field possessing a primitive eight root of unity, then $2 = 1 + 1$ is a square in F. [Hint: Let ζ be a primitive eighth root. Consider the square of $\zeta + \zeta^{-1}$.]

2. Let F be a field with a primitive twelfth root of unity. Show that the number 3 is a square in this field. [Hint: Consider the cube of $\zeta + \zeta^{-1}$ where ζ is a primitive 12th root.][16]

3. Suppose K is the splitting field for an irreducible polynomial $p(x) \in F[x]$. Suppose L is an intermediate field—that is $F \subseteq L \subseteq K$. Show that if $p(x)$ is solvable by radicals relative to the field F, then $p(x)$ is solvable by radicals relative to the intermediate field L. (Remark: At first sight it seems a triviality. If solvability by radicals over F is seen as the ability to express the roots of $p(x)$ in terms of some compound formula involving successive field operations and (multiple) root-extractions whose input variables belong to F, then the result is a triviality, for any such formula with input variables from F is a formula with input variables from the intermediate field L. But that is not exactly how we defined the property that "$p(x)$ is solvable by radicals". Our criterion was that the splitting field K was a radical extension of F, that is, there is a tower of fields

$$F_0 \subseteq F_1 \subseteq \cdots \subseteq F_k = K,$$

such that F_{i+1} was obtained for F_i by adjoining the r_ith root of an element in the latter field. Since L may not be one of the F_i, some work is needed to show that K is a radical extension of L. The point of this remark is to motivate the following hint.) [Hint: Quoting the correct Theorems and Lemmas, observe that $p(x)$ is solvable by radicals if and only the Galois group $G(K/F)$ is solvable. Note that $G(K/L)$ is a subgroup of the former, and apply Galois' criterion.]

4. Suppose $p(x)$ and $r(x)$ are two irreducible polynomials in $F[x]$, and let K be the splitting field of $p(x)$ over F. Suppose the field K contains a root of $r(x)$.

   (a) Show that K contains a copy of the splitting field L of $r(x)$.
   (b) By the "Primitive Element Theorem" (Theorem 11.10.2, p. 407), $L = F(\theta)$ for some element θ whose irreducible polynomial is $s(x) := \text{Irr}(\theta) \in F[x]$. Using the theorems of the sections cited above for this exercise set, show the following:

   $p(x)$ is solvable by radicals over F if and only if (i) $p(x)$ is solvable by radicals over L and (ii) $s(x)$ is solvable by radicals over F.

   [Hint: Use the fact that L is a normal extension of F, Galois' criterion, and an elementary fact about solvable groups.]

5. The final exercise of this section is due to Prof. Michael Artin, and uses such a combination of facts that it cannot easily be assigned to any one subsection of this chapter. On the other hand, it is not difficult if one uses the full symphony of facts available. A perfect test question!

---

[16]In the case that F is $\mathbf{Z}/(p)$, where p is a rational prime, the conclusions of these two exercises follow from the more extensive theory of quadratic reciprocity. A beautiful account of this, using Gauss sums, can be found in the book *Elements of Number Theory* by Ireland, K. and Rosen, M.I. Bogden & Quigley, New York, 1972.

Suppose F is a finite field of characteristic 2 and suppose K is an extension of F of degree 2.

(a) Show $K = F(\alpha)$ for some element α whose irreducible monic polynomial $p(x) \in F[x]$ has the specific form

$$x^2 + x + a, \text{ where, of course } a \text{ belongs to the subfield } F.$$

(b) Show that necessarily $\alpha + 1$ is also a root of the irreducible quadratic polynomial $p(x) \in F[x]$.

(c) Find an explicit formula for an automorphism of K that takes α to $\alpha + 1$.

11.12.8 Exercises for Sect. 11.11

1. Let G be a finite subgroup of the general linear group $GL(n, F)$. We suppose each element $g \in G$ to be an invertible matrix $(\alpha_{ij}^{(g)})$. Let K be the field of quotients of the polynomial ring $F[x_1, \ldots x_n]$. (In other words, $K = F(x_1, \ldots, x_n)$ is a transcendental extension of F by algebraically independent transcendental elements x_i.)

(a) For each (transcendental) element x_i, and $g \in G$, let

$$x_i^g := \alpha_{i1}^{(g)} x_1 + \cdots + \alpha_{in}^{(g)} x_n.$$

For each rational function $f(x_1, \ldots, x_n)$, define

$$f^g(x_1, \ldots x_n) := f(x_1^g, \ldots, x_n^g).$$

   i. Show that the permutation $f \mapsto f^g$, for all $f \in K$ is a field automorphism $\phi(g)$ of K.

   ii. Show that $\phi : G \to \text{Aut}(K)$ is an embedding of groups.

(b) Assume G to be embedded (that is, with a little abuse of notation we write G for $\phi(G)$). Let K_G be the subfield of elements of K which are fixed by all elements of G. (In this particular context, the elements of K_G are called *invariants of G.*) Show that $G = \text{Gal}(K/K_G)$.

(c) Show that any $n+1$ rational functions $f_1, \ldots, f_{n+1} \in K$ are connected by an algebraic equation—precisely, there is a polynomial $p \in F[y_1, \ldots, y_{n+1}]$ such that

$$p(f_1, \ldots, f_{n+1}) = 0 \in K.$$

[Hint: Consider the transcendence degree of the subfield generated by $f_1, \ldots f_{n+1}.$]

(d) Is it true that there exists a set X of n algebraically independent invariants in the subfield K_G such that every element of K_G is a rational function in the elements of X?

11.12.9 Exercises Associated with Appendix 1 of Chap. 10

1. Prove Lemma 11.A.1.
2. Prove the uniqueness of the limit of a a convergent sequence with respect to a valuation ϕ of a field F. (See the paragraph following Eq. 11.38.).
3. Prove that every convergent sequence is a Cauchy sequence.

Appendix 1: Fields with a Valuation

Introduction

Let $\mathbb{R}$ denote the field of real numbers and let F be any field. A mapping $\phi : F \to \mathbb{R}$ from F into the field of real numbers is a *valuation on F* if and only if, for each α and β in F:

1. $\phi(\alpha) \geq 0$, and $\phi(\alpha) = 0$ if and only if $\alpha = 0$.
2. $\phi(\alpha\beta) = \phi(\alpha)\phi(\beta)$, and
3. $\phi(\alpha + \beta) \leq \phi(\alpha) + \phi(\beta)$.

This concept is a generalization of certain "distance" notions which are familiar to us in the case that F is the real or complex field. For example, in the case of the real field $\mathbb{R}$, the *absolute value function*

$$|r| = \begin{cases} r & \text{if } r \geq 0 \\ -r & \text{if } r < 0 \end{cases}$$

is easily seen to be a valuation. Similarly, for the complex numbers, the function "$\| \ \|$", defined by

$$\|a + bi\| = \sqrt{a^2 + b^2},$$

where $a, b \in \mathbb{R}$ and the square root is non-negative, is also a valuation. In this case, upon representing complex numbers by points in the complex plane, $\|z\|$ measures the distance of z from 0. The "subadditive property" (3.) then becomes the triangle inequality.

A valuation $\phi : F \to \mathbb{R}$ is said to be *archimedean* if and only if there is a natural number n such that if

$$s(n) := 1 + 1 + \cdots + 1 \ \{ \text{ with exactly } n \text{ summands}\},$$

in F, then $\phi(s(n)) > 1$.

These valuations of the real and complex numbers are clearly archimedean.

An example of a non-archimedian valuation is the *trivial valuation* for which $\phi(0) = 0$, and $\phi(\alpha) = 1$ for all non-zero elements α in F.

Here is another important example. Let F be the field of quotients of some unique factorization domain D which contains a prime element p. Evidently, $F \neq D$. Then any nonzero element α of F can be written in the form

$$\alpha = \frac{a}{b} \cdot p^k,$$

where $a, b \in D$, p does not divide either a or b, and k is an integer. The integer k and the element $a/b \in F$ are uniquely determined by the element α. Then, if $\alpha \neq 0$, set

$$\phi(\alpha) = e^{-k},$$

where e is any fixed number larger than 1, such as $2.71\ldots$. Otherwise we set $\phi(0) = 0$. The student may verify that ϕ is a genuine valuation of F. (Only the subadditive property really needs to be verified.) From this construction, it is clear that ϕ is *non-archimedean*—that is, it is not archimedean.

Lemma 11.A.1 *If ϕ is a valuation of F, then the following statements hold:*

1. $\phi(1) = 1$.
2. *If ζ is a root of unity in F, then $\phi(\zeta) = 1$. In particular, $\phi(-1) = 1$ and so $\phi(-\alpha) = \phi(\alpha)$, for all α in F.*
3. *The only valuation possible on a finite field is the trivial valuation*
4. *For any $\alpha, \beta \in F$,*
$$|\phi(\alpha) - \phi(\beta)| \leq \phi(\alpha - \beta).$$

The proof of this lemma is left as an exercise for the student on p. 424.

Lemma 11.A.2 *If ϕ is a non-archimedean valuation of F, then for any $\alpha, \beta \in F$,*

$$\phi(\alpha + \beta) \leq \max(\phi(\alpha), \phi(\beta)).$$

Proof From the subadditivity, and that fact that for any number of summands, $\phi(1 + 1 + \cdots + 1) \leq 1$, we have

$$\phi((\alpha + \beta)^n) = \phi\left(\sum_{j=0}^{n} \binom{n}{j} \alpha^{n-j} \beta^j\right)$$
$$\leq \sum_{j=0}^{n} \phi(\alpha)^{n-j} \phi(\beta)^j$$
$$\leq (n+1)\max(\phi(\alpha)^n, \phi(\beta)^n).$$

Since the taking of positive nth roots in the reals is monotone, we have

$$\phi(\alpha + \beta) \le (n + 1)^{1/n}\max\,(\phi(\alpha), \phi(\beta)).$$

Now since n is an arbitrary positive integer, and $(n + 1)^{1/n}$ can be made arbitrarily close to 1, the result follows. $\square$

The Language of Convergence

We require a few definitions. Throughout them, ϕ is a valuation of the field F.

First of all, a *sequence* is a mapping $\mathbb{N} \to F$, from the natural numbers to the field F, but as usual it can be denoted by a displayed indexed set, for example

$$\{\alpha_0, \alpha_1, \ldots\} = \{\alpha_k\} \text{ or } \{\beta_0, \beta_1, \ldots\} = \{\beta_k\}, \text{ etc.}.$$

A sequence $\{\alpha_k\}$ is said to be a *Cauchy sequence* if and only if, for every positive real number ϵ, there exists a natural number $N(\epsilon)$, such that

$$\phi(\alpha_m - \alpha_n) < \epsilon,$$

for all natural numbers n and m larger than $N(\epsilon)$.

A sequence $\{\alpha_k\}$ is said to *converge relative to* ϕ if and only if there exists an element $\alpha \in F$ such that for any real $\epsilon > 0$, there exists a natural number $N(\epsilon)$, such that

$$\phi(\alpha - \alpha_k) < \epsilon, \tag{11.38}$$

for all natural numbers k exceeding $N(\epsilon)$.

It is an easy exercise to show that an element α satisfying (11.38) for each $\epsilon > 0$ is unique. [Note that in proving uniqueness, the choice function $N : \mathbb{R}^+ \to \mathbb{N}$ taking ϵ to $N(\epsilon)$ is posited for the assertion that α is a limit. To assert that β is a limit, an entirely different choice function $M : \epsilon \to M(\epsilon)$ may be posited. That $\alpha = \beta$ is the thrust of Exercise 2 in Sect. 11.12.9.] In that case we say that α is *the limit of the convergent sequence* $\{\alpha_k\}$ or that *the sequence* $\{\alpha_k\}$ *converges to* α.

A third easy exercise is to prove that every convergent sequence is a Cauchy sequence. (See Exercise 3 in Sect. 11.12.9.)

A sequence which converges to 0 is called a *null sequence*. Thus $\{\alpha_k\}$ is a null sequence if and only if, for every real $\epsilon > 0$, however small, there exists a natural number $N(\epsilon)$, such that $\phi(\alpha_k) < \epsilon$, for all $k > N(\epsilon)$.

A field F with a valuation ϕ is said to be *complete with respect to* ϕ if and only if every Cauchy sequence converges (with respect to ϕ) to an element of F.

Now, finally, a *completion of F (with respect to ϕ)* is a field $\bar{F}$ with a valuation $\bar{\phi}$ having these three properties:

1. The field $\bar{F}$ is an extension of the field F.
2. $\bar{F}$ is complete with respect to $\bar{\phi}$.
3. Every element of $\bar{F}$ is the limit of a ϕ-convergent sequence of elements of F. (We capture this condition with the phrase "F is dense in $\bar{F}$".

So far our glossary entails

- Convergent sequence
- Limit
- Null sequence
- Cauchy sequence
- Completeness with respect to a valuation ϕ
- A completion of a field with a valuation

Completions Exist

The stage is set. In this subsection our goal is to prove that a completion $(\bar{F}, \bar{\phi})$ of a field with a valuation (F, ϕ) always exists.

In retrospect, one may now realize that the field of real numbers $\mathbb{R}$ is the completion of the field of rational numbers $\mathbb{Q}$ with respect to the "absolute" valuation, $|\ |$. The construction of such a completion should not be construed as a "construction" of the real numbers. That would be entirely circular since we used the existence of the real numbers just to define a valuation and to deduce some of the elementary properties of it.

One may observe that the set of all sequences over F possesses the structure of a ring, namely the direct product of countably many copies of the field F. Thus

$$S := F^{\mathbb{N}} = \prod_{k \in \mathbb{N}} F_k \text{ where each } F_k = F.$$

Using our conventional sequence notation, addition and multiplication in S are defined by the equations

$$\{\alpha_k\} + \{\beta_k\} = \{\alpha_k + \beta_k\}$$
$$\{\alpha_k\} \cdot \{\beta_k\} = \{\alpha_k \beta_k\}.$$

For each element $\alpha \in F$, the sequence $\{\alpha_k\}$ for which $\alpha_k = \alpha$, for every natural number k is called a *constant sequence* and is denoted $\bar{\alpha}$. The zero element of the ring S is clearly the constant sequence $\bar{0}$. The multiplicative identity of the ring S is the constant sequence $\bar{1}$ (where "1" denotes the multiplicative identity of F).

Obviously, the constant sequences form a subring of S which is isomorphic to the field F. It should also be clear that every constant sequence is also a Cauchy sequence. We shall show that the collection Ch of all Cauchy sequences also forms a subring of S.

In order to do this, we first make a simple observation. A sequence $\{\alpha_k\}$ is said to be *bounded* if and only if there exists a real number b such that $\phi(\alpha_k) < b$ for all natural numbers k. Such a real number b is called an *upper bound* of the sequence $\{\alpha_k\}$. The observation is

Lemma 11.A.3 *Every Cauchy sequence is bounded.*

Proof Suppose $\{\alpha_k\}$ is a Cauchy sequence. Then there is a natural number $N = N(1)$ such that $\phi(\alpha_N - \alpha_m) < 1$, for all natural numbers m exceeding N. Now set

$$b := 1 + \max\ \{\phi(\alpha_0), \phi(\alpha_1), \ldots, \phi(\alpha_N)\}.$$

If $k \leq N$, $\phi(\alpha_k) < b$, by the definition of b. If $k > N$ then $\phi(\alpha_k) < b$ by the definition of N and the fact that $\phi(\alpha_N) \leq b - 1$. So b is an upper bound. $\square$

Lemma 11.A.4 *The collection Ch of all Cauchy sequences of S is closed under addition and multiplication of sequences. Thus Ch is a subring of S, the ring of all sequences over F.*

Proof Recall that if $\{\alpha_k\}$ is a Cauchy sequence, there must exist an auxiliary function $N : \mathbb{R}^+ \to \mathbb{N}$ such that if $\epsilon \in \mathbb{R}^+$, then $\phi(\alpha_n - \alpha_m) < \epsilon$ for all numbers n and m larger than $N(\epsilon)$. This auxiliary function N is not unique; it is just that at least one such function is available for each Cauchy sequence. If we wish to indicate an available auxiliary function N, we simply write $(\{\alpha_k\}, N)$ instead of $\{\alpha_k\}$.

Now suppose $(\{\alpha_k\}, N)$ and $(\{\beta_k\}, M)$ are two Cauchy sequences. For each positive real number ϵ set $K(\epsilon) := \max(N(\epsilon/2), M(\epsilon/2))$. Then for any natural numbers n and m exceeding $K(\epsilon)$, we have

$$\phi(\alpha_n + \beta_n - \alpha_m - \beta_m) \leq \phi(\alpha_n - \alpha_m) + \phi(\beta_n - \beta_m) \leq \epsilon/2 + \epsilon/2 = \epsilon.$$

Thus $(\{\alpha_k + \beta_k\}, K)$ is a Cauchy sequence.

Again suppose $(\{\alpha_k\}, N)$ and $(\{\beta_k\}, M)$ are two Cauchy sequences. By the preceding Lemma 11.A.3, we may assume that these sequences possess positive upper bounds b and c, respectively. For each positive real number ϵ set

$$K(\epsilon) := \max\ (N(\epsilon/2c), M(\epsilon/2b)).$$

Now suppose n and m are natural numbers exceeding $K(\epsilon)$. Then

$$\begin{aligned}
\phi(\alpha_m\beta_m - \alpha_n\beta_n) &= \phi(\alpha_m\beta_m - \alpha_m\beta_n + \alpha_m\beta_n - \alpha_n\beta_n) \\
&\leq \phi(\alpha_m)(\phi(\beta_m - \beta_n) + \phi(\beta_n)\phi(\alpha_n - \alpha_m) \\
&< b(\epsilon/2b) + c(\epsilon/2c) = \epsilon.
\end{aligned}$$

Thus $(\{\alpha_k\beta_k\}, K)$ is a Cauchy sequence.

Finally, it should be obvious that $\{-\alpha_k\}$ is the additive inverse in S of $\{\alpha_k\}$ and, from an elementary property of ϕ, that the former is Cauchy if and only if the latter is. Finally one notices that the constant sequence $\bar{1}$ is a Cauchy sequence that is the multiplicative identity element of Ch.

It follows that the Cauchy sequences form a subring of S. $\square$

The null sequences of S are simply the sequences that converge to zero. We have already remarked that every convergent sequence is a Cauchy sequence. Thus the set Z of all null sequences lies in Ch. Note that any sequence whose coordinates are nonzero at only finitely many places (that is, the direct sum $\bigoplus_N F$) form a subset Z_0 of Z.

Clearly the sum or difference of two null sequences is null. Suppose $\{\alpha_k\}$ is a null sequence so that for every positive real ϵ, there is a natural number $N(\epsilon)$ such that $\phi(\alpha_m) \leq \epsilon$ for all $m > N(\epsilon)$ Next let $\{\gamma_k\}$ be any Cauchy sequence with upper bound b. Then, for every real $\epsilon > 0$, and $n > N(\epsilon/b)$, we see that

$$\phi(\alpha_n \gamma_n) \leq b \cdot (\epsilon/2) = \epsilon.$$

Thus the product $\{\alpha_k\}\{\gamma_k\}$ is a null sequence. So we have shown the following:

Lemma 11.A.5 Z *is an ideal of the ring Ch.*

A sequence $\{\alpha_k\}$ is said to possess a *lower bound* ℓ if and only if for some real $\ell > 0$, one has $\phi(\alpha_n) > \ell$ for *every* natural number n. The next lemma, though mildly technical, is important.

Lemma 11.A.6 *The following assertions hold:*

(i) *Suppose a sequence $\{\gamma_k\}$ in S has the property that for every real $\epsilon > 0$, $\phi(\gamma_k) > \epsilon$ for only finitely many natural numbers k. Then $\{\gamma_k\}$ is a null sequence.*

(ii) *Suppose $\{\beta_k\}$ is a Cauchy sequence which is not a null sequence. Then there exists a real number $\lambda > 0$, and a natural number N such that $\phi(\beta_k) > \lambda$ for all $k > N$.*

(iii) *If $\{\alpha_k\}$ is a Cauchy sequence having a non-zero lower bound ℓ, then it is a unit in the ring Ch.*

Proof (i). The reader may want to do this as an exercise. The hint is that the hypotheses allow us to define a mapping $K : \mathbb{R}^+ \to \mathbb{N}$ where

$$K(\epsilon) := \max \{k \in \mathbb{N} | \phi(\gamma_k) > \epsilon\}.$$

Then K serves the desired role in the definition of a null sequence.

(ii). Using the contrapositive of the statement in part (i), we see that if $\{\beta_k\}$ is a Cauchy sequence which is not null, there must exist some $\epsilon > 0$ such that $\phi(\beta_k) > \epsilon$ infinitely often. Also, since the sequence is Cauchy, there exists a natural number $N(\epsilon/2)$ such that

$$\phi(\beta_n - \beta_m) \leq \epsilon/2,$$

for all natural numbers n and m exceeding $N(\epsilon/2)$. So there is an natural number N exceeding $N(\epsilon/2)$ such that $\phi(\beta_M) > \epsilon$. Then for $m > N$,

$$\phi(\beta_m) = \phi(\beta_M - (\beta_M - \beta_m))$$
$$\geq \phi(\beta_M) - \phi(\beta_M - \beta_n)$$
$$> \epsilon - (\epsilon/2) = \epsilon/2.$$

Thus $\lambda := \epsilon/2$ and N satisfy the conclusion of part (ii).

(iii). For each $\epsilon > 0$, let $K(\epsilon) := N(\epsilon \cdot \ell^2)$. Then for any natural numbers n and m exceeding $K(\epsilon)$, we have

$$\phi\left(\frac{1}{\alpha_n} - \frac{1}{\alpha_m}\right) = \phi\left(\frac{1}{\alpha_n \alpha_m}\right) \phi(\alpha_m - \alpha_n)$$
$$\leq \ell^{-2} \cdot (\epsilon \ell^2) = \epsilon.$$

Thus the sequence of reciprocals $\{\alpha_k{}^{-1}\}$ is a Cauchy sequence and is a multiplicative inverse to $\{\alpha_k\}$ in Ch. The proof is complete. $\square$

Corollary 11.A.1 *The factor ring* $\bar{F} := Ch/Z$ *is a field.*

Proof It suffices to show that every element of $Ch\backslash Z$ is congruent to a unit modulo the ideal Z. Suppose $\{\beta_k\}$ is a non-null Cauchy sequence. Then by Lemma 11.A.6, there is a positive real number λ and a natural number N such that $\phi(\beta_n) > \lambda$ for all natural numbers n greater than N. Now let $\{\zeta_k\}$ be a sequence for which

$$\zeta_k = \begin{cases} 1 & \text{if } k \leq N \\ 0 & \text{if } k > N. \end{cases}$$

Then $\{\zeta_k + \beta_k\}$ is bounded below by $\ell := \min(1, \lambda)$ and so by Lemma 11.A.6, part 3., is a unit. Since $\{\zeta_k\}$ belongs to Z, we have shown the sufficient condition announced at the beginning of this proof. $\square$

Note that we have a natural embedding $F \to \bar{F}$ taking each element α to the constant sequence $\bar{\alpha}$, all of whose terms are equal to α.

Now we need to extend the valuation ϕ on (the embedded copy of) F to a valuation $\bar{\phi}$ of $\bar{F}$. The key observation is that whenever we consider a Cauchy sequence $\{\alpha_k\}$, then $\{\phi(\alpha_k)\}$ is itself a Cauchy sequence of real numbers with respect to the absolute value. For, given real $\epsilon > 0$, there is a number $N(\epsilon)$ such that $\phi(\alpha_n - \alpha_m) < \epsilon$ for all n and m exceeding $N(\epsilon)$. But

$$|\phi(\alpha_m) - \phi(\alpha_m)| \leq \phi(\alpha_n - \alpha_m) < \epsilon$$

with the same conditions on n and m.

Now Cauchy sequences $\{r_n\}$ of real numbers tend to a real limit which we denote by the symbol

$$\lim_{k \to \infty} r_k.$$

Thus, for any Cauchy sequence $\{\alpha_k\}$ in Ch, we may define

$$\bar{\phi}(\{\alpha_k\}) := \lim_{k \to \infty} \phi(\alpha_k).$$

Now note that if $\{\alpha_k\}$ happens to be a sequence that already converges to α in F, then, by the definition of convergence, $\lim_{k \to \infty} \phi(\alpha_k) = \phi(\alpha)$. In short, on convergent sequences, $\bar{\phi}$ can be evaluated simply by taking ϕ of the existing limit. In particular this works for constant sequences, whereby we we have arrived at the fact that $\bar{\phi}$ extends the mapping ϕ defined on the embedded subfield F.

Now observe

Lemma 11.A.7 *If $\{\alpha_k\}$ and $\{\beta_k\}$ are two Cauchy sequences with the property that*

$$\phi(\alpha_k) \le \phi(\beta_k),$$

for all natural numbers k larger than some number N, then

$$\bar{\phi}(\{\alpha_k\}) \le \bar{\phi}(\{\beta_k\}).$$

Proof The student should be able to prove this, either by assuming the result for real Cauchy sequences, or from first principles by examining what it would mean for the conclusion to be false. □

Corollary 11.A.2 *If $\{\alpha_k\}$ and $\{\beta_k\}$ are two Cauchy sequences,*

$$\bar{\phi}(\{\alpha_k + \beta_k\}) \le \bar{\phi}(\{\alpha_k\}) + \bar{\phi}(\{\beta_k\}).$$

Proof Immediate from Lemma 11.A.7. □

Corollary 11.A.3 *The following statements are true.*

(i) $\{\alpha_k\}$ is a null sequence if and only if $\bar{\phi}(\{\alpha_k\}) = 0 \in \mathbb{R}$.
(ii) $\bar{\phi}$ assumes a constant value on each coset of Z in Ch.

Proof Another exercise in the meaning of the definitions. □

Now for any coset $\gamma := \{\beta_k\} + Z$ of Z in Ch we can let $\bar{\phi}(\gamma)$ be the constant value of $\bar{\phi}$ on all Cauchy sequences in this coset. In effect $\bar{\phi}$ is now defined on the factor ring $\bar{F} = Ch/Z$, which, as we have noted, is a field. Moreover, from the preceding Corollary $\bar{\phi}$ assumes the value 0 only on the zero element $\bar{0} = 0 + Z = Z$ of $\bar{F}$.

At this stage—and with a certain abuse of notation—$\bar{\phi}$ is a non-negative-real-valued function defined on two distinct domains, Ch and $\bar{F}$. However the ring epimorphism $f : Ch \to \bar{F}$ preserves the $\bar{\phi}$-values. That is important for three reasons.

First it means that if $\bar{\phi}$ is multiplicative on Ch, then it is on $\bar{F}$. But it is indeed the case that $\bar{\phi}$ is multiplicative on Ch, since, for $\{\alpha_k\}$ and $\{\beta_k\}$ in Ch,

$$\lim_{k \to \infty} \phi(\alpha_k \beta_k) = \lim_{k \to \infty} \phi(\alpha_k)\phi(\beta_k)$$
$$= (\lim_{k \to \infty} \phi(\alpha_k)) \cdot (\lim_{k \to \infty} \phi(\beta_k)).$$

Second, the fact that the homomorphism f preserves $\bar{\phi}$, means that if $\bar{\phi}$ is subadditive on Ch, then it is also subadditive on $\bar{F}$. But the subadditivity on Ch is asserted in Corollary 11.A.2.

Third, since $F \cap Z = 0 \in Ch$ (that is, the only constant sequence which is null is the zero constant sequence), the homomorphism f embeds F (in its transparent guise as the field of constant sequences) into $\bar{F}$ without any change in the valuation.

These observations, taken together, yield

Corollary 11.A.4 *The function $\bar{\phi} : \bar{F} \to \mathbb{R}^+$ is a valuation of $\bar{F}$, which extends the valuation ϕ on its subfield F.*

It remains to show that $\bar{F}$ is complete with respect to the valuation $\bar{\phi}$ and that F dense in $\bar{F}$.

Recall that the symbol $\bar{\alpha}$ denotes the constant sequence $\{\alpha, \alpha, \ldots\}$. Its image in $\bar{F}$ is then $f(\bar{\alpha}) = \bar{\alpha} + Z$. Now suppose $\{\alpha_k\}$ is any Cauchy sequence, and let $\gamma = \{\alpha_k\} + Z$ be its image in $\bar{F}$. We claim that the sequence

$$\{f(\bar{\alpha}_1), f(\bar{\alpha}_2), \ldots\}$$

(of constant sequences mod Z) is a sequence of elements in $\bar{F}$ converging to γ relative to $\bar{\phi}$. Choose a positive real number ϵ. Then there exists a natural number $N(\epsilon)$ such that

$$\phi(\alpha_{N(\epsilon)} - \alpha_m) < \epsilon \text{ for all } m > N(\epsilon).$$

Then as f preserves $\bar{\phi}$,

$$\bar{\phi}\left(f(\bar{\alpha}_{N(\epsilon)}) - \bar{\gamma}\right) = \bar{\phi}\left(\bar{\alpha}_{N(\epsilon)} - \{\alpha_k\}\right)$$
$$= \lim_{n \to \infty} \phi\left(\alpha_{N(\epsilon)} - \alpha_n\right) < \epsilon.$$

So our claim is true. F is indeed dense in $\bar{F}$.

Now suppose $\{\gamma_\ell\}$ is a Cauchy sequence of elements of $\bar{F}$ with respect to $\bar{\phi}$. For each index ℓ, we can write

$$\gamma_\ell = \{\beta_{\ell k}\},$$

a Cauchy sequence over F indexed by k. So there exists a natural number $N(\ell)$ such that

$$\phi(\beta_{\ell N(\ell)} - \beta_{\ell m}) < 2^{-\ell} \tag{11.39}$$

for all m exceeding $N(\ell)$. This allows us to define the following set of elements of F:

$$\delta_\ell := \beta_{\ell N(\ell)}, \text{ for } \ell = 0, 1, \ldots.$$

Then taking limits in Eq. (11.39), we have

$$\bar{\phi}(f(\bar{\delta}_\ell) - \gamma_\ell) < 2^{-\ell},$$

where, as usual, $f(\bar{\delta}_k)$ is the image of a constant sequence, namely the coset

$$\{\delta_k, \delta_k, \delta_k, \ldots\} + Z.$$

Now for any natural numbers n and m, we have

$$\begin{aligned}
\phi(\delta_n - \delta_m) &= \bar{\phi}\left(f(\bar{\delta}_m) - f(\bar{\delta}_n)\right) \\
&= \bar{\phi}\left(f(\bar{\delta}_n) - \gamma_m + \gamma_m - \gamma_n + \gamma_n - f(\bar{\delta}_m)\right) \\
&\leq \bar{\phi}(f(\bar{\delta}_n - \gamma_m) + \bar{\phi}(\gamma_n - \gamma_m) + \bar{\phi}(\gamma_m - f(\bar{\delta}_m)) \\
&\leq \bar{\phi}(\gamma_n - \gamma_m) + 2^{-m} + 2^{-n}
\end{aligned}$$

But since $\{\gamma_\ell\}$ is a Cauchy sequence, the above equations tell us that $\delta := \{\delta_k\}$ is a Cauchy sequence in Ch. It is now clear that

$$\bar{\phi}(\delta - \gamma_\ell) = \lim_{k \to \infty} (\delta_k - \beta_{\ell k}) < 2^{-\ell}.$$

Thus $\{\gamma_\ell\}$ converges to δ with respect to $\bar{\phi}$. It follows that $\bar{F}$ is a complete field.

Summarizing all of the results of this subsection, one has the following

Theorem 11.A.1 *The field $\bar{F} = Ch/Z$ with valuation $\bar{\phi}$ is a completion of (F, ϕ).*

The Non-archimedean p-Adic Valuations

Our classical example of a non-archimedean valuation was derived from the choice of a prime p in a unique factorization domain D. The valuation ϕ is defined on the field F of quotients of D by the rule that $\phi(r) = e^{-k}$, whenever we write $r \in F \backslash \{0\}$ in the canonical form $r = (a/b)p^k$ where a and b are elements of D which are not divisible by p, and k is an integer. Of course to make this function subadditive, we must take e to be a fixed real number greater than 1. (Many authors choose p itself. That way, the values never overlap from one prime to another.)

Now notice that from Lemma 11.A.2, the collection of all elements α of F for which $\phi(\alpha) \leq 1$ is closed under addition as well as multiplication, and so form a subring O of F called the *valuation ring* of (F, ϕ). From the multiplicative properties of ϕ, it is clear that every non-zero element of F either lies in O, or has its multiplicative inverse in O.

A somewhat smaller set is $P := \{\alpha \in F | \phi(\alpha) < 1\}$. It is easy to see that this is an ideal in O. In fact the set $O - P = \{u \in F | \phi(u) = 1\}$ is closed under taking multiplicative inverses in F and so consists entirely of units of O. Conversely, since

$\phi(1) = 1$, every unit of O must have ϕ-value 1, so in fact

$$O \backslash P = U(O), \tag{11.40}$$

the (multiplicative) group of units of O. Thus P is the unique maximal ideal of O and so O is a *local ring*.

Note that O and P have been completely determined by F and ϕ. The factor ring O/P is a field called the *residue field of F with respect to ϕ*. The same sort of argument used to show (11.40) can be used to show that there exists an element $\pi \in P$, such that for any natural number k,

$$P^k \backslash P^{k+1} = \pi^k U(O).$$

Taking the disjoint union of these set differences, one concludes that P is the principal ideal $P = O\pi$.

Suppose p is a prime integer, $D = \mathbb{Z}$, so $F = \mathbb{Q}$, the field of rational numbers. Then O is the ring of all fractions $\frac{a}{b}$ where a and b are integers and p does not divide b. Terribly odd things happen! For example $\{1, p, p^2, \ldots\}$ is a null sequence (since $\phi(p^k) = e^{-k}$, for all positive integers k), and

$$1 + p + p^2 + \cdots$$

is a "convergent series"—that is its sequence of partial sums is a Cauchy sequence.

Let us return to our classic example. There F is the field of fractions of a unique factorization domain D, p is a prime element of D and $\phi(r) = e^{-k}$ when $r \in F$ is written in the form $(a/b)p^k$, $a, b \in D$, k an integer. Let O be the valuation ring and let P be its unique prime ideal. Similarly, let $\bar{F}, \bar{\phi}, \bar{O}$ and $\bar{P}$ be the completion of F, its extended valuation, valuation ring and uniqe maximal ideal, respectively.

Theorem 11.A.2 *The following statements hold:*

 (i) $D \subseteq O$, *and* $D \cap P = pD$.
 (ii) $D + P = O$.
 (iii) $F \cap \bar{O} = O$, $O \cap \bar{P} = P$.
 (iv) $O + \bar{P} = \bar{O}$.
 (v) *The residue class fields of F and $\bar{F}$ are both D/pD.*

Proof (i) Clearly every element of D has valuation at most 1 so the first statement holds. Those elements of D of value less than 1 are divisible by p, and vice versa. So the second statement holds.

(ii) Suppose $r \in O \backslash P$. Then r has the form a/b where a and b are elements of D which are not divisible by p. Since p is a prime and D is a UFD, this means there are elements c and d of D such that $cp + db = a$.

Then

$$r = \frac{a}{b} = \left(\frac{c}{b}\right) \cdot p + d. \tag{11.41}$$

Now p does not divide d, otherwise it would divide $a = cp + db$, against our assumption about r. Thus d is an element of D also in $O - P$, and by Eq. (11.41), $r \equiv (a/b)$ modulo P. This proves (ii).

(iii) Both statements here just follow from that fact the $\bar{\phi}$ extends ϕ.

(iv) Now let $\bar{\alpha}$ be any element in $\bar{O} - \bar{P}$ so $\bar{\phi}(\bar{\alpha}) = 1$. Since F is dense in $\bar{F}$, there is an element $r \in F$ such that $\bar{\phi}(\bar{\alpha} - r) < 1$. This means $r - \bar{\alpha}$ is an element of P. Since O is an additive subgroup of F containing P, r belongs to $\bar{O} \cap F = O$, and we can say $r \equiv \bar{\alpha} \bmod \bar{P}$. This proves (iv).

(v) First using (iii) and (iv), and then (i) and (ii),

$$\bar{O}/\bar{P} = (O + \bar{P})/\bar{P}$$
$$\simeq O/(O \cap \bar{P}) = O/P, \text{ while}$$
$$O/P = (D + P)/P$$
$$\simeq D/(D \cap P) = D/pD.$$

The proof is complete. $\square$

Now fix a system Y of coset representatives of pD in the additive group $(D, +)$ and suppose $0 \in Y$. For example, if $D = \mathbb{Z}$, then p is a rational prime, the residue class field is $\mathbb{Z}/p\mathbb{Z}$, and we may take Y to be the set of integers,

$$\{0, 1, 2, \ldots, p - 1\}.$$

For another example, suppose $D = K[x]$ where K is some field, and $p = x + a$ is a monic polynomial of degree one. Then the residue class field is $K[x]/(x + a) \simeq K$. But as K is a subring of D, it makes perfect sense to take $Y = K$ (the polynomials of degree zero).

Now in the notation of Theorem 11.A.2, $D + \bar{P} = (D + P) + \bar{P} = O + \bar{P} = \bar{O}$. So, for a fixed $\bar{\alpha} \in \bar{O}$, there is a unique element $a_0 \in Y$ such that $\bar{\alpha} - a_0 \in \bar{P}$. (If $\bar{\alpha}$ is already in $\bar{P}$, this element is 0.) Now $\bar{P} = p\bar{O}$, so

$$\bar{\alpha}_1 = \frac{1}{p}(\bar{\alpha} - a_0) \in \bar{O}.$$

Then we repeat this procedure to obtain a unique element $a_1 \in Y$ such that $\bar{\alpha}_1 - a_1 \in p\bar{O}$. Then

$$\bar{\alpha} \equiv a_0 + pa_1 \bmod p^2 \bar{O}.$$

Again, one can find $a_2 \in Y$ such that if $\bar{\alpha}_2 := (1/p)(\bar{\alpha}_1 - a_1)$, then $\bar{\alpha}_2 - a_2 \in p\bar{O}$. So, after k repetitions of this procedure, we obtain a sequence of $k + 1$ unique elements of Y, $(a_0, a_1, \ldots, a_k)$ such that

$$\bar{\alpha} \equiv a_0 + a_1 p + a_2 p^2 + \cdots + a_k p^k \bmod \bar{P}^{k+1}.$$

Thus we can write any $\bar{\alpha} \in \bar{F}$ in the "power series" form,

$$\bar{\alpha} = p^k[a_0 + a_1 p + a_2 p^2 + \cdots],$$

where the a_i are uniquely determined elements of Y, and $a_0 \neq 0$, so that k is determined by $e^{-k} = \bar{\phi}(\bar{\alpha})$.

For example, if $p = x \in K[x] = D$, then, taking $Y = K$, we see that $\bar{F}$ is the field of *Laurent series over K*—that is, the ring of power series of the form

$$b_{-k}x^{-k} + \cdots + b_{-1}x^{-1} + b_0 + b_1 x + b_2 x^2 + \cdots, \ k \in \mathbb{N}$$

where the b_i are in K and one follows the usual rules of adding and multiplying power series.

The ring of Laurent series is very important in the study of power series generating functions. Be aware that the field K here is still arbitrary. It can even be a finite field.

Note that if $D = \mathbb{Z}$ and p is a rational prime, the system of coset representatives Y is not closed under either addition or multiplication and so, in order to restore the coefficients of the powers of p to their proper location in Y, some local adjustments are needed after addition or multiplication. For example if $p = 5$, then we must rewrite

$$(1 + 3(5))(1 + 4(5)) = 1 + 7(5) + 12(5^2)$$

as

$$1 + 2(5) + 3(5^2) + 2(5^3).$$

Appendix 2: Finite Division Rings

In 1905 J.H.M. Wedderburn proved that every finite division ring is in fact a field.

On pp. 104–105 of van Der Waerden's classic book (1930–31), (cited at at the end of this appendix) one finds a seemingly short proof that depends only on the easy exercise which proves that no finite group can be the union of the conjugates of a proper subgroup. But the proof is short only in the sense that the number of words required to render it is not large. Measured from actual elementary first principles, the proof spans a considerable "logical distance". It uses the notion of a splitting field for a division ring, and a certain structure theorem involving a tensor product of algebras. Both of these notions are beyond what has been developed in this and the preceding chapters.

Several other more elementary proofs are available. The first proof given below is due to T.J. Kaczynski (cited at the end of this appendix) and has several points of interest.

We begin with a small lemma.

Lemma 11.A.8 *Every element in a finite field* $\mathbf{GF}(q)$ *is the sum of two squares. As a consequence, when* p *is a prime number, there exist numbers* t *and* r *such that* $t^2 + r^2 \equiv -1$, *mod* p, *with* $t \not\equiv 0$.

Proof If q is even, all elements are squares and hence is the sum of 0^2 and a square. So suppose q is odd and let S be the set of non-zero squares. Then, for any non-square $g \in \mathbf{GF}(q)$, we have a partition $\mathbf{GF}(q)^* = S + Sg$. Now using 0 as one of the summand squares, we see that every element of $\{0\} \cup S$ is the sum of two squares, and if $(S + S) \cap gS \neq \emptyset$ (where now "$S + S$" indicates all possible sums of two squares), then this is true of gS as well and the result is proved. Otherwise $S + S \subseteq \{0\} \cup S$ and so $\{0\} \cup S$ is a subfield of $\mathbf{GF}(q)$. That is impossible since $|\{0\} \cup S| = (q + 1)/2$ is not a prime power dividing q.

In particular, -1 can be represented as the sum of two squares in $\mathbb{Z}/(p)$, and one of them at least (say t) is non-zero. $\square$

Theorem 11.A.3 (Wedderburn) *Every finite division ring is a field.*

Proof ([1]) Let D be a finite division ring and let P be the prime subfield in its center. Also let D^* be the multiplicative group of non-zero elements of D. For some prime r let S be an r-Sylow subgroup of D^*, of order r^a, say. Now S cannot contain an abelian subgroup A of type $Z_r \times Z_r$ otherwise the subfield generated by $P \cup A$ would have too many roots of $x^r - 1 = 0$. Thus S is either cyclic or else $r = 2$ and S is a generalized quaternion group. We shall show that the latter case is also forbidden.

Suppose S is generalized quaternion. Then the characteristic p is not 2 and S contains a subgroup S_0 which is quaternion of order 8, generated by two elements a and b or order 4 for which $bab = a^{-1}$ and $z = a^2 = b^2$ is its central involution. Then the extension $P(z)$ is a subfield containing the two roots of $x^2 - 1 = 0$, namely 1 and -1. It follows that

$$a^2 = -1 = z = b^2$$

so $a^3 = -a$. Thus

$$ba = a^{-1}b = -ab.$$

Now let $p = |P|$, the characteristic of D. Then by the lemma we can find elements t and r in P (and hence central in D) such that $t^2 + r^2 = z = -1$. Then

$$\begin{aligned} ((ta + b) + r)((ta + b) - r) &= (ta + b)^2 - r^2 \\ &= t^2 a^2 + r(ab + ba) + b^2 - r^2 \\ &= -t^2 + b^2 - r^2 = 0. \end{aligned}$$

So one of the factors on the left is zero. But then $b = -ta \pm r$ and so as $t \neq 0$, b commutes with a, a contradiction.

Thus, regardless of the parity of p, every r-Sylow subgroup of D^* must be cyclic. If follows that D^* must be solvable. Now let $Z^* = Z(D^*)$ be the center of the group

D^* and suppose $Z^* \neq D^*$. Then we can find a normal subgroup A of D^* containing Z^* such that A/Z^* is a minimal normal subgroup of D^*/Z^*. Since the latter is solvable with all Sylow subgroups cyclic, A/Z^* is cyclic. Now Z^* is in the center of A, and since no group has a non-trivial cyclic factor over its center, A is abelian.

Now we shall show that in general, D^* cannot contain a normal abelian subgroup A which is not in the center Z^*. Without loss of generality we can take (as above) $Z^* < A \lhd D^*$. Choose $y \in A - Z^*$. Then there exists an element x in D^* which fails to commute with y. Then also $x + 1$ is not in A as it fails to commute with y as well. Then, as A is normal,

$$(1 + x)y = a(1 + x),$$

for some element $a \in A$. Then

$$y - a = ax - xy = (a - xyx^{-1})x. \tag{11.42}$$

Now if $y - a = 0$, then $a = y$ and $a = xyx^{-1}$ by (11.42), so $y = xyx^{-1}$ against the fact that x and y do not commute. Thus $(a - xyx^{-1})$ is invertible, and from (11.42),

$$x = (a - xyx^{-1})^{-1}(y - a). \tag{11.43}$$

But a, y and xyx^{-1} all belong to A and so commute with y. Thus every factor on the right side of (11.43) commutes with y, and so x commutes with y, the same contradiction as before. $\square$

Remark As noted, the last part of Kaczynski's proof did not use finiteness at all. As a result, one can infer

Corollary 11.A.5 (Kaczynski) *Suppose D is any division ring with center Z. Then any abelian normal subgroup of D^* lies in Z.*

One might prefer a proof that did not use the fact that a finite group with all Sylow subgroups cyclic is solvable. This is fairly elementary to a finite group theorist (one uses a transfer argument on the smallest prime and applies induction) but it would not be as transparent to the average man or woman on the street.

Another line of argument due to I.N. Herstein (cited below) develops some truly beautiful arguments from first principles. Consider:

Lemma 11.A.9 *Suppose D is a division ring of characteristic p, a prime, and let P be its prime subfield, which, of course, is contained in the center Z. Suppose a is any element of $D - Z$ which is algebraic over P. Then there exists an element $y \in D$ such that*

$$yay^{-1} = a^i \neq a$$

for some integer i.

Proof Fix element a as described. The commutator mapping $\delta : D \to D$, which takes an arbitrary element x to $xa - ax$, is clearly an endomorphism of the additive group $(D, +)$. Also, for any element λ of the finite subfield $P(a)$ obtained by adjoining a to P, the fact that λ commutes with a yields

$$\delta(\lambda x) = \lambda xa - a\lambda x = \lambda(xa - ax) = \lambda\delta(x).$$

Thus, regarding D as a left vector space over the field $P(a)$, we see that δ is a linear transformation of V.

Now, for any $x \in D$, and positive integer k, an easy computation reveals

$$\delta^k(x) = \sum_{j=0}^{k}(-1)^j \binom{k}{j} a^j x a^{k-j}. \tag{11.44}$$

Now $P(a)$ is a finite field of order p^m, say. Then $a^{p^m} = a$. Since D has characteristic p, putting $k = p^m$ in (11.44) gives

$$\delta^{p^m}(x) = xa^{p^m} - a^{p^m}x = xa - ax = \delta(x).$$

Thus the linear transformation δ of V satisfies

$$\delta^{p^m} = \delta.$$

Thus, in the commutative subalgebra of Hom (V, V) generated by δ, we have

$$0 = \delta^{p^m} - \delta = \prod_{\lambda \in P(a)} (\delta - \lambda I),$$

where the λI are the scalar transformations $V \to V, \lambda \in P(a)$.

Now $\ker \delta$ is the subalgebra $C_D(a)$ of all elements of D which commute with a. (This is a proper subspace of V since a is not in the center Z.) Thus δ acts as a non-singular transformation of $W := V/C_D(a)$ while $\prod(\delta - \lambda I)$ (the product taken over all non-zero $\lambda \in P(a)$) vanishes on it. It follows that for some $\lambda \in P(a) - \{0\}$, $\delta - \lambda I$ is singular on W. This means there is a coset $b_0 + C_D(a) \neq C_D(a)$ such that

$$\delta(b_0) = \lambda b_0 + c, \quad c \in C_D(a).$$

Setting $y = b_0 + \lambda^{-1}c$, we have $\delta(y) = \lambda y \neq 0$. This means

$$yay^{-1} = a + \lambda \in P(a).$$

Now if m is the order of a in the multiplicative group $P(a)^*$, then yay^{-1} is one of the roots of $X^m - 1 = 0$ lying in $P(a)$. But since these roots are just the powers of a, we have $yay^{-1} = a^i \neq a$, for some i, $1 < m \leq m - 1$, as desired. $\square$

Now we proceed with a proof of Wedderburn's theorem. We may assume that D is a non-commutative finite division ring in which all proper subalgebras are fields. In particular, for any element $v \in D - Z$, the centralizing subalgebra $C_D(v)$ is a field. It follows that the commuting relation on $D - Z$ is an equivalence relation. For each commuting equivalence class A_i, and element $a_i \in A_i$, we have

$$F_i := Z \cup A_i = C_D(a_i). \tag{11.45}$$

Let D^* be the multiplicative group of non-zero elements of D, so $Z^* := Z - \{0\}$ is its center. Choose a coset aZ^* so that its order in D^*/Z^* is the smallest prime s dividing the order of D^*/Z^*. The actual representative element a can then be chosen to have s-power order.

Now by Herstein's lemma, there is an element $y \in D$ such that

$$yay^{-1} = a^t \neq a.$$

From (11.45) it follows that conjugation by y leaves the subfield $F := C_D(a)$ invariant, and induces on it an automorphism whose fixed subfield is

$$C_D(y) \cap C_D(a) = Z.$$

But the same conclusion must hold when y is replaced by a power y^k which is not in Z. It follows that $\mathrm{Gal}(F/Z)$ is cyclic of prime order r and conjugation by y induces an automorphism of order r. Thus $y^r \in C_D(a) \cap C_D(a) = Z$, so r is the order of yZ^* in the group D^*/Z^*. We can then choose the representative y to be in the r-Sylow subgroup of $Z(y)^*$, so that it has r-power order.

Now, by the "(very) little Fermat Theorem" $i^{s-1} \equiv 1 \bmod s$. Thus

$$y^{(s-1)} a y^{-(s-1)} = a^{i^{s-1}}$$
$$\equiv a \bmod Z^*.$$

So, setting $b := y^{s-1}$, we have
$$bab^{-1} = az,$$

for some element z in Z which is also a power of a. Now from this equation, $z = 1$ if and only if $b \in C_D(y) \cap C_D(a) = Z$.

Case 1: $z = 1$ and b is in the center. Then $y^{s-1} \in Z$, so r, the order of yZ^* in D^*/Z^*, divides $s - 1$. That is impossible since s was chosen to be the smallest prime divisor of $|D^*/Z^*|$.

Case 2: $z \neq 1$ and b has order r mod Z^*. Then $b^r a^{-r} = a = az^r$. But since z is an element of s-power order, we now have that $r = s$ and is the order of z. Now at this stage, the multiplicative subgroup $H := \langle a, b \rangle$ of D^* generated by a and b is a nilpotent group generated by two elements of r-power order, and so is an r group. Since it is non-abelian without a subgroup of type $Z_r \times Z_r$, it is generalized quaternion and so contains the quaternion subgroup of order 8. Herstein goes on to eliminate the presence of Q_8 by an argument that essentially anticipates by three years the argument given in the first part of Kaczynski's proof above.[17] □

There is another fairly standard proof which comes closer to being a proof from elementary principles. It uses an argument due to Witt (cited below) concerning cyclotomic polynomials and the partition of the group D^* into conjugacy classes. One may consult Jacobson's book (cited below), pp. 431–2, for a presentation of this proof.

Reference

1. Kaczynski TJ (1964) Another proof of Wedderburn's theorem. Am Math Mon 71:652–653

---

[17] Actually Herstein inadvertently omitted Case 1 by taking $s = r$ from a misreading of the definition of s. As we see, this presents no real problem since his lemma applies when s is the smallest possible prime.

Chapter 12
Semiprime Rings

Abstract As was the case with groups, a ring is said to be *simple* if it has no proper homomorphic images (equivalently, it has no proper 2-sided ideals). On the other hand, a right (or left) R-module without proper homomorphic images is said to be *irreducible*. A right (or left) module is said to be *completely reducible* is it is a direct sum of irreducible modules. Similarly, a ring R is said to be *completely reducible* if and only if the right module R_R is completely reducible. A ring is *semiprimitive* if and only if the intersection of its maximal ideals is zero, and is *semiprime* if and only if the intersection of *all* its prime ideals is the zero ideal. Written in the presented order, each of these three properties of rings implies its successor—that is, the properties become weaker. The goal here is to prove the Artin-Wedderburn theorem, basically the following two statements: (1) A ring is completely reducible if and only if it is a direct sum of finitely many full matrix algebras, each summand defined over its own division ring. (2) If R is semiprimitive and Artinian (i.e. it has the descending chain condition on right ideals) then the same conclusion holds. A corollary is that any completely reducible simple ring is a full matrix algebra.

12.1 Introduction

In this chapter we develop the material necessary to reach the famous Artin-Wedderburn theorem which—in the version given here—classifies the Socle of a prime ring. The reader should probably be reminded here that *all rings in this chapter have a multiplicative identity element.*

We begin with the important concept of complete reducibility of modules. When transferred to rings, this notion and its equivalent incarnations give us quick access to the Socle of a ring and its homogeneous components. Most of the remaining development takes place in a general class of rings called the semiprime rings. It has several important subspecies, the prime, primitive, and completely reducible rings, where the important role of idempotents unfolds.

We remind the reader that R-module homomorphisms are applied on the *left*. Thus if M and M' are right R-modules and if $\phi : M \to M'$ is an R-module homomorphism, then one has

© Springer International Publishing Switzerland 2015

E. Shult and D. Surowski, *Algebra*, DOI 10.1007/978-3-319-19734-0_12

$$\phi(mr) = (\phi m)r, \text{ for all } m \in M \text{ and all } r \in R. \tag{12.1}$$

Next, if we set $E = \operatorname{End}_R(M) = \operatorname{Hom}_R(M, M)$, the *endomorphism ring* of M, then (12.1) shows that M has the structure of an (E, R)-*bimodule*, a formalism that will be useful in the sequel.

12.2 Complete Reducibility

Given a right R-module M, the concept of *minimal* or *submodule* refers to minimal or maximal elements in the poset of all *proper* submodules of M. Thus 0 is not a minimal submodule and M itself is not a maximal submodule of M.

One defines the **radical** rad(M) of a right R-module M as the intersection of all maximal submodules of M. If there are no maximal submodules—as in the $\mathbb{Z}$-module

$$\mathbb{Z}(p^\infty) = \langle 1/p, 1/p^2, \ldots \rangle$$

—then rad$(M) = M$, by definition.

There is a dual notion: One defines the *socle* Soc(M) as the submodule generated by all minimal submodules (that is, all *irreducible* submodules) of M. Similarly, if there are no irreducible submodules—as in the case of the free $\mathbb{Z}$-module $\mathbb{Z}_{\mathbb{Z}}$—then by convention one sets Soc$(M) = 0$, the zero submodule.

Examples If D is any principle ideal domain (PID), and if p is a prime element in D, then the D module D/p^2D has radical rad$(M) = $ Soc$(M) = pD/p^2D \simeq D/pD$. (See Exercise (1) in Sect. 12.6.1.) In a similar vein, if F is a field, and if R is the matrix ring

$$R = \left\{ \begin{bmatrix} x & 0 \\ y & z \end{bmatrix} \mid x, y, z \in F \right\},$$

consisting of lower-triangular matrices over F, and if V is the right R-module consisting of 1×2 row vectors with entries in F:

$$V = \left\{ [\, x \; y \,] \mid x, y \in F \right\},$$

it also follows that
$$\operatorname{rad} V = \operatorname{Soc}(V) = \left\{ [\, x \; 0 \,] \mid x \in F \right\}.$$

(See Exercise (4) in Sect. 12.6.1.)

Theorem 12.2.1 *Let M be a right R-module having at least one irreducible submodule, so that Soc$(M) \neq 0$. Then*

(i) $Soc(M) = \oplus_{i \in J} M_i$, where $\{M_i\}_{i \in J}$ is a family of irreducible submodules of M, and the sum is direct.

(ii) In addition, $Soc(M)$ is invariant under the endomorphism ring, $E = \text{Hom}_R (M, M)$.

Proof Note that the second statement (ii) is obvious, for if $L \subseteq M$ is any irreducible R-module and ϕ is an R-module endomorphism, then the restriction of ϕ to L will have as kernel either 0 or all of L (as L is irreducible). Therefore, either $\phi(L) = 0$ or $\phi(L) \cong_R L$. In either case, $\phi(L) \subseteq Soc(M)$.

We now prove (i). Let the set I index the family of all irreducible modules of M. Trivially, we may assume I is non-empty. We shall call a subset J of I *a direct set* if and only if $\sum_{j \in J} M_j = \oplus_{j \in J} M_j$ is a direct sum. (Recall that this means that if a finite sum $\sum_{k \in K} a_k = 0$ where $a_k \in M_k$ and $K \subseteq J$, then each $a_k = 0$. We note also that singleton subsets of I are direct, and so, as irreducible modules are assumed to exist, direct subsets of I really exist.)

Suppose

$$I_0 \subset I_1 \subset \cdots \subset \cdots$$

is a properly ascending tower of direct sets, and let J be the union of the sets in this tower. If J were not direct, there would be a finite subset $K \subseteq J$ together with a sum $\sum_{k \in K} a_k = 0$ with not all a_k equal to zero. But since K is a finite subset, it clearly must be contained in I_n for some index n, which then says that I_n cannot have been direct, a contradiction. This says that in the poset of all direct subsets of I, partially ordered by inclusion, all simply ordered chains possess an upper bound. Thus by Zorn's Lemma, this poset contains a maximal member, say H.

We now claim that $M_H := \oplus_{j \in H} M_j = Soc(M)$. Clearly, $M_H \leq Soc(M)$. Suppose M_H were properly contained in $Soc(M)$. Then there would exist an irreducible submodule M_0 not in M_H. This forces $M_H \cap M_0 = 0$ from which we conclude that $M_H + M_0 = M_H \oplus M_0$, and so $H \cup \{0\}$ is a direct subset if I, properly containing H, contrary to the maximality of H. Thus $Soc(M) = \oplus_{j \in H} M_j$, and the proof is complete. $\square$

Corollary 12.2.2 *The following conditions on an R-module M are equivalent:*

(i) $M = Soc(M)$.

(ii) M is a sum of minimal submodules.

(iii) M is isomorphic to a direct sum of irreducible submodules.

An R-module M satisfying any of the above three conditions is said to be *completely reducible*.

Next, call a submodule L of a right R-module M *large* (some authors use the term *essential*) if it has a non-zero intersection with every non-zero submodule. (Note that if $M \neq 0$, M is a large submodule of itself.) We can produce (but not actually construct) many large submodules using the following Lemma.

Lemma 12.2.3 *If B is a submodule of M and C is maximal among submodules of M meeting B at zero, then $B + C$ is large.*

Proof Assume B and C chosen as above and let $D \neq 0$ be any non-trivial submodule of M. Suppose $(B + C) \cap D = 0$. If b, c, and d are elements of B, C and D, respectively, and $b = c + d$, then $d = b - c$ and this is zero by our supposition. Thus $b = c \in B \cap C = 0$, so $b = c = 0$. Thus $B \cap (C + D) = 0$. By maximality of C, $D \subseteq C$. Then $D = (B + C) \cap D = 0$, a contradiction. Thus always $(B + C) \cap D$ is non-zero, and so $B + C$ is large. $\square$

That such a submodule C exists which is maximal with respect to meeting B trivially, as in Lemma 12.2.3, is an easy application of Zorn's Lemma.

If M is a right R-module, let $L(M)$ be the lattice of all submodules of M (sum is the lattice "join", intersection is the lattice "meet"). We say $L(M)$ is *complemented* (or that M is *semisimple*) if every submodule B of M has a *complement* in $L(M)$—i.e. there exists a submodule B' such that $B \cap B' = 0$ and $B + B' = M$ (this means $M = B \oplus B'$).

Lemma 12.2.4 *If $L(M)$ is complemented, then so is $L(N)$ for any submodule N of M.*

Proof Let C be a submodule of N. By hypothesis there is a complement C' to C in $L(M)$. Then setting $C'' := C' \cap N$, we see

$$C \cap C'' \subseteq C \cap C' = 0, \text{ and}$$
$$C + C'' = C + (C' \cap N) = (C + C') \cap N, \text{ (the modular law)}$$
$$= M \cap N = N.$$

Thus C'' is a complement of C in N. $\square$

Remark Of course the above is actually a lemma about any modular lattice: if it is complemented, then so are any of its intervals above zero.

Lemma 12.2.5 *If M is a right R-module for which $L(M)$ is complemented, then $rad(M) = 0$.*

Proof Let m be a non-zero element of M and let A be maximal among the submodules which do not contain m. (A exists by Zorn's Lemma.) Suppose now $A < B < M$ so A is not maximal in $L(M)$. Let B' be a complement to B in $L(M)$. Then by the modular law (Dedekind's Lemma)

$$A = (B \cap B') + A = B \cap (B' + A).$$

Now as $B < M$ implies $B' \neq 0$ and $A < B$ implies $B' \cap A = 0$, we see that A is properly contained in $B' + A$. Then the maximality of A implies $m \in B' + A$ and also $m \in B$. Thus m lies in $B \cap (B' + A) = A$, contrary to the choice of A.

Thus no such B exists and A is in fact a maximal submodule of M. Then we see that every non-zero module element is avoided by some maximal submodule of M, and hence is outside $rad(M)$. It follows that $rad(M) = 0$. $\square$

Theorem 12.2.6 *The following conditions on an R-module M are equivalent:*

(i) *M is completely reducible.*

(ii) *M contains no proper large submodule.*

(iii) *M is semisimple.*

Proof Assume condition (i), i.e., that $M = \text{Soc}(M)$ and that $A \subsetneq M$ is a proper submodule of M. If A contained every irreducible submodule of M, then it would clearly contain $\text{Soc}(M) = M$. But then A couldn't be proper, contrary to our assumption on A. Thus there is an irreducible submodule U not in A, and so $A \cap U = 0$, since U is minimal. Thus A is not large, proving condition (ii).

Next, assume that (ii) holds, i.e., M contains no proper large submodules, and let $A \subseteq M$ be a submodule. Using Zorn's lemma, there exists a submodule $B \subseteq M$ which is maximal with respect to the condition $A \cap B = 0$. But then Lemma 12.2.3 asserts that $A + B$ is large and so by part (ii), $A + B = M$. This means that $M = A \oplus B$, i.e., that B is a complement to A in M, proving that M is semisimple.

Finally, assume condition (iii). Since M is semisimple, $\text{Soc}(M)$ has a complement C in $L(M)$. Next, by Lemma 12.2.5, we know that $\text{rad}(C) = 0$ and so the set of maximal submodules of C must be nonempty (else $C = \text{rad}(C)$ would be the intersection of the empty collection of maximal submodules). Letting $L \subseteq C$ be a maximal submodule, we now apply Lemma 12.2.4 to infer the existence of a submodule $A \subseteq C$ with $C = A \oplus L$. But then, $A \cong C/L$ and so A is an irreducible R-module with $A \cap \text{Soc}(M) = 0$. This is clearly impossible unless $C = 0$, in which case $\text{Soc}(M) = M$, proving (i). $\square$

12.3 Homogeneous Components

12.3.1 The Action of the R-Endomorphism Ring

Let M be a right R-module and let $E = \text{Hom}_R(M, M)$ be its R-endomorphism ring. As remarked earlier, we may regard M as a left E-module, giving M the structure of an (E, R)-bimodule.

Let M be any right R-module, and let $\mathcal{F}$ be a family of irreducible R-modules. We set $M[\mathcal{F}] = \sum A$, where the sum is taken over all irreducible submodules $A \subseteq M$ such that A is R-isomorphic to at least one module in $\mathcal{F}$. Note that if M contains no submodules isomorphic with any member of $\mathcal{F}$, then $M[\mathcal{F}] = 0$. If $\mathcal{F} = \{N\}$, the set consisting of a single irreducible right R-module N, then we write $M[N] := M[\{N\}]$, and call it the N-*homogeneous component* of M; again, it is entirely possible that $M[N] = 0$. Note that by Corollary 12.2.2, $M[\mathcal{F}]$ is completely reducible for any family $\mathcal{F}$ of irreducible submodules of M that we may choose. Next, suppose that $\mathcal{F}$ is a family of irreducible submodules of M, and that $B \subseteq M$ is a submodule of M isomorphic with some irreducible submodule $A \in \mathcal{F}$. Then for any $e \in E = \text{Hom}_R(M, M)$ we have that $eB = 0$ or $eB \cong_R \cong_R A$, proving that also

$eB \subseteq M[A] \subseteq M[\mathcal{F}]$. This proves already that $M[A]$ (as well as $M[\mathcal{F}]$) is invariant under the left action of every element of E.

We can internalize this discussion to the ring R itself. In particular, if $M = R_R$, the regular right R-module R, and if $A \subseteq R$ is a minimal right ideal, then $R[A] = M[A]$ is obviously also a right ideal of R. But it's also a left ideal since left multiplication by an element $r \in R$ gives, by the associative multiplication in R, an R-module endomorphism $R \to R$. By what was noted above, this proves that $rR[A] \subseteq R[A]$, and so $R[A]$ is a left ideal, as well. We shall capitalize on this simple observation shortly.

Lemma 12.3.1 *Assume that M is a completely reducible R-module and set $E = \mathrm{Hom}_R(M, M)$. If $A \subseteq M$ is any irreducible submodule of M, then $EA = M[A]$.*

Proof We have already observed above that $M[A]$ is invariant under every R-module endomorphism of M, so $EA \subseteq EM[A] \subseteq M[A]$. Conversely, assume that $A' \subseteq M$ is a submodule with $A \cong_R A'$. Since $\mathrm{rad}(M) = 0$ we may find an irreducible R-submodule $L \subseteq M$ with $A \cap L = 0$. Therefore it follows that $M = A \oplus L$. If $\pi : M = A \oplus L \to A$ is the projection onto the first coordinate, and if $\phi : A \xrightarrow{\cong} A'$ is an R-module isomorphism, then the composite

$$M \xrightarrow{\pi} A \xrightarrow{\phi} A' \hookrightarrow M$$

defines an R-module endomorphism carrying A to A'. $\square$

12.3.2 The Socle of a Module Is a Direct Sum of the Homogeneous Components

Our first lemma below informs us that inside the modules $M[\mathcal{F}]$ we won't encounter any unexpected submodules.

Lemma 12.3.2 *If M is an R-module and if $\mathcal{F}$ is a family of irreducible R-modules, then any irreducible submodule of $M[\mathcal{F}]$ is isomorphic with some member of $\mathcal{F}$.*

Proof Set $N = M[\mathcal{F}] \subseteq M$, and let $A \subseteq N$ be an irreducible submodule. By Lemma 12.2.5 we have $\mathrm{rad}(N) = 0$ from which we infer the existence of a maximal submodule $L \subseteq N$ with $A \cap L = 0$. Since N is generated by submodules each of which is isomorphic with a member of $\mathcal{F}$, we may choose one, call it B, where $B \cap L = 0$. Since L is maximal, we have $N = A \oplus L = B \oplus L$. Therefore,

$$A \cong (A \oplus L)/L = N/L = (B \oplus L)/L \cong B,$$

proving the result. $\square$

As a consequence of the above, we may exhibit the socle of a module as the direct sum of its *homogeneous components*, that is, the direct sum of its A-homogeneous components $M[A]$ as A ranges over a set of representatives of the R-isomorphisms classes of irreducible submodules of M.

Corollary 12.3.3 *Soc(M) is the direct sum of the homogeneous components of M.*

Proof Let $\mathcal{F}$ be a family of pairwise non-isomorphic irreducible R-submodules of M such that each irreducible submodule of M is isomorphic with a member of $\mathcal{F}$. Obviously $\mathrm{Soc}(M) = M[\mathcal{F}] = \sum_{A \in \mathcal{F}} M[A]$. By Lemma 12.3.2 we see that for all $A \in \mathcal{F}$, one has $M[A] \cap \sum_{B \in \mathcal{F}_A} M[B] = 0$, where $\mathcal{F}_A = \mathcal{F} \backslash \{A\}$. Now by Theorem 8.1.6 on internal direct sums, it follows immediately that

$$\mathrm{Soc}(M) = \sum_{A \in \mathcal{F}} M[A] = \bigoplus_{A \in \mathcal{F}} M[A]$$

as claimed. $\square$

The next several sections concern rings rather than modules. But we can already apply a few of the results of this section to the module R_R. Of course, the first statement below has already been observed above.

Corollary 12.3.4 *Every homogeneous component of Soc(R_R) is a 2-sided ideal of R. Also Soc(R_R) is itself a 2-sided ideal of R (which we write as Soc(R)).*

12.4 Semiprime Rings

12.4.1 Introduction

We now pass to rings. The basic class of rings that we shall deal with are the semiprime rings and certain subspecies of these. In order to state the defining properties, a few reminders about 2-sided ideals are in order.

If A and B are 2-sided ideals of the ring R, the *product AB of ideals* is the additive group generated by the set of all products $\{ab \,|\, (a, b) \in A \times B\}$. This additive group is itself already a 2-sided ideal which the ring-theorists denote by the symbol AB. Of course it lies in the ideal $A \cap B$. (From this definition, taking products among ideals is associative.)

We say that a (2-sided) ideal A is *nilpotent* if and only if there exists some positive integer n such that

$$A^n := A \cdot A \cdots A \ (n \text{ factors}) = 0 := \{0_R\},$$

the zero ideal.

An ideal P is said to be a *prime ideal* if and only if, whenever A and B are ideals such that $AB \subseteq P$, then either $A \subseteq P$ or $B \subseteq P$. The intersection of all prime ideals of the ring R is called the *prime radical* of R and is denoted $\mathrm{Prad}(R)$. We say that the ring R is *semiprime* if the intersection of its prime ideals is 0, i.e., if $\mathrm{Prad}(R) = 0$. Clearly $\mathrm{Prad}(R)$ is an ideal of R. A moment's thought should be enough to reveal that $R/\mathrm{Prad}(R)$ is semiprime. If the 0-ideal of R is itself a prime ideal, the ring R is called a *prime* ring.

12.4.2 The Semiprime Condition

Theorem 12.4.1 *The following conditions on a ring R are equivalent:*

(i) R is semiprime.
(ii) For any ideals A and B of R, $AB = 0$ implies $A \cap B = 0$.
(iii) 0 is the only nilpotent ideal.
(iv) For each non-zero $x \in R$, $xRx \neq 0$.

Proof (i)$\Longrightarrow$(ii) Assume A and B are ideals for which $AB = 0$. Then $AB \subseteq P$ for any prime ideal P so either A lies in P or B lies in P. In any event $A \cap B \subseteq P$ for each prime ideal P. Thus $A \cap B$ is contained in the intersection of all prime ideals, and so, since R is semiprime, is zero.

(ii)$\Longrightarrow$(iii) Suppose, by way of contradiction, that A is a non-zero nilpotent ideal of R. Let k be the least positive integer for which $A^k = 0$. Then $A^{k-1} \cdot A = 0$, which implies by condition (ii) that $A^{k-1} \subseteq A^{k-1} \cap A = 0$, contrary to the choice of exponent k.

(iii)$\Longrightarrow$(iv) If, for some $0 \neq x \in R$ we had $xRx = 0$, then as $x \in RxR$, we conclude that RxR is a non-zero ideal of R. But then

$$RxR \cdot RxR = RxRxR = R \cdot 0 \cdot R = 0,$$

forcing RxR to be a non-zero nilpotent ideal, against (iii).

(iv)$\Longrightarrow$(i) Let $a \neq 0$; we shall prove that there exists a prime ideal $P \subseteq R$ with $a \notin P$, proving that R is semiprime. Set $a_0 := a$; by (iv), $a_0Ra_0 \neq 0$. Therefore, there exists a non-zero element $a_1 \in a_0Ra_0$. We may continue this process to generate a sequence of non-zero elements $a_0, a_1, \ldots$, such that $a_{i+1} \in a_iRa_i$. Set $T = \{a_n | n \in \mathbb{N}$. Before proceeding further, we note an important property of this sequence:

(S) If the element a_i is in some ideal $J \subseteq R$, then so are all of its successors in T.

Let P be an ideal maximal with respect to not containing any element of T. (It exists by Zorn's Lemma.) Now suppose A and B are ideals of R neither of which is contained in P but with $AB \subseteq P$. By the maximality of P the ideals $A+P$ and $B+P$ both meet T non-trivially. So there are subscripts i and j such that $a_i \in A + P$, and $a_j \in B + P$. Set $m = \max(i, j)$. Then by the property (S), a_m lies in both $P + A$ and $P + B$, so

$$a_{m+1} \in a_m R a_m \subseteq (A + P)(B + P) \subseteq AB + P \subseteq P$$

against $P \cap T = \emptyset$. Thus no such ideals A and B exist, so P is a prime ideal. Since P doesn't contain $a = a_0$, the proof that R is semiprime is complete. $\square$

12.4.3 Completely Reducible and Semiprimitive Rings Are Species of Semiprime Rings

At the beginning of this chapter we defined the radical of a right R-module M to be the intersection of all maximal submodules of M, or, if there are no maximal submodules, to be M itself. Similarly, we define the *radical*, rad(R), of the ring R to be rad(R_R)—the radical of the right module R_R. Accordingly, rad(R) is the intersection of all maximal right ideals of R and is called the *Jacobson radical of R*.

While rad(R) is, a priori, a right ideal of R, the reader may be surprised to learn that it is also a left ideal. As this is relatively easy to show, we shall pause long enough to address this issue. Assume first that M is a non-zero irreducible right R-module. Then for each $0 \neq m \in M$, we have that (by irreducibility) $M = mR$. Furthermore, the mapping $R \to M$, $r \mapsto mr$ is easily verified to be a surjective R-module homomorphism. This implies immediately that if $J = \text{Ann}_R(m) := \{r \in R \mid mr = 0\}$, then J is the kernel of the above R-module homomorphism: $M \cong_R R/J$. Clearly J is a maximal right ideal of R. Next, recall that any R-module M determines a ring homomorphism $R \to \text{End}_{\mathbb{Z}}(M)$, the ring of abelian group homomorphisms of M. Furthermore, it is clear that $\text{Ann}_R(M) := \{r \in R \mid Mr = 0\}$ is the kernel of the above ring homomorphism, and hence $\text{Ann}_R(M)$ is a 2-sided ideal of M. Finally, note that

$$\text{Ann}_R(M) = \bigcap_{0 \neq m \in M} \text{Ann}_R(m), \tag{12.2}$$

i.e., $\text{Ann}_R(M)$ is the intersection of the maximal right ideals $\text{Ann}_R(m)$, $m \in M$. Our proof will be complete as soon as we can show that

$$\text{rad}(R) = \bigcap_{J \in \mathcal{M}(R)} \text{Ann}_R(R/J),$$

where $\mathcal{M}(R)$ denotes the collection of all maximal right ideals of R. By Eq. (12.2) we see that each $\text{Ann}_R(R/J)$ is an intersection of maximal right ideals of R; therefore if $x \in \text{rad}(R)$ then x is contained in $\text{Ann}_R(R/J)$ for every maximal right ideal $J \subseteq R$. This proves that

$$\text{rad}(R) \subseteq \bigcap_{J \in \mathcal{M}(R)} \text{Ann}_R(R/J).$$

Conversely, if $x \in \mathrm{Ann}_R(R/J)$, then Eq. (12.2) implies that $x \in J = \mathrm{Ann}_R(1+J)$, and so it is now clear that

$$\mathrm{rad}(R) \supseteq \bigcap_{J \in \mathcal{M}(R)} \mathrm{Ann}_R(R/J),$$

proving that, as claimed $\mathrm{rad}(R)$ is a 2-sided ideal of R.

We note in passing that the Jacobson radical annihilates every irreducible right R-module.

A ring is said to be *semiprimitive* if and only its Jacobson radical is trivial. A ring is *primitive* if and only if there exists an irreducible right R-module M with $\mathrm{Ann}(M) = 0$, i.e., if and only if R has a *faithful* irreducible module. It follows, of course, that every primitive ring is semiprimitive as is $R/\mathrm{rad}(M)$ for any ring R. Finally we say that a ring R is *completely reducible* if and only if the right module R_R is a completely reducible module. A hierarchy of these concepts is displayed in the following Theorem:

Theorem 12.4.2 *The following implications hold among properties of a ring:*

$$\text{completely reducible} \Rightarrow \text{semiprimitive} \Rightarrow \text{semiprime}$$

Proof The first implication is just Lemma 12.2.5, together with Theorem 12.2.6 applied to the module $M = R_R$.

For the second implication, suppose R is a ring which is semiprimitive but not semiprime. Then $\mathrm{rad}(R) = 0$ and there exists a non-zero nilpotent ideal N. By replacing N by an appropriate power if necessary, we may assume $N^2 = 0 \neq N$. Since $\mathrm{rad}(R) = 0$, there exists a maximal right ideal M not containing N. Then $R = N + M$ by maximality of M. Thus we may write $1 = n + m$, where $n \in N$ and $m \in M$. Then

$$n = 1 \cdot n = n^2 + mn = 0 + mn \in M.$$

Thus $1 = n + m \in M$, which is impossible. Thus we see that N must lie in every maximal right ideal, so $N \subseteq \mathrm{Rad}(R) = 0$. This contradicts $N \neq 0$. Thus no such ring R exists, and so the implication holds. $\square$

12.4.4 Idempotents and Minimal Right Ideals of Semiprime Rings

The importance of semiprimness for us is that it affects the structure of the minimal right ideals of R.

We pause, however, to define an *idempotent* in a ring R to be a non-zero element e satisfying $e^2 = e$. Such elements play a central role in this and the following subsection. Furthermore, the role that they play is analogous to that played by projections in linear algebra.

Lemma 12.4.3 (Brauer's lemma) *Let K be a minimal right ideal of a ring R. Then either*

(i) $K^2 = 0$, *or*
(ii) $K = eR$, *for some idempotent $e \in R$.*

Proof If $K^2 \neq 0$, then by the minimality of K, $K = K^2 = kK$ for some element $k \in K$. From this we infer the existence of an element e such that $ke = k$. Next, since the right annihilator, $\mathrm{Ann}_K(k) = \{x \in K \mid kx = 0\}$, is a right ideal of R lying properly in K, we have $\mathrm{Ann}_K(k) = 0$. But as $ke = k$,

$$k(e^2 - e) = (ke)e - ke = ke - ke = 0,$$

so $e^2 - e \in \mathrm{Ann}_K(k) = 0$, forcing $e^2 = e$. Finally, $0 \neq eR \subseteq K$, so $eR = K$ as K was minimal. $\square$

Corollary 12.4.4 *If R is semiprime, then every minimal right ideal of R has the form eR for some idempotent $e \in R$.*

Proof Let K be a minimal right ideal of R. Then by Brauer's Lemma, the conclusion holds or else $K^2 = 0$. But in the latter case, RK is a nilpotent 2-sided ideal containing K, as

$$(RK)^2 = R(KR)K \subseteq RK^2 = 0.$$

This contradicts the fact that R is semiprime. $\square$

Lemma 12.4.5 *Suppose e and f are non-zero idempotent elements of a ring R. Then there is an additive group isomorphism $\mathrm{Hom}_R(eR, fR) \simeq fRe$. Moreover, if $f = e$, then this isomorphism can be chosen to be a ring isomorphism.*

Proof Consider the mapping $\psi : fRe \to \mathrm{Hom}_R(eR, fR)$ defined by

$$\psi(fre) : es \mapsto (fre)es = fres, \text{ where } r, s \in R$$

In other words, for each $r \in R$, $\psi(fre)$ is left multiplication of eR by fre. This is clearly a homomorphism of right R-modules $eR \to fR$. Note that $\psi(fre)(e) = fre$, so the image of $\psi(fre)$ can be the zero homomorphism if and only if $fre = 0$. So the mapping ψ is injective.

The mapping is also surjective, for if λ is a right module homomorphism $eR \to fR$, then $\lambda(e) = fs$, for some $s \in R$. Then, for any $r \in R$,

$$\lambda(er) = \lambda(e^2 r) = \lambda(e) \cdot er = (fse)er = \psi(fse)(er),$$

so the R-homomorphism λ is an image of the mapping ψ.

Of course for elements r_1, r_2 in R, $f(r_1 + r_2)e = f(r_1 + r_2)e$ has a ψ-image that is the sum of the R-morphisms $\psi(fr_1e) + \psi(fr_2e)$, so ψ preserves addition. Thus ψ is an isomorphism $fRe \to \mathrm{Hom}_R(eR, fR)$ of additive groups.

Now suppose $e = f$. Then eRe inherits a ring multiplication from R that is distributive with respect to addition. We must show that ψ preserves multiplication in this case. For two arbitrary elements er_1e, er_2e of eRe, the ψ-image of their product is left multiplication of eR by er_1eer_2, which by the associative law, is the composition $\psi(er_1e) \circ \psi(er_2e)$ in $Hom_R(eR, fR) = End_R(eR)$. $\square$

Lemma 12.4.6 *If R is a semiprime ring, and $e^2 = e \in R$, then eR is a minimal right ideal if and only if eRe is a division ring.*

Proof ($\Longrightarrow$) If eR is a minimal ideal, then $End_R(eR)$ is a division ring by Schur's Lemma (Lemma 8.2.6). By the preceding Lemma 12.4.5, eRe is isomorphic to this division ring.

($\Longleftarrow$) Assume eRe is a division ring. Let us assume, by way of contradiction, that eR is not a minimal ideal. Then there is a right ideal N, such that $0 < N < eR$. Then $Ne \subseteq eRe \cap N$. Since a division ring has no proper right ideals, $eRe \cap N$ is either 0 or eRe.

Suppose $Ne \neq 0$. Then there exists an element $n \in N$ such that $ne \neq 0$. Then ne, being a non-zero element of a division ring, possesses a multiplicative inverse in eRe, say ese. Then

$$ne(ese) = nese = e.$$

Since N is a right ideal in R, it follows that e lies in N. Thus $eR \subseteq N$, contrary to our choice of N. Thus we must have $Ne = 0$.

Now eN is a right ideal of R such that $(eN)^2 = e(Ne)N = 0$. Since R is semiprime, it possesses no non-trivial nilpotent right ideals (Theorem 12.4.1). Thus $eN = 0$. But left multiplication by e is the identity endomorphism $eR \to eR$, yet it annihilates the submodule N of eR. It follows that $N = 0$, also against our choice of N.

Thus no such N, as hypothesized exists, and so eR is an irreducible R-module— i.e. a minimal right ideal. $\square$

Corollary 12.4.7 *Let e be an idempotent element of the semiprime ring R. Then eR is a minimal right ideal if and only if Re is a minimal left ideal.*

Proof In Lemma 12.4.6 a fact about a right ideal is equivalent with a fact about a ring eRe which possesses a complete left-right symmetry. Thus, applying this symmetry, eRe is a division ring if and only if the principal left ideal Re is minimal—that is, an irreducible R-module. $\square$

Lemma 12.4.8 *Let e and f be idempotent elements of the ring R. Then the right ideals eR and fR are isomorphic as R-modules if and only if there exist elements u and v in R such that $vu = e$ and $uv = f$.*

Proof ($\Longrightarrow$) Assume there exists an isomorphism $\alpha : eR \to fR$. Here we exploit the isomorphism $\psi^{-1} : Hom_R(eR, fR) \to eRf$ of Lemma 12.4.5, to see that α is achieved by left multiplication by an element $u = fae \in fRe$. Note that $u = fue$, since e and f are idempotent elements. Similarly, α^{-1} is achieved by left multiplication by $v = ebf \in eRf$. Clearly $v = evf$, since e and f are idempotents. The left multiplications compose as $\alpha \circ \alpha^{-1} = 1_{eR}$, which is represented as left-multiplication by e. Thus $vu = e$. Similarly, $\alpha^{-1} \circ \alpha = 1_{fR}$ is left multiplication of fR by uv and by f. Thus $uv = f$.

($\Longleftarrow$) Now assume that there exist elements u and v in R such that $vu = e$ and $uv = mf$. Then $ue = uvu = fu$, so left multiplication by u effects an R-homomorphism $eR \to fR$ as right R modules. Similarly, left multiplication by v effects another R-homomorphism of right R-modules $fR \to eR$. Now since $vu = e$ and $vu = f$, these morphisms are inverses of each other, and so are R-isomorphisms of R-modules. Hence $eR \simeq fR$. $\square$

Now consider a homogeneous component A of the semiprime ring R. We have already observed that left multiplication of a minimal right ideal eR by an element $r \in R$ produces a module reR which, being a homomorphic image of an irreducible module, is either 0 or is an irreducible right module isomorphic to eR. It follows that $RA \subseteq A$, so A is a 2-sided ideal. But there is more.

First, A, being the sum of irreducible right R-modules, is a completely reducible module. By Lemma 12.3.2 every irreducible right submodule in A is of the same isomorphism type. Because of this, Lemma 12.4.8 tells us that we can pass from one irreducible right submodule of A to any other by multiplying on the left by an appropriate element of R. Thus A is a completely reducible left R-module spanned by irreducible left R-modules of the same isomorphism type. These observations imply the following:

Corollary 12.4.9 *Let R be a semiprime ring, and let A be any homogeneous component of R_R. Then A is an irreducible left R-module, and so is a minimal 2-sided ideal of R. We also see from Lemma 12.4.6, that A is also a homogeneous component of $_R R$.*

12.5 Completely Reducible Rings

The left-right symmetry that has emerged in the previous section can be applied to the characterization of completely reducible rings.

Theorem 12.5.1 *The following four statements are equivalent for any ring R.*

  (i) *Every right R-module is completely reducible.*
 (ii) *R_R is completely reducible.*
(iii) *Every left R-module is completely reducible.*
(iv) *$_R R$ is completely reducible.*

Proof That (i) implies (ii) is true *a fortiori*. Now assume (ii), and let M be a right R-module. We may write $R = \sum_{i \in I} A_i$ where the A_i are minimal right ideals in R. We clearly have

$$M = \sum_{m \in M} mR = \sum_{m \in M} \sum_{i \in I} mA_i.$$

Note that the mapping $A_i \to mA_i$, $a \mapsto ma$ is a surjective R-module homomorphism; therefore, it follows that either $mA_i = 0$ or mA_i is an irreducible submodule of M. By Corollary 12.2.2, we see that M is completely reducible. This proves that (i) and (ii) are equivalent. In an entirely similar fashion (iii) and (iv) are equivalent. Finally, by Corollary 12.4.7, (ii)$\Longleftrightarrow$(iv), which clearly finishes the proof. $\square$

There is an immediate application:

Corollary 12.5.2 *Let D be a division ring. Then any right vector space V over D is completely reducible. In particular V is a direct sum of copies of D_D.*

Proof Since D is a division ring, D_D is an irreducible module, so any right D module V (for that is what a right vector space is!) is completely reducible. $\square$

Before stating the following, recall that R is a *prime ring* if and only if 0 is a prime ideal.

Lemma 12.5.3 *Let R be a prime ring, and assume that $S = Soc(R_R) \neq 0$. If eR is a minimal right ideal of R (where e is an idempotent element), then $\operatorname{End}_R(S) := \operatorname{Hom}_R(S, S)$ is isomorphic to the ring of all linear transformations $\operatorname{End}_{eRe}(Re) := \operatorname{Hom}_{eRe}(Re, Re)$. (Note that Re is a right eRe-vector space.)*

Proof Since $S = Soc(R_R)$ is a 2-sided ideal and is the direct sum of its homogeneous components, each of which is a 2-sided ideal, the prime hypothesis implies that S is itself a homogeneous component. Let S be written as a direct sum $\sum e_i R$ of minimal right ideals with $e_i^2 = e_i$, and fix one of these, say $e = e_1$. Then each $e_i R \simeq eR$ so, by Lemma 12.4.8, there exist elements u_i and v_i such that $v_i u_i = e$ and $u_i v_i = e_i$. Then

$$u_i e v_i = u_i v_i u_i v_i = e_i^2 = e_i.$$

Now let $\phi \in \operatorname{Hom}_R(S, S)$. We define

$$\tau(\phi) : Re \to Re$$

by the rule

$$\tau(\phi)(re) = \phi(re) = \phi(re)e, \text{ for all } r \in R.$$

Note that $\tau(\phi)$ is right eRe-linear, for if $r_1 e, r_2 e \in Re$, and if $es_1 e, es_2 e \in eRe$, then,

$$\begin{aligned}
\tau(\phi)(r_1 e \cdot es_1 e + r_2 e \cdot es_2 e) &= \phi(r_1 e \cdot es_1 e + r_2 e \cdot es_2 e) \\
&= \phi(r_1 e \cdot es_1 e) + \phi(r_2 e \cdot es_2 e) \\
&= \phi(r_1 e)es_1 e + \phi(r_2 e)es_2 e \\
&= \tau(\phi)(r_1 e)es_1 e + \tau(\phi)(r_2 e)es_2 e.
\end{aligned}$$

Also, if $\phi_1, \phi_2 \in \mathrm{Hom}_R(S, S)$, it is clear that

$$\tau(\phi_1 + \phi_2) = \tau(\phi_1) + \tau(\phi_2)$$

and that $\tau(\phi_1 \circ \phi_2) = (\tau(\phi_1)) \circ (\tau(\phi_2))$, proving that τ is a ring homomorphism.
Clearly, $\tau(\phi) = 0$ if and only if $\phi(re) = 0$ for all $r \in R$. So

$$\phi(e_i) = \phi(u_i e v_i) = \phi(u_i e) v_i = 0 \cdot v_i = 0.$$

for all indices i. Thus $\phi(S) = 0$, and hence is the zero mapping. It follows that τ is injective.

We shall now show that $\tau : \mathrm{Hom}_R(S, S) \to \mathrm{Hom}_{eRe}(Re, Re))$ is surjective. Let $\psi \in \mathrm{Hom}_{eRe}(Re, Re)$, and define $\phi \in \mathrm{Hom}_R(S, S)$ as follows. If $s \in S$, then since $S = \sum e_i R$ is a direct sum, we may write s as $s = \sum e_i r_i$, with unique summands $e_i r_i$, for suitable elements $r_i \in R$. Here we warn the reader that the elements r_i are *not* necessarily unique, as the idempotents e_i do not necessarily comprise a basis of S (that is, S need not be a free R-module). We set

$$\phi(s) = \sum \psi(u_i e) v_i r_i.$$

To show that ϕ is well defined, we need only show that if $e_i r_i = 0$, then $\psi(u_i e) v_i r_i = 0$.
However, as $\psi \in \mathrm{Hom}_{eRe}(Re, Re)$, and since $e \in eRe$, we have that

$$\begin{aligned}
\psi(u_i e) v_i r_i = \psi(u_i e \cdot e) v_i r_i &= \psi(u_i e) e v_i r_i \\
&= \psi(u_i e) v_i u_i v_i r_i \\
&= \psi(u_i e) v_i e_i r_i \\
&= \psi(u_i e) \cdot 0 = 0,
\end{aligned}$$

and so ϕ is well defined. Checking that ϕ is R-linear is entirely routine.

The proof will be complete as soon as we show that $\tau(\phi) = \psi$. Thus, let $re \in Re$ and write $re = \sum e_i r_i \in S$. Then

$$\begin{aligned}
\tau(\phi)(re) = \phi(re) &= \phi(re)e \\
&= \sum \psi(u_i e) v_i r_i e \\
&= \sum \psi(u_i e^2) v_i r_i e \\
&= \sum \psi(u_i e) e v_i r_i e \\
&= \sum \psi(u_i e \cdot e v_i r_i e) \\
&= \sum \psi(u_i e v_i r_i e) \\
&= \sum \psi(u_i v_i u_i v_i r_i e)
\end{aligned}$$

$$= \sum \psi(e_i^2 r_i e)$$
$$= \sum \psi(e_i r_i e)$$
$$= \psi\left(\sum e_i r_i e\right)$$
$$= \psi(re \cdot e) = \psi(re),$$

Where all sums on the right are over the parameter i. Comparing first and last terms, we see that $\tau(\phi) = \psi$, proving that τ is a surjective mapping. $\square$

Before passing to our main result, we pause long enough to observe that *for any ring R, we have $R \cong \operatorname{End}_R(R_R)$.* This isomorphism is given via $r \mapsto \lambda_r$, where, as usual, $\lambda_r(s) = rs$, for all $r, s \in R$. Furthermore, the associative and distributive laws in R guarantee not only that each λ_r is a homomorphism of the right R-module R but also that the mapping $r \mapsto \lambda_r$ defines a ring homomorphism $R \to \operatorname{End}_R(R_R)$. This homomorphism is injective since $\lambda_r = 0$ implies $0 = \lambda_r(1) = r \cdot 1 = r$. Finally, if $\phi \in \operatorname{End}_R(R)$, set $r = \phi(1)$; then for all $s \in R$,

$$\lambda_r(s) = rs = \phi(1)s = \phi(1 \cdot s) = \phi(s).$$

Theorem 12.5.4 (Wedderburn-Artin)

 (i) *A ring R is completely reducible if and only if it is isomorphic to a finite direct product of completely reducible simple rings.*
(ii) *A ring R is completely reducible and simple if and only if it is the ring of all linear transformations of a finite dimensional vector space.*

Proof First we prove part (i). Let R be completely reducible, so R is a direct sum $R = \bigoplus_{i \in I} A_i$, where the A_i are minimal right ideals. Now write $1 = \sum e_i$, where each $e_i \in A_i$. Note that for each $j \in I$, we have

$$e_j = 1 \cdot e_j = \left(\sum_{i \in I} e_i\right) \cdot e_j = \sum_{i \in I} e_i e_j.$$

Since $e_i e_j \in A_i e_j \subseteq A_i$, we conclude that $e_j^2 = e_j$ and that when $i \neq j$, $e_i e_j = 0$. Next, since $1 = \sum e_i$ is necessarily a finite sum, we may as well write this as

$$1 = \sum_{i=1}^{n} e_i,$$

for some n. Therefore, it follows that

$$R = 1 \cdot R = \sum_{i=1}^{n} e_i R. \tag{12.3}$$

Since $e_i R \subseteq A_i$, we infer that the sum in Eq. (12.3) is necessarily finite. Furthermore, since each A_i is a minimal right ideal, we must have $e_i R = A_i$, and so conclude that R is a *finite* direct sum of minimal right ideals:

$$R = e_1 R \oplus e_2 R \oplus \cdots \oplus e_n R. \tag{12.4}$$

Next, we gather the above minimal right ideals into their constituent homogeneous components, giving rise to the direct sum

$$R = C_1 \oplus C_2 \oplus \cdots \oplus C_m,$$

where the $C_1, \ldots, C_m$ are the distinct homogeneous components of R. If we write the identity element of R according to the above sum:

$$1 = \sum_{j=1}^m c_j,$$

we easily infer that $c_j^2 = c_j$, and that $c_j c_k = 0$ whenever $j \neq k$. Furthermore, if $x_j \in C_j$, and if $k \in \{1, 2, \ldots, m\}$, we see that $x_j c_k$ and $c_k x_j$ are both elements of $C_j \cap C_k$. If $j \neq k$, then $C_j \cap C_k = 0$ so $x_j c_k = c_k x_j = 0$. From this it follows that

$$x_j = x_j \cdot 1 = x_j c_j = 1 \cdot x_j = c_j x_j.$$

This proves that each of the 2-sided ideals C_j has a multiplicative identity, viz., c_j. That C_j is a simple ring is already contained in Corollary 12.4.9. That C_j is completely reducible follows from the definition of homogeneous component.

Conversely, if R is a finite direct product of completely reducible rings, this finite direct product can be regarded as a finite direct sum, which already implies that R_R is completely reducible.

We turn now to part (ii). Let R be a completely reducible and simple ring. Let eR be a minimal right ideal of R. By Lemma 12.5.3 together with the discussion above, we have

$$R \simeq \operatorname{End}_R(R) \simeq \operatorname{End}_{eRe}(Re).$$

Our proof will be complete as soon as we show that Re is a finite-dimensional vector space over the division ring eRe. Note first of all that the right R-module R_R, being a direct sum of finitely many irreducible modules as in Eq. (12.4) is certainly a Noetherian module. Therefore, any collection of right ideals has a maximal member. To show that Re is finite dimensional over eRe, it suffices to show that any collection of subspaces of Re has a maximal member (i.e., that Re is a Noetherian eRe-module). Thus, let $\{L_\mu\}$ be a collection of eRe-subspaces of Re, and note that $\{L_\mu R\}$ is a family of right ideals of R. Since R_R is Noetherian, it has a maximal member, say $L_\sigma R$. Suppose L_ν is a subspace with $L_\nu \supseteq L_\sigma$. Then we have $L_\nu R = L_\sigma R$. Furthermore, as $LeRe = L$ for any subspace $L \subseteq Re$, we have

$$L_\nu = L_\nu eRe = L_\nu Re = L_\sigma Re = L_\sigma eRe = L_\sigma,$$

and so L_σ is a maximal subspace.

Finally, we must prove that the ring of endomorphisms of a finite-dimensional vector space forms a simple ring. Assume that $R \cong \mathrm{End}_D(V)$ the ring of linear transformations of a finite-dimensional vector space V over a division ring D. In order to make the arguments more transparent, we shall show first that $\mathrm{End}_D(V) \cong M_n(D)$, the ring of $n \times n$ matrices over D, where n is the dimension of V over D. Fix an ordered basis $(v_1, v_2, \ldots, v_n)$ and let $\phi \in \mathrm{End}_D(V)$. Write

$$\phi(v_j) = \sum_{i=1}^{n} v_i a_{ij}, \quad j = 1, 2, \ldots, n,$$

where $a_{ij} \in D$. It is entirely routine to check that the mapping $\phi \mapsto A = [a_{ij}] \in M_n(D)$ is a ring isomorphism. Next, one argues that for $i = 1, 2, \ldots, n$, the sets

$$L_i = \left\{ \begin{bmatrix} 0 & 0 & \cdots & 0 \\ \vdots & \vdots & \cdots & \vdots \\ a_{i1} & a_{i2} & \cdots & a_{in} \\ \vdots & \vdots & \cdots & \vdots \\ 0 & 0 & \cdots & 0 \end{bmatrix} \;\middle|\; a_{i1}, a_{i2}, \ldots, a_{in} \in D \right\} \subseteq M_n(D)$$

are isomorphic minimal right ideals in $M_n(D)$. This makes all of $M_n(D)$ a homogeneous component and so is a simple ring. $\square$

We turn, finally, to an alternative formulation of the Wedderburn-Artin theorem. In this version, we assume the ring R to be semiprimitive (so $\mathrm{rad}(R) = 0$), as well as being (right) Artinian. The object shall be to prove that R_R is completely reducible.

Theorem 12.5.5 *Let R be a semiprimitive Artinian ring. Then R can be represented as a finite direct product of matrix rings over division rings.*

Proof It clearly suffices to prove that R_R is completely reducible. Since R is right Artinian, we may find a minimal right ideal $I_1 \subseteq R$. Also, since $\mathrm{rad}(R) = 0$ we may find a maximal right ideal, say, M_1 with $I_1 \not\subseteq M_1$. Clearly

$$R = I_1 + M_1.$$

Next, using again the fact that R is right Artinian, we may extract a minimal right ideal $I_2 \subseteq M_1$. Since $\mathrm{rad}(R) = 0$ there is a maximal right ideal, say M_2 such that $I_2 \nsubseteq M_2$. Since $M_1 = R \cap M_1$, and since $R = I_2 + M_2$, we may use the modular law to obtain

$$
\begin{aligned}
M_1 &= R \cap M_1 \\
&= (I_2 + M_2) \cap M_1 \\
&= I_2 + (M_2 \cap M_1),
\end{aligned}
$$

from which we conclude that

$$R = I_1 + M_1 = I_1 + I_2 + (M_1 \cap M_2).$$

We may continue in this fashion to produce a sequence $M_1, M_2, \ldots$ of maximal right ideals and a sequence $I_1, I_2, \ldots$ of minimal right ideals such that

$$R = I_1 + I_2 + \cdots I_k + (M_1 \cap M_2 \cap \cdots \cap M_k).$$

Since R is right Artinian, the strictly decreasing sequence

$$M_1 \supsetneq M_1 \cap M_2 \supsetneq \cdots \supsetneq \bigcap_{i=1}^{t} M_i \supsetneq \cdots$$

must terminate. Since R is semiprime, it must terminate at 0. This means that for a suitable positive integer s, we will have found finitely many minimal right ideals $I_1, I_2, \ldots, I_s$ such that

$$R = I_1 + I_2 + \cdots + I_s.$$

The proof is complete. $\square$

12.6 Exercises

12.6.1 General Exercises

1. Let D be a PID and let $0 \neq a \in D$ be a non-unit. Compute the radical and the socle of the cyclic D-module D/aD.
2. Let D be a PID.

   (a) Show that every nonzero submodule of the free right D-module D_D is large.
   (b) Let $p \in D$ be a prime. Show that for every integer $n \geq 1$, every nonzero submodule of the cyclic D-module $D/p^n D$ is large.

3. Show that every nonzero $\mathbb{Z}$-submodule of $\mathbb{Z}(p^\infty)$ is large.
4. Let F be a field and let $A \subseteq M_n(F)$ be the ring of lower triangular $n \times n$ matrices over F. Let V be the A-module of $1 \times n$ matrices

$$V = \{[\, x_1\ x_2\ \dots\ x_n \,] \mid x_1, x_2, \dots, x_n \in F\}.$$

Compute $\mathrm{Soc}(V)$ and $\mathrm{rad}(V)$ and show that they are both large submodules.
5. Let M be a right R-module. Show that the socle of M is the intersection of the large submodules of M.
6. Let $R = M_2(F)$, where F is a field. Show that

$$L = \left\{ \begin{bmatrix} x & y \\ 0 & 0 \end{bmatrix} \mid x, y \in F \right\},$$

is a minimal right ideal in R and hence is an irreducible R-module. Show that any other minimal right ideal of R is isomorphic to L as a right R-module. [Hint: If

$$L' = \left\{ \begin{bmatrix} 0 & 0 \\ x & y \end{bmatrix} \mid x, y \in F \right\},$$

then $L' \cong_R L$.) Conclude that $R_R = R[L]$.]
7. Let F be a field and let x be indeterminate over F. Let A be the 2×2 matrix

$$A = \begin{bmatrix} 1 & 0 \\ 1 & 1 \end{bmatrix}.$$

Make $V = F_2 = \{[a\ b] \mid a, b \in F\}$ into an $F[x]$-module in the usual way and compute $\mathrm{Soc}(V)$ and $\mathrm{rad}(V)$.
8. Let F be a field, and let $A \in M_n(F)$. Let M be the set of $1 \times n$ matrices over F regarded as a right $F[x]$-module in the usual way. If A is diagonalizable over F, show that M is a completely reducible $F[x]$-module. (This strong hypothesis isn't necessary. In fact all one really requires is that the minimal polynomial of A factor into distinct irreducible factors in $F[x]$.)
9. A right R-module M is called a *prime module* if for any submodule $N \subseteq M$, and any ideal $I \subseteq R$ we have that $NI = 0$ implies that either $N = 0$ or that $MI = 0$. Show that the annihilator in R of a prime module is a prime ideal.
10. Define a *primitive ideal* in the ring R to be the annihilator of an irreducible right R-module. Now show that we have the following hierarchy for ideals in the ring R:

$$\text{maximal ideal} \implies \text{primitive ideal} \implies \text{prime ideal}.$$

11. Let R and S be rings, and let ρ and σ be antiautomorphisms of R and S, respectively. Given an (R, S)-bimodule M, define new multiplications

$$S \times M \times R \to M$$

by the rules that

$$sm = m(s^\sigma) \text{ and } mr = (r^\rho)m,$$

for all $(s, m, r) \in S \times M \times R$.

(a) Show that with respect to these new multiplications, the additive group $(M, +)$ is endowed with the structure of an (S, R)-bimodule, which we denote as $^\sigma M^\rho$.

(b) Show that if $\mathbb{Z}$ is the ring of integers, any abelian additive group $(M, +)$ is a $(\mathbb{Z}, \mathbb{Z})$-bimodule where the action of an integer n on module element m (from the right or left) is to add m or $-m$ to itself $|n|$ times, according as n is positive or negative, and to set $0m = m0 = 0$. With this interpretation, any left R-module is an $(R, \mathbb{Z})$-bimodule, and any right S-module is already a $(\mathbb{Z}, S)$-bimodule.

(c) By applying part 1 with one of (R, ρ) or (S, σ) set equal $(\mathbb{Z}, 1_\mathbb{Z})$, the ring of integers with the identity antiautomorphism, show that any left R-module M can be converted into a right R-module M^ρ. Similarly, any right S-module N can be converted into a left S-module, $^\sigma N$, by the recipe of part 1.

(d) Suppose N is a monoid with anti-automorphism μ. Let K be a field and let KN denote the monoid ring—that is, a K-vector space with the elements of N as a basis, and all multiplications of K-linear combinations of these bases elements determined (via the distributive laws) by the multiplication table of N. Define $\hat{\mu} : KN \to KN$ by applying μ to each basis element of N in each K-linear combination, while leaving the scalars from K unaffected. Show that $\hat{\mu}$ is an anti-automorphism of KN.

(e) Suppose now $N = G$, a group. Show that every anti-automorphism σ of G has the form $g \to (g^{-1})^\alpha$, where α is a fixed automorphism of G.

12.6.2 Warm Up Exercises for Sects. 12.1 and 12.2

1. An element x of a ring R is *nilpotent* if and only if, for some positive integer n, $x^n = 0$. Show that no nilpotent element is right or left invertible.
2. Show that if x is a nilpotent element, then $1 - x$ is a unit of R.
3. A right ideal A of R is said to be *nilpotent* if and only if, for some positive integer n (depending on A) $A^n = 0$. Show that if A and B are nilpotent right ideals of R, then so is $A + B$.
4. (An interesting example) Let F be any field. Let $F\langle x, y \rangle$ denote the ring of all polynomials in two "non-commuting" variables x and y. Precisely $F\langle x, y \rangle$ is the monoid-ring FM where M is the free monoid generated by the two-letter alphabet $\{x, y\}$. (We shall see later on that this ring is the tensor algebra $T(V)$ where V is a two-dimensional vector space over F.) Let I be the 2-sided ideal generated by $\{x^2, y^2\}$, and form the factor ring $R := F\langle x, y \rangle / I$.

(a) For each positive integer k, let

$$(x)_k := xyx \cdots (k \text{ factors }) + I,$$
$$(y)_k := yxy \cdots (k \text{ factors }) + I.$$

Show that the set $\{1, (x)_k, (y)_k | 0 < k \in \mathbf{Z}\}$ is a basis for R as a vector space over F. Thus each element of R is a finite sum

$$r = \alpha_0 \cdot 1 + \sum_{i \in J} \alpha_i (x)_i + \beta_i (y)_i,$$

where J is a finite set of positive integers.

(b) Show that the sum $(x + I)R + (y + I)R$ of two right ideals is direct, and is a maximal two-sided ideal of R.

(c) An element $(x)_k$ is said to be x-*palindromic* if it begins and ends in x—i.e., if k is odd. Similarly $(y)_k$ is y-palindromic if and only if k is odd: Show that any linear combination of x-palindromic elements is nilpotent.

(d) Let P_x (P_y) denote the vector subspace spanned by all x-palindromic (y-palindromic) elements. Then the group of units of R contains $(1 + P_x) \cup (1 + P_y)$ and hence all possible products of these elements.

(e) Show that for any positive integer k, $(x + y)^k = (x)_k + (y)_k$. (Here is a sum of two nilpotent elements which is certainly not nilpotent.)

5. A right ideal is said to be a *nil right ideal* if and only all of its elements are nilpotent. (If the nil right ideal is 2-sided, we simply call it a *nil ideal*.)

(a) For any family $\mathcal{F}$ of nilpotent right ideals, show that the sum $\sum_{\mathcal{F}}$ of all ideals in $\mathcal{F}$ is a nil ideal.

(b) Show that if A is a nilpotent right ideal, then so is rA.

(c) Define $N(R)$ to be the sum of all nilpotent right ideals of R. (This invariant of R can be thought of as a certain kind of "radical" of R.) Show that $N(R)$ is a 2-sided nil ideal.

12.6.3 Exercises Concerning the Jacobson Radical

Recall that Jacobson radical of a ring R (with identity) was defined to be the intersection of all the maximal right ideals of R, and was denoted by the symbol rad(R).

1. Show that an element r of the ring R is right invertible if and only if it belongs to no maximal right ideal.

2. Let "1" denote the multiplicative identity element of R. Show that the following four conditions on an element $r \in R$ are equivalent:

(a) r belongs to the Jacobson radical of R.

(b) For each maximal right ideal M, 1 does not belong to $M + rR$.

(c) For every element s in R, $1 - rs$ belongs to no maximal right ideal of M.

(d) For every element s in R, $1 - rs$ is right invertible.

3. Show that if an element $1 - rt$ is right invertible in R, then so is $1 - tr$. [Hint: By hypothesis there exists an element u such that $(1 - rt)u = 1$. So we have :
$u = 1 + rtu$. Use this to show that the product

$$(1 - tr)(1 + tur) = 1 + tur - t(1 + rtu)r,$$

is just 1.] Prove that $\text{Rad}(R)$ is a 2-sided ideal. [Hint: Show that if $t \in \text{rad}(R)$, then so is rt.]

4. Show that $\text{Rad}(R)$ contains every nilpotent 2-sided ideal. [Hint: If false, there is a nilpotent 2-sided ideal A and a maximal ideal M such that $A^2 \subseteq M$ while $R = A + M$. Then $1 = a + m$ where $a \in A$ and $m \in M$. Thus $1 - a$ is on the one hand an element of M, and on the other hand it is a unit by Exercise (2) in Sect. 12.6.2.]

12.6.4 The Jacobson Radical of Artinian Rings

A ring whose poset of right ideals satisfies the descending chain condition is called an *Artinian ring*. The next group of exercises will lead the student to a proof that for any Artinian ring, the Jacobson radical is a nilpotent ideal.

1. We begin with something very elementary. Suppose b is a non-zero element of the ring R. If $b = ba$, then there is a maximal right ideal not containing a—in particular a is not in the Jacobson radical. [Hint: Just show that $1 - a$ has no right inverse.]

2. Suppose R is an Artinian ring, and A is a non-zero ideal for which $A^2 = A$. Show that there exists a non-zero element b and an element $a \in A$ such that $ba = b$. [Hint: Consider $\mathcal{F}$, the collection of all non-zero right ideals C such that $CA \neq 0$. $A \in \mathcal{F}$, so this family is non-empty. By the DCC there is a minimal element B in $\mathcal{F}$. Then for some $b \in B$, $bA \neq 0$. Then as $bA = bA^2$, $bA \in \mathcal{F}$. But minimality of B shows $bA = B$, and the result follows.]

3. Show that if R is Artinian, then $\text{rad}(R)$ is a nilpotent ideal. [Hint: $J := \text{rad}(R)$ is an ideal (Exercise (3) in Sect. 12.6.3). Suppose by way of contradiction that J is not nilpotent. Use the DCC to show that for some positive integer k, $A := J^k = J^{k+1}$, so $A^2 = A \neq 0$. Use Exercise (2) in Sect. 12.6.3 to obtain a non-zero element b with $ba = b$ for some $a \in A$. Use Exercise (1) in Sect. 12.6.3 to argue that a is not in J. It is apparent that the last assertion is a contradiction.]

4. Show that an Artinian ring with no zero divisors is a division ring. [No hints this time. Instead a challenge.]

12.6.5 Quasiregularity and the Radical

Given a ring R, we define a new binary operation "$\circ$" on R, by declaring

$$a \circ b := a + b - ab, \text{ for all } a, b \in R.$$

The next few exercises investigate invertibility properties of elements with respect to this operation, and use these facts to describe a new poset of right ideals for which the Jacobson radical is a *supremum*. One reward for this will be a dramatic improvement of the elementary result in Exercise (4) in Sect. 12.6.3: *the Jacobson radical actually contains every nil right ideal.*

1. Show that $(R, \circ)$ is a monoid with monoid identity element being the zero element, 0, of the ring R.
2. An element x of R is said to be *right quasiregular* if it has a right inverse in $(R, \circ)$—that is, there exists an element r such that $x \circ r = 0$. Similarly, x is *left quasiregular* if $\ell \circ x = 0$ for some $\ell \in R$. The elements r and ℓ are called *right* and *left quasi-inverses of* x, respectively. (They need not be unique.) If x is both left and right quasiregular, then x is simply said to be *quasiregular*. You are asked here to verify again a fundamental property of all monoids: Show that if x is quasiregular, then any left quasi-inverse is equal to any right quasi-inverse, so there is a unique element x° such that

$$x^\circ \circ x = x \circ x^\circ = 0.$$

   Show also that $(x^\circ)^\circ = x$, for any $x \in R$.
3. This exercise has four parts, all elementary.

   (a) Show that for any elements $a, b \in R$,

   $$a \circ b = 1 - (1 - a)(1 - b).$$

   (b) Show that x is right (left) quasiregular if and only if $1 - x$ is right (left) invertible in R. From this, show that if x is quasiregular, then $1 - x$ has $1 - x^\circ$ for a two-sided multiplicative inverse.
   (c) Conclude that x is quasiregular if and only if $1 - x$ is a unit.
   (d) Prove that any nilpotent element a of R is quasiregular. [Hint: Think about the geometric series in powers of a.]

4. Now we apply these notions to right ideals. Say that a *right ideal is quasiregular* if and only if each of its elements is quasiregular. Prove the following

Lemma 12.6.1 *If every element of a right ideal K is right quasiregular, then each of its elements is also quasiregular—that is, K is a quasiregular right ideal.*

[Hint: First observe that if $x \in K$ and y is a right quasi-inverse of x, then $y \in K$ as well. So now y has both left and right quasi-inverses which must be x (why?). Why does this make x quasiregular?]

5. Next prove the following:

Lemma 12.6.2 *Suppose K is a quasiregular right ideal and M is a maximal right ideal. Then $K \subseteq M$. (Thus every quasiregular right ideal lies in the Jacobson radical.)*

[Hint: If K does not lie in M, $R = M + K$ so $1 = m + k$ where $m \in M$ and $k \in K$, Now k has a right quasi-inverse z. As above, $z \in K$. Then

$$0 = 1 \cdot z - z = (m + k)z - z = mz + (kz - z) = mz - k,$$

to force $k \in M$, an absurdity.]

6. Show that the Jacobson radical is itself a right quasiregular ideal. [Hint: Show each element $y \in \mathrm{rad}(R)$ is right quasiregular. If not, $(1 - y)R$ lies in some maximal right ideal M.] We have now reached the following result:

Lemma 12.6.3 *The radical $\mathrm{rad}(R)$ is the unique supremum in the poset of right quasiregular ideals of R.*

7. Prove that every nil right ideal lies in the Jacobson radical $\mathrm{Rad}(R)$. [Hint: It suffices to show that every nil right ideal is quasiregular. There are previous exercises that imply this.]
8. Are there rings R (with identity) in which some nilpotent elements do not lie in the Jacobson radical? Try to find a real example.)

12.6.6 Exercises Involving Nil One-Sided Ideals in Noetherian Rings

1. Show that if B is a nil right ideal, then for any element $b \in B$, Rb is a nil left ideal.
2. For any subset X of ring R, let $X^{\perp} := \{r \in R | Xr = 0\}$, the *right annihilator of* X.

    (a) Show that always $X^{\perp}R \subseteq X^{\perp}$.
    (b) Show that $X^{\perp} \subseteq (RX)^{\perp}$.
    (c) Show that $X^{\perp}$ is a right ideal.
    (d) Show that if X is a left ideal of R, then $X^{\perp}$ is a 2-sided ideal.

3. Now suppose L is a left ideal, and P is the poset of right ideals of the form $y^{\perp}$ as y ranges over the non-zero elements of L (partially ordered by the containment relation).

(a) Show that R cannot be a member of P.

(b) Show that if $y^\perp$ is a maximal member of P, then $(ry)^\perp = y^\perp$ for all $r \in R$.

4. Suppose R is Noetherian—that is, it has the ascending chain condition on right ideals.

   (a) Show that if L is a left ideal, then there exists an element $u \in L - \{0\}$ such that $u^\perp = (ru)^\perp$, for all $r \in L$.

   (b) Suppose L is a nil left ideal and that u is chosen as in Part (a) of this exercise. Show that $uRu = 0$. Conclude from this that $(RuR)^2 = 0$, so u generates a nilpotent 2-sided ideal. [Hint: Note that if k is the least positive integer such that $u^k = 0$, then $u \in (u^{k-1})^\perp = u^\perp$, so $Ru \subseteq u^\perp$.]

   (c) Because R is Noetherian, there exists an ideal M which is maximal among all nilpotent 2-sided ideals. Show that M contains every nil left ideal. [Hint: If L is a nil left ideal, so is $L + M$, and so $(L + M)/M$ is a nil left ideal of the Noetherian ring R/M. Now if $L + M$ is not contained in M, Part (b) of this exercise applied to R/M will yield a non-zero nilpotent 2-sided ideal in R/M. Why is this absurd?]

   (d) Show that every nil right ideal lies in a nilpotent 2-sided ideal. [Hint: Let N be a nil right ideal and let M be the maximal nilpotent 2-sided ideal of part (c) of this exercise. (M exists by the Noetherian condition.) By way of contradiction assume $n \in N - M$. Select an element $r \in R$. Since N is a right nil ideal there exists an integer $k \geq 2$ such that $(nr)^k = 0$. Then $(rn)^{k+1} = r \cdot (nr)^k \cdot n = 0$. Thus Rn is a left nil ideal, and so, by part (c) of this exercise, lies in M against our choice of n.]

   (e) In a Noetherian ring, the sum of all nilpotent right ideals, $N(R)$, is a nil 2-sided ideal (Exercise (5) in Sect. 12.6.2). Prove the following:

Theorem 12.6.4 (Levitsky) *In a Noetherian ring, $N(R)$ is a nilpotent ideal and it contains all left and right nil ideals.*

Remarks The reader is advised that most of the theory exposed in these exercises also exist in "general ring theory" where (unlike this limited text) rings are not required to possess a multiplicative identity element. The proofs of the results without the assumption of an identity element—especially the results about radicals—require a bit of ingenuity. There are three basic ways to handle this theory without a multiplicative identity. (1) Embed the general ring in a ring with identity (there is a rather 'minimal' way to do this) and apply the theory for these rings.. Only certain sorts of statements about general rings can extracted in this way. For example, quasiregularity can be defined in any general ring. But to establish the relation connecting quasiregularity of x with $1 - x$ being a unit, the embedding is necessary. (2) Another practice is to exploit the plentitude of right ideals by paying attention only to right ideals possessing properties they would have had if R had an identity element. Thus a *regular right ideal* in a general ring, is an ideal I for which there is an element e, such that $e - er \in I$ for all R in the ring. The Jacobson radical is then defined to be the intersection of all such regular right ideals. With this modified definition, nearly all of the previous exercises regarding the Jacobson radical can be emulated in this more general context. But even this is just "scratching the surface". There is an extensive literature on all kinds of radicals of general rings. (3) In Artinian general rings, one can often find sub-(general)-rings containing an idempotent serving as a multiplicative identity for that subring. Then one can apply any relevant theorem about rings with identity to such subrings.

Two particularly good sources extending the material in this chapter are the classic *Rings and Modules* by J. Lambeck [2] and the excellent and thorough book of J. Dauns, *Modules and Rings* [1].[1]

References

1. Dauns J (1994) Modules and rings. Cambridge University Press, Cambridge
2. Lambek J (1966) Rings and modules. Blaisdell Publishing Co., Toronto

---

[1] As indicated by the titles of these two books, they have this subject covered "coming and going".

Chapter 13
Tensor Products

Abstract No algebra course would be complete without introducing the student to the language of category theory. Some properties of the objects of algebra are defined by their internal structure, while other properties describe how the object sits in a morphism-closed environment. Universal mapping properties are of the latter sort. Their relation to initial and terminal objects of another suitably-chosen category is emphasized. The tensor product in the category of right R-modules is defined in two ways: as a constructed object, and as a unique solution to a universal mapping problem. From the tensor product one derives functors which are adjoint to the "Hom" functors. Another feature is that tensor products can also be defined for F-algebras. The key facts that tensor products "distribute" over direct sums and that there is a uniform way to define multiple tensor products, allows one to define the *tensor algebra*. In the category of F-algebras generated by n elements, this algebra becomes an initial object. This graded algebra, $T(V)$, is uniquely determined by an F-vector space V and has two important homomorphic offspring: the *symmetric algebra*, $S(V)$ (modeled by polynomial rings), and the *exterior algebra*, $E(V)$, (modeled by an algebra on poset chains). In the category of vector spaces, T, S and E, and their restrictions to the homogenous summands, are all functors—that is, morphisms among vector spaces induce morphisms among the algebras and their components of fixed degree. Herein lie the basic theorems concerning multilinear forms.

13.1 Introduction

In this chapter we shall present the notion of tensor product. The exposition will emphasize many of its "categorical" properties (including its definition), as well as its ubiquity in multilinear algebra (and algebra in general!). In so doing, we shall find it convenient to weave some rather general discussions of "category theory" into this chapter, which will also help to explain the universal nature of many of the constructions that we've given thus far.

© Springer International Publishing Switzerland 2015   471
E. Shult and D. Surowski, *Algebra*, DOI 10.1007/978-3-319-19734-0_13

13.2 Categories and Universal Constructions

13.2.1 Examples of Universal Constructions

In past chapters, we have encountered "universal mapping properties". They always seem to say that given a certain situation, there exists a special collection of maps such that for every similar collection of maps, there is a unique collection of morphisms to or from that special collection. So, far we have discussed such "universal mapping properties" for special realms such as rings, or R-modules. But it is time to give this notion a concrete setting. The natural home for describing "universal mapping properties" in a general way is a "category". In this section, we shall define the notion of a *category*, which will not only unify some of the concepts that we've considered thus far, but also give meaning to the concept of a "universal construction"—that is, the construction of an object satisfying a "universal mapping property". Examples of universal constructions given so far include

1. the kernel of a group (or ring or module) homomorphism;
2. the commutator subgroup $[G, G]$ of a group G;
3. the direct product of groups (rings, modules, ...);
4. the direct sum of modules;
5. the free group on a set;
6. the free module on a set.

We shall encounter more examples. The reader is likely to wonder what the above examples have in common. The objective of this discussion is to make that clear.

Before turning to the formal definitions, let us indicate here the "universality" of kernels. Indeed, let $\phi : G \rightarrow H$ be a homomorphism of groups, with kernel $\mu : K \hookrightarrow G$ where μ is the inclusion mapping.[1] From this, it is a trivial fact that the composition $\phi \circ \mu : K \rightarrow H$ is the trivial homomorphism. With $\phi : G \rightarrow H$ fixed, the "universal" property of the pair (K, μ) is as follows: Suppose that (K', μ') is another pair consisting of a group K' and a homomorphism $\mu' : K' \rightarrow G$ such that $\phi \circ \mu' : K' \rightarrow H$ is also the trivial homomorphism. It follows from Theorem 3.4.5 that there exists a *unique* homomorphism $\theta : K' \rightarrow K$ such that $\mu \circ \theta = \mu' : K' \rightarrow G$. In other words, the homomorphism $\mu' : K' \rightarrow G$ satisfying the given property (viz., that $\phi \circ \mu'$ is the trivial homomorphism) must occur through the "courtesy" of the homomorphism $\mu : K \rightarrow G$. Perhaps a commutative diagram depicting the above would be helpful:

---

[1]The student may be more accustomed to saying that $K = \{g \in G | \phi(g) = 1_H\}$ *is a subgroup of* G. But these "category people" like to think in terms of morphisms and their compositions.

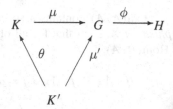

As another example, we consider the direct sum $D := \bigoplus_{\alpha \in \mathcal{A}} M_\alpha$ of a family $\{M_\alpha \mid \alpha \in \mathcal{A}\}$ of right R-modules (where R is a fixed ring). Thus D is a right R-module having, for each $\alpha \in \mathcal{A}$, an injective homomorphism $\mu_\alpha : M_\alpha \to D$. The universal property of the direct sum is as follows: if $\{(D', \mu'_\alpha) \mid \alpha \in \mathcal{A}\}$ is another family consisting of an R-module D' and R-module homomorphisms $\mu'_\alpha : M_\alpha \to D'$, then there must exist a *unique* homomorphism $\theta : D \to D'$ making all of the relevant triangles commute, i.e., for each $\alpha \in \mathcal{A}$, we have $\theta \circ \mu_\alpha = \mu'_\alpha$. Thus, in analogy with our first example (that of the kernel), all homomorphisms must factor through the "universal object." The relevant picture is depicted below:

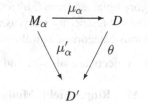

13.2.2 Definition of a Category

In order to unify the above examples as well as to clarify the "universality" of the given constructions, we start with a few key definitions. First of all, a *category* is a triple $\mathcal{C} = (\mathcal{O}, \mathrm{Hom}, \circ)$ where

1. $\mathcal{O}$ is a class (of *objects* of $\mathcal{C}$),
2. Hom assigns to each pair (A, B) of objects a class $\mathrm{Hom}(A, B)$ (which we sometimes write as $\mathrm{Hom}_\mathcal{C}(A, B)$ when we wish to emphasize the category $\mathcal{C}$), whose elements are called *morphisms* from A to B, and
3. There is a law of composition,

$$\circ : \mathrm{Hom}(B, C) \times \mathrm{Hom}(A, B) \to \mathrm{Hom}(A, C),$$

defined for all objects A, B, and C for which $\mathrm{Hom}(A, B)$ and $\mathrm{Hom}(B, C)$ are non-empty. (The standard convention is to write composition as a binary operation—thus if $(f, g) \in \mathrm{Hom}(B, C) \times \mathrm{Hom}(A, B)$, one writes $f \circ g$ for $\circ(f, g)$.) Composition is subject to the following rules:

(a) Where defined, the (binary) composition on morphisms is associative.

(b) For each object A, $\mathrm{Hom}(A, A)$ contains a special morphism, 1_A, called the *identity morphism at* A, such that for all objects B, and for all $f \in \mathrm{Hom}(A, B)$, $g \in \mathrm{Hom}(B, A)$,

$$f \circ 1_A = f, \quad 1_B \circ g = g.$$

NOTE: When object A is a set, we often write id_A for 1_A, since the identity mapping $\mathrm{id}_A : A \to A$, taking each element of A to itself, is the categorical morphism 1_A in that case.

Occasionally one must speak of subcategories. If $\mathcal{C} = (\mathcal{O}, \mathrm{Hom}, \circ)$ is a category, then any subcollection of objects, $\mathcal{O}_0 \subseteq \mathcal{O}$, determines an *induced subcategory* $\mathcal{C}_0 := (\mathcal{O}_0, \mathrm{Hom}_0, \circ)$, where Hom_0 denotes those morphisms in Hom which connect two objects in $\mathcal{O}_0$. There is also another kind of subcategory in which the collection of morphisms is restricted. Let $\mathcal{C} = (\mathcal{O}, \mathrm{Hom}, \circ)$, as before, and let Hom_1 be a subcollection of Hom containing 1_A, for all $A \in \mathcal{O}$, with the property that if a composition of two morphisms in Hom_1 exists in Hom, then that composition lies in Hom_1. Then $(\mathcal{O}, \mathrm{Hom}_1, \circ)$ is a *morphism restricted subcategory*. In general a *subcategory* of $\mathcal{C}$ is any category obtained from $\mathcal{C}$ by a sequence of these processes: forming induced subcategories and restricting morphisms.

There are some rather obvious categories, as follows.

1. The category **Set**, whose objects are all sets and whose morphisms are just mappings of sets.
2. The categories **Group**, **Ab**, **Ring**, **Field**, **Mod**$_R$ of groups, abelian groups, rings, fields, and (right) R-modules and their relevant homomorphisms. Note that **Ab** is an induced subcategory of **Group**.
3. The category **Top** of all topological spaces and continuous mappings.
   Here are a few slightly less obvious examples.
4. Fix a group G and define the category $(\mathcal{O}, \mathrm{Hom}, \circ)$ where $\mathcal{O} = G$, and where, for $x, y \in G$, $\mathrm{Hom}(x, y) = \{g \in G \mid gx = y\}$. (Thus $\mathrm{Hom}(x, y)$ is a singleton set, viz., $\{yx^{-1}\}$). Here, $\circ$ is defined in the obvious way and is associative by the associativity of multiplication in G.
5. Let M be a fixed monoid. We form a category with $\{M\}$ as its sole object. Hom will be the set of monoid elements of M, and composition of any two of these is defined to be their product under the binary monoid operation. (Conversely, note that in *any* category with object A, $\mathrm{Hom}(A, A)$ is always a monoid with respect to morphism composition.)
6. Let G be a *simple graph* (V, E). Thus, V is a set (of *vertices*) and E (the set of *edges*) is a subset of the set of all 2-element subsets of V. If x and y are vertices of G, we define a *walk* from x to y to be a finite sequence $w = (x = x_0, x_1, \ldots, x_n = y)$ where each successive pair $\{x_i, x_{i+1}\} \in E$ for $i = 0, \ldots, n - 1$. The graph is said to be *connected* if, for any two vertices, there is a walk beginning at one of the vertices and terminating at the other. For a connected graph (V, E), we define a category as follows:

   (a) The objects are the vertices of V.

(b) If x and y are vertices, then $\text{Hom}(x, y)$ is the collection of all walks that begin at x and end at y. (This is often an infinite set.)

Suppose now $u = (x = x_0, x_1, \ldots, x_n = y)$ is a walk in $\text{Hom}(x, y)$, and $v = (y = y_0, y_1, \ldots, y_m = z)$ is a walk in $\text{Hom}(y, z)$. Then the *composition* of the two morphisms is defined to be their *concatenation*, that is, the walk $v \circ u := (x = x_0, \ldots, x_n = y_0, \ldots, y_m = z)$ from x to z in $\text{Hom}(x, z)$.

7. If $G_1 := (V_1, E_1)$ and $G_2 = (V_2, E_2)$ are two simple graphs, a *graph homomorphism* $\phi : G_1 \rightarrow G_2$ is a set-mapping $f : V_1 \rightarrow V_2$ such that if $\{x, y\}$ is an edge in E_1, then either $f(x) = f(y) \in V_2$, or else $\{f(x), f(y)\}$ is an edge in E_2. The category of simple graphs, denoted **Graphs**, has all simple graphs and their graph homomorphisms as objects and morphisms. [By restricting morphisms or objects, this category has all sorts of *subcategories* for which each object possesses a unique "universal cover".]

We conclude with two examples of categories that are especially relevant to the constructions of kernel and direct sum given above.

8. Let $\phi : G \rightarrow H$ be a homomorphism of groups. Define the category whose objects are the pairs (L, η), where L is a group and where $\eta : L \rightarrow G$ is a homomorphism such that $\phi \circ \eta : L \rightarrow H$ is the trivial homomorphism. If (L, η), (L', η') are two such pairs, a morphism from (L, η) to (L', η') is simply a group homomorphism $\theta : L \rightarrow L'$ such that the diagram below commutes:

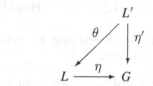

It should be clear that the axioms of a category are satisfied for this example.

9. Let $\{M_\alpha \mid \alpha \in \mathcal{A}\}$ be a family of right R-modules and form the category whose objects are of the form $(M, \{\phi_\alpha \mid \alpha \in \mathcal{A}\})$, where M is a right R-module and where each ϕ_α is an R-module homomorphism from M_α to M. In this case a morphism from $(M, \{\phi_\alpha \mid \alpha \in \mathcal{A}\})$ to $(M', \{\phi'_\alpha \mid \alpha \in \mathcal{A}\})$ is simply an R-module homomorphism $\theta : M \rightarrow M'$ such that for all $\alpha \in \mathcal{A}$, the following triangle commutes:

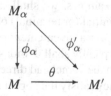

13.2.3 Initial and Terminal Objects, and Universal Mapping
Properties

We continue with a few more definitions. Let $C = (\mathcal{O}, \text{Hom}, \circ)$ be a category and
let A and B be objects of C. As usual, let 1_A and 1_B be the identity morphisms of
$\text{Hom}(A, A)$ and $\text{Hom}(B, B)$. A morphism $f \in \text{Hom}(A, B)$ is called an *isomorphism*
if there exists a morphism $g \in \text{Hom}(B, A)$ such that $g \circ f = 1_A$ and $f \circ g = 1_B$. In
this case, we write $f = g^{-1}$ and $g = f^{-1}$. An *initial object* in C is an object I of C
such that for all objects A, $\text{Hom}(I, A)$ is a singleton set. That is to say, I is an initial
object of C if and only if, for each object A of C, there exists a unique morphism
from I to A. Likewise, a *terminal object* is an object T such that for all objects A,
$\text{Hom}(A, T)$ is a singleton set.

The following should be clear, but we'll pause to give a proof anyway.

Lemma 13.2.1 ("Abstract Nonsense") *If the category C contains an initial object,
then it is unique up to isomorphism in C. Similarly, if C contains a terminal object,
it is unique up to isomorphism in C.*

Proof Let I and I' be initial objects of C. Let $f \in \text{Hom}(I, I')$, $f' \in \text{Hom}(I', I)$ be
the unique morphisms. Then

$$f' \circ f \in \text{Hom}(I, I) = \{1_I\}, \quad f \circ f' \in \text{Hom}(I', I') = \{1_{I'}\}.$$

Thus, if C has a terminal object, then it is unique. Essentially the same proof shows
that if C has a terminal object, it is unique. $\square$

At first blush, it might seem that the notion of an initial object in a category
cannot be very interesting. Indeed, the category **Group** of groups and group homomorphisms contains an initial object, viz., the trivial group $\{e\}$, which is arguably not
very interesting. At the same time the trivial group is also a terminal object in **Group**.
Similarly, $\{0\}$ is both an initial and a terminal object in $\textbf{Mod}_R$. As a less trivial example, note that if K is the kernel of the group homomorphism $\phi : G \to H$, then by our
discussions above, (K, μ) is a terminal object in the category of example 8 above.
Likewise, $(\bigoplus_{\alpha \in \mathcal{A}} M_\alpha, \{\mu_\alpha \mid \alpha \in \mathcal{A}\})$ is an initial object in the category of example
9, above. In light of these examples, we shall say that an object *has a universal
mapping property* if it is an initial (or terminal) object (in a suitable category).

In general, we cannot expect a given category to have either initial or terminal
objects. When they do, we typically shall demonstrate this via a direct construction.
This is certainly the case with the kernel and direct sum; once the constructions were
given, one then proves that they satisfy the appropriate universal mapping properties,
i.e., that they are terminal (resp. initial) objects in an appropriate category.

13.2.4 Opposite Categories

For each category $C = (\mathcal{O}, \text{Hom}_C, \circ)$, there is an *opposite category* $C^{\text{opp}} = (\mathcal{O}, \text{Hom}^{\text{opp}}, \circ')$, having the same set of objects $\mathcal{O}$; but we now turn the arrows that depict morphisms around. More precisely, $f \in \text{Hom}(A, B)$, if and only if $f^{\text{opp}} \in \text{Hom}^{\text{opp}}(B, A)$. In category C, if $f \in \text{Hom}(A, B)$, $g \in \text{Hom}(B, C)$, so that $h := g \circ f \in \text{Hom}(A, C)$, then in category C^{opp} we have $h^{\text{opp}} = f^{\text{opp}} \circ' g^{\text{opp}} \in \text{Hom}^{\text{opp}}(C, A)$. One notices that an identity mapping 1_A in C becomes an identity mapping in C^{opp} and that an initial (terminal) object in C becomes a terminal (initial) object in C^{opp}.

A universal mapping property in one category corresponds to an "opposite" mapping property in its opposite category. Let us illustrate this principle with the notion of the *kernel*. Earlier we discussed the universal mapping property of the kernel in the category of right R-modules. Let us do it again in a much more general categorical context.

Suppose $C = (\mathcal{O}, \text{Hom}, \circ)$ is a category in which a terminal object T exists and that T is also the initial object I—that is $T = I$. Now for any two objects A and B in $\mathcal{O}$, consider the class $\text{Hom}(A, B)$, in category C. This class always contains a unique morphism which factors through $I = T$. Thus there is a unique mapping $\tau_A : A \to T$, since T is a terminal object, and a unique mapping $\iota_B : T = I \to B$, since T is also the initial object. Then $0_{AB} := \iota_B \circ \tau_A$ is the unique mapping in $\text{Hom}(A, B)$ which factors though $I = T$. This notation, 0_{AB} is exploited in the next paragraphs.

Now, suppose $\phi : A \to B$ is a fixed given morphism. Then the *kernel* of ϕ, is a morphism $\kappa : (\ker \phi) \to A$ such that $\phi \circ \kappa = 0_{(\ker \phi)B}$, with this universal property:

If $\gamma : \quad X \to A$ is a morphism of C such that $\phi \circ \gamma = 0_{XB}$, then there is a morphism $\theta_X : X \to \ker \phi$ such that $\kappa \circ \theta_X = \gamma$.

Now in the opposite category this property becomes the defining property of the *cokernel*. Again suppose $C = (\mathcal{O}, \text{Hom}, \circ)$ is a category in which a terminal object T exists and that T is also the initial object I. Fix a morphism $\phi : A \to B$. Then the *cokernel* (if it exists) is a morphism $\epsilon : B \to C := \text{coker}\phi$ such that $\epsilon \circ \phi = 0_{AC}$, with the following universal property:

If $\epsilon' : \quad B \to X$ is a morphism of C such that $\gamma \circ \phi = 0_{AX}$, then there exists a unique morphism $\theta_X : C \to X$ such that $\theta_x \circ \epsilon = \epsilon'$. That is, the following diagram commutes:

$$A \xrightarrow{\phi} B \xrightarrow{\epsilon} C$$

with ϵ' from B down to X and θ_X from C to X.

Again, it is easy to describe the cokernel as an initial object in a suitable category of diagrams that depend on the morphism ϕ.

13.2.5 Functors

Let $C_1 = (\mathcal{O}_1, \mathrm{Hom}_1, \circ_1)$ and $C_2 = (\mathcal{O}_2, \mathrm{Hom}_2, \circ_2)$ be two categories. We may write $\mathrm{Hom}_i = \cup\{\mathrm{Hom}_i(A, B), |A, B \in \mathcal{O}_i\}$, the full collection of morphisms of category C_i, $i = 1, 2$.

A *covariant functor* $F : C_1 \to C_2$ is a pair of mappings,

$$F_{\mathcal{O}} : \mathcal{O}_1 \to \mathcal{O}_2,$$
$$F_H : \mathrm{Hom}_1 \to \mathrm{Hom}_2,$$

such that

(*) *If* $f \in \mathrm{Hom}_1(A, B)$ *and* $g \in \mathrm{Hom}_1(B, C)$, *then*

1. $F_H(f) \in \mathrm{Hom}_2(F(A), F(B))$, $F_H(g) \in \mathrm{Hom}_2(F(B), F(C))$, and
2. $F_H(g \circ_1 f) = F_H(g) \circ_2 F_H(f) \in \mathrm{Hom}_2(F(A), F(C))$.

In other words: (1) F_H is compatible with $F_{\mathcal{O}}$ in the sense that if $f : A \to B$ is a morphism of C_1, then $F_H(f)$ is a morphism in $\mathrm{Hom}_2(F(A), F(B))$, and (2) F_H preserves composition of mappings.

Here is an example : Let C be the category of example 4 of p. 597, whose objects were the elements of a group G and whose morphisms were the maps induced by left multiplication by elements of G. Then any endomorphism $G \to G$ is a a covariant functor of this category into itself. The reader can easily devise a similar functor using a monoid endomorphism for example 5 of p. 597.

Here is a classic example we have met before (see p. 494). Let M be a fixed right R-module. Now for any right R-module A, the set $\mathrm{Hom}_R(M, A)$ is an abelian group. If $f : A \to B$ is a morphism, composing all the elements of $\mathrm{Hom}(M, A)$ with f yields a mapping $F(f) : \mathrm{Hom}(M, A) \to \mathrm{Hom}(M, B)$ preserving the additive structure. Thus $(\mathrm{Hom}, -) : \mathbf{Mod}_R \to \mathbf{Ab}$ is a covariant functor from the category of right R-modules to the category of abelian groups.

Similarly, $\mathrm{Hom}_R(M, -)$ is a covariant functor $\mathbf{Mod}_R \to_R \mathbf{Mod}$, from the category of right R-modules to the category of left R-modules (again see p. 494).

Now suppose F is a covariant functor from category $C_1 = (\mathcal{O}_1, \mathrm{Hom}_1, \circ_1)$ to $C_2 = (\mathcal{O}_2, \mathrm{Hom}_2, \circ_2)$ such that $F_{\mathcal{O}} : \mathcal{O}_1 \to \mathcal{O}_2$ and $F_H : \mathrm{Hom}_1 \to \mathrm{Hom}_2$ are bijections. Let $F_{\mathcal{O}}^{-1}$ and F_H^{-1} be the corresponding inverse mappings. Then $F^{-1} = (F_{\mathcal{O}}^{-1}, F_H^{-1})$ preserves composition and so is a functor $F^{-1} : C_2 \to C_1$. In this case the covariant functor F is called an *isomorphism of categories*.

There is another kind of functor—a *contravariant functor*—which reverses the direction and order of compositions. Again it consists of two maps: $F_{\mathcal{O}} : \mathcal{O}_1 \to \mathcal{O}_2$, $F_H : \mathrm{Hom}_1 \to \mathrm{Hom}_2$, except now we have:

(1*) *If* $f : A \to B$ *is in* Hom_1, *then* $F_H(f) \in \mathrm{Hom}_2(B, A)$.

(2*) If $f : A \to B$ and $g : B \to C$ are in Hom_1, then $F_H(g \circ f) = F_H(f) \circ F_H(g) \in \text{Hom}_2(C, A)$.

Now, fixing the right R-module M, the functor $\text{Hom}(-, M) : \mathbf{M}_R \to \mathbf{Ab}$ is a contravariant functor from the category of right R-modules to the category of abelian groups.

There are further versions of functors of this sort, where M is a bimodule. A classical example is the functor which, in the category of vector spaces, takes a vector space to it's dual space. Specifically, let F be a field and let **Vect** be the category whose objects are the (F, F)-bimodules—that is, F-vector spaces where $\alpha v = v \alpha$ for any vector v and scalar α. The morphisms are the linear transformations between vector spaces. The dual space functor $\text{Hom}_F(-, F) : \mathbf{Vect} \to \mathbf{Vect}$ takes a linear transformation $T : V \to W$, to the mapping $\text{Hom}_F(W, F) \to \text{Hom}_F(V, F)$ which takes the functional $f \in \text{Hom}_F(W, F)$ to the functional $f \circ T \in \text{Hom}_F(V, F)$.

One final issue is that functors between categories can be composed: If $F : \mathcal{C}_1 \to \mathcal{C}_2$ and $G : \mathcal{C}_2 \to \mathcal{C}_3$ are functors, then, by composing the maps on objects and on the morphism classes: $G_\mathcal{O} \circ F_\mathcal{O}$ and $G_H \circ F_H$, one obtains a functor $G \circ F : \mathcal{C}_1 \to \mathcal{C}_3$. Obviously

If $F : \mathcal{C}_1 \to \mathcal{C}_2$ and $G : \mathcal{C}_2 \to \mathcal{C}_3$ are functors, then the composite functor $G \circ F$ is covariant if and only if F and G are either both covariant or both contravariant.

Of course you know where this is leading: to a "category of all categories". Let's call it $\mathbf{Cat}$. It's objects would be all categories (if this is even conceivable), and its morphisms would be the covariant functors between categories. This is a dangerous neighborhood, for as you see, $\mathbf{Cat}$ is also one of the objects of $\mathbf{Cat}$. Clearly, we have left the realm of sets far behind. The point of this meditation is that actually we have left sets behind long ago: the collection of objects in many of our favorite categories $\mathbf{Ab}$, $\mathbf{Ring}$, $\mathbf{Mod}_R$, etc. are not sets. Of course this does not preclude them from being collections that possess morphisms. These things are still amazingly useful since they provide us with a language in which we can describe universal mapping properties. We just have to be very careful not to inadvertently insert inappropriate axioms of set-theory to the collections of objects of these categories.[2]

13.3 The Tensor Product as an Abelian Group

13.3.1 The Defining Mapping Property

Let M be a right R-module, let N be a left R-module, and let A be an abelian group. By a *balanced mapping* (or *R-balanced mapping* if we wish to specify R), we mean

---

[2]For a lucid superbly-written account of the difference between sets and classes, the authors recommend the early chapters of *Set Theory and the Continuum Hypothesis* by Raymond Smullyan and Melvin Fitting [3].

a function

$$f : M \times N \longrightarrow A,$$

such that

(i) $f(m_1 + m_2, n) = f(m_1, n) + f(m_2, n),$
(ii) $f(m, n_1 + n_2) = f(m, n_1) + f(m, n_2),$
(iii) $f(mr, n) = f(m, rn)$

where $m, m_1, m_2 \in M$, $n, n_1, n_2 \in N$, $r \in R$.

Define a *tensor product* of M and N to mean an abelian group T, together with a balanced map $t : M \times N \to T$ such that given any abelian group A, and any balanced map $f : M \times N \to A$ there exists a unique abelian group homomorphism $\phi : T \to A$, making the diagram below commute.

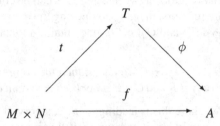

Notice that a tensor product has been defined in terms of a universal mapping property. Thus, by our discussions in the previous section, we should be able to express a tensor product of M and N as an initial object of some category. Indeed, define the category having as objects ordered pairs (A, f) where A is an abelian group and $f : M \times N \to A$ is a balanced mapping. A morphism from (A, f) to (A', f') is an abelian group homomorphism $\theta : A \to A'$ such that the following diagram commutes:

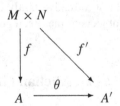

Then the preceding diagram defining the tensor product T displays its role as an initial object in this category. The following corollary is an immediate consequence of "abstract nonsense" (see Lemma 13.2.1).

Corollary 13.3.1 *The tensor product of the right R-module M and the left R-module N (if it exists) is unique up to abelian group isomorphism.*

13.3.2 Existence of the Tensor Product

Thus, there remains only the question of existence of the tensor product of the respective right and left R-modules M and N. To this end, let F be the free $\mathbb{Z}$-module with basis $M \times N$ (often called "the free abelian group on the set $M \times N$"). Let B be the subgroup of F generated by the set of elements of the form

$$(m_1 + m_2, n) - (m_1, n) - (m_2, n),$$
$$(m, n_1 + n_2) - (m, n_1) - (m, n_2),$$
$$(mr, n) - (m, rn),$$

where $m, m_1, m_2 \in M$, $n, n_1, n_2 \in N$, $r \in R$. Write $M \otimes_R N := F/B$ and set $m \otimes n := (m, n) + B \in M \otimes_R N$. Therefore, the defining relations in $M \otimes_R N$ are precisely

$$(m_1 + m_2) \otimes n = m_1 \otimes n + m_2 \otimes n,$$
$$m \otimes (n_1 + n_2) = m \otimes n_1 + m \otimes n_2,$$
$$mr \otimes n = m \otimes rn.$$

for all $m, m_1, m_2 \in M$, $n, n_1, n_2 \in N$, $r \in R$.

The elements of the form $m \otimes n$ where $m \in M$ and $n \in N$ are called *pure tensors* (or sometimes "*simple tensors*"). Furthermore, from its definition, $M \otimes_R N$ is generated by all of its pure tensors $m \otimes n$, $m \in M, n \in N$.

Finally, define the mapping $t : M \times N \to M \otimes_R N$ by setting $t(m, n) = m \otimes n$, $m \in M$, $n \in N$. Then, *by construction*, t is a balanced map. In fact:

Theorem 13.3.2 *The abelian group $M \otimes_R N$, together with the balanced map $t : M \times N \to M \otimes_R N$, is a tensor product of M and N.*

Proof Let A be an abelian group and let $f : M \times N \to A$ be a balanced mapping. If the abelian group homomorphism $\theta : M \otimes_R N \to A$ is to exist, then it must satisfy the condition that for all $m \in M$ and for all $n \in N$, $\theta(m \otimes n) = \theta \circ t(m, n) = f(m, n)$. Since $M \otimes_R N$ is generated by the simple tensors $m \otimes n$, $m \in M$, $n \in N$, it follows that $\theta : M \otimes_R N \to A$ is already uniquely determined. It remains, therefore, to show that θ is an abelian group homomorphism. To do this, we need only verify that the defining relations are satisfied. We have, since t and f are balanced,

$$\theta((m_1 + m_2) \otimes n) = f((m_1 + m_2), n)$$
$$= f(m_1, n) + f(m_2, n)$$
$$= \theta(m_1 \otimes n) + \theta(m_2 \otimes n),$$

for all $m_1, m_2 \in M$, $n \in N$. The remaining relations are similarly verified, proving that θ is, indeed, a group homomorphism $M \otimes_R M \to A$, satisfying $\theta \circ t = f$. We have already noted that θ is uniquely determined by this condition, and so the proof is complete. $\square$

A few very simple applications are in order here. The first is cast as a Lemma.

Lemma 13.3.3 *If M is a right R-module, then $M \otimes_R R \cong M$ as abelian groups.*

To show this, one first easily verifies that the mapping $t : M \times R \to M$ defined by setting $t(m, r) = mr$, $m \in M$, $r \in R$ is balanced. Next, let A be an arbitrary abelian group and let $f : M \times R \to A$ be a balanced mapping. Define $\theta : M \to A$ by setting $\theta(m) := f(m, 1_R)$, $m \in M$. (Here 1_R is the multiplicative identity element of the ring R.) Using the fact that f is balanced, we have that $\theta(m_1 + m_2) = f(m_1 + m_2, 1_R) = f(m_1, 1_R) + f(m_2, 1_R) = \theta(m_1) + \theta(m_2)$, $m_1, m_2 \in M$, and so θ is a homomorphism. Finally, let $m \in M$, $r \in R$; then $\theta \circ t(m, r) = \theta(mr) = f(mr, 1_R) = f(m, r)$, and so $\theta \circ t = f$, proving that (M, t) satisfies the defining universal mapping property for tensor products. By Corollary 13.3.1, the proof is complete. $\square$

Remark Of course in the preceding proof one could define $e : M \to M \otimes R$ by setting $e(m) := m \otimes 1_R$ and then observe that the mappings $t \circ e$ and $e \circ t$ are identity mappings of M and $M \otimes R$ respectively—but that proof would not illustrate the use of the uniqueness of of the tensor product. Later in this chapter we shall encounter other proofs of an isomorphism relation which also exploit the uniqueness of an object defined by a universal mapping property.

As a second application of our definitions the consider the following:

If A is any torsion abelian group and if D is any divisible abelian group, then $D \otimes_{\mathbb{Z}} A = 0$. If $a \in A$, let $0 \neq n \in \mathbb{Z}$ be such that $na = 0$. Then for any $d \in D$ there exists $d' \in D$ such that $d'n = d$. Therefore $d \otimes a = d'n \otimes a = d' \otimes na = d' \otimes 0 = 0$. Therefore every simple tensor in $D \otimes_{\mathbb{Z}} A$ is zero; since $D \otimes_{\mathbb{Z}} A$ is generated by simple tensors, we conclude that $D \otimes_{\mathbb{Z}} A = 0$.

Using the simple observation in the preceding paragraph, the reader should have no difficulty in proving that if m and n are positive integers with greatest common divisor d, then

$$\mathbb{Z}/m\mathbb{Z} \otimes_{\mathbb{Z}} \mathbb{Z}/n\mathbb{Z} \cong \mathbb{Z}/d\mathbb{Z}$$

(see Exercise (2) in Sect. 13.13.2).

13.3.3 Mapping Properties of the Tensor Product

The following is an important mapping property of the tensor product.

Theorem 13.3.4 *Let R be a ring, let $\phi : M \to M'$ be a right R-module homomorphism and let $\psi : N \to N'$ be a left R-module homomorphism. Then there exists a unique abelian group homomorphism $\phi \otimes \psi : M \otimes_R N \to M' \otimes_R N'$ such that for all $m \in M$, $n \in N$, $(\phi \otimes \psi)(m \otimes n) = \phi(m) \otimes \psi(n)$.*

Proof We start by defining the mapping $\phi \times \psi : M \times N \to M' \otimes_R N'$, by the rule

$$(\phi \times \psi)(m, n) := \phi(m) \otimes \psi(n) \text{ for all } m \in M, \ n \in N.$$

Using the fact that ϕ, ψ are both module homomorphisms, $\phi \times \psi$ is easily checked to be balanced. By the universality of $M \otimes_R N$, there exists a unique mapping—which we shall denote $\phi \otimes \psi$—from $M \otimes_R N$ to $M' \otimes_R N'$ such that $(\phi \otimes \psi) \circ t = \phi \times \phi$, where, as usual $t : M \times N \to M \otimes_R N$ is the balanced mapping $t(m, n) = m \otimes n$, $m \in M$, $n \in N$. Therefore, $(\phi \otimes \psi)(m \otimes n) = (\phi \otimes \psi) \circ t(m, n) = (\phi \times \psi)(m, n) = \phi(m) \otimes \psi(n)$, and we are done. $\square$

The following is immediate from Theorem 13.3.4:

Corollary 13.3.5 (Composition of tensored maps) *Let R be a ring, let $M \xrightarrow{\phi} M' \xrightarrow{\phi'} M''$ be a sequence of right R-module homomorphisms, and let $N \xrightarrow{\psi} N' \xrightarrow{\psi'} N''$ be a sequence of left R-module homomorphisms. Then*

$$(\phi' \otimes \psi') \circ (\phi \otimes \psi) = (\phi' \circ \phi) \otimes (\psi' \circ \psi) : M \otimes_R N \to M'' \otimes_R N''.$$

Corollary 13.3.6 (The Distributive Law of Tensor Products) *Let $\sum_{\sigma \in I} A_\sigma$ be a direct sum of right R-modules $\{A_\sigma | \sigma \in I\}$, and let N be a fixed left R-module. Then $(\sum_{\sigma \in I} A_\sigma) \otimes N$ is a direct sum of its submodules $\{A_\sigma \otimes N\}$.*

Proof By Theorem 8.1.6 on identifying internal direct sums, it is sufficient to demonstrate two things: (i) that $(\sum_{\sigma \in I} A_\sigma) \otimes N$ is spanned by its submodules $A_\sigma \otimes N$ and (ii) that for any index $\tau \in I$, that

$$(A_\tau \otimes N) \cap \sum_{\sigma \neq \tau} A_\sigma \otimes N = 0. \tag{13.1}$$

We begin with (i). Any element $a \in \sum A_\sigma$, has the form $a = \sum_{\tau \in S} a_\tau$ where S is a finite subset of I, and $a_\tau \in A_\tau$. From the elementary properties of "$\otimes$" presented at the beginning of Sect. 13.3.2, for any $n \in N$, we have $a \otimes n = \sum_S a_\tau \otimes n$, the summands of which lie in submodules $A_\tau \otimes N$. Since $(\sum_{\sigma \in I} A_\sigma) \otimes N$ is spanned by such elements $a \otimes n$, it is also spanned by its submodules $A_\sigma \otimes N$. Thus (i) holds.

Since the sum $\sum_{\sigma \in I} A_\sigma$ is direct, there exists a family of projection mappings $\{\pi_\tau : \sum_{\sigma \in I} A_\sigma \to A_\tau\}$ with the property that the restriction of π_τ to the submodule A_σ is the identity mapping if $\sigma = \tau$, while it is the zero mapping $A_\sigma \to 0$, whenever $\sigma \neq \tau$ (see Sect. 8.1.8).

Now, letting 1_N denote the identity mapping on the submodule N, Theorem 13.3.4 gives us mappings $\pi_\tau \otimes 1_N : (\sum A_\sigma) \otimes N \to A_\tau \otimes N$ whose value at a pure element $a \otimes n$ is $\pi_\tau(a) \otimes n$, for $(a, n) \in (\sum A_\sigma) \times N$. Thus $\pi_\tau \otimes 1_N$ restricted to the submodule $A_\tau \otimes N$ is the identity mapping $1_{A_\tau} \otimes 1_N$ on that submodule. In contrast, for $\sigma \neq \tau$, the mapping $\pi_\tau \otimes 1_N$ restricted to $A_\sigma \otimes N$ has image $0 \otimes N = 0$.

Now consider an element $b \in (A_\tau \otimes N) \cap \sum_{\sigma \neq \tau}(A_\sigma \otimes N)$. Then $(\pi_\tau \otimes 1_N)(b) = b$, as $b \in A_\tau \otimes N$, while on the other hand this value is 0, since, $b \in \sum_{\sigma \neq \tau}(A_\sigma \otimes N)$. Thus $b = 0$ and so Eq. (13.1) holds. Thus (ii) has been established. $\square$

Of course, a "mirror image" of this proof will produce the following:

Corollary 13.3.7 *If N is a right R-module and $\sum_{\sigma \in I} A_\sigma$ is a direct sum of a family of left R-modules $\{A_\sigma | \sigma \in I\}$, then $N \otimes (\sum_{\sigma \in I} A_\sigma)$ is a direct sum of its submodules $\{N \otimes A_\sigma | \sigma \in I\}$. Thus one can write*

$$N \otimes (\bigoplus_{\sigma \in I} A_\sigma) \simeq \bigoplus_{\sigma \in I} (N \otimes A_\sigma).$$

From Corollary 13.3.5, we see that for any ring R and for each left R-module N, $F : \mathbf{Mod}_R \to \mathbf{Ab}$ given by

$$F(M) := M \otimes_R N, \quad F(\phi) := \phi \otimes \mathrm{id}_N : M \otimes_R N \to M' \otimes_R N$$

(where $\phi' : M \to M'$ is a right R-module homomorphism and as usual id_N is the identity mapping on N) defines a functor.

Similarly, Theorem 13.3.4 and Corollary 13.3.5 tell us that for each right R-module M the assignment $G : \mathbf{Mod}_R \to \mathbf{Ab}$ given by

$$G(N) := M \otimes_R N, \quad F(\phi) := \mathrm{id}_M \otimes \phi : M \otimes_R N \to M \otimes_R N'$$

(where $\phi : N \to N'$ is a left R-module homomorphism) also defines a functor. These two functors are often denoted $- \otimes_R N$ and $M \otimes_R -$, respectively.

We shall continue our discussion of mapping properties of the tensor product, with the existence of these functors in mind.

The next theorem shows how the tensor product behaves with respect to exact sequences. (In the language just introduced, this result describes the right exactness of the tensor functors just introduced.)

Theorem 13.3.8 *(i) Let $M' \xrightarrow{\mu} M \xrightarrow{\epsilon} M'' \to 0$ be an exact sequence of right R-modules, and let N be a left R-module. Then the sequence*

$$M' \otimes_R N \xrightarrow{\mu \otimes \mathrm{id}_N} M \otimes_R N \xrightarrow{\epsilon \otimes \mathrm{id}_N} M'' \otimes_R N \to 0$$

is exact.

(ii) Let $N' \xrightarrow{\mu} N \xrightarrow{\epsilon} N'' \to 0$ be an exact sequence of left R-modules, and let M be a right R-module. Then

$$M \otimes_R N' \xrightarrow{\mathrm{id}_M \otimes \mu} M \otimes_R N \xrightarrow{\mathrm{id}_M \otimes \epsilon} M \otimes_R N'' \to 0$$

is exact.

Proof We shall be content to prove (i) as (ii) is entirely similar. First of all, if $m'' \in M''$, then there exists an element $m \in M$ with $\epsilon(m) = m''$. Therefore, it follows that for any $n \in N$, $(\epsilon \otimes \mathrm{id}_N)(m \otimes n) = (m) \otimes \mathrm{id}_N(n) = m'' \otimes n$. Since $M'' \otimes_R N$ is generated by the simple tensors of the form $m'' \otimes n$, we infer that $\epsilon \otimes \mathrm{id}_N : M \otimes_R N \to M'' \otimes_R N$ is surjective.

Next, since $(\epsilon \otimes \mathrm{id}_N)(\bar{\ } \otimes \mathrm{id}_N) = \bar{\ } \otimes \mathrm{id}_N = 0 : M' \otimes_R N \to M'' \otimes_R N$, we infer that $\mathrm{im}\,(\mu \otimes \mathrm{id}_N) \subseteq \ker(\epsilon \otimes \mathrm{id}_N)$. Next, let A be the subgroup of $M \otimes_R N$ generated by simple tensors of the form $m \otimes n$, where $m \in \ker \epsilon$. Note that it is clear that $A \subseteq \ker(\epsilon \otimes \mathrm{id}_N)$. As a result, $\epsilon \otimes \mathrm{id}_N$ factors through a homomorphism

$$\overline{\epsilon \otimes \mathrm{id}_N} : (M \otimes_R N)/A \to M'' \otimes_R N$$

satisfying

$$\overline{\epsilon \otimes \mathrm{id}_N}(m \otimes n + A) = \epsilon(m) \otimes n \in M'' \otimes_R N, \ m \in M, n \in N.$$

If we can show that $\overline{\epsilon \otimes \mathrm{id}_N}$ is an isomorphism, we will have succeeded in showing that $A = \ker \epsilon \otimes \mathrm{id}_N$. To do this we show that $\overline{\epsilon \otimes \mathrm{id}_N}$ has an inverse. Indeed. define the mapping

$$f : M'' \times N \to (M \otimes_R N)/A, \quad f(m'', n) := m \otimes n + A,$$

where $m \in M$ is any element satisfying $\epsilon(m) = m''$. Note that if $m_1 \in M$ is any other element satisfying $\epsilon(m_1) = m''$, then $m \otimes n - m_1 \otimes n = (m - m_1) \otimes n \in A$, which proves that $f : M'' \times N \to (M \otimes_R N)/A$ is well defined. As it is clearly balanced, we obtain an abelian group homomorphism $\theta : M'' \otimes_R N \to (M \otimes_R N)/A$ satisfying $\theta(m'' \otimes n) = (m \otimes n) + A$, where $\epsilon(m) = m''$. As it is clear that θ is inverse to $\overline{\epsilon \otimes \mathrm{id}_N}$, we conclude that $A = \ker(\epsilon \otimes \mathrm{id}_N)$.

Finally, note that since $\ker \epsilon = \mathrm{im}\,\mu$ we have that $A \subseteq \mathrm{im}\,(\mu \otimes \mathrm{id}_N)$. Since we have already shown that $\mathrm{im}\,(\mu \otimes \mathrm{id}_N) \subseteq \ker(\epsilon \otimes \mathrm{id}_N) = A$, it follows that $\mathrm{im}\,(\mu \otimes \mathrm{id}_N) = \ker(\epsilon \otimes \mathrm{id}_N)$, and the proof is complete. $\square$

We hasten to warn the reader that in Theorem 13.3.8 (i) above, even if $M' \xrightarrow{\mu} M$ is injective, it need not follow that $M' \otimes_R N \xrightarrow{\mu \otimes \mathrm{id}_N} M \otimes_R N$ is injective. (A similar comment holds for part (ii).) Put succinctly, the tensor product does not take short exact sequences to short exact sequences. In fact a large portion of "homological algebra" is devoted to the study of functors that do not preserve exactness. As an easy example, consider the short exact sequence of abelian groups (i.e. $\mathbb{Z}$-modules):

$$\mathbb{Z} \xrightarrow{\mu_2} \mathbb{Z} \to \mathbb{Z}/2\mathbb{Z} \to 0,$$

where $\mu_2(a) = 2a$. If we tensor the above short exact sequence on the right by $\mathbb{Z}/2\mathbb{Z}$, we get the sequence

$$\mathbb{Z}/2\mathbb{Z} \xrightarrow{0} \mathbb{Z}/2\mathbb{Z} \xrightarrow{\cong} \mathbb{Z}/2\mathbb{Z}.$$

Thus the exactness breaks down.

Motivated by Theorem 13.3.8 and the above example, we call a left R-module N *flat* if for any short exact sequence of the form

$$0 \to M' \xrightarrow{\mu} M \xrightarrow{\epsilon} M'' \to 0,$$

the following sequence is also exact:

$$0 \to M' \otimes_R N \xrightarrow{\mu \otimes \mathrm{id}_N} M \otimes_R N \xrightarrow{\epsilon \otimes \mathrm{id}_N} M'' \otimes_R N \to 0.$$

Note that by Theorem 13.3.8, one has that the left R-module is flat if and only if for every injective right R-module homomorphism $\mu : M' \to M$, the induced homomorphism $\mu \otimes \mathrm{id}_N : M' \otimes_R N \to M \otimes_R N$ is also injective.

The following lemma will prove useful in the discussion of flatness.

Lemma 13.3.9 *Let R be a ring, and let F be a free left R-module with basis $\{f_\beta \mid \beta \in \mathcal{B}\}$. Let M be a right R-module. Then every element of $M \otimes_R F$ can be uniquely expressed as $\sum_{\beta \in \mathcal{B}} m_\beta \otimes f_\beta$, where only finitely many of the elements m_β, $\beta \in \mathcal{B}$ are nonzero.*

Proof First of all, it is obvious that each element of $M \otimes_R F$ admits such an expression. Now assume that the element $\sum_{\beta \in \mathcal{B}} m_\beta \otimes f_\beta = 0$. We need to show that each of the elements m_β, $\beta \in \mathcal{B}$ are 0. Since F is free with basis $\{f_\beta \mid \beta \in \mathcal{B}\}$, there exist, for each $\beta \in \mathcal{B}$, a so-called projection homomorphism $\pi_\beta : F \to_R R$ such that

$$\pi_\beta(f_\gamma) = \begin{cases} 0 \text{ if } \gamma \neq \beta \\ 1 \text{ if } \gamma = \beta. \end{cases}$$

Next, we recall that $M \otimes_R R \cong M$ via the isomorphism $\sigma : M \otimes_R R \to M$, $m \otimes r \mapsto mr \in M$ (see Lemma 13.3.3). Combining all of this, we have the following

$$\begin{aligned} 0 &= \sigma(1_M \otimes \pi_\beta) \sum m_\beta \otimes f_\beta \\ &= \sigma(m_\beta \otimes 1) \\ &= m_\beta. \end{aligned}$$

The result follows. $\square$

Remark Of course one could easily devise a proof of Lemma 13.3.9 based on right distributive properties (see Corollary 13.3.6). However here we are interested in retrieving a "flatness" result.

Theorem 13.3.10 *Let R be a ring. If F is a free left R-module, then it is flat.*

Proof Let $\{f_\beta \mid \beta \in \mathcal{B}\}$ be a basis of F and let $\mu : M' \to M$ be an injective homomorphism of right R-modules. We shall show that the induced homomorphism $\mu \otimes 1_F : M' \otimes_R F \to M \otimes_R F$ is also injective. Using Lemma 13.3.9 we may express a typical element of $M' \otimes_R F$ as $\sum m'_\beta \otimes f_\beta$, where only finitely many of the elements m'_β, $\beta \in \mathcal{B}$ are nonzero. Thus, if $(\mu \otimes 1_F)(\sum m'_\beta \otimes f_\beta) = 0$, then $\sum \mu(m'_\beta) \otimes f_\beta = 0 \in M \otimes_R F$. But then the uniqueness statement of Lemma 13.3.9 guarantees that each $\mu(m'_\beta) = 0$. Since $\mu : M' \to M$ is injective, we infer that each $m'_\beta = 0$, and so the original element $\sum m'_\beta \otimes f_\beta = 0$, proving the result. $\square$

13.3.4 The Adjointness Relationship of the Hom and Tensor Functors

In this section, we use some of the elementary language of category theory introduced in Sect. 13.2. Let R be a ring and let $_R\mathbf{Mod}$, $\mathbf{Ab}$ denote the categories of left R-modules and abelian groups, respectively. Thus, if M is a fixed right R-module, then we have a functor

$$M \otimes_R - \; :_R \mathbf{Mod} \longrightarrow \mathbf{Ab}.$$

In an entirely similar way, for any fixed left R-module N, there is a functor

$$- \otimes_R N : \mathbf{Mod}_R \to \mathbf{Ab},$$

where $\mathbf{Mod}_R$ is the category of right R-modules. Next we consider a functor $\mathbf{Ab} \to_R \mathbf{Mod}$, alluded to in Sect. 8.4.2, Chap. 8. Indeed, if M is a fixed right R-module, we may define

$$\mathrm{Hom}_\mathbb{Z}(M, -) : \mathbf{Ab} \to_R \mathbf{Mod}.$$

Indeed, note that if A is an abelian group, then $\mathrm{Hom}_\mathbb{Z}(M, A)$ is a left R-module via $(r \cdot f)(m) = f(mr)$. For the fixed right R-module M, the functors $M \otimes_R -$ and $\mathrm{Hom}_\mathbb{Z}(M, -)$ satisfy the following important adjointness relationship:

Theorem 13.3.11 (Adjointness Relationship) *If M is a right R-module, N is a left R-module, and if A is an abelian group, there is a natural equivalence of sets:*

$$\mathrm{Hom}_\mathbb{Z}(M \otimes_R N, A) \cong_{\mathbf{Set}} \mathrm{Hom}_R(N, \mathrm{Hom}_\mathbb{Z}(M, A)). \tag{13.2}$$

Proof One defines a mapping θ from the left side of (13.2) to the right side in the following way. For each $f \in \mathrm{Hom}_\mathbb{Z}(M \otimes_R N, A)$ let

$$\theta(f) : N \to \mathrm{Hom}_\mathbb{Z}(M, A)$$

be defined by

$$\theta(f)(n) : M \to A, \text{ where } [\theta(n)](m) = f(m \otimes n)$$

for all $n \in N$ and $m \in M$.

For the inverse consider an arbitrary abelian group morphism $\psi : N \to \mathrm{Hom}_\mathbb{Z}(M, A)$. Observe that $\theta^*(\psi) : M \times N \to A$ defined by $\theta^*(\psi)(m, n) = [\psi(m)](n)$ is a balanced mapping, and so factors through $M \otimes_R N$ to yield a mapping $\theta^{-1}(\psi) : M_R \otimes N \to A$. The fact that θ^{-1} really is an inverse (that is, $\theta^{-1} \circ \theta$ and $\theta \circ \theta^{-1}$ are respectively the identity mappings on the left and right sides of (13.2)) is easily verified. $\square$

In general if $\mathcal{C}, \mathcal{D}$ are categories, and if $F : \mathcal{C} \to \mathcal{D}, G : \mathcal{D} \to \mathcal{C}$ are functors, we say that F is *left adjoint* to G (and that G is *right adjoint* to F) if there is a natural equivalence of sets

$$\mathrm{Hom}_\mathcal{D}(F(X), Y) \cong_{\mathbf{Set}} \mathrm{Hom}_\mathcal{C}(X, G(Y)),$$

where X is an object of $\mathcal{C}$ and Y is an object of $\mathcal{D}$. Thus, we see that the functor $M \otimes_R -$ is left adjoint to the functor $\mathrm{Hom}_\mathbb{Z}(M, -)$.

13.4 The Tensor Product as a Right S-Module

In the last section we started with a right R-module M and a left R-module N and constructed the abelian group $M \otimes_R N$ satisfying the universal mapping properties relative to balanced mappings. In this section, we shall discuss conditions that will enable $M \otimes_R N$ to carry a module structure.

To this end let S, R be rings, assume that M is a right R-module, and assume that N is an (R, S)-bimodule. In order to give $M \otimes_R N$ the structure of a right S-module, it suffices to construct a ring homomorphism

$$\phi : S \to \mathrm{End}_\mathbb{Z}(M \otimes_R N)^*;$$

where the "*" is there to indicate that the endomorphisms are to act on the right, in this abselian group. This will allow for the definition of an S-scalar multiplication: $a \cdot s := a\phi(s)$, $a \in M \otimes_R N$. For each $s \in S$ define $f_s : M \times N \to M \otimes_R N$ by setting $f_s(m, n) := m \otimes ns$, $s \in S$, $m \in M$, $n \in N$. Then f_s is easily checked to be a balanced map. By the universal mapping property of the tensor product, there exists a uniqsue abelian group homomorphism $\phi_s : M \otimes_R N \to M \otimes_R N$ satisfying $\phi_s(m \otimes n) = m \otimes ns$. Note that if $m \in M$ and $n \in N$, then

$$
\begin{aligned}
\phi_{s_1+s_2}(m \otimes n) &= m \otimes n(s_1 + s_2) \\
&= m \otimes (ns_1 + ns_2) \\
&= m \otimes ns_1 + m \otimes ns_2 \\
&= \phi_{s_1}(m \otimes n) + \phi_{s_2}(m \otimes n) \\
&= (\phi_{s_1} + \phi_{s_2})(m \otimes n).
\end{aligned}
$$

It follows, therefore, that $\phi_{s_1+s_2} = \phi_{s_1} + \phi_{s_2}$. Similarly, one verifies that $\phi_{s_1 s_2} = (\phi_{s_1}) \cdot (\phi_{s_2})$, where, for composition of right operators, the "dot" indicates that s_1 is applied first and s_2 second. In turn, this immediately implies that the mapping $\phi : S \to \text{End}_{\mathbb{Z}}(M \otimes_R N)$, $\phi(s) := \phi_s$ is the desired ring homomorphism. In other words, we have succeeded in giving $M \otimes_R N$ the structure of a right S-module.

The relevant universal property giving rise to a module homomorphism is the following:

Theorem 13.4.1 *Let R and S be rings, let M be a right R-module, and let N be an (R, S)-bimodule. If K is a right S-module and if $f : M \times N \to K$ is a balanced mapping which also satisfies*

$$
f(m, ns) = f(m, n)s, \ s \in S, \ m \in M, \ n \in N,
$$

then the uniquely induced abelian group homomorphism $\theta : M \otimes_R N \to K$ is also a right S-module homomorphism.

Proof Let $m \in M$, $n \in N$, and let $s \in S$. Then

$$
\begin{aligned}
\theta((m \otimes n)s) &= \theta(m \otimes (ns)) \\
&= \theta \circ t(m, ns) \\
&= f(m, ns) \\
&= f(m, n)s \\
&= (\theta \circ t(m, n))s \\
&= \theta(m \otimes n)s.
\end{aligned}
$$

Since $M \otimes_R N$ is generated by the simple tensors $m \otimes n$, $m \in M$, $n \in N$, we conclude that θ preserves S-scalar multiplication, and hence is an S-module homomorphism $M \otimes_R N \to K$. $\square$

Corollary 13.4.2 (Exchange of Rings) *Let R be a subring of S. If M is a right R-module, then $M \otimes_R S$ is a right S-module.*

Proof Since S is an (R, S)-bimodule, it can replace the module N in Theorem 13.4.1. $\square$

A frequent example of such an exchange is the enlargement of the ground field of a vector space: here an F-vector space V becomes $V \otimes_F E$, where E is an extension field of F.[3] In fact one has the following:

Corollary 13.4.3 *Suppose E is a field containing F as a subfield. Let 1_E denote the multiplicative identity element of E. If V is an F-vector space, then $V \otimes_F E$ is a right vector space over E. If $X = \{x_\sigma\}$ is a basis for V then $X \otimes 1_E := \{x_\sigma \otimes 1_E\}$ is an E-basis for $V \otimes E$. Hence*

$$\dim_F V = dim_E(V \otimes E).$$

The proof is left as Exercise (1) in Sect. 13.13.4.

Of particular importance is the following special case. Assume that R is a commutative ring. In this case, any right R-module M can also be regarded as a left R-module by declaring that $rm := mr$, $m \in M$, $r \in R$. Furthermore, this specification also gives M the structure of an (R, R)-bimodule. On p. 234 we called such a module M a *symmetric bimodule*. For example, vector spaces over a field are normally treated as symmetric bimodules.

suppose M and N are such symmetric (R, R)-bimodules. Then for all $m \in M$, $n \in N$, and $r \in R$, we can define left and right scalar multiplication of pure tensors by the first and last entries in the following equations:

$$(m \otimes n)r = m \otimes (nr) = m \otimes (rn) = (mr) \otimes n = (rm) \otimes n = r(m \otimes n).$$

Of course, this endows $M \otimes_R N$ with the structure of a symmetric (R, R)-bimodule.

Note that Theorem 13.4.1 says in particular that if F is a field and if V and W are vector spaces over F, then $V \otimes_F W$ is automatically a vector space over F. In fact, we can say more:

Theorem 13.4.4 *Let F be a field, and let V and W be F-vector spaces with bases $\{v_\sigma | \sigma \in I\}$, and $\{w_\tau | \tau \in J\}$, respectively. Then $V \otimes_F W$ has basis $\{v_\sigma \otimes w_\tau | (\sigma, \tau) \in I \times J\}$. In particular,*

$$dim_F(V \otimes_F W) = dim_F V \cdot dim_F W,$$

as a product of cardinal numbers.

Proof Since F-vector spaces are free (F, F)-bimodules, we may write them as direct sums of their 1-dimension subspaces. Thus

$$V = \bigoplus_{\sigma \in I} v_\sigma F , \ W = \bigoplus_{\tau \in J} F w_\tau.$$

---

[3]Of course for over a century mathematicians (perhaps in pursuit of eigenvalues of transformations) have been replacing F-linear combinations of a basis with E-linear combinations without feeling the need of a tensor product.

Now by the "distributive laws" (Corollaries 13.3.6 and 13.3.7) we have

$$V \otimes_F W = \bigoplus_{(\sigma,\tau)\in I\times J} v_\sigma F \otimes F w_\tau.$$

Since each summand is an (F, F)-bimodule, we may write $v_\sigma \otimes w_\tau F$ in place of $v_\sigma F \otimes F w_\tau$, and the direct sum of these summands is also an (F, F)-bimodule. Thus $V \otimes_F W$ is an F-vector space with basis $\{v_\sigma \otimes w_\tau | (\sigma, \tau) \in I \times J\}$. $\square$

13.5 Multiple Tensor Products and R-Multilinear Maps

Everything that we did in defining a tensor product can also be done in defining *multiple tensor products*. This important topic is the basis of what is called multilinear algebra. There are two reasons that it should be introduced at this stage: (1) it is needed in Sect. 13.7 to define the tensor product of algebras, and (2) it is needed again in a construction of the tensor algebra in Sect. 13.9, and in our discussions of the symmetric and exterior algebras. The main idea here is to avoid having to justify any kind of associative law of the tensor product.

Throughout this section, R is a commutative ring and all modules considered are (not neccessarily symmetric) (R, R)-bimodules.

Let $\{M_i | i = 1, \ldots, n\}$ be a sequence of (R, R)-bimodules, and let A be any abelian group (with its operation denoted by addition). A homomorphism of additive groups

$$\alpha : M_1 \oplus \ldots \oplus M_n \to A$$

is said to be *balanced* if, for each index $i = 1, \ldots, n$, for each n-tuple $(a_1, \ldots, a_n) \in \bigoplus M_i$, and for any $r \in R$, one has

$$\alpha((a_1, \ldots, a_i r, a_{i+1}, \ldots, a_n)) = \alpha(a_1, \ldots, a_i, r a_{i+1}, a_{i+2}, \ldots, a_n). \quad (13.3)$$

This simply extends our previous definition of balanced mapping where the parameter n was 2.

Generalizing the recipe for the usual tensor product, we form the free $\mathbb{Z}$-module F with basis consisting of the set of all n-tuples in $M_1 \times \ldots M_n$,[4] and let B be the subgroup of F generated by all elements of the form:

$$(a_1, \ldots, a_i + a_i', \ldots, a_n) - (a_1, \ldots, a_n) - (a_1, \ldots, a_i', \ldots, a_n)$$
$$(a_1, \ldots, a_i r, a_{i+1}, \ldots, a_n) - (a_1, \ldots, a_i, r a_{i+1}, \ldots, a_n),$$

where $i = 1, 2, \ldots, n$, $a_i, a_i' \in M_i$ and $r \in R$.

---

[4]Again, the reader is warned that addition in F is not addition in $\bigoplus M_i$ when the elements being added lie in the latter sum.

Let $a_1 \otimes \cdots \otimes a_n := (a_1, \ldots, a_n) + B$, the image of $(a_1, \ldots, a_n)$ under the canonical projection $\pi : F \to F/B$ (such an element is called a *pure multitensor*) and we write

$$F/B := M_1 \otimes M_2 \otimes \cdots \otimes M_n.$$

Then for each i,

$$a_1 \otimes \cdots \otimes (a_i + a_i') \otimes \cdots \otimes a_n = (a_1 \otimes \cdots \otimes a_n) + (a_1 \otimes \cdots \otimes a_i' \otimes \cdots \otimes a_n)$$
$$a_1 \otimes \cdots \otimes a_i r \otimes a_{i+1} \otimes \cdots \otimes a_n = a_1 \otimes \cdots \otimes a_i \otimes r a_{i+1} \otimes \cdots \otimes a_n$$

for all $a_j \in M_j$, $a_i' \in M_i$ and $r \in R$. Thus the mapping

$$\theta : M_1 \times \cdots \times M_n \to F/B$$

(a restriction of π) is balanced.

Conversely, suppose β is a balanced mapping

$$M_1 \times \cdots \times M_n \to A$$

for some abelian group A. Using the universal property of free Z-modules, β determines a unique group homomorphism $\hat{\beta} : F \to A$, and, since β is balanced, ker $\hat{\beta}$ contains B. Thus by the Composition Theorem of homomorphisms and the two applications of the Fundamental Theorems of homomorphisms we have the following sequence of homomorphisms:

$$\pi : F \to F/B = M_1 \otimes \cdots \otimes M_n, \tag{13.4}$$

$$\pi_\beta : F/B \to (F/B)/(\ker \hat{\beta}/B), \tag{13.5}$$

$$iso_1 : (F/B)/(\ker \hat{\beta}/B) \to F/(\ker \hat{\beta}), \tag{13.6}$$

$$iso_2 : F/(\ker \hat{\beta}) \to \hat{\beta}(F), \tag{13.7}$$

$$inc : \hat{\beta}(F) \to A \tag{13.8}$$

Here, (13.4) is the afore-mentioned projection, (13.5) is a projection using $B \subseteq \ker \hat{\beta}$, (13.6) is a classical isomorphism, (13.7) is another classical isomorphism, and (13.8) is the inclusion mapping.

Then

$$\hat{\beta} = inc \circ iso_2 \circ iso_1 \circ \pi_\beta \circ \pi,$$

which is a detailed way of saying that $\hat{\beta}$ *factors through the tensor product*:

$$F \to M_1 \otimes \cdots \otimes M_n \to A.$$

Recalling that β is a restriction of $\hat{\beta}$ one concludes the following:

Theorem 13.5.1 *Suppose* $M_1, \ldots, M_n, A$ *are* (R, R)-*bimodules Then there is a unique balanced map* $t : M_1 \times \cdots \times M_n \to M_1 \otimes \cdots \otimes M_n$ *such that for any balanced mapping*

$$\beta : M_1 \times \cdots \times M_n \to A,$$

There exists a balanced mapping

$$\theta(\beta) : M_1 \otimes \cdots \otimes M_n \to A$$

such that $\beta = \theta(\beta) \circ t$—*that is, "$\beta$ factors through the tensor product".*

The theorem is only about R-balanced mappings—certain morphisms of additive groups. If the component M_i's are symmetric (R, R) bimodules, one naturally realizes that such a theorem can hold so that the mappings involved, t, β and $\theta(\beta)$ are all morphisms of symmetric (R, R)-bimodules.

Theorem 13.5.2 *Suppose* $M_1, \ldots, M_n, A$ *are* (R, R)-*bimodules where* R *is a commutative ring and* $rm_i = m_i r$ *and* $ra = ar$ *for all* $(r, , m_i) \in R \times M_i \times A, i \in \{1, 2, \ldots, n\}$. *Then there is a unique balanced map* $t : M_1 \times \cdots \times M_n \to M_1 \otimes \cdots \otimes M_n$ *such that for any balanced mapping*

$$\beta : M_1 \times \cdots \times M_n \to A,$$

There exists an (R, R)-*bimodule homomorphism*

$$\theta(\beta) : M_1 \otimes \cdots \otimes M_n \to A$$

such that $\beta = \theta(\beta) \circ t$—*that is, "$\beta$ factors through the tensor product".*

Remark It is important for the student to realize that in this development, the multiple tensor product is not approached by a series of iterations of the binary tensor product. Rather, it is directly defined as a factor group F/B—a definition completely free of parentheses.

13.6 Interlude: Algebras

Let A be a right R-module, where, for the moment, R is an arbitrary ring (with identity 1_R). We say that A is an R- *algebra* if and only if A is both a ring and that the R-scalar multiplication satisfies the condition

$$(ab)r = a(br) = (ar)b, \quad \text{for all } a, b \in A, r \in R. \tag{13.9}$$

Note that as a consequence of the above requirement, we have, for all $a, b \in A$, $r, s \in R$, that

$$(ab)(rs) = (a(br))s = ((as(br) = ((as)rb = ((a(sr))b.$$

Moreover if we set $b = 1_A$, the multiplicative identity of A, in the above equation, then we obtain the fact that $a(rs) = a(sr)$ for all $a \in A$ and all $r, s R$. Therefore the annihilator $I := \mathrm{Ann}_R(A)$ contains all commutators $rs - sr \in R$, from which we conclude that A is an R/I-module. Noting that R/I is a commutative ring, we see that in our definition of R-algebra, we may as well assume—and we shall—that R is commutative from the outset. Note that with this convention in force, the R-algebra A is a symmetric (R, R)-bimodule via the definition $ra := ar$, $a \in A$, $r \in R$.

We consider some familiar examples:

Example 57 Let F be a field and let $\mathrm{M}_n(F)$ be the ring of $n \times n$ matrices over F. Then the usual F-scalar multiplication with matrices is easily checked to provide $M_n(F)$ the structure of an F-algebra.

Example 58 Let F-be a field, let V be an F-vector space and let $\mathrm{End}_F(V)$ be the set of linear transformations $V \to V$. This is already an F-vector space by the definitions

$$(T + S)(v) := T(v) + S(v), \ (T\alpha)(v) := T(v\alpha),$$

$T, S \in \mathrm{End}_F(V)$, $v \in V$, $\alpha \in F$. Multiplication of linear transformations is just composition: $(TS)(v) := T(S(v))$, $T, S \in \mathrm{End}_F(V)$, $v \in V$, which is again a linear transformation. This gives $\mathrm{End}_F(V)$ a ring structure. Finally, one checks that for all $T, S \in \mathrm{End}_F(V)$ and for all $\alpha \in F$, $(TS)\alpha = T(S\alpha) = (T\alpha)S$ by showing that these three linear transformations have the same effect at all $v \in V$. Therefore, $\mathrm{End}_F(V)$ is an F-algebra.

Example 59 Let R be a commutative ring, let X be a set and let R^X be the set of functions $X \to R$ with ring structure given by point-wise multiplication:

$$(f + g)(x) := f(x) + g(x), \ (fg)(x) := f(x)g(x), \quad \text{for all} \ f, g \in R^X, \ x \in X.$$

An R-scalar multiplication can again be given point-wise:

$$(f \cdot \alpha)(x) := f(x)\alpha, \ f \in R^X, \ x \in X, \ \alpha \in R.$$

The reader should have no difficulty in verifying that the above definition endows R^X with the structure of an R-algebra.

Example 60 The Dirichlet Algebra was defined as a certain collection of complex-valued functions on the set of positive integers. Multiplication was a sort of convolution (see Example 43 on p. 216 for details). Of course one can replace the complex numbers by any commutative ring R in the definition of "Dirichlet multiplication" on the R-valued functions on the positive integers, to obtain an R-algebra.

Example 61 Let F be a field, and M be a monoid. Then the F-monoid ring FM admits an F-scalar multiplication via pointwise scalar multiplication: if $f \in FM$, $\alpha \in F$, and $m \in M$, set $(f\alpha)(m) := f(m)\alpha$. Perhaps this is made more transparent by writing elements of M in the form $f : \sum_{m \in M} \alpha_m m$, where the usual convention that $\alpha_m \neq 0$ for only finitely many $m \in M$, and defining

$$f\alpha := \sum_{m \in M} \alpha \alpha_m m.$$

This gives FM the structure of an F-algebra. We hasten to emphasize two very important special cases of this construction: the polynomial rings and group rings (over a finite group). Thus, we can—and often shall—refer to such objects as *polynomial algebras* and *group algebras*. Note that the F-group algebra FG over the finite group G is finite dimensional over F.

Note that if R is a commutative ring and A is an R-algebra, then any ideal $I \subseteq A$ of the *ring* A must also be an R-submodule of A. Indeed, we realize that if 1_A is the identity of A, and if $\alpha \in R$, then we have, for all $x \in I$, that $x\alpha = (x1_A)\alpha = x(1_A\alpha) \in I$, as I is an ideal. From this it follows immediately that quotients of A by right ideals are also R-modules. Furthermore, a trivial verification reveals that quotients of A by two-sided ideals of A are also R-algebras.

We shall conclude this short section with a tensor product formulation of R-algebras. Thus, we continue to assume that R is a commutative ring and that A is an R-module. First of all, notice that if A is an R-algebra, then by Eq. (13.9) the multiplication $A \times A \to A$ in A is R-balanced, and hence factors through the tensor product, giving an R-module homomorphism $\mu : A \otimes_R A \to A$. The associativity of the multiplication in A translates into the commutativity of the diagram of Fig. 13.1.

At the same time, the identity element $1 \in A$ determines an R-linear homomorphism $\eta : R \to A$, defined by $\eta(\alpha) := 1_A \cdot \alpha \in A$. The fact that 1 is the identity of A translates into the commutativity of the diagrams of the following figure, where $\epsilon : R \otimes A \to A$ and $\epsilon' : A \otimes R \to A$ are the isomorphisms given by

$$\epsilon(\alpha \otimes a) = \alpha a, \quad \epsilon'(a \otimes \alpha) = a\alpha,$$

$a \in A$, $\alpha \in R$.

Fig. 13.1 Diagram for associativity

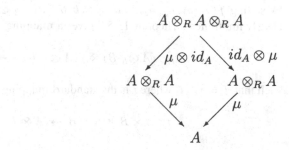

Fig. 13.2 Diagrams for a
two-sided identity

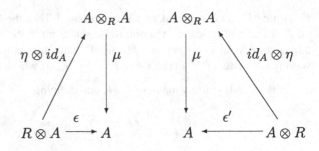

Note that the distributive laws are already contained in the above. Indeed, if $a, b, c \in A$, then using the fact that $\mu : A \otimes_R A \to A$ is an abelian group homomorphism, we have

$$a(b + c) = \mu(a \otimes (b + c)) = \mu(a \otimes b + a \otimes c) = \mu(a \otimes b) + \mu(a \otimes c) = ab + ac.$$

Conversely, if A is an algebra over the commutative ring R, and if a multiplication $\mu : A \otimes_R A \to A$ is given which makes the diagrams of Figs. 13.1 and 13.2 commute, then μ gives A the structure of an R-algebra (Exercise (1) in Sect. 13.13.5).

13.7 The Tensor Product of R-Algebras

Throughout this section, we continue to assume that R is a commutative ring. If A, B are both R-algebras, we shall give a natural R-algebra structure on the tensor product $A \otimes_R B$. Recall from Sect. 13.4 that $A \otimes_R B$ is already an R-module with scalar multiplication satisfying

$$(a \otimes b)\alpha = a \otimes b\alpha = a\alpha \otimes b,$$

$a \in A, \ b \in B, \ \alpha \in R.$

To obtain an R-algebra structure on $A \otimes_R B$, we map

$$f : A \times B \times A \times B \longrightarrow A \otimes_R B,$$

by setting $f(a_1, b_1, a_2, b_2) = a_1 a_2 \otimes b_1 b_2, \ a_1, a_2 \in A, \ b_1, b_2 \in B$. Then, as f is clearly multilinear, Theorem 13.5.2 gives a mapping

$$\mu : (A \otimes_R B) \otimes_R (A \otimes_R B) \longrightarrow A \otimes_R B$$

such that $f = \mu \circ t$, where t is the standard mapping

$$t : A \times B \times A \times B \to A \otimes B \otimes A \otimes B.$$

Thus, we define the multiplication on $A \otimes_R B$ by setting $xy := \mu(x \otimes y)$, $x, y \in A \otimes_R B$. Note that on simple tensors, this multiplication satisfies $(a \otimes b)(a' \otimes b') = aa' \otimes bb'$, $a, a' \in A$, $b, b' \in B$. One now has the desired result:

Theorem 13.7.1 *Let A, B be R-algebras over the commutative ring R. Then there is an R-algebra structure on $A \otimes_R B$ such that $(a \otimes b) \cdot (a' \otimes b') = aa' \otimes bb'$, $a, a' \in A, b, b' \in B$.*

Proof It suffices to prove that the Figs. 13.1 and 13.2 are commutative. In turn, to prove that the given diagrams are commutative, it suffices to verify the commutativity when applied to simple tensors. Thus, let $a, a', a'' \in A$, $b, b', b'' \in B$. We have

$$
\begin{aligned}
\mu(\mu \otimes \mathrm{id}_{A \otimes_R B})(a \otimes b \otimes a' \otimes b' \otimes a'' \otimes b'') &= \mu(aa' \otimes bb' \otimes a'' \otimes b'') \\
&= (aa')a'' \otimes (bb')b'' \\
&= a(a'a'') \otimes b(b'b'') \\
&= \mu(a \otimes b \otimes a'a'' \otimes b'b'') \\
&= \mu(\mathrm{id}_{A \otimes_R B} \otimes \mu)(a \otimes b \otimes a' \otimes b' \otimes a'' \otimes b''),
\end{aligned}
$$

proving that Fig. 13.1 commutes. Proving that the Fig. 13.2 also commute is even easier, so the result follows. $\square$

13.8 Graded Algebras

Let $(M, \cdot)$ be a (not necessarily commutative) monoid, let R be a commutative ring, and let A be an R-algebra, as described in the previous sections. The algebra A is said to possess an M-*grading* if and only if

(i) $A = \bigoplus_{\sigma \in M} A_\sigma$. a direct sum of R-modules, A_σ, indexed by M.
(ii) For any $\sigma, \tau \in M$,

$$
A_\sigma A_\tau = \{a_\sigma a_\tau | (a_\sigma, a_\tau) \in A_\sigma \times A_\tau\} \subseteq A_{\sigma \cdot \tau}.
$$

Any R-algebra A that possesses a grading with respect to some monoid M is called a *graded algebra*.

The elements of A_σ are said to be *homogeneous of degree σ with respect to the M-grading*, and the submodules A_σ themselves will be called *homogeneous summands*.[5] Note that by this definition, the additive identity element $0 \in A$ is homogeneous of every possible degree chosen from M.

Example 62 We have already met such algebras in Sect. 7.3.2. The monoid algebras FM, where F is a field, for example, have homogeneous summands that are

---

[5] Although "homogenous summand" is not a universally used term, it is far better than "homogeneous component" which has already been assigned a specific meaning in the context of completely reducible modules.

1-dimensional over the field F. (The converse is not true. There are M-graded F-algebras with 1-dimensional summands, which are not monoid rings.)

We also should include in this generic example the polynomial rings $D[x]$ and $D[X]$ where D is an integral domain and the respective multiplicative monoids are respectively the powers of x, or all monomials $x_1^{a_1} x_2^{a_2} \cdots x_n^{a_n}, a_i, n \in \mathbb{N}$, in commuting indeterminates $x_i \in X^6$ Note that we must use an integral domain in order to maintain property (i), that $A_\sigma A_\tau \subseteq A_{\sigma \cdot \tau}$.

But this genus of examples also includes examples where the monoid $(M, \cdot)$ is not commutative:

1. The polynomial ring $D\{X\}$ in non-commuting indeterminates, where the relevant grading is provided by the free monoid on X (consisting of two or more symbols) whose elements are words in the alphabet X.
2. The group ring DG where G is a non-commutative group.

In these examples, which were discussed in Sects. 7.3.2 and 7.3.5, the homogeneous summands are of the form Dm (one-dimensional if D is a field).

The following observation provides examples of graded D-algebras whose homogeneous summands are *not* of this form.

Lemma 13.8.1 *Let A be an D-algebra graded with respect to a monoid $(M, \cdot)$. Suppose $\phi : (M, \cdot) \to (M', \cdot)$ is a surjective homomorphism of monoids. For each element m' of the image monoid M', define*

$$A_{m'} := \sum_{\phi(\sigma)=m'} A_\sigma.$$

Then the direct decomposition $A = \bigoplus_{m' \in M'} A_{m'}$, defines a grading of A with respect to the image monoid M'.

Proof We need only show that if m'_1 and m'_2 are elements of M', then $A_{m'_1} A_{m'_1} \subseteq A_{m'_1 \cdot m'_2}$. But if $\phi(m_i) = m'_i, i = 1, 2$, then $\phi(m_1 \cdot m_2) = m'_1 m'_2$, since ϕ is a monoid morphism. Thus $A_{m_1} A_{m_2} \subseteq A_{m'_1 \cdot m'_2}$ for all m_i such that $\phi(m_i) = m'_i, i = 1, 2$. Since $A_{m'_i} = \sum_{\phi(m)=m'_i} A_m$, the result follows. $\square$

In effect, the monoid homomorphism ϕ produces a grading that is more "course" than the original M-grading.

Example 63 The (commutative) polynomial ring $R[X]$, where R is any commutative ring, is graded by the multiplicative monoid $\mathcal{M}^*(X)$ of finite monomials chosen from X. As noted in Sect. 7.3.5, p. 205, there is a monoid morphism

$$\deg : \mathcal{M}^*(X) \to (\mathbb{N}, +)$$

---

[6]Recall that these monoids are respectively isomorphic to $(\mathbb{N}, +)$ and the additive monoid of all finite multisets chosen from X. (Once again, we remind the beginning reader that the set of natural numbers, $\mathbb{N}$, includes 0.).

which is defined by

$$\deg : \prod_{i=1}^{n} x_i^{a_i} \mapsto \sum_{i=1}^{n} a_i$$

for any finite subset $\{x_1, \ldots, x_n\}$ of X. The image $\deg m$ is called the *degree* of m, for monomial $m \in \mathcal{M}^*(X)$.

Assume that A, A' are both M-graded algebras over the commutative ring R. An algebra homomorphism $\phi : A \to A'$ is said to be a *graded algebra homomorphism* if for each $m \in M$, $\phi(A_m) \subseteq A'_m$. When A is an M-graded algebra over R, we shall be interested in ideals $I \subseteq A$ for which R/I is also M-graded and such that the projection homomorphism $A \to A/I$ is a homomorphism of M-graded algebras.

For example, suppose $A = D[x]$, the usual polynomial ring, and the ideal I is $Ax^n, n > 1$. Then is one writes $A_i := x^i D, 0 \le i \le n - 1$, then

$$A/I = (A_0 + I)/I \oplus (A_1 + I)/I \oplus \cdots \oplus (A_{n-1} + I)/I$$

is a grading on A/I by the monoid $\mathbb{Z}/n\mathbb{Z}$.

On the other hand assume that $D = F$ is a field, and let I be the ideal of $A = F[x]$ generated by the inhomogeneous polynomial $x + x^2$. By the division algorithm, we see that the quotient algebra A/I has dimension 2 over the field F.

Assume, however, that A/I is graded by the nonnegative integers, and that $A \to A/I$ is a homomorphism of graded algebras. This would yield

$$A/I = \sum_{m=0}^{\infty} (A_m + I)/I.$$

Since, for each nonnegative integer m, $A_m = F \cdot x^m$, we would have $(A_m + I)/I = F(x^m + I) \neq 0$, since x^m cannot be a multiple of $x + x^2$. In turn, this would clearly imply that A/I is infinite dimensional, contradicting our prior inference that A/I has F-dimension 2. Thus A/I is not a graded algebra in this case.

Motivated by the above, we call the ideal I of the M-graded R-algebra A *homogeneous* if

$$A/I = \bigoplus_{m \in M} (A_m + I)/I.$$

Note that this implies both that A/I is M-graded via $(A/I)_m = (A_m + I)/I$ and that the projection homomorphism $A \to A/I$ is a homomorphism of M-graded R-algebras.

Theorem 13.8.2 *Let A be an M-graded R-algebra, and let I be an ideal of A. Then, the following are equivalent:*

(i) *I is homogeneous,*
(ii) *$I = \bigoplus_{m \in M} (A_m \cap I)$,*

(iii) I is generated by homogeneous elements.

Proof Assume (i), i.e., that I is homogeneous; thus

$$A/I = \bigoplus_{m \in M} (A_m + I)/I.$$

Let $x \in I$ and write x as $x = \sum_{m \in M} x_m$, where each $x_m \in A_m$. We shall show that, in fact, each $x_m \in I$. For otherwise, we would have

$$0 + I = x + I = \sum_{m \in M} x_m + I = \sum_{m \in M} (x_m + I).$$

Since the last sum is direct, we conclude that each $x_m + I = 0 + I$, i.e., that each $x_m \in I$. This implies that $I = \sum_{m \in M} (A_m \cap I)$. Since this sum is obviously direct, we infer (ii).

If we assume (ii), and if elements $a_m \in A_m$ are given with

$$\sum_{m \in M} (a_m + I) = 0 + I \in A/I,$$

then

$$\sum_{m \in M} a_m \in I = \sum_{m \in M} (A_m \cap I).$$

This clearly forces each $a_m \in (A_m \cap I) \subseteq I$, and so the sum $\sum_{m \in M} (A_m + I)/I$ is direct. As it clearly equals A/I, we infer condition (i). Thus, conditions (i) and (ii) are equivalent.

If we assume condition (ii), then it is clear that I is generated by homogeneous elements, as each $A_m \cap I$, $m \in M$ consists wholly of homogeneous elements, so (iii) holds.

Finally, assume condition (iii). The homogeneous elements that lie in the ideal I form a set $S = \cup_{\sigma \in M} (I \cap A_\sigma)$, which, by hypothesis, generates I as a 2-sided ideal. Let

$$H := \bigoplus_{\sigma \in M} H_\sigma \text{ where } H_\sigma = I \cap A_\sigma, \tag{13.10}$$

so that H is the additive group generated by S. Clearly $H \subseteq I$. Since part (ii) of the definition of an M-grading requires that homogeneous elements be closed under multiplication, we see that if h is an arbitrary homogeneous element of A, then

$$hS \cup Sh \subseteq S,$$

since the products to the left of the containment symbol are homogeneous elements of A lying in the ideal I. It follows that hH and Hh both lie in H for any homogeneous element h chosen from A. Since the additive span of these homogeneous elements is A itself, the distributive law of the ring yields $AHA \subseteq H$, so H is itself a two-sided ideal of A. Since H contains S the ideal-generators of I, we have $I \subseteq H$. Since $H \subseteq I$ we also have $H = I$ and now Eq. (13.10) produces

$$I = \bigoplus_{\sigma \in M} (I \cap A_\sigma),$$

proving (ii).

Thus all three conditions (i), (ii), and (iii) are equivalent. $\square$

13.9 The Tensor Algebras

13.9.1 Introduction

In this section and the next we hope to introduce several basic algebras with important universal properties. There are two ways to do this: (1) first define the object by a construction, and then show it possesses the desired universal property, or (2) first base the definition upon a universal mapping property, and then utilize our categorical arguments for uniqueness to show that it is something we know or can construct. We think that the second strategy is particularly illustrative in the case of the tensor algebra.

13.9.2 The Tensor Algebra: As a Universal Mapping Property

Fix a vector space V over a field F. The *tensor algebra of V* is defined to be a pair $(\iota, T(V))$ where $\iota : V \to T(V)$ is an injection of V into an F-algebra, $T(V)$ such that every F-linear transformation $t : V \to A$ into an F-algebra A uniquely factors through ι—that is, there exists a unique F-algebra homomorphism $\theta(t) : T(V) \to A$, such that $\theta(t)$ extends t. In other words $t = \theta(T) \circ \iota$ and we have the commutative triangle below:

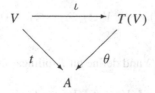

Theorem 13.9.1 *Define the category whose objects are the pairs (f, A), where A is an F-algebra, and $f : V \to A$ is an F-linear transformation, for a fixed vector space V. In this category, a morphism from object (f, A) to object (f', A') is an F-algebra homomorphism $\phi : A \to A'$ such that $\phi \circ f = f'$. Then $(\iota, T(V))$ (if it exists) is an initial object in this category.*

Proof This is just a restatement of the universal property described above.

But now the "abstract nonsense", has a wonderful consequence!

The tensor algebra $(\iota, T(V))$ (if it exists) is unique up to isomorphism.

But it *does* exist and we have met it before. Let X be any basis of the vector space V. We may regard V as a vector subspace of the polynomial ring $F\{X\}$ in non-commutting indeterminates X—namely as the subspace spanned by the monomials of total degree one. We let $\iota_X : V \hookrightarrow F\{X\}$ denote this containment. Let A be any F-algebra, and let $t : V \to A$ be any linear transformation. Now we note that if 1_A is the multiplicative identity element of A, then the subfield $1_A F$ lies in the center of A. Now we apply Exercise (6) in Sect. 7.5.3, with $(t, t(X), F1_A, A)$ in the roles of (α, B, R, S) to conclude that the restricted mapping $t : X \to A$ extends to a ring homomorphism (actually an algebra homomorphism in this case) $E_t^* : F\{X\} \to A$, which we called the "evaluation homomorphism". It follows that it induces $t : V \to A$ when restricted to its degree-one subspace V (see p. 205). We also note that the evaluation mapping $E_t^* : F\{X\} \to A$ was uniquely determined by t.

Thus the pair $(\iota_X, F\{X\})$ satisfies the definition of a tensor algebra given above. We record this as

Corollary 13.9.2 *Let V be a vector space with basis X. Then the tensor algebra is isomorphic to the polynomial algebra $F\{X\}$ in non-commuting indeterminates X.*

13.9.3 The Tensor Algebra: The Standard Construction

In this construction we utilize the (parenthesis free) construction of the multiple tensor product defined in Sect. 13.5.

Let F be a field and let V be an F-vector space. We define a sequence $T^r(V)$ of F-vector spaces by setting $T^0(V) = F$, $T^1(V) = V$, and in general,

$$T^r(V) = \bigotimes_{i=1}^{r} V = V \otimes_F \otimes \cdots \otimes_F V \ (r \text{ factors }).$$

Set $T^*(V) := \bigoplus_{r=0}^{\infty} T^r(V)$ and define an F-bilinear mapping

$$\mu : T^*(V) \times T^*(V) \to T^*(V)$$

by setting

$$\mu\left(\sum_{r=0}^{\infty}\tau_r, \sum_{s=0}^{\infty}\tau_s'\right) := \sum_{k=0}^{\infty}\sum_{r+s=k}\tau_r \otimes \tau_s' \in T^{r+s}(V) \subseteq T^*(V).$$

This is easily verified to endow $T^*(V)$ with the structure of an F-algebra, called the *tensor algebra* over F. When V has a basis X, then $T^*(V)$ has an F-basis consisting all pure tensors of the form $x_1 \otimes \cdots \otimes x_d$, where $d \in \mathbb{N}$, the x_i are in X and an empty product is taken to be the identity element e of $T^*(V)$. In terms of this basis of pure tensors the multiplication follows this rule:

$$\mu(v_1 \otimes \cdots \oplus v_n, v_{n+1} \otimes \cdots \otimes v_{n+m}) = v_1 \otimes \cdots \otimes v_{n+m}. \tag{13.11}$$

Since this is simply a concatenation of pure multitensors, the operation is associative.

Let $\iota : V = T^1(V) \to T^*(V)$, be the (injective) containment mapping. We shall show that the pair $(\iota, T^*(V))$ satisfies the universal mapping property of our previous definition of a tensor algebra $T(V)$.

Theorem 13.9.3 *Fix an F-vector space V. Define the category whose objects are the pairs (A, f), where A is any F-algebra, and $f : V \to A$ is any F-linear transformation. A morphism from (A, f) to (A', f') in this category is an F-algebra homomorphism $\phi : A \to A'$ such that $\phi \circ f = f'$. Then $(\iota, T^*(V))$ (the embedding of V into it's tensor algebra) is an initial object in this category.*

Proof Let (A, f) be an object in this category. Note that if $\phi : T(V) \to A$ is an F-algebra homomorphism such that $\phi \circ \iota = f$, then for all nonnegative integers r, we must have $\phi(v_1 \otimes v_2 \otimes \cdots \otimes v_r) = f(v_1)f(v_2)\cdots f(v_r)$. Since any element of $T(V)$ is a sum of such simple tensors, we must have that $\phi : T(V) \to A$ must be unique. It remains to prove the existence of such a mapping ϕ. First define, $\phi_0 : T^0(V) = Fe \to F1_A$, to be the F-linear mapping which takes the identity element e of $T^*(V)$ to 1_A, the identity element of the algebra A. Now, by Theorem 13.5.2, for each positive integer $r > 0$, the F-balanced mapping

$$f_r : V \times V \times \cdots \times V \to A, \text{ where } f_r(v_1, v_2, \ldots, v_r) = f(v_1)f(v_2)\cdots f(v_r).$$

induces a unique F-linear mapping

$$\phi_r : T^r(V) := V \otimes_F V \otimes_F \otimes_F \cdots \otimes_F V \to A,$$

where $\phi_r(v_1 \otimes v_2 \otimes \cdots \otimes v_r) = f(v_1)f(v_2) \cdots f(v_r)$. In turn, by the universality of the direct sum $T^*(V) = \bigoplus_{r=0}^{\infty} T^r(V)$, we then get a unique F-linear mapping

$$\phi : T^*(V) = \bigoplus_{r=0}^{\infty} T^r(V) \to A$$

which is an F-algebra homomorphism. This proves the existence of ϕ, and the result follows. $\square$

At this point, we know from the uniqueness of initial objects that the three objects $T(V)$, $F\{X\}$ (where X is a basis of V) and $T^*(V)$ are all three the same object.[7] In fact, the isomorphism $T^*(V) \to F\{X\}$ is induced by the following mapping connecting their basis elements:

$$e \mapsto 1, \text{ the monic polynomial of degree zero}$$
$$x_1 \otimes \cdots \otimes x_n \mapsto x_1 x_2 \cdots x_n, \text{ a monomial of degree n}$$

(Here n is any positive integer and the x_i lie in X, a basis for V.)

From now on we will write $T(V)$ for $T^*(V)$.

13.10 The Symmetric and Exterior Algebras

13.10.1 Definitions of the Algebras

Inside the tensor algebra $T(V)$ over the field F, we define the following homogeneous ideals (cf. Theorem 13.8.2):

 (i) $I \subseteq T(V)$, generated by all elements of the form $v \otimes w - w \otimes v$, $v, w \in V$;
(ii) $J \subseteq T(V)$, generated by all simple tensors of the form $v \otimes v$, $v \in V$.

In terms of the above ideals, we now define the *symmetric algebra* over V to be $S(V) := T(V)/I$. Likewise, we define the *exterior algebra* by setting $E(V) := T(V)/J$. By Theorem 13.8.2, these are both graded algebras. Thus,

$$S(V) = \bigoplus_{r=0}^{\infty} S^r(V), \quad \text{where each } S^r(V) = (T^r(V) + I)/I,$$

---

[7]Up to isomorphism, of course.

and

$$E(V) = \bigoplus_{r=0}^{\infty} E^r(V), \quad \text{where each } E^r(V) = (T^r(V) + J)/J.$$

(Many authors write $\bigwedge V$ for $E(V)$ and then write $\bigwedge^r V$ for $E^r(V)$.)

We denote by $\iota_S : V \to S(V)$ and by $\iota_E : V \to E(V)$ the F-linear mappings

$$\iota_S(v) := v + I \in S(V), \quad \iota_E(v) := v + J \in E(V), \quad v \in V.$$

Since it is clear that $V \cap I = V \cap J = 0$, we see that ι_S and ι_E are both injective F-linear transformations. We shall often identify V with their respective image subspaces $\iota_S(V) \subseteq S(V)$ and $\iota_E(V) \subseteq E(V)$ and write $V \subseteq S(V)$ and $V \subseteq E(V)$.

Multiplication in $S(V)$ is typically written simply as juxtaposition: if $a, b \in S(V)$, then the product of a and b is denoted $ab \in S(V)$. We note that $S(V)$ is a commutative algebra. Indeed, any element of $S(V)$ is an F-linear combination of elements of the form $v_1 v_2 \cdots v_r$ (that is, the product $v_1 \otimes v_2 \otimes \cdots \otimes v_r + I$), where $v_1, v_2, \ldots, v_r \in V$. Now write

$$a := v_1 v_2 \cdots v_r, \quad b := w_1 w_2 \cdots w_s,$$

where $v_1, \ldots, v_r, w_1, \ldots, w_s \in V$. Since for any vectors $v, w \in V$ we have $vw = wv$, it follows immediately that $ab = ba$ and so it follows easily that $S(V)$ is commutative, as claimed.

In the exterior algebra, products are written as "wedge products": if $c, d \in E(V)$, their product is denoted $c \wedge d$. It follows immediately that for all vectors $v, w \in V$, one has $v \wedge v = 0$, and from $(v + w) \wedge (v + w) = 0$ one obtains $v \wedge w = -w \wedge v$.

Like the tensor algebra, the universal properties satisfied by the symmetric and exterior algebras involve an F-linear mapping $V \to A$ into an F-algebra A. Such an F-linear mapping is said to be *commuting* if and only if $f(u)f(v) = f(v)f(u)$ for all $u, v \in V$—that is, the image $f(V)$ generates a commutative subring. Similarly such a mapping is said to be *alternating* if and only if $f(v)^2 = 0$, for all $v \in V$.

Theorem 13.10.1 *Let F be a field, let V be a vector space over F, and let $S(V)$, $E(V)$ be the symmetric and exterior algebras over V, respectively.*

(i) *Consider the category of pairs (A, f) of F-algebras and commuting F-linear mappings $f : V \to A$. A morphism from the pair (A, f) to the pair (A', f') is an F-algebra homomorphism $\phi : A \to A'$ such that $\phi \circ f = f'$. Then $(S(V), \iota_S)$ is an initial object in this category.*

(ii) *Consider the category of pairs (A, f) of F-algebras and alternating F-linear mappings $f : V \to A$. A morphism from the pair (A, f) to the pair (A', f') is an F-algebra homomorphism $\phi : A \to A'$ such that $\phi \circ f = f'$. Then $(E(V), \iota_E)$ is an initial object in this category.*

We shall leave the proof to the motivated reader in Exercise (2) in Sect. 13.13.7.

The uniqueness of these algebras follows. Whenever one hears the word "uniqueness" in a categorical setting, one knows that isomorphisms are being implied, and so

a number of corollaries can be expected to result from the preceding Theorem 13.10.1.
We produce some of them in the next subsection.

13.10.2 Applications of Theorem 13.10.1

A Model for $S(V)$

We begin with this example: From the uniqueness assertion in Theorem 13.10.1,
one can easily exploit the evaluation morphism of polynomial rings (in commuting
indeterminates) to prove the following:

Corollary 13.10.2 *Let X be an F-basis for the vector space V. Then $S(V) \cong F[X]$
as F-algebras.*

The proof is simply that the universal property of the evaluation mapping E_α
for polynomial rings in commuting indeterminates proves that the pair $(\iota_X, F[X])$
satisfies the universal mapping property of the tensor algebra. (See Sect. 7.3 for
background and Exercise (8) in Sect. 7.5.3 for the universal property. A formal proof
is requested in Exercise (3) in Sect. 13.13.7.)

A Model for $E(V)$

Let $(X, \leq)$ denote a totally ordered poset bijective with a basis X of the vector space
V. Let C be the collection of all finite ascending subchains (including the empty
chain, ϕ) of $(X, \leq)$. (Clearly this collection is in one-to-one-correspondence with
the finite subsets of X. It forms a boolean lattice $(C, \leq)$ under the refinement relation
among chains. Thus the "join" $c_1 \vee c_2$, is the smallest chain refining both c_1 and c_2.)
For every non-empty ascending chain $c = \{x_1 < \cdots < x_r\} \in C$, let w_c be the
word $x_1 x_2 \cdots x_r$ (viewed either as a word in the free monoid over X or simply as a
sequence of elements of X). For the empty chain ϕ we denote w_ϕ by the symbol e.
(For convenience, we shall refer to these w_c as "chain-words".)

Let A be the F-vector space defined as the formal direct sum

$$E = \bigoplus_{c \in C} F w_c.$$

We convert A into an F-algebra by defining multiplication between basis elements
and extending this to a definition of multiplication ("$*$") on A by the usual distributive
laws. In order to accomplish this we consider two chains $a = \{x_1 < \cdots < x_r\}$ and
$b = \{y_1 < \cdots < y_s\}$. We let $m(a, b)$ be the number of pairs (x_k, y_l) such that $x_k > y_l$.
Then if the two chains a and b possess a common element, we set $a*b := 0$. But if the
chains a and b possess no common element, their disjoint union can be reassembled

into a properly ascending chain $c = a \vee b$, the smallest common refinement of the chains a and b. In that case we set

$$a * b := (-1)^{m(a,b)} w_c.$$

(Note that $(-1)^{m(a,b)}$ is simply the sign of the permutation $\pi(a, b)$ of indices that permutes the concatenated sequence $w_a \cdot w_b$ to $w_{a \vee b}$.)

Two make sure that the student gets the sense of this, note that when $(X, \leq) = \{1 < 2 < \cdots < 9\}$, we have

$$(1 < 2 < 4 < 7) * (3 < 6 < 7) = 0 \text{ while}$$
$$(1 < 2 < 4 < 7) * (3 < 6 < 8) = -(1 < 2 < 3 < 4 < 6 < 7 < 8).$$

To show that "$*$" is an associative binary operation for A it is enough to check it on the basis elements—the chains. Thus we wish to compare two products $(a * b) * c$ and $a * (b * c)$ where w_a, w_b, w_c are the respective ascending sequences $(a_1, \ldots, a_r)$, $(b_1 \ldots, b_s)$, $(c_1, \ldots, c_t)$. We may assume r, s, t are all positive integers and that a, b and c pairwise share no common element (otherwise $(a * b) * c = 0 = a * (b * c)$). Form the concatenation of sequences

$$w_a \cdot w_b \cdot w_c = (a_1, \ldots, a_r, b_1, \ldots, b_s, c_1, \ldots, c_t) = (y_1, \ldots, y_{r+s+t}) := y,$$

and let π be the permutation of indices so that $\{y_{\pi(i)}\}_{i=1}^{r+s+t}$ is in ascending order, thus realizing the chain $a \vee b \vee c$. (We view the group of permutations of sequences as acting on sequences from the right, so that a factorization of permutations $\pi_1 \pi_2$ means π_1 is applied first.) Now the permutation π has two factorizations: (i) $\pi = \pi(a, b) \cdot \pi(a \vee b, c)$ and (ii) $\pi = \pi(b, c) \pi(a, b \vee c)$. Here $\pi(a, b)$ fixes the last t indices to effect $w_a \cdot w_b \cdot w_c \to w_{a \vee b} \cdot w_c$, while $\pi(a \vee b, c)$ rearranges the entries of the latter sequence to produce $w_{a \vee b \vee c}$. So (i) holds. Similarly, we could also apply $\pi(b, c)$ to the last $s + t$ entries in $y = w_a \cdot w_b \cdot w_c$ while fixing the first r entries to get $w_a * w_{b \vee c}$ and then apply $\pi(a, b \vee c)$ to obtain the ascending sequence $w_{a \vee b \vee c}$. Thus the factorization (ii) holds.

Now $(a * b) * c = a * (b * c) = sgn(\pi)(a \vee b \vee c)$ follows from the associativity of the "join" operation in $(C, \leq)$, the definition of "$*$", and

$$sgn(\pi) = sgn(\pi(a, b)) sgn(\pi(a \vee b, c)) = sgn(\pi(b, c)) sgn(\pi(a, b \vee c)),$$

given to us by the factorizations (i) and (ii).

Now, extending the binary operation $*$ to all of A by linearity from its definition on the basis C, we obtain an F-vector space A with an associative binary operation "$*$" which distributes with respect to addition and respects scalar multiplication from F on either side. One easily checks that $e * a = a * e = a$ for each $a \in A$, so e is a multiplicative identity. Thus $(A, *)$ is an algebra.

Notice that the lengths of the chains produce a grading $A = \bigoplus A_d$ where, for every natural number d, A_d is the subspace spanned by all w_c as c ranges over the chains of length d. Thus $A_0 = Fe$ and $A_r * A_s \subseteq A_{r+s}$.

At this stage we see that $(A, *)$ is a genuine graded F-algebra which we call the *algebra of ascending chains*. Its dimension is clearly the number of finite subchains of $(X, \leq)$, and if V has finite dimension n then A has dimension 2^n.

There is a natural injection $\iota_A : V \to A$ mapping linear combinations of basis elements x_i of V to the corresponding linear combination of chain-words $w_i := w_{\{x_i\}}$ of length one.

If $x_\sigma < x_\tau$ is a chain, and $w_\sigma = \iota_A(x_\sigma)$ and $w_\tau = \iota_A(x_\tau)$ are the length-one chain words $\{x_\sigma\}$ and $\{x_\tau\}$, we have

$$w_\sigma^2 = w_\tau^2 = 0 \text{ (writing } a^2 \text{ for } a * a \text{) and}$$

$$w_\sigma * x_\tau = -w_\tau * x_\sigma.$$

This gives us $w_v^2 = 0$ for every vector $v \in V$, and so $\iota_A : V \to A$ is an alternating mapping as defined above. Thus the universal property of $E(V)$ requires that there is a unique algebra epimorphism $\epsilon : E(V) \to A$ extending ι_A. If V had finite dimension, we would know at once that ϵ is an isomorphism. Instead we obtain the isomorphism by showing that (A, ι_A) possesses the desired universal mapping property.

Suppose $f : V \to (B, \circ)$ is an alternating mapping into an F-algebra B having identity element 1_B. Then of course

$$f(v)^2 = 0, \text{ and} \tag{13.12}$$
$$f(u) \circ f(v) = -f(v) \circ f(u) \tag{13.13}$$

for all vectors $u, v \in V$.

We then define an F-linear mapping $\alpha_f : A \to B$ by describing its values on a basis for A. First we set $\alpha_f(e) = 1_B$. For each finite ascending chain $c = (x_1, \ldots, x_n)$ in $(X, \leq)$ we set

$$\alpha_f(w_c) = f(x_1) \cdots f(x_n).$$

This can be extended to linear combinations of the x_i to yield

$$\alpha_f(\iota_A(v)) = f(v), \text{ for all } v \in V.$$

We also note that for any two finite ascending chains a and b in $(X, \leq)$, we have

$$\alpha_f(w_a * w_b) = f(w_a) \circ f(w_b).$$

So once again extending by linearity, we see that

$$\alpha_f : A \to B$$

is an algebra homomorphism and that $\alpha_f \circ \iota_A = f$. Thus A possesses the universal property which defines the exterior algebra. By the uniqueness asserted in Theorem 13.10.1, we have a companion to Corollary 13.10.2:

Corollary 13.10.3 *Suppose V is any F-vector space. Then the exterior algebra $E(V)$ is isomorphic to the algebra $(A, *)$ of finite ascending chains of a total ordering of a basis X of V.*

In particular, if V is an F-vector space of finite dimension n, then the exterior algebra $E(V)$ is isomorphic to the algebra of ascending chains, $(A, *)$ based on the chain $N = \{1 < \cdots < n\}$. It is a graded algebra of dimension 2^n.

Morphisms Induced on $S(V)$ and $E(V)$

Suppose $t : V \to V$ is any F-linear transformation of V. Then, $\iota_S \circ t$ is a commuting mapping $V \to S(V)$ and so, by the universal mapping property of $S(V)$ (placing $(\iota_S \circ t, S(V))$ in the role of (t, A)), we obtain an algebra homomorphism

$$S(t) : S(V) \to S(V)$$

which forces the commutativity of the following diagram:

$$
\begin{array}{ccc}
S(V) & \xrightarrow{\ S(t)\ } & S(V) \\
{\scriptstyle \iota_S}\big\uparrow & & \big\uparrow{\scriptstyle \iota_S} \\
V & \xrightarrow{\ \ t\ \ } & V
\end{array}
$$

The algebra homomorphism $S(t) : S(V) \to S(V)$ preserves the grading and so induces morphisms

$$S^r(t) : S^r(V) \to S^r(V)$$

taking any product of elements of V, $v_1 v_2 \cdots v_n$, to $v_1^t \cdots v_n^t$ where v^t denotes the image of the vector $v \in V$ under the linear transformation t.

It is apparent that if $s, t \in \hom_F(V, V)$, then

$$S^r(s \circ t) = S^r(s) \circ S^r(t).$$

The same journey can be emulated for the exterior algebra $E(V)$ using its particular universal mapping property. Given F-linear $t : V \to V$ one obtains the *alternating mapping* $\iota_E \circ t \to E(V)$ and so by the universal mapping property, a commutative diagram

$$E(V) \xrightarrow{\; E(t) \;} E(V)$$

$$\iota_E \uparrow \qquad\qquad \uparrow \iota_E$$

$$V \xrightarrow{\quad t \quad} V$$

where the algebra homomorphism $E(t)$ at the top of the diagram preserves degrees and so induces mappings

$$E^r(t) : S^r(V) \to E^r(V)$$

taking a product of vectors in V, $v_1 \wedge v_2 \wedge \cdots \wedge v_n$ to $v_1^t \wedge \cdots \wedge v_n^t$ where v^t denotes the image of an arbitrary vector v under the linear transformation t. Again

$$E^r(s \circ t) = E^r(s) \circ E^r(t).$$

We collect these observations in this way:

Corollary 13.10.4 *Suppose $t \in \hom_F(V, V)$. Let*

$$\iota_S : V \to S(V), \text{ and}$$
$$\iota_E : V \to E(V)$$

be the canonical embeddings of V into the symmetric and exterior algebras, respectively.

Then there exist F-algebra homomorphisms

$$S(t) : \; S(V) \to S(V) \text{ and}$$
$$E(t) : \; E(V) \to E(V)$$

such that one has

$$S(t) \circ \iota_S = \iota_S \circ t, \text{ and}$$
$$E(t) \circ \iota_E = \iota_E \circ t,$$

respectively. Both $S(t)$ and $E(t)$ preserve the algebra grading and so induce mappings

$$S^r(t) : \; S^r(V) \to S^r(V) \text{ and}$$
$$E^r(t) : \; E^r(V) \to E^r(V).$$

For any r-tuple $(v_1, \ldots, v_r) \in V^{(r)}$ one has

$$S^r(t)(v_1 \cdots v_n) = v_1^t v_2^t \cdots v_r^t, \text{ and}$$
$$E^r(t)((v_1 \wedge \cdots \wedge v_r) = a_1^t \wedge \cdots \wedge v_r^t.$$

Finally if the symbol Φ denotes either $T, T^r, S, S^r, E,$ or E^r, one has

$$\Phi(s \circ t) = \Phi(s) \circ \Phi(t). \tag{13.14}$$

Remark Of course by now even the beginning student has noticed that T, T^r, S, S^r, E, and E^r—all of them—are functors $\mathbf{V} \to \mathbf{V}$ where $\mathbf{V}$ is an appropriate sub-category of **Vect**, the category of all F-vector spaces.

There are other consequences of Corollary 13.10.4. For example, if $t : V \to V$ is an invertible linear transformation, then so are $S(t)$, $S^r(t)$, $E(t)$ and $E^r(t)$. Thus

Corollary 13.10.5 *If $t : V \to V$ is invertible, then $S(t)$ and $E(t)$ are automorphisms of the F-algebras $S(V)$ and $E(V)$, respectively.*

In particular, there are injective homomorphisms

$$\rho_S : GL(V) \to Aut(S(V))$$
$$\rho_E : GL(V) \to Aut(E(V))$$

whose images elements preserve the grading.

As a result, for any subgroup G of $GL(V)$ one obtains group representations

$$\rho_S^r : G \to GL(S^r(V))$$
$$\rho_E^r : G \to GL(E^r(V))$$

Remark All of the results above were consequences of just one result: Theorem 13.10.1. There is a lesson here for the student.[8] It is that universal mapping properties—as abstract as they may seem—possess a real punch! Suppose you are in a situation where something is defined in the heavenly sunlight of a universal mapping property. Of course that does not mean it exists. But it does mean, that any two constructions performed in the shade which satisfy this universal property (whether known or new) are isomorphic (a consequence of the "nonsense" about initial and terminal objects in a category). So we prove existence. What more is there to do, one might ask? The answer is that one does not so easily depart from a gold mine they are standing on. Here are three ways the universal characterizations produce results: (1) It is a technique by which one can prove that two constructed algebraic objects are isomorphic—simply show that they satisfy the same universal mapping property within the same category. (2) One can prove that an object defined in one context has a specific property by recasting it in the context of another "solution" of the same universal mapping property. (3) One can prove the existence of derived endomorphisms and automorphisms of the universal object.

---

[8]Of course such a phrase should be a signal that the teacher is about to become engulfed by the urge to present a sermon and that perhaps the listener should surreptitiously head for an exit! But how else can a teacher portray a horizon beyond a mere list of specific theorems?.

13.11 Basic Multilinear Algebra

In Sect. 13.5 we proved that R-multibalanced mappings from a finite direct sum of symmetric (R, R)-bimodules to a target (R, R)-bimodule factored through the multiple tensor product (see Theorem 13.5.2). In this brief subsection we wish to extend this theorem to alternating and symmetric forms (the key theorems of multilinear algebra), but we shall do this in the case that R is a field so that we can exploit universal properties of the symmetric and exterior algebras developed in the previous section. As the student shall observe, the device for transferring statements about algebras to statements about multilinear mappings depends critically upon our discussion of graded algebras and homogeneous ideals.

First the context: Let V and W be vector spaces over a field F. We regard both V and W as symmetric (F, F)-bimodules—that is $\alpha v = v\alpha$ for all $(\alpha, v) \in F \times V$ or $F \times W$.

A mapping

$$f : V \times \cdots \times V (n \text{ factors}) \to W$$

is F-*multilinear* if and only if (1) it is an F-linear mapping of the n-fold direct product (and so preserves addition in the direct product) and (2) that for any n-tuple of vectors of V, $(v_1, \ldots, v_n)$, for any field element $\alpha \in F$, and for any index $j < n$, one has

$$f(v_1, \ldots, v_j\alpha, v_{j+1}, \ldots, v_n) = f(v_1, \ldots, v_j, \alpha v_{j+1}, \ldots, v_n) \quad (13.15)$$
$$= (f(v_1, \ldots, v_n))\alpha. \quad (13.16)$$

Such an F-multilinear mapping is said to be   *alternating* if and only if $f(v_1, \ldots, v_n) = 0$ if at least two distinct entries among the v_i are equal. By considering a sequence whose ith and $(i + 1)$st entries are both $v_i + v_{i+1}$, one sees that for an alternating F-multilinear mapping f, the sign of $f(v_1, \ldots, v_n)$ is changed by the transposition of entries in the ith and $(i + 1)$st positions. It follows that the sign is in fact changed by any odd permutation of the entries, but the value of f is preserved by any even permutation of the entries.

An F-multilinear mapping $f : V \times \cdots \times V (n \text{ factors}) \to W$ is said to be *symmetric* if $f(v_1, \ldots, , \ldots, v_n)$ is unchanged under the transposition (v_i, v_{i+1}) interchanging the two entries in the ith and $i + 1$st positions. It then follows that the value of f is unchanged by any permutation of its arguments.

Theorem 13.11.1 (Fundamental Theorem of Multilinear Mappings) *Let V be a vector space over a field F, viewed as an (F, F)-bimodule so that $\alpha v = v\alpha$ for all $(\alpha, v) \in F \times V$. Throughout, we assume that f is an F-multilinear mapping*

$$f : V \times \cdots \times V (n \text{ factors}) \to W$$

into another (F, F)-bimodule W. The following three statements hold:

1. $f = \bar{f} \circ \iota$ where $\bar{f}$ is a homomorphism

$$\bar{f} : T^n(V) = V \otimes \ldots \otimes V \to W,$$

uniquely determined by f. Specifically, the mapping f "factors through $\bar{f}$" so that

$$f(v_1, \ldots, \ldots, v_n) = \bar{f}(v_1 \otimes \cdots \otimes v_n).,$$

for all n-tuples of vectors $v_i \in V$.

2. Now suppose the mapping f is alternating. Then the mapping f factors through a unique mapping

$$f_E : E^n(V) = V \bigwedge \cdots \bigwedge V (n \text{ factors}) \to W$$

so that

$$f(v_1, \ldots, \ldots, v_n) = f_E(v_1 \wedge v_2 \wedge \cdots \wedge v_n),$$

for all n-tuples of vectors $v_i \in V$.

3. Finally suppose the multilinear mapping f is symmetric. Then f factors through a unique $f_S : S^n(V) \to W$ where $S^n(V)$ is the space of all homogeneous elements of degree n in the graded symmetric algebra $S(V)$. Thus $f(v_1, \ldots, v_n) = f_S(v_1 \cdots v_n)$.

Proof Part 1 is just Theorem 13.5.2 with the commutative ring R replaced by a field.

Part 2. By definition $E(V) = T(V)/J$ where J is the homogeneous ideal given at the beginning of Sect. 13.10. We have these direct decompositions:

$$T(V) = F \oplus V \oplus T^2(V) \oplus \cdots \oplus T^r(V) \oplus \cdots \qquad (13.17)$$

$$J(V) = F \cap J \oplus (V \cap J) \oplus \cdots \oplus (T^r(V) \cap J) \oplus \cdots \qquad (13.18)$$

$$E(V) = F \oplus V \oplus \bigoplus_{r=2}^{\infty} (T^r(V)/(T^r(V) \cap J)) \qquad (13.19)$$

$$= F \oplus V \oplus V \wedge V \oplus \cdots \bigwedge^r (V) \oplus \cdots \qquad (13.20)$$

The first two summands in Eq. (13.18) are of course zero. Comparing degree components in the last two equations yields

$$\bigwedge^n (V) \cong T^n(V)/(T^n(V) \cap J). \qquad (13.21)$$

Now, since f is multilinear, it factors through $\bar{f} : T^n(V) \to W$ as in part 1. But since f is alternating, we see that ker $\bar{f}$ contains $(T^n(V) \cap J)$. Thus it follows that

$\bar{f}$ factors through this chain:

$$T^n(V) \to T^n(V)/(T^n(V) \cap J) \to [T^n(V)/(T^n(V) \cap J)]/(\ker \bar{f}/(T^n(V) \cap J))$$
$$\cong T^n(V)/\ker \bar{f} \cong \bar{f}(T^n(V) \hookrightarrow W.$$

(The first two morphisms are natural projections, the next two isomorphisms are from the classical homomorphisms theorems, and the last mapping is the inclusion relation into W.) Folding in the isomorphism of Eq. (13.21) at the second term and composing with the remaining mappings we obtain the desired morphism

$$f_E^n : \bigwedge^n (V) \to W$$

through which $\bar{f}$ (and hence f) factors. That

$$f(v_1, \ldots, v_n) = \bar{f}(v_1 \otimes \cdots \otimes v_n) = f_E^n(v_1 \wedge \cdots \wedge v_n)$$

follows upon applying the mappings.

Part 3 follows the same pattern except that we use the ideal I, given at the beginning of Sect. 13.10, in place of J. Again one exploits the fact that I is a homogeneous ideal to deduce that $S^n(V) \cong T^n(V)/(T^n(V) \cap I)$. We leave the student to fill in the details in Exercise (1) in Sect. 13.13.8. $\square$

13.12 Last Words

From here the subject of multilinear algebra begins to blend with another subject called Geometric Algebra. Although space does not allow us to pursue the latter subject, a hint of its flavor may be suggested by the following.

Suppose V is both a left vector space, as well as a right vector space over a division ring D. Do *not* assume that V is an (D, D)-bimodule. (This would virtually make D a field.) A 2-*form* is a bi-additive mapping $f : V \times V \to D$ such that

$$f(\alpha u, v\beta) = \alpha f(u, v)\beta \text{ for all } \alpha, \beta \in D, u, v, \in V.$$

If $g = \alpha f$ for a fixed scalar α in D, then g is said to be *proportional* to f. Such a 2-form f is said to be *reflexive* if and only if $f(u, v) = 0$ implies $f(v, u) = 0$, and is *non-degenerate* if and only if $f(v, V) = 0$, implies $v = 0, u, v \in V$.

Theorem 13.12.1 *Suppose* dim $V \geq 2$ *and* $f : V \times V \to F$ *is a reflexive non-degenerate 2-form, where* F *is a field. Then* f *is proportional to a 2-form* g *such that for all* $u, v \in V$ *exactly one of the following holds:*

    (i) $g(u, v) = g(v, u)$, *(symmetric form)*

(ii) $g(u, u) = 0$, (alternating of symplectic form)

(iii) $g(u, v) = g(v, u)^\sigma$, *where σ is a field automorphism of order* 2 (Hermitian form).

The first case can be derived from quadratic forms (see Exercise (9) in Sect. 13.13.7). All three cases are the children of a description of 2-forms over division rings—the so-called (σ, ϵ)-*Hermitian forms* (see [2], pp. 199–200 for versions of J. Tits' proof that reflexive forms must take this shape).

There are many beautiful results intertwining 2-forms and multilinear forms as well as widespread geometric applications of such forms. The interested student is referred to the classic book *Geometric Algebra* by E. Artin [1], which, after more than fifty years, now begs for an update.

13.13 Exercises

13.13.1 Exercises for Sect. 13.2

1. Let $\phi : M \to N$ be a homomorphism of right R-modules. Recall that the *cokernel* of ϕ was defined by a universal mapping property on p. 477, in any category for which an initial object exists and is also a terminal object. (i) In the category of right R-modules, show that the projection $\pi : N \to N/\mathrm{im}\,\phi$ satisfies this mapping property. (ii) In the category of groups and their homomorphisms, do cokernels of homomorphisms exist? If so define it. [Hint: Examine the universal mapping property of a cokernel in this category.]

2. Given the morphism $M \overset{\phi}{\to} N$ in the category of right R-modules, define a category within which the cokernel of ϕ appears as an initial object.

3. Let $\{M_\alpha \mid \alpha \in \mathcal{A}\}$ be a family of right R-modules. Recall from p. 260 that if $P := \prod_{\alpha \in \mathcal{A}} M_\alpha$, then P satisfies the following universal mapping property: First of all, there are projection homomorphisms $\pi_\alpha : P \to M_\alpha$, $\alpha \in \mathcal{A}$; if P' is another right R-module with homomorphisms $\pi'_\alpha : P' \to M_\alpha$, $\alpha \in \mathcal{A}$, then there exists a unique R-module homomorphism $\theta : P' \to P$, making the following diagram commute for every $\alpha \in \mathcal{A}$.

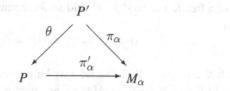

Now define a category within which the direct product P appears as a terminal object.

4. Let X be a fixed set and define the following category. The objects are the pairs (G, μ), where G is a group and $\mu : X \to G$ is a mapping of the set X into G. A morphism from the (G, μ) to the pair (G', μ') is a group homomorphism $\phi : G \to G'$ such that $\mu' = \phi \circ \mu$. Show how the free group $F(X)$ on the set X affords an initial object in this category.

5. Rephrase and repeat the above exercise with "groups" replaced with "right R-modules". [Hint: The free module with basis X replaces the free group $F(X)$.]

6. Suppose $F : \mathcal{C}_1 \to \mathcal{C}_2$ is a contravariant functor. Show that there are also covariant functors $F' : \mathcal{C}_1^{opp} \to \mathcal{C}_2$ and $F'' : \mathcal{C}_1 \to \mathcal{C}_2^{opp}$.

7. Show that there exists an isomorphism $\mathcal{C} \to (\mathcal{C}^{opp})^{opp}$ which induces the identity mapping on objects. (It may move maps, however.)

13.13.2 Exercises for Sect. 13.3

1. Let A be an abelian group. If n is a positive integer, prove that

$$\mathbb{Z}/n\mathbb{Z} \otimes_{\mathbb{Z}} A \cong A/nA.$$

2. Let m, n be positive integers and let $d = \text{g.c.d}(m, n)$. Prove that

$$\mathbb{Z}/m\mathbb{Z} \otimes_{\mathbb{Z}} \mathbb{Z}/n\mathbb{Z} \cong \mathbb{Z}/d\mathbb{Z}.$$

3. Let

$$0 \to M' \overset{\mu}{\to} M \overset{\epsilon}{\to} M'' \to 0$$

be a split exact sequence of right R-modules, and let N be a left R-module. Show that the induced sequence

$$0 \to M' \otimes_R N \overset{\mu \otimes 1_N}{\longrightarrow} M \otimes_R N \overset{\epsilon \otimes 1_N}{\longrightarrow} M'' \otimes_R N \to 0$$

is also split exact. Prove the corresponding statement if the split exact sequence occurs as the right hand factor of the tensor product sequence.

4. Prove that if P is a projective left R-module, then P is flat. [Hint: Let $\mu : M' \to M$ be an injective homomorphism of right R-modules. We know that P is the direct summand of a free left R-module F, and so there must be a split exact sequence of the form

$$0 \to P \overset{i}{\to} F \overset{j}{\to} N \to 0,$$

for some left R-module N. By Exercise 3 we know that both $1_{M'} \otimes i : M' \otimes_R P \to M' \otimes_R F$ and $1_M \otimes i : M \otimes_R P \to M \otimes_R F$ are injective. We have the commutative square

$$M' \otimes_R P \xrightarrow{\mu \otimes 1_P} M \otimes_R P$$

$$1_{M'} \otimes l \downarrow \qquad\qquad \downarrow i_M \otimes l$$

$$M' \otimes_R F \xrightarrow{\mu \otimes 1_F} M \otimes_R F$$

Now what?]

5. Generalize the proof of Theorem 13.3.8 so as to prove the following. Let R be a ring, let M, M' be right R-modules, and let N, N' be left R-modules. Let $\phi : M \to M'$ and let $\psi : N \to N'$ be module homomorphisms. Define $K \subseteq M \otimes_R N$ to be the subgroup generated by simple tensors $m \otimes n$ where either $m \in \ker \phi$ or $n \in \ker \psi$. Show that, in fact, $K = \ker (\phi \otimes \psi : M \otimes_R N \to M' \otimes_R N')$.

13.13.3 Exercises for Sect. 13.3.4

1. Let $\mathcal{C}$ be a category and let $\mu : A \to B$ be a morphism. We say that μ is a *monomorphism* if whenever A' is an object with morphisms $f : A' \to A$, $g : A' \to A$ such that $\mu \circ f = \mu \circ g : A' \to B$, then $f = g : A' \to A$. In other words, monomorphisms are those morphisms that have "left inverses." Similarly, epimorphisms are those morphisms that have right inverses. Now assume that $\mathcal{C}, \mathcal{D}$ are categories, and that $F : \mathcal{C} \to \mathcal{D}$, $G : \mathcal{D} \to \mathcal{C}$ are functors, with F *left adjoint* to G. Prove that F preserves epimorphisms and that G preserves monomorphisms.

2. Let $i : \mathbb{Z} \hookrightarrow \mathbb{Q}$ be the inclusion homomorphism. Prove that in the category of rings, i is an epimorphism. Thus an epimorphism need not be surjective.

3. Let V, W be $\mathbb{F}$-vector spaces and let V^* be the $\mathbb{F}$-dual of V. Prove that there is a vector space isomorphism $V^* \otimes_{\mathbb{F}} W \cong \mathrm{Hom}_{\mathbb{F}}(V, W)$.

4. Let G be a group. Exactly as in Sect. 7.3.2, we may define the *integral group ring* $\mathbb{Z}G$. (These are formal $\mathbb{Z}$-linear combinations of group elements in G.) In this way a group gives rise to a ring. Correspondingly, given a ring R we may form its *group of units* $U(R)$. This time, a ring is giving rise to a group. Show that these correspondences define functors

$$Z : \textbf{Groups} \longrightarrow \textbf{Rings}, \quad U : \textbf{Rings} \longrightarrow \textbf{Groups}.$$

Prove that $\mathbb{Z}$ is left adjoint to U.

5. Below are some further examples of adjoint functors. In each case you are to prove that F is left adjoint to G.

   (a)

   $$\textbf{Groups} \xrightarrow{F} \textbf{Abelian Groups} \xrightarrow{G} \textbf{Groups};$$

   F is the commutator quotient map.

(b)
$$\textbf{Sets} \xrightarrow{F} \textbf{Groups} \xrightarrow{G} \textbf{Sets},$$

where $F(X) =$ free group on X and $G(H)$ is the underlying set of the group H.

(c)
$$\textbf{Integral Domains} \xrightarrow{F} \textbf{Fields} \overset{G}{\hookrightarrow} \textbf{Integral Domains};$$

$F(D)$ is the field of fractions of D. (Note: for this example we consider the morphisms of the category **Integral Domains** to be restricted only to injective homomorphisms.)

(d) Fix a field K.

$$K - \textbf{Vector Spaces} \xrightarrow{F} K - \textbf{Algebras} \xrightarrow{G} K - \textbf{Vector Spaces};$$

$F(V) = T(V)$, the tensor algebra of V and $G(A)$ is simply the underlying vector space structure of algebra A.

(e)
$$\textbf{Abelian Groups} \xrightarrow{F} \textbf{Torsion Free Abelian Groups} \overset{G}{\hookrightarrow} \textbf{Abelian Groups};$$

$F(A) = A/T(A)$, where $T(A)$ is the torsion subgroup of A.

(f)
$$\textbf{Left } R\textbf{-modules} \xrightarrow{F} \textbf{Abelian Groups} \xrightarrow{G} \textbf{Left } R\textbf{-modules}.$$

F is the forgetful functor, $G = \text{Hom}_{\mathbb{Z}}(R, -)$.

(g) If G is a group, denote by G **Set** the category of all sets X acted on by G. If X_1, X_2 are objects in G **Set** then the morphisms from X_1 to X_2 are the mappings in $\text{Hom}_G(X_1, X_2)$. Let H be a subgroup of G. Then an adjoint functor pair is given by

$$G \textbf{ Set} \overset{F}{\to} H \textbf{ Set} \overset{I^G}{\to} G \textbf{ Set}$$

where, for G-set X, and H-set Y, $F(X) = \text{Res}_H^G(X)$, $I^G(Y) = \text{Ind}_H^G(Y)$. [The induced action of G on $Y \times G/H$, denoted by the symbol $\text{Ind}_H^G(Y)$, is defined in Exercise 7 of the following Sect. 13.13.4 on p. 519.]

13.13.4 Exercises for Sect. 13.4

1. Give a formal proof of Corollary 13.4.3, which asserts that if $F \subseteq E$ is an extension of fields, and V is an F-vector space with basis $\{x_i\}$, then as an E-vector

space, $V \otimes_F E$ has basis $\{x_i \otimes 1_E\}$. [Hint: Write $V = \sum x_i F$ and use Corollary 13.3.6.]

2. Let $T : V \to V$ be a linear transformation of the finite-dimensional F-vector space V. Let $m_{T,F}(x)$ denote the minimal polynomial of T with respect to the field F. If $F \subseteq E$ is a field extension, prove that $m_{T,F}(x) = m_{T \otimes 1_V, E}(x)$. [Hint: Apply the previous Exercise 1, above.]

3. Let W be an F-vector space and let $T : V_1 \to V_2$ be an injective linear transformation of F-vector spaces. Prove that the sequence $T \otimes 1_W : V_1 \otimes_F W \to V_2 \otimes_F W$ is injective. (Note that by Theorem 13.3.8, we're really just saying that every object in the category of F-vector spaces is flat.)

4. Let F be a field and let $A \in M_n$, $B \in M_m$ be square matrices. Define the *Kronecker* (or *tensor*) product $A \otimes B$ as follows. If $A = [a_{ij}]$, $B = [b_{kl}]$, then $A \otimes B$ is the block matrix $[D_{pq}]$, where each entry D_{pq} (in row p and column q of D) is the $m \times m$ matrix $D_{pq} = a_{pq}B$. Thus, for instance, if

$$A = \begin{bmatrix} a_{11} & a_{12} \\ a_{21} & a_{22} \end{bmatrix}, \quad B = \begin{bmatrix} b_{11} & b_{12} \\ b_{21} & b_{22} \end{bmatrix}.$$

then

$$A \otimes B = \begin{bmatrix} a_{11}b_{11} & a_{11}b_{12} & a_{12}b_{11} & a_{12}b_{12} \\ a_{11}b_{21} & a_{11}b_{22} & a_{12}b_{21} & a_{12}b_{22} \\ a_{21}b_{11} & a_{21}b_{12} & a_{22}b_{11} & a_{22}b_{12} \\ a_{21}b_{21} & a_{21}b_{22} & a_{22}b_{21} & a_{22}b_{22} \end{bmatrix}.$$

Now Let V, W be F-vector spaces with ordered bases $\mathcal{A} = (v_1, v_2, \ldots, v_n)$, $\mathcal{B} = (w_1, w_2, \ldots, w_m)$, respectively. Let $T : V \to V$, $S : W \to W$ be linear transformations with matrix representations $T_\mathcal{A} = A$, $S_\mathcal{B} = B$. Assume that $\mathcal{A} \otimes \mathcal{B}$ is the ordered basis of $V \otimes_\mathbb{F} W$ given by $\mathcal{A} \otimes \mathcal{B} = (v_1 \otimes w_1, v_1 \otimes w_2, \ldots, v_1 \otimes w_m; v_2 \otimes w_1, \ldots, v_2 \otimes w_m; \ldots; v_n \otimes w_m)$. Show that the matrix representation of $T \otimes S$ relative to $\mathcal{A} \otimes \mathcal{B}$ is given by $(T \otimes S)_{\mathcal{A} \otimes T_\mathcal{B}} = T_\mathcal{A} \otimes T_\mathcal{B}$ (the Kronecker product of matrices given earlier in this exercise).

5. Let V be a two-dimensional vector space over the field F, and let $T, S : V \to V$ be linear transformations. Assume the minimal polynomials of S and T are given by: $m_T(x) = (x - a)^2$, $m_S(x) = (x - b)^2$. (Therefore T and S can be represented by 2×2 Jordan blocks, $J_2(a)$, $J_2(b)$, respectively.) Compute the invariant factors of $T \otimes S : V \otimes V \to V \otimes V$. (See Sect. 10.7.2, p. 349, for notation and definitions concerning minimal polynomials and Jordan blocks.)

6. Let M be a right R-module, and let $I \subseteq R$ be a 2-sided ideal in R. Prove that, as right R-modules,

$$M \otimes_R (R/I) \cong M/MI.$$

7. (Induced representations) Here is an important application of the tensor product. Let G be a finite group, let F be a field, and let FG be the F-group ring

(see Sect. 7.3). Note that FG is clearly an F-vector space via scalar multiplication

$$\alpha \sum_{g \in G} \alpha_g g := \sum_{g \in G} (\alpha \alpha_g) g, \ \alpha \in F.$$

Likewise, if M is any right FG-module, then M naturally carries the structure of an F-vector space by setting $\alpha \cdot m := m(\alpha e)$, $\alpha \in F$, $m \in M$, where e is the identity of the group G. Now let H be a subgroup of G and regard FH as a subring of FG in the obvious way. Let V be a right FH-module, finite dimensional over F, and form the *induced module*

$$\operatorname{Ind}_H^G(V) := V \otimes_{FH} FG.$$

Since FG is an (FH, FG)-bimodule, we infer that, in fact, $\operatorname{Ind}_H^G(V)$ is actually a right FG-module. Now show that

$$\dim_F \operatorname{Ind}_H^G(V) = [G : H] \cdot \dim V.$$

8. Let A be an abelian group. Prove that a ring structure on A is equivalent to an abelian group homomorphism $\mu : A \otimes_{\mathbb{Z}} A \to A$, together with an element $e \in A$ such that $\mu(e \otimes a) = \mu(a \otimes e) = a$, for all $a \in A$, and such that

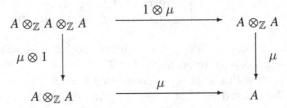

commutes. (The above diagram, of course, stipulates that multiplication is associative.)

9. The following theorem is an application of the tensor product to ideal classes in the field of fractions of a Dedekind Domain.

Theorem 13.13.1 *Let I and J be fractional ideals in the field of fractions E of a Dedekind domain R. If $[I] = [J]$ (that is, I and J belong to the same ideal class in E) then I and J are isomorphic as right R-modules.*

Provide a proof of this theorem. (The converse is easy: see Exercise (5) in Sect. 9.13.5.) [Hint: First, since I has the form $\alpha I'$ for some $\alpha \in E$ and ideal I' in R, we have $[I] = [I']$. Thus, without loss of generality, we may assume that I is an ideal of R. Note that since R is an integral domain, the multiplicative identity of R is the multiplicative identity 1_E of E. Define the injections $i_I : I \to I \otimes_R E$ and $i_J : J \to J \otimes_R E$ by

$$i_I(r) \rightarrow r \otimes 1_E, \text{ and}$$
$$i_J(s) \rightarrow s \otimes 1_E$$

for all $(r, s) \in I \times J$.

Now consider the commutative diagram below:

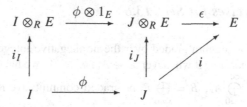

where i_I, i_J are the injections, given above, $\epsilon : J \otimes_R E \rightarrow E$ is given by $\epsilon(b \otimes \lambda) = b\lambda$, and where $i : J \hookrightarrow E$ is the containment mapping. Note also that $\phi \otimes 1_E : I \otimes_R E \rightarrow J \otimes_R E$ is an E-linear transformation.

Next, note that if $0 \neq a_0 \in I$, then $(a_0 \otimes a_0^{-1})a = (a_0 \otimes 1)a_0^{-1}a = (a_0 \otimes a)a_0^{-1} = (aa_0 \otimes 1)a_0^{-1} = (a \otimes a_0)a_0^{-1} = (a \otimes 1)a_0^{-1}a_0 = a \otimes 1$. Therefore, set $\alpha_0 := \epsilon(\phi \otimes 1)(a_0 \otimes a_0^{-1}) \in E$ and obtain $\phi(a) = \epsilon(\phi \otimes 1)(a \otimes 1) = \epsilon(\phi \otimes 1)((a_0 \otimes a_0^{-1})a) = \epsilon(\phi \otimes 1)(a_0 \otimes a_0^{-1})a = \alpha_0 a \in J$. Since $\phi : I \rightarrow J$ is an isomorphism, the result follows.]

13.13.5 Exercises for Sects. 13.6 and 13.7

1. Show that if R is a commutative ring, and if A is an R-module, then a multiplication $\mu : A \otimes_R A \rightarrow A$ gives A the structure of an R-algebra if and only if the diagrams in Figs. 13.1 and 13.2 (see p. 495) are commutative.
2. Let A_1, A_2 be commutative R-algebras. Prove that $A_1 \otimes_R A_2$ satisfies a universal condition reminiscent of that for direct sums of R-modules. Namely, there exist R-algebra homomorphisms $\mu_i : A_i \rightarrow A_1 \otimes_R A_2$, $i = 1, 2$ satisfying the following. If B is any commutative R-algebra such that there exist R-algebra homomorphisms $\phi_i : A_i \rightarrow B$, then there exists a unique R-algebra homomorphism $\theta : A_1 \otimes_R A_2 \rightarrow B$ such that for $i = 1, 2$, the diagram

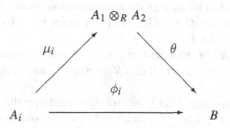

commutes.

3. Prove that $R[x] \otimes_R R[y] \cong R[x, y]$ as R-algebras.
4. Let F be a field and form F-group algebras for the finite groups G_1, G_2 as in Sect. 7.3. Prove that $F[G_1 \times G_2] \cong FG_1 \otimes_F FG_2$ as F-algebras.

13.13.6 Exercises for Sect. 13.8

1. Let A be an R-algebra graded over the nonnegative integers. We say that A is *graded-commutative* if whenever $a_r \in A_r$, $a_s \in A_s$ we have $a_r a_s = (-1)^{rs} a_s a_r$. Now let $A = \bigoplus_{r=0}^{\infty} A_r$, $B = \bigoplus_{s=0}^{\infty} B_s$ be graded-commutative R-algebras. Prove that there is a graded-commutative algebra structure on $A \otimes_R B$ satisfying

$$(a_r \otimes b_s) \cdot (a_p \otimes b_q) = (-1)^{sp}(a_r a_p \otimes b_s b_q),$$

$a_r \in A_r$, $a_p \in A_p$, $b_s \in B_s$, $b_q \in B_q$. (This is usually the intended meaning of "tensor product" in the category of graded-commutative R-algebras.)

13.13.7 Exercises for Sect. 13.10

1. Assume that the $\mathbb{F}$-vector space V has dimension n. For each $r \geq 0$, compute the $\mathbb{F}$-dimension of $S^r(V)$.
2. Prove Theorem 13.10.1.
3. Prove Corollary 13.10.2.
4. (Determinants) Let $T : V \to V$ be a linear transformation, and assume that dim $V = n$. Show that there exists a scalar $\det(T)$, such that

$$E^n(T) = \det T \cdot \mathrm{id}_{E^n(V)} : E^n(V) \to E^n(V).$$

Cite the results in Sect. 13.9 that show, for $S, T \in \mathrm{Hom}(V, V)$, that

$$\det(S) \cdot \det(T) = \det(S \circ T) = \det(T \circ S).$$

5. Let $G \to GL_F(V)$ be a group representation on the F-vector space V. Show that the mapping $G \to \mathrm{GL}_F(E^r(V))$ given by $g \mapsto E^r(g)$ defines a group representation on $E^r(V)$, $r \geq 0$.
6. Let V be a vector space and let $v \in V$. Define the linear map $\cdot \wedge v : E^r(V) \to E^{r+1}(V)$ by $\omega \mapsto \omega \wedge v$. If dim $V = n$, compute the dimension of the kernel of $\cdot \wedge v$.
7. (Boundary operators) Let V be n-dimensional over the field F, and having basis $\{v_1, v_2, \ldots, v_n\}$. For each integer r, $1 \leq r \leq n$, define the linear transformation $\partial_r : E^r(V) \to E^{r-1}(V)$, from its value on basis elements by setting

$$\partial_r(v_{i_1} \wedge v_{i_2} \wedge \cdots \wedge v_{i_r}) = \sum_{j=1}^{r}(-1)^{j-1}v_{i_1} \wedge v_{i_2} \wedge \cdots \wedge \widehat{v_{i_j}} \wedge \cdots \wedge v_{i_r},$$

where $\widehat{(\cdot)}$ means *delete the factor* $(\cdot)$. The mapping $\partial_0 : E^1(V) = V \to E^0(V) = F$ is the mapping determined by $\partial_0(v_i) = 1$, $i = 1, 2, \ldots, n$.

(a) Show that if $r \geq 1$, then $\partial_{r-1}\partial_r = 0 : E^r(V) \to E^{r-2}(V)$.
(b) Define a mapping $h_r : E^r(V) \to E^{r+1}(V)$, $0 \leq r \leq n - 1$ by setting
$h_r(\eta) = v_1 \wedge \eta$, where $\eta \in E^r(V)$.
Show that if $1 \leq r \leq n - 1$, then $\partial_{r+1}h_r + h_{r-1}\partial_r = \mathrm{id}_{E^r(V)}$, and that
$h_{n-1}\partial_n = \mathrm{id}_{E^n(V)}$, $\partial_1 h_0 = \mathrm{id}_{E^0(V)}$.
(c) Conclude that the sequence

$$0 \to E^n(V) \xrightarrow{\partial_n} E^{n-1} \xrightarrow{\partial_{n-1}} E^{n-2} \to \cdots \xrightarrow{\partial_2} E^1(V) \xrightarrow{\partial_1} E^0(V) \to 0$$

is exact.

8. Let V be an F-vector space. An F-linear mapping $\delta : E(V) \to E(V)$ is called an *antiderivation* if for all $\omega \in E^r(V)$, $\eta \in E^s(V)$ we have

$$\delta(\omega \wedge \eta) = \delta(\omega) \wedge \eta + (-1)^r \omega \wedge \delta(\eta).$$

Now let $f : V \to F$ be a linear functional, and show that f can be extended uniquely to an antiderivation $\delta : E(V) \to E(V)$ satisfying $\delta(v) = f(v) \cdot 1_{E(V)}$, for all $v \in V$. In addition, show that δ satisfies

(a) $\delta : E^r(V) \to E^{r-1}(V)$, $r \geq 1$, $\delta : E_0(V) \to \{0\}$.
(b) $\delta^2 = 0 : E(V) \to E(V)$.

9. (The *Clifford Algebra*) Let V be an F-vector space and let $Q : V \to F$ be a function. We call Q a *quadratic form* if Q satisfies

(a) $Q(\alpha v) = \alpha^2 Q(v)$, for all $\alpha \in F$, $v \in V$, and
(b) the mapping $B : V \times V \to F$ given by

$$(v, w) \mapsto Q(v + w) - Q(v) - Q(w)$$

defines a (clearly symmetric) bilinear form on V.

Given the quadratic form Q on V, we define a new algebra, $C(Q)$, the *Clifford algebra*, as follows. Inside the tensor algebra $T(V)$ define the ideal I to be generated by elements of the form $v \otimes v - Q(v) \cdot 1_F$, $v \in V$. (Note that I is *not* a homogeneous ideal of $T(V)$.) The Clifford algebra is the quotient algebra $C(Q) = T(V)/I$.

Show that $C(Q)$ satisfies the following universal criterion. Assume that C' is an F-algebra and that there exists a linear mapping $f : V \to C'$ such that for all $v \in V$, $(f(v))^2 = Q(v) \cdot 1_{C'}$. Prove that there exists a unique F-algebra homomorphism $\gamma : C(Q) \to C'$ making the following triangle commute:

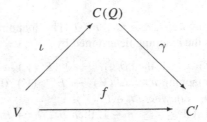

(In the above diagram, $\iota : V \to C(Q)$ is just the natural homomorphism $v \mapsto v + I$.)

10. Let V be an n-dimensional F-vector space. If $d \leq n$, define the (n, d)-*Grassmann space*, $\mathcal{G}_d(V)$ as the set of all d-dimensional subspaces of V. In particular, if $d = 1$, the set $\mathcal{G}_1(V)$ is more frequently called the *projective space* on V, and is denoted by P(V).[9] We define a mapping

$$\phi : \mathcal{G}_d(V) \longrightarrow \mathrm{P}(E^d(V)),$$

as follows. If $U \in \mathcal{G}_d(V)$, let $\{u_1, \dots, u_d\}$ be a basis of U, and let $\phi(U)$ be the 1-space in P($E^d(V)$) spanned by $u_1 \wedge \cdots \wedge u_d$. Prove that $\phi : \mathcal{G}_d(V) \to \mathrm{P}(E^d(V))$ is a well-defined injection of $\mathcal{G}_d(V)$ into P($E^d(V)$). (This mapping is called the *Plücker embedding*.)

11. Let V be an n-dimensional vector space over the field F.

   (a) Show that if $1 < d < n - 1$, then the Plücker embedding $\phi : \mathcal{G}_{n-1}(V) \longrightarrow \mathrm{P}(E^{n-1}(V))$ is never surjective. [Hint: Why is it sufficient to consider only $(d, n) = (2, 4)$?]

   (b) If F is a finite field F_q, show that the Plücker embedding $\phi : \mathcal{G}_{n-1}(V) \longrightarrow \mathrm{P}(E^{n-1}(V))$ is surjective. This implies that every element of $z \in E^{n-1}(V)$ can be written as a "decomposable element" of the form $z = v_1 \wedge v_2 \wedge \cdots \wedge v_{n-1}$ for *suitable* vectors $v_1, v_2, \dots, v_{n-1} \in V$. [An obvious counting argument.] .

12. Let V, W be F-vector spaces. Prove that there is an isomorphism

$$\bigoplus_{i+j=r} E^i(V) \oplus E^j(W) \longrightarrow E^r(V \oplus W).$$

---

[9] In geometry, the terms "Grassmann space", and "Projective space" not only include the "points", $\mathcal{G}_d(V)$ and $\mathcal{G}_1(V)$, respectively, but also include collections of "lines" as well. In the case of the projective space, the lines are $\mathcal{G}_2(V)$, while, for the Grassmann space $\mathcal{G}_d(V)$, $1 < d < n$, the lines are all pairs $(X, Y) \in \mathcal{G}_{d-1}(V) \times \mathcal{G}_{d+1}(V)$ such that $X \subset Y$. The "points" belonging to (X, Y) are those $Z \in \mathcal{G}_d(V)$ such that $X \subset Z \subset Y$. These lines correspond to 2-dimensional subspaces of $E^d(V)$.

13. Let V be an F-vector space, where char $F \neq 2$. Define the linear transformation $S : V \otimes V \to V \otimes V$ by setting $S(v \otimes w) = w \otimes v$.

 (a) Prove that S has minimal polynomial $m_S(x) = (x - 1)(x + 1)$.
 (b) If $V_1 = \ker(S - I)$, $V_{-1} = \ker(S + I)$, conclude that $V \otimes V = V_1 \oplus V_{-1}$.
 (c) Prove that $V_1 \cong S^2(V)$, $V_{-1} \cong E^2(V)$.
 (d) If $T : V \to V$ is any linear transformation, prove that V_1 and V_{-1} are $T \otimes T$-invariant subspaces of $V \otimes V$.

14. Let V be an n-dimensional F-vector space.

 (a) Prove that $E(V)$ is graded-commutative in the sense of Exercise (1) in Sect. 13.13.6.
 (b) If $\{L_i\}$, $i = 1, \ldots, n$, is a collection of one-dimensional subspaces of V which span V, prove that as graded-commutative algebras,

$$E(V) \cong E(L_1) \otimes E(L_2) \otimes \cdots \otimes E(L_n)$$

15. Let $G = \mathrm{GL}(V)$ act naturally on the n-dimensional F-vector space V so that V becomes a right FG-module..

 (a) Show that the recipe $g(f) = \det g \cdot f \circ g^{-1}$, $g \in G$, $f \in V^* = \mathrm{Hom}(V, F)$, defines a representation of G on V^*, the dual space of V.
 (b) Show that in the above action, G acts transitively on the non-zero vectors of V^*.
 (c) Fix any isomorphism $E^n V \cong F$; show that the map $E^{n-1}(V) \to V^*$ given by $\omega \mapsto \omega \wedge \cdot$ is a morphism of right FG-modules, i.e., the map commutes with the action of G.
 (d) A vector $z \in E^d(V)$ is said to be *decomposable* or *pure* if and only if it has the form $z = v_1 \wedge \cdots \wedge v_d$ for suitable vectors $v_i \in V$. Since G clearly acts on the set of decomposable vectors in $E^{n-1}(V)$, conclude from (b) that every vector in E^{n-1} is decomposable.

16. (Veronesean action) Again fix a vector space V with finite basis $X = \{x_1, \ldots, x_n\}$ with respect to the field F.

 (a) Lef $f : V \to F$ be a *functional* (that is, a vector of the dual space V^*). Show that $S^d(f)$ is a linear mapping $S^d(V) \to F$, and so is a functional of $S^d(V)$, $1 \le d \le n$.
 (b) Set $G = \mathrm{GL}(V)$. First let us view V as a *left* G-module. Thus for all $S, T \in G$ and $v \in V$, we have $(ST)v = S(T(v))$. If $f \in \mathrm{Hom}(V, F)$ and $T \in G$, then $f \circ T$ is also a functional. Show that for all $(f, T) \in V^* \times G$, the mapping

$$f \to f \circ T$$

defines a *right* action of G on V^* (i.e. V^* becomes a right FG-module). [Hint: One must show that $f \circ (ST) = f \circ S \circ T$. One can check these maps at arbitrary $v \in V$ recalling that the operators T, S and f operate on V from the left.] (By way of comparison with part (a) of the previous exercise, note that we have defined a representation of G on the dual space V^*, without employing the determinant.)

(c) For each $f \in \mathrm{Hom}(V, F)$, and integer $d \geq 1$, $S^d(f) : S^d(V) \to F$ is a functional of $S^d(V)$ called a *Veronesean vector*. With respect to the basis $X = \{x_i\}$, we may regard $S^d(V)$ as the space of homogeneous polynomials of degree d in $F[x_1, \ldots, x_n]$. If $f(x_i) = \epsilon_i \in F$, show that at each homogeneous polynomial $p \in S^d(V)$,

$$f(p(x_1, \ldots, x_n)) = p(\epsilon_1, \ldots, \epsilon_n) \in F.$$

conclude from this, that if $(\alpha, f) \in F \times V^*$, then

$$S^d(\alpha f) = \alpha^d S^d(f). \tag{13.22}$$

Describe how this produces a bijection between the 1-dimensional subspaces of V^* (the points of the projective space $P(V^*)$) and the 1-spaces of $S^d(V)^*$ spanned by the individual Veronesean vectors. (The latter collection of 1-dimensional subspaces is called *the (projective) Veronesean variety of degree d*.) [Hint: Use Eq. (13.22).]

(d) Prove that the (right) action of $G = GL(V)$ on V^* described in part (b) of this exercise transfers to an action on G on $S^d(V)^*$ which is transitive on the non-zero Veronesean vectors it contains. It also induces a permutation isomorphism between the action of G on the projective points of $P(V^*)$ and the action of G on the projective Veronesean variety of degree d. [Hint: Citing relevant theorems, show that if $(f, T) \in V^* \times G$ then $S^d(f \circ T) = S^d(f) \circ S^d(T)$, so that

$$\rho^d(T) : S^d(f) \to S^d(f) \circ S^d(T)$$

defines a mapping
$$\rho^d : G \to \mathrm{End}_F(S^d(V)^*)$$

which describes this action.]

17. Attempt to emulate the development of the previous exercise with $E(V)$ replacing $S(V)$ everywhere. Thus for $1 \leq d \leq n$, one desires an injective mapping

$$\phi : P(V^*) \to P(E^d(V^*))$$

by transferring functional f of V^* to $E^d(f) : E^d(V) \to F$. Explain in detail what goes wrong.

13.13.8 Exercise for Sect. 13.11

1. Write out a proof of part 3 of Theorem 13.11.1

References

1. Artin E (1957) Geometric algebra. Interscience Publishers, Inc., New York
2. Shult E (2011) Points and lines. Universitex series. Springer, New York
3. Smullyan R, Fitting M (2010) Set theory and the continuum problem. Dover Publications, Mineola

Bibliography

1. Aigner M (1997) Combinatorial theory. Springer, Berlin
2. Alperin J, Lyons R (1971) On conjugacy classes of p-elements. J Algebra 19:536–537
3. Artin E (1957) Geometric algebra. Interscience Publishers, Inc., New York
4. Aschacher M (1993) Finite group theory. Cambridge studies in advanced mathematics, vol 10. Cambridge University Press, Cambridge
5. Aschbacher M, Smith S (2004) The structure of strong quasithin K-groups. Mathematical surveys and monographs, vol 112. American Mathematical Society, Providence
6. Aschbacher M, Smith S (2004) The classification of quasithin groups II, main theorems: the classification of QTKE-groups. Mathematical surveys and monographs, vol 172. American Mathematical Society, Providence
7. Aschbacher M, Lyons R, Smith S, Solomon R (2004) The classification of finite simple groups of characeristic 2 type. Mathematical surveys and monographs, vol 112. American Mathematical Society, Providence
8. Birkhoff G, MacLane S (1965) A survey of modern algebra, 3rd edn. Macmillan, New York
9. Cameron P (1999) Permutation groups. London mathematical society student texts, vol 45. Cambridge University Press, Cambridge
10. Cohn PM (1977) Algebra, vol 2. Wiley, London
11. Collins MJ (1990) Representations and characters of finite groups. Cambridge studies in advanced mathematics, vol 22. Cambridge University Press, Cambridge
12. Conway JH, Curtis RT, Norton SP, Parker RA, Wilson RA (1985) Atlas of finite groups. Oxford University Press, Eynsham
13. Coxeter HMS (1934) Discrete groups generated by reflections. Ann Math 35:588–621
14. Curtis CW, Reiner I (1962) Representation theory of finite groups and associative algebras. Pure and applied mathematics, vol IX. Interscience Publishers, New York, pp xiv+685
15. Dauns J (1994) Modules and rings. Cambridge University Press, Cambridge
16. Devlin K (1997) The joy of sets. Springer, New York
17. Feit W (1967) Characters of finite groups. W. A. Benjamin Inc., New York
18. Feit W, Thompson J (1963) Solvability of groups of odd order. Pac J Math 13:775–1029
19. Gorenstein D (1968) Finite groups. Harper & Row, New York
20. Greene C (1982) The Möbius function of a partially ordered set. In: Rival I (ed) Appeared in ordered sets. D. Reidel Publishing Company. Dordrecht, pp 555–581
21. Griess R (1998) Twelve sporadic groups. Springer monographs in mathematics. Springer, Berlin
22. Grove LC (1997) Groups and characters. Pure and applied mathematics interscience series. Wiley, New York

23. Ireland K, Rosen M (1972) Elements of number theory. Bogden and Quigley Inc., New York
24. Isaacs IM (1994) Character theory of finite groups. Dover Publications Inc., New York
25. Jacobson N (1985) Basic algebra I, 2nd edn. W. H. Freeman and Co., New York
26. Jacobson N (1989) Basic algebra II, 2nd edn. W. H. Freeman and Co., New York
27. Kleidman P, Liebeck M (1990) The subgroup structure of the finite classical groups. London mathematical society lecture note series, vol 129. Cambridge University Press, Cambridge
28. Kung JP (ed) (1995) Gian-Carlo Rota on combinatorics. Birkhaüser, Boston
29. Lambek J (1966) Rings and modules. Blaisdell Publishing Co., Toronto
30. Oxley J (2011) Matroid theory. Oxford graduate texts, vol 21. Oxford University Press, Oxford
31. Passman D (1968) Permutation groups. W. A. Benjamin Inc., New York
32. Rota G-C (1964) On the foundations of combinatorial theory I. Theory of Mobius functions Zeitschrift für Wahrscheinlichkeitstheorie, Band 2, Heft 4. Springer, Berlin, pp 340–368
33. Rotman J (1979) An introduction to homological algebra. Academic Press, Boston
34. Smullyan R, Fitting M (2010) Set theory and the continuum problem. Dover Publications, Mineola
35. Shult E (2011) Points and lines. Universitex series. Springer, New York
36. Stanley RP (1997) Enumerative combinatorics, vol 1, Cambridge studies in advanced mathematics, vol 49. Cambridge University Press, Cambridge
37. van der Waerden BL (1991) Algebra, vols I–II. Springer, New York
38. Wielandt H (1964) Finite permutation groups. Academic Press, New York

Index

A
Abelian group, 88
    divisible, 268
    torsion subgroup, 362
Absolute value, 424
ACC, *see* ascending chain condition
Adjoint
    left, 488
    right, 488
Algebra, 493
    Clifford, 523
    exterior, 504
    graded commutative, 522
    monoid ring, 199
    of ascending chains, 508
    symmetric, 504
    tensor, 503
Algebraic independence, 412
Algebraic integer, 255, 316
Algebraic interval, 239
Alternating group, 111
Annihilator
    of a module, 271
    right, 196
Antiautomorphism
    of a ring, 190
Antichain, 32
Antiderivation, 523
Ascending chain condition, 42
Association class, 281
Associative law, 74
Atom
    in a poset, 47, 69
Automorphism
    of a group G, 81
    of a graph, 76
    of a ring, 190
Automorphism group, of a group, 81

B
Baer's criterion, 277
Balanced map
    multiple arguments, 491
Balanced mapping, 479
Basis
    of a free module, 335
    of a module, 243
Bell numbers, 36, 136
Bilinear form
    symmetric, 315
Bimodule, 234
    symmetric, 234

C
Canonical form, rational, 348
Cardinal number, 16, 18
Cartesian product
    of n sets, 5
    of two sets, 5
Category, 473
    initial object, 476
    isomorphism, 476, 478
    morphism, 473
    object, 473
    opposite, 477
    subcategory, 474
    terminal object, 476
Cauchy sequence, 426
    convergence of, 426
Cayley graph, 164

© Springer International Publishing Switzerland 2015

E. Shult and D. Surowski, *Algebra*, DOI 10.1007/978-3-319-19734-0

Printed in the United States
By Bookmasters